Lecture Notes in Mathematics

Volume 2376

This series reports on new developments in all areas of mathematics and their applications - quickly, informally and at a high level. Mathematical texts analysing new developments in modelling and numerical simulation are welcome. The type of material considered for publication includes:

1. Research monographs
2. Lectures on a new field or presentations of a new angle in a classical field
3. Summer schools and intensive courses on topics of current research.

Texts which are out of print but still in demand may also be considered if they fall within these categories. The timeliness of a manuscript is sometimes more important than its form, which may be preliminary or tentative. Please visit the LNM Editorial Policy (https://drive.google.com/file/d/1MOg4TbwOSokRnFJ3ZR3ciEeKs9hOnNX_/view?usp=sharing)

Titles from this series are indexed by Scopus, Web of Science, Mathematical Reviews, and zbMATH.

Preface

The notion of relative rearrangement is based on monotone rearrangement (Chap. 2) but is not a symmetrization as defined in previous books (see [103] or [17]). Indeed, monotone rearrangement (or symmetrization) involves only one function u on a domain Ω that u is rearranged or transformed to a function u_* on another domain of the same measure as Ω. The fundamental property of such transformation is the invariance of the measures of level sets. While in relative rearrangement, as those words indicate, the transformation depends on two functions (v, u). Firstly, the word "rearrangement" means that we transform a function, here we denote by v, and secondly, the word "relative" means that the transformation depends on another function, we denote by u, here the transformations are made according to the level sets of a second function u . So since it involves a couple of functions (u, v), relative rearrangement is denoted by v_{*u}. The rigorous definition is given in the second chapter. The definition can be described as: Look at a point u the computation of the directional derivative in a direction v, say

$$\Big[(u + tv)_* - u_*\Big]/t$$

as t tend to zero, then we have a function called relative rearrangement of v with respect to u. Therefore, the two notions, relative rearrangement and rearrangement, do not have the same properties, for instance the invariance of integrals cannot be true in general for the relative rearrangement. A numerical construction is proposed in the second chapter, Sect. 2.8, to help visualize this notion. Relative rearrangement is also called pseudo-rearrangement by the Alvino-Trompetti group in Naples (see A. Ferone and R. Volpicelli [73], Alvino and Trompetti [7]) but was obtained in a different manner. G. Talenti and V. M. Maz'ja also used pseudo-rearrangement in their pioneering works [164], [112], but in a different form than that defined here and under some regularity conditions. While some applications of this notion concern the same topics as the Schwartz symmetrization, the techniques presented here allow us to consider more general situations and results.

For instance, when we talk about Sobolev inclusions, most of the time we think of those associated with Lebesgue spaces. The techniques for obtaining the associated or corresponding inequalities are based either on an integral representation of the function to be estimated (this is the original Sobolev method (see the book by Gilbard and Trudinger [94])), or on the use of the Gagliardo-Nirenberg inequality (see the books by Brézis [29], Adams [1], and J.E. Rakotoson and J.M. Rakotoson [136]). When we try to move to more general spaces like Lorentz spaces, Birnbaum–Orlicz spaces, Zygmund spaces $L^{\beta}(\operatorname{Log} L)^{\alpha}$, or any other normed space like the one recently introduced by Alberto Fiorenza [80]: small Lebesgue spaces, these usual techniques for obtaining classical Sobolev inclusions are no longer easily adaptable to prove Sobolev inclusions associated with the latter spaces.

Here, we propose *pointwise inequalities* that link a function and its gradient. These inequalities turn out to be *the root* of many Sobolev-type inequalities associated with normed spaces, that is, if $\rho,\ \rho_0$ are two norms, $W^1L(\Omega, \rho)$ a Sobolev space associated with the norm ρ, and $L(\Omega, \rho_0)$ the normed space associated with ρ_0, we give conditions (called inclusion indices) so that $W^1L(\Omega, \rho) \subset L(\Omega, \rho_0)$ with a "continuous" inclusion.

In addition to the Sobolev inclusions, we will show that these inequalities also lead to *interpolation inequalities* of the Gagliardo–Nirenberg type.

Another advantage of the method we are going to present is that we can estimate the constants that appear in these inequalities. Here is an example of a frequently used inequality that we will re-prove: if Ω is a bounded open subset of $\mathbb{R}^2$, then for any element $u \in H_0^1(\Omega)$ we have

$$|u|_{L^4(\Omega)} \leqslant \left(\frac{1}{\pi}\right)^{\frac{1}{4}} |u|^{\frac{1}{2}}_{L^2(\Omega)} |\nabla u|^{\frac{1}{2}}_{L^2(\Omega)},$$

or again, we will show that for spaces $W^1L(\Omega, \rho)$ which are injected into the set of continuous functions $\big($e.g., if $L(\Omega, \rho) \subsetneqq L^{N,1}(\Omega)\big)$ we have

$$\operatorname*{osc}_{B(x,r)\cap\Omega} u \leqslant \frac{\alpha_N^{1-\frac{1}{N}}}{\alpha_{N-1}} \rho_{B(x,r)}(\nabla u),$$

where α_m denotes the measure of the unit ball of $\mathbb{R}^m$,

$\rho_{B(x,r)}$ denotes the restriction of the norm ρ to maps defined on the ball $B(x,r)$ with center x and radius $r > 0$.

In the particular case of Lorentz spaces, such a result was demonstrated by E.M. Stein [157], but without estimating the constant. His method is based on an integral representation of the function u and the use of the theory of weak type operators (see Bennett-Sharpley's book [21]).

To refine these estimates we have classified the functions that we study into three categories according to their boundary conditions on the domain Ω:

1. Functions with zero trace
2. Functions with partially zero trace
3. Functions with arbitrary trace

These *pointwise inequalities* can be obtained, in *boundary value problems*, which leads to regularity theorems for normed spaces, i.e., we can partially answer the question: if f belongs to a normed space $L(\Omega, \rho)$, what can we say about the solution u?

For the sake of clarity, we have often illustrated these results by choosing the Lorentz and Lebesgue spaces according to the boundary conditions 1. or 2. or 3. above. They can be replaced by the spaces mentioned above or by any other normed space whose norm satisfies certain properties.

These pointwise inequalities are obtained by the intermediary of monotone and relative rearrangement whose definitions and properties are introduced in Chaps. 1 and 2. One of the main advantages of these rearrangements is that they are transformations that send the set of measurable functions $L^0(\Omega)$ into the set of measurable functions on *an interval* of the same measure $\Omega_* \ =]0, |\Omega|[$ while preserving certain "qualities" of the original function.

These conservation properties are illustrated by the equimeasurability properties for monotone rearrangements, the inequalities as we have emphasized the techniques of rearrangement allow us to give simple proofs, at least easily reproducible for various estimates both for the inclusion theorems and for boundary value problems. We have essentially shown this when working with spaces of Lebesgue type and their generalization like Lorentz spaces.

Moreover, we show that the techniques of rearrangement can be adapted to obtain analogous results when the spaces on which we work are not invariant under rearrangement, as is the case for BMO and VMO spaces or Lebesgue spaces with variable exponents. We have thus devoted Chap. 10 to Lebesgue spaces with variable exponents.

To obtain estimates of the gradient of a function or to use the gradient in inclusion problems, we have exploited in more depth the ideas that we used in the article [144–146] using the relative rearrangement of the gradient. In particular, by taking up a result of Cianchi and Pick [45], we show here that the Sobolev–Lorentz space $W^1L^{N,\infty}(\Omega)$ is contained in $bmo_r(\Omega)$, by giving in addition an estimate of the inclusion constant. We do this by proving, among other things, that for a ball B contained in Ω of measure $a = |B|$, for all v in $W_0^{1,1}(\Omega)$, if u denotes the restriction of v to the ball B, then we have

$$\int_B |v(x) - v_B|dx \leqslant D_N \int_0^a \operatorname{Max}\left(t^{\frac{1}{N}}; (a-t)^{\frac{1}{N}}\right) |\nabla u|_{\star u}(t)|dt$$

where $v_B = |B|^{-1} \displaystyle\int_B v(y)dy$ and $D_N = \dfrac{2^{\frac{1}{N}}\alpha_N^{1-\frac{1}{N}}}{\alpha_{N-1}}$.

This estimate allows us to find other inclusions by replacing $L^{N,\infty}(\Omega)$ by a more general space $L(\Omega, \rho)$, where ρ is a Fatou norm invariant under rearrangement. These results are given in Chap. 4.

In Chap. 5, we added in the same spirit, by adapting a result of Cianchi–Maz'ja [44], and by using the techniques of relative rearrangement, a result on the regularity of the gradient. If f is in $L^1(\Omega) \cap W^{-1,p'}(\Omega)$, and $u \in W_0^{1,p}(\Omega)$ is a solution of $-\Delta_p(u) = f$ in the sense of distributions, then the gradient of the solution is in a Lorentz–Marcienkiewicz space and we have the estimate

$$|\nabla u|_{L^{\frac{N(p-1)}{(N-1)},\infty}} \leqslant c_N \|f|_1^{1/(p-1)}.$$

We also give an estimate of the constant c_N.

Other results concerning Sobolev-type inclusions, but associated with more general spaces called G-Γ spaces, are added in Chap. 4. They generalize inequalities such as Trudinger's, but also give inclusions for "limit" spaces, such as BMO-spaces, Zygmund spaces, and small or grand Lebesgue spaces.

We have chosen not to include applications to Peetre-type interpolation spaces. However, we have added to Chap. 11 exercises about the computation of K-functionals between Lebesgue spaces (including large Lebesgue spaces) because their computation involves rearrangement.

In this same chapter, we have grouped the statements of the exercises and their solutions (or indications).

In Chap. 1, we shall start by recalling some properties of the monotone rearrangement. Alternative proofs of some of the results of this chapter can be found in the books by Kesavan [103], J. Mossino [115], Bennet and Sharpley [21], and Chong and Rice [39].

Chapter 2 explains the purpose of this book, providing a description of the notion of relative rearrangement which is more developed than in the book by J. Mossino [115].

The aim of Chap. 3 is to present some fundamental pointwise inequalities, such as the Polyà–Szegö inequalities. Note that in previous books (see, for instance, [103], [115]) only classical Polyà–Szegö inequalities in the usual Sobolev spaces are given.

However, to establish those pointwise inequalities, we need the regularity of the first derivative of the monotone rearrangement.

In Chap. 4, we exploit the fundamental pointwise inequality established in Theorem 3.4.1 in order to present a "simple" way of getting Sobolev embeddings in a general setting, with explicit estimates of the Sobolev constants. In particular, we give some non-standard embeddings, as we have mentioned before.

In Chap. 5, we pursue the applications of the fundamental inequality Theorem 4. This chapter uses the same arguments as Talenti's works but instead of using the distribution function $m_u(t) = meas\{u > t\}$ it directly uses the monotone rearrangement and its first derivative. So, the argument requires the regularity of the first derivative of the monotone rearrangement. The gradient estimates for quasilinear PDEs are given as in the works of Cianchi et al. Note that in Talenti's work (see also [103], [115]) the estimates are derived from a comparison of the solutions two PDEs and in general do not give any information on the gradient.

Chapter 6 is based on the works of Coron–Almgren–Lieb and studies the continuity of the mapping $u \to u'_*$ (the derivative of the monotone rearrangement). It presents some useful extensions of the regularity of u'_*.

Chapter 7 is devoted to the study of the continuity of the mapping $u \to b_{*u}$ (the relative rearrangement of b with respect to u). It happens that this chapter is related to Chap. 6 and describes the strong convergence of the directional derivative $u \to u_*$.

In Chaps. 8 to 10, we present different applications of the relative rearrangement, in particular for evolutions problems that have not been considered in previous books on monotone rearrangement and their applications (see [103], [115], [17]).

Chapter 11 ends the book with a set of exercises related to relative rearrangement, but also to monotone rearrangement. An overview of an application of the monotone rearrangement is given for the computation of the so-called K-functional appearing in interpolation theory.

The bibliography includes a nice and complete survey by G. Talenti [164] concerning monotone rearrangement, pseudo-rearrangement and their applications in which he gives some historical facts but also some graphical illustrations of rearrangements. Many references are also given.

Poitiers, France Jean Michel Rakotoson

Acknowledgements The concept of *relative rearrangement* was introduced by my former teachers Professor Roger Temam and Jacqueline Mossino, when they computed for the first time the directional derivative of the monotone rearrangement. I would like to thank them to give me the opportunity to develop this concept, and to write this first book in English about this issue. They were my co-authors in the study of several properties given in Chaps. 2, 3, and 9.

May my various collaborators find here my deep gratitude, in particular Ildefonso Díaz from the Complutense University of Madrid, who brought a wonderful problem from plasma physics. This model, related to the so-called Stellarator machines, raised several open questions about relative rearrangement. This led me to develop several properties of relative rearrangement, in particular the results of Chap. 7. Also all his students, namely, J.F. Padial, Lourdes Tello, and Lerena Belen, brought me to some questions which led to several improvements in Chap. 7.

The multi-constraint problem given as an example in the introduction to Chap. 8 was proposed to me by Professor Denis Serre of the Ecole Normale Supérieure de Lyon; I am grateful to him. This problem led to the development of a nice application of the notion of relative rearrangement exposed in this chapter. A special thought in the memory of Professor Robert Janin, with whom the discussion on this problem were fruitful.

I wish to thank Professor Marisa Seoane of the University of Santiago de Compostela (Saint Jacques de Compostelle), with whom I wrote several articles, some of the results of which are in this book, and to all my Italian friends and collaborators, including Professor Alberto Fiorenza of Naples, who introduced the small Sobolev spaces that gave rise to applications of the pointwise inequalities for relative rearrangement.

And recently, I worked with Maria Rosaria Formica, El Hamidi Abdallah, Amiran Gogatishvilii, and Ahmed Irshaad to the study of interpolation spaces in the sense of Peetre. These studies led us to some applications of monotone rearrangement. I am grateful to all them. My gratitude goes in particular to Professor Mara Rosaria Formica not only for our joint works but for having read this book entirely.

I would like to say a special thanks to my wife Catherine Falaise for the typing and formatting of this document, also for doing the translation via Google translate.

Finally, I extend my sincere acknowledgments to the anonymous reviewers for the additional references and insightful remarks that helped improve the content of this book. The Language Editor of Springer who polishes my English.

Last but not the least, I thank Dr. Ute McCrory and Dr. Remi Loth, the Springer editors, for considering this new book in English concerning relative rearrangement.

Competing Interests The author has no competing interests to declare that are relevant to the content of this manuscript.

Contents

Chapter 1
Motivations and Generalities on the Monotone Rearrangement

To show Sobolev inequalities, we can use potential theory (Riesz potential), which is a way of using integral representations, or the Gagliardo–Nirenberg inequality,

$$\text{for } u \in C_c^\infty(\mathbb{R}^N) \qquad |u|_{L^{\frac{N}{N-1}}(\mathbb{R}^N)} \leqslant \prod_{j=1}^{N} |\partial_j u|_{L^1(\mathbb{R}^N)}^{\frac{1}{N}}.$$

The proofs are often very technical and are difficult to adapt to spaces that are invariant under rearrangement.

The tool that we are going to present in this book has been used by several schools of mathematics. The tools that they use are in different forms than ours, but they are all based on the use of the monotone rearrangement of a function defined on a domain Ω (most often here: a subset of $\mathbb{R}^N$ of finite measure). The monotone rearrangement mapping reduces a function u to a function of a real variable on an interval $[0, |\Omega|]$ while preserving good properties of the initial function u and moving the singularities of u to the origin (or to the boundary $|\Omega|$).

Our work consists in adding another transformation, which to two functions (u, v) associates a function of a single variable v_{*u}. By a varied choice of v, this leads to interesting relations between u_*, u'_* and other quantities. This has the consequence of giving simple and practical proofs of Sobolev inequalities, while providing precise estimates of the inclusion constants. Among the results that we will present, we will explain, how to show that $W^1L^{N,1}(\Omega) \hookrightarrow C(\overline{\Omega})$ (by specifying the oscillations of the function u) and $W^1L^{N,\infty}(\Omega) \hookrightarrow bmo_r(\Omega)$ (functions with bounded "mean oscillation "). All this will be discussed in Chap. 4 using monotone and relative rearrangement.

J. M. Rakotoson, *Relative Rearrangement*, Lecture Notes in Mathematics 2376,
https://doi.org/10.1007/978-3-032-02228-8_1

The same tool that we will develop will allow us:

- to solve non-classical optimization problems, such as the following minimization:

$$\text{Inf}\left\{\frac{1}{2}\int_\Omega |\nabla v|^2\,dx - \int_\Omega f v\,dx,\quad v \in K(h)\right\},\ \text{where } h \in L^\infty(\Omega)$$

and $K(h) = \Big\{v \in H_0^1(\Omega)$

$$\int_\Omega \Phi(v)dx \leqslant \int_\Omega \Phi(h)dx\ \ \forall\,\Phi : \mathbb{R} \to \mathbb{R}\,\text{convex Lipschitzian}\Big\}.$$

- to find the Euler equation associated with this problem. We will see that the set of constraints can be expressed in monotone rearrangement terms, while obtaining the Euler equations requires relative rearrangement.

Nonlocal partial differential equations will be discussed as applications of these tools.

Other multi-constrained problems can be reformulated in terms of relative rearrangement and monotone rearrangement, see Chap. 2.

1.1 Notation and Reminders

Throughout this book, we will denote by $\mathbb{R}^N$ the Euclidean space equipped with the norm $|x| = \left(\sum_{i=1}^N x_i^2\right)^{1/2}$ associated with the scalar product

$$x \cdot y = \sum_{i=1}^N x_i y_i (with x = (x_1, \cdots, x_N),\ y = (y_1, \cdots, y_N));$$

$B(x, r)$ the open ball with center x and radius r.

For simplicity, we will use the Lebesgue measure, and if E is measurable in $\mathbb{R}^N$ we denote by of the notation $|E|$ the measure of E or by abuse of the notation the volume of E. We denote by Ω the set (often bounded) on which we work.

The characteristic function of E will be denoted by χ_E i.e.

$$\chi_E = \begin{cases} 1 & \text{if } x \in E, \\ 0 & \text{otherwise.} \end{cases}$$

The m-dimensional Haussdorf measure of E will sometimes be used and will be denoted by $H_m(E)$. Note that

$$H_N(E) = |E| \equiv \mathcal{L}_N(E).$$

In short, the rearrangement of a function $u : \Omega \to \mathbb{R}$ is the function u_* which is inverse to the volume function along

$$t \to \mathrm{vol}\Big\{u > t\Big\} = |u > t|\,.$$

It is therefore a decreasing function. Its main interest is summarized by the conservation of the volume:

$$\mathrm{vol}\Big\{u > t\Big\} = \mathrm{vol}\Big\{u_* > t\Big\} \qquad (\forall t)$$

(level set volume). This implies the conservation of norms in different normed spaces.

If $u : \Omega \to \mathbb{R}$ is a measurable function, we will simply write:
$\Big\{u > t\Big\} = \Big\{x \in \Omega : u(x) > t\Big\}$, $\Big\{u = t\Big\} = \Big\{x \in \Omega : u(x) = t\Big\}$ and similarly for sets $\{u < t\}$, $\{u \leqslant t\}$, $\{u \geqslant t\}$. Their measure is denoted by $\Big|\Big\{u > t\Big\}\Big| = |u > t|$, and likewise $|u < t|$, $|u \geqslant t|$, $|u = t|$. If $E \subset \Omega$ measurable, we denote by $v\big|_E$ the restriction of v to the set E. Its Lebesgue measure is denoted by $|E| = \mathrm{meas}\,E$.

We will denote by $L^p(\Omega)$ the classical Lebesgue spaces,

$$1 \leqslant p \leqslant +\infty, \text{ equipped with the norm } |\cdot|_p\,.$$

Definition 1.1.1 (Plateau in a Value t) Let $t \in \mathbb{R}$, *and let* $u : \Omega \to \mathbb{R}$ be measurable.

We will say that u has a plateau at point t if $|u = t| > 0$.

We call the plateaus of u is the set $P(u) = \bigcup_{t \in D_u} \Big\{u = t\Big\}$ where D_u is the set of points where u has a plateau.

Proposition 1.1.1 *Let* $u : \Omega \to \mathbb{R}$ *be measurable,* Ω *of finite measure. Then* D_u *(given in the previous definition) is at most countable.*

Proof For $n \in \mathbb{N}^*$, we set $D_{u,n} = \Big\{t \in D_u : |u = t| > \dfrac{1}{n}\Big\}$.

Then $D_u = \bigcup_{n\in\mathbb{N}^*} D_{u,n}$. Let us show that $D_{u,n}$ is at most enumerable. If this is not the case, then $D_{u,n}$ is an infinite non-countable set and by the axiom of choice, there would exist a countable infinite subset $D'_{u,n} \subset D_{u,n}$. So

$$\frac{1}{n}\text{card}(D'_{u,n}) \leqslant \sum_{t\in D'_{u,n}} |u=t| = \left|\bigcup_{t\in D'_{u,n}} \{x : u(x)=t\}\right| \leqslant |\Omega|,$$

which is absurd.

Consequently D_u itself is at most countable. ◇

Definition 1.1.2 (Distribution Function) Let $u : \Omega \to \mathbb{R}$ measurable, $|\Omega| < +\infty$. We call The distribution function of u is the decreasing function

$$m_u = m : \begin{array}{l} \mathbb{R} \to I\!R \\ t \ \to m(t) \end{array}$$

defined by $m(t) = m_u(t) = \text{measure}\Big\{u > t\Big\}$.

Definition 1.1.3 (Decreasing Rearrangement) Let $u : \Omega \to \mathbb{R}$ be measurable, $|\Omega| < +\infty$. We call The decreasing rearrangement of u is the function u_*: $\Omega_* = \big]0, |\Omega|\big[\to \mathbb{R}$ defined by

$$u_*(s) = \text{Inf }\{t \in \mathbb{R},\ m(t) \leqslant s\}\ .$$

If $s = 0$, $then\ u_*(0) = \text{ess sup}_{\,\Omega} u$, $if\ s = |\Omega|$, $then\ u_*(|\Omega|) = \text{ess inf}_{\,\Omega} u$.

Property 1.1.1

(a) m is right continuous,
(b) $u_*\big(m(t)\big) \leqslant t \quad \forall t$,
(c) $m\,(u_*(s)) \leqslant s \quad \forall s$,
(d) u_* is right continuous on $\big[0, |\Omega|\big)$,
(e) if $u \leqslant v$ a.e. then $u_*(s) \leqslant v_*(s)\ \forall s \in \overline{\Omega_*}$.

Proof

(a) Let $E(h) = \Big\{t < u \leqslant t + h\Big\}$. Then

$$E(h_1) \subset E(h_2) \text{ if } h_1 \leqslant h_2 \text{ and } \bigcap_{h>0} E(h) = \emptyset : \lim_{h\to 0} |E(h)| = 0 \text{ i.e.}$$

$\lim_{h\to 0} m(t+h) = m(t)$.

(b) $u_*\big(m(t)\big) = \text{Inf}\Big\{\theta : m(\theta) \leqslant m(t)\Big\} \leqslant t$.

(c) Let $s \in \Omega_*$. Then there exists $t_n \in \mathbb{R}$, s.t.

$$m(t_n) \leqslant s,\ t_n > u_*(s),\ t_n \xrightarrow[n\to+\infty]{} u_*(s).$$

By right continuity, we have

$$m(t_n) \xrightarrow[n\to+\infty]{} m\big(u_*(s)\big) : m\big(u_*(s)\big) = \lim_n m(t_n) \leqslant s.$$

For $s = |\Omega|$, $m(\text{ess inf}_{\,\Omega} u) \leqslant |\Omega|\, and for\ s = 0$ we have $m(\text{ess sup}_{\,\Omega}) = 0$.

(d) Suppose there exists $s \in \big[0, |\Omega|\big[$ s.t. $\lim_{h\searrow 0} u_*(s+h) \neq u_*(s)$, then there exists γ s.t. $u_*(s) > \gamma > u_*(s+h)$, $\forall h > 0$. Hence

$$m(\gamma) \leqslant m\big(u_*(s+h)\big) \leqslant s + h : m(\gamma) \leqslant s\,.$$

By definition of u_*, $u_*(s) \leqslant \gamma$, a contradiction.

(e) If $u \leqslant v$ then $m_u(t) \leqslant m_v(t)\ \forall t$, which entails by definition $u_*(s) \leqslant v_*(s)$, $\forall s \in \Omega_*$ (by right continuity for $s = 0$, by left continuity for $s = |\Omega|$).

◇

Property 1.1.2 (Fundamentals of Equimeasurability) For any real t, we have:

1. $|u > t| = |u_* > t|$,
2. $|u \geqslant t| = |u_* \geqslant t|$, $|u \leqslant t| = |u_* \leqslant t|$, $|u = t| = |u_* = t|$.

Proof We only show the first statement because: $|u \leqslant t| = |\Omega| - |u > t| = |\Omega| - |u_* > t| = |u_* \leqslant t|$,

$$\lim_{h\searrow 0} |u > t - h| = \lim_{h\searrow 0} |u_* > t - h| \text{ implies } |u \geqslant t| = |u_* \geqslant t|\,.$$

To show 1., let $I(t) = \Big\{\sigma \in \overline{\Omega_*} :\ u_*(\sigma) > t\Big\}$.

- If $t \geqslant \text{ess sup}_{\,\Omega} u$, then $I(t) = \emptyset : \text{meas}\big(I(t)\big) = |u > t| = 0$.

- If $t < \operatorname{ess\,sup}_{\Omega} u$, then $0 \in I(t)$ and as u_* is decreasing, $I(t)$ is an interval. Hence $I(t) = \Big[0, \text{meas}\Big\{u_* > t\Big\}\Big)$. As $u_*\big(m(t)\big) \leqslant t$, $m(t) \notin I(t)$. Therefore $m(t) \geqslant \text{length}\big(I(t)\big) = |u_* > t|$. Thus $\gamma = |u > t| \geqslant |u_* > t|$.

  Let $s > |u_* > t|$. Then $s \notin I(t)$, i.e. $u_*(s) \leqslant t$, $m(t) \leqslant m\big(u_*(s)\big) \leqslant s$. Letting s tend towards, $|u_* > t|$ we then have $|u > t| \leqslant |u_* > t|$. ◇

Corollary 1.1.1 *Let $F : \mathbb{R} \to \mathbb{R}_+$ be Borel function. Then*

$$\int_\Omega F(u)dx = \int_{\Omega_*} F(u_*)ds\ .$$

In particular, $|u|_{L^p(\Omega)} = |u_*|_{L^p(\Omega_*)} \qquad 1 \leqslant p \leqslant +\infty.$

Proof

- If $F(t) = \chi_{(a,b)}(t)$, $-\infty \leqslant a < b \leqslant +\infty$ characteristic function of an interval, then the equality is true.
- If $\mathcal{O}$ is an open subset of $\mathbb{R}$, then $\mathcal{O} = \bigcup_{j \in D}]a_j, b_j[$, $]a_i, b_i[$, $]a_j, b_j[$ are two disjoint intervals whenever $i \neq j$, and D is at most countable. We have

$$\int_\Omega \chi_{\mathcal{O}}(u)dx = \sum_{j \in D} \int_\Omega \chi_{]a_j,b_j[}(u)dx = \sum_{j \in D} \int_{\Omega_*} \chi_{]a_j,b_j[}(u_*)ds = \int_{\Omega_*} \chi_{\mathcal{O}}(u_*)ds.$$

- If A is a closed subset of $\mathbb{R}$, then since $\chi_A = 1 - \chi_{\mathbb{R}\backslash A}$ and $\mathbb{R}\backslash A = \mathcal{O}$ is an open set, then we have the equality. We define

$$\mathcal{M}_1 = \{E \text{ Borel set} : \int_\Omega \chi_E(u)dx = \int_{\Omega_*} \chi_E(u_*)ds\}.$$

  We check that $\mathcal{M}_1$ is a σ-algebra containing all open sets (Carathèodory extension principle), hence it contains the Borel σ-algebra.
- If $F : \mathbb{R} \to \mathbb{R}_+$ is a bounded Borel function then there exist $(a_i^n)_{i \leqslant n}$ and $\{E_i\}_{i \leqslant n}$ with $F_n(\sigma) = \sum_{i=0}^{n} a_i^n \chi_{E_i}(\sigma)$ which is a simple function, E_i a Borel set, $\lim_{n \to +\infty} F_n(\sigma) = F(\sigma)\ \forall \sigma \in \mathbb{R}$, and the family F_n is uniformly bounded in n. Then, by the theorem of dominated convergence we have:

$$\int_\Omega F_n(u)dx = \int_{\Omega_*} F_n(u_*)ds \Longrightarrow \int_\Omega F(u)dx = \int_\Omega F(u_*)ds.$$

- If F is an unbounded Borel function, then we consider the sequence of functions, called *truncations*: $T_k(\sigma) = \begin{cases} \sigma & |\sigma| \leqslant k \\ k \operatorname{sign}(\sigma) & \text{otherwise.} \end{cases}$

$$F_k(\sigma) = T_k \circ F(\sigma) \; satisfies \; 0 \leqslant F_k(\sigma) \leqslant F_{k+1}(\sigma) \leqslant F(\sigma)$$

  and we have

$$\int_\Omega F_k(u) = \int_{\Omega_*} F_k(u_*)ds \overset{\text{(Beppo Levi)}}{\underset{k\to+\infty}{\Longrightarrow}} \int_\Omega F(u) = \int_{\Omega_*} F(u_*).$$

◇

Note 1 We can replace the Lebesgue measures by the weighted measure $|E|_a = \int_E a(x)dx$ if $a > 0$, $a \in L^1(\Omega)$ and define the decreasing rearrangement u_*^a associated with u (See Exercise 11.1.4, Chap. 11, or the articles Rakotoson–Simon [141, 142], Mercaldo A. [113], Brock F. et al. [31]).

Lemma 1.1.1 *Let* $\psi : \mathbb{R} \to \mathbb{R}$ *be increasing. Then, for any measurable* u,

$$(\psi \circ u)_* = \psi(u_*).$$

Proof By equimeasurability, we have:

$$|\psi \circ u > t| = \int_\Omega \chi_{]t,+\infty[}(\psi \circ u)du = \int_{\Omega_*} \left(\chi_{]t,+\infty[} \circ \psi\right)(u_*)ds = |\psi \circ u_* > t|,$$

which means that $(\psi \circ u)_* = (\psi \circ u_*)_*$. Indeed, let $f = \psi \circ u_*$, then f is decreasing. If $s \in \Omega_*$ then the interval $\{f > f(s)\}$ does not contain s but contains 0 if it is non-empty, so we always have $|f > f(s)| \leqslant s$, thus $f_*(s) \leqslant f(s)$ (by definition). As for all k: $\int_{\Omega_*} T_k\big(f(s)\big)ds = \int_{\Omega_*} T_k\big(f_*(s)\big)ds$ and $T_k\left(f_*(s)\right) \leqslant T_k\left(f(s)\right)$, we have:

$$\forall k : T_k\left(f_*(s)\right) = T_k\left(f(s)\right) \text{ a.e. in s.} \Longrightarrow f_*(s) = f(s)$$

a.e., hence $(\psi \circ u)_* = (\psi \circ u_*)_* = \psi \circ u_*$. ◇

1.2 Hardy–Littlewood Inequalities

We will need the following Lyapounov lemma:

Lemma 1.2.1 (Lyapounov) *Let $s \in \Omega_*$ so that there exists $t \in \mathbb{R}$ such that $|u > t| \leqslant s \leqslant |u \geqslant t|$. Then there exists a E measurable set such that:*
$$\begin{cases} \bullet \{u > t\} \subset E \subset \{u \geqslant t\}, \\ \bullet |E| = s. \end{cases}$$

Proof Let P be a polynomial (example $P(x) = \sum_{i=1}^{N} x_i^2 = |x|^2$) and v be the restriction of P to $\{u \geqslant t\} \backslash \{u > t\} = \{u = t\}$ (note that if $|u = t| = 0, s = |u > t|$, $E = \{u > t\}$). Consider

$$E = \Big\{x \in \Omega : u(x) > t\Big\} \cup \Big\{x : u(x) = t,\ v(x) > v_*(s - |u > t|)\Big\} .$$

Since v is stepless (i.e without a plateau), we have

$$|E| = |u > t| + |v_* > v_* (s - |u > t|)| = s .$$

◇

Lemma 1.2.2 (1st Hardy–Littlewood Inequality) *Let $E \subset \Omega$ be measurable, and $u : \Omega \to \mathbb{R}$ be measurable $\geqslant 0$ or integrable. Then,*

$$\int_E u(x)dx \leqslant \int_0^{|E|} u_*(s)ds .$$

Similarly we have

$$\int_E u(x)a(x)dx \leqslant \int_0^{|E|_a} u_*^a(s)ds,\ \textit{for}\ a > 0,\ a \in L^1(\Omega),$$

u_*^a being the decreasing rearrangement of u with respect to the weighted measure adx. If E satisfies $\{u > t_0\} \subset E \subset \{u \geqslant t_0\}$, then we have an equality.

Proof Let $v = u|_E$, the restriction of u to E. So by equimeasurability, we deduce:

$$\int_E u(x)dx = \int_0^{|E|} v_*(\sigma)d\sigma \ .$$

(Which is true if $u \geqslant 0$ or $u \in L^1(E)$.)

But $m_v(t) \leqslant m_u(t)$, so $v_*(s) \leqslant u_*(s),\ s \leqslant |E|$.

Hence the inequality.

If $\{u > t_0\} \subset E \subset \{u \geqslant t_0\}$, then we have:

$$\int_E u\,dx = \int_{E\backslash\{u>t_0\}} u\,dx + \int_{\{u>t_0\}} u\,dx = t_0\left|E\backslash\{u > t_0\}\right| + \int_{u_*>t_0} u_* ds$$

$$\int_0^{|E|} u_* d\sigma = \int_0^{|u>t_0|} u_* d\sigma + \int_{|u>t_0|}^{|E|} u_* d\sigma \leqslant \int_{u_*>t_0} u_* d\sigma + t_0\Big[\,|E| \setminus |u > t_0|\,\Big],$$

indeed $\sigma > m(t_0) \Longrightarrow u_*(\sigma) \leqslant u_*\left(m(t_0)\right) \leqslant t_0$.

$$\int_0^{|E|} u_* d\sigma \leqslant \int_E u\,dx.$$

Further Proof

For this last case, let us show that

$$u_*(\sigma) = v_*(\sigma) \text{ if } \sigma < |E| \,.$$

Let $\sigma \in [0, |E|[$ and $t \in \mathbb{R}$, if $t < t_0$, $m_u(t) \geqslant |E| > \sigma$, if $t \geqslant t_0$, $m_u(t) = m_v(t)$ because

$$\Big(\Big\{x \in \Omega : u(x) > t\Big\} = \Big\{x \in E : u(x) > t\Big\}\Big).$$

So $u_*(\sigma) = \text{Inf}\,\Big\{t \geqslant t_0 : m_u(t) \leqslant \sigma\Big\} = v_*(\sigma)$. ◇

Corollary 1.2.1 (Of the 1st Inequality)

(a) For all $s \in \overline{\Omega_*}$ *if* $u \geqslant 0$ *is measurable or u is integrable:*

$$\int_0^s u_*(\sigma)d\sigma = \text{Max}\,\left\{\int_E u(x)dx, \quad |E| = s\right\}.$$

(b) If $u \geqslant 0$, $\displaystyle\int_0^s u_*(\sigma)d\sigma = \text{Max}\,\left\{\int_E u(x)dx, \quad |E| \leqslant s\right\}.$

Proof

(a) If $|u = u_*(s)| = 0$, then $E = \Big\{u > u_*(s)\Big\}$ satisfies

$$\int_{\{u>u_*(s)\}} u(x)dx = \int_{\Big\{u_*>u_*(s)\Big\}} u_*(\sigma)d\sigma \ .$$

If $|u = u_*(s)| > 0$, then $|u > u_*(s)| \leqslant s \leqslant |u \geqslant u_*(s)|$, there exists a measurable $E \subset \Omega$ s.t. $|E| = s$ and

$$\Big\{u > u_*(s)\Big\} \subset E \subset \Big\{u \geqslant u_*(s)\Big\}.$$

From the previous lemma, we deduce that:

$$\int_E u(x)dx = \int_0^{|E|} u_*(\sigma)d\sigma \ .$$

(b) If $u \geqslant 0$, then we have:

$$\int_0^s u_*(\sigma)d\sigma = \text{Max}\left\{\int_E u(x)dx, \quad |E| = s\right\}$$

and

$$\text{Max}\left\{\int_0^{|E|} u_*(\sigma)d\sigma, \quad |E| \leqslant s\right\} \leqslant \int_0^s u_*(\sigma)d\sigma.$$

◇

Corollary 1.2.2 *Under the same conditions as Lemma 1.2.2:*

$$\int_0^s u_*(\sigma)d\sigma = \text{Max}\left\{\int_\Omega u(x)z(x)dx,\ 0 \leqslant z \leqslant 1,\ \int_\Omega z(x)dx = s\right\}\ \forall s \in \overline{\Omega_*}.$$

Proof Let $0 \leqslant z \leqslant 1$, $\displaystyle\int_\Omega z(x)dx = s > 0$

$$\int_\Omega u(x)z(x)dx = \int_{\{z>0\}} u(x)z(x)dx = \int_0^{|z>0|_z} u_*^z(\sigma)d\sigma$$

(where u_*^z is the decreasing rearrangement with respect to the measure $z(x)dx$ on $\{z > 0\}$, $|z > 0|_z = \displaystyle\int_{\{z>0\}} z(x)dx = s$.)

Now, $0 \leqslant z \leqslant 1 \Longrightarrow u_*^z(\sigma) \leqslant u_*(\sigma)$, hence

$$\int_\Omega u(x)z(x)dx \leqslant \int_0^s u_*(\sigma)d\sigma.$$

As: $\displaystyle \text{Max}\left\{\int_E u(x)dx,\ |E| = s\right\}$

$$\leqslant \text{Max}\left\{\int_\Omega u(x)z(x)dx,\ 0 \leqslant z \leqslant 1,\ \int_\Omega z(x)dx = s\right\}$$

we deduce the result.

See Chap. 11, (Exercise 11.1.18) for another proof of this corollary. ◇

Proposition 1.2.1 *Let $u : \Omega \to \mathbb{R}$ be integrable. For all $s \in \overline{\Omega_*}$, let us set: $F(u) = \min\limits_{t\in\mathbb{R}}\left\{ts + \displaystyle\int_\Omega (u-t)_+dx\right\}$. Then*

$$F(u) = \int_0^s u_*(\sigma)d\sigma\ .$$

Proof Let $F(t,u) = ts + \displaystyle\int_\Omega (u-t)_+dx,\ t \in \mathbb{R}$. So,

$$\begin{aligned}
F(t,u) &= ts + \int_\Omega (u-t)_+(\sigma)d\sigma \\
&= ts + \int_0^{|u\geqslant t|} (u_* - t)(\sigma)d\sigma \\
&= ts + \int_0^s (u_* - t)(\sigma)d\sigma + \int_s^{|u\geqslant t|} (u_* - t)(\sigma)d\sigma \\
&= \int_0^s u_*(\sigma)d\sigma + \int_s^{|u\geqslant t|} (u_* - t)(\sigma)d\sigma.
\end{aligned}$$

We have

$$\int_s^{|u\geqslant t|} (u_* - t)(\sigma)d\sigma \geqslant 0.$$

Indeed:

- if $t \leqslant u_*(s)$ then $|u_* \geqslant t| \geqslant |u_* \geqslant u_*(s)| \geqslant s$ and if $\sigma < |u \geqslant t|$ then $u_*(\sigma) - t \geqslant 0$: $\displaystyle\int_s^{|u\geqslant t|} (u_*(\sigma) - t)\, d\sigma \geqslant 0$;
- if $t > u_*(s)$ then $|u \geqslant t| \leqslant |u > u_*(s)| \leqslant s$ and if $\sigma > |u \geqslant t|$ then $u_*(\sigma) - t < 0$. Consequently, the integral is positive;
- if $t = u_*(s)$, on the interval $(s, |u \geqslant u_*(s)|)$ the function u_* is constant say $u_* = u_*(s)$ then

$$F(u_*(s), u) = \int_0^s u_*(\sigma)d\sigma.$$

Hence $\displaystyle\int_0^s u_*(\sigma)d\sigma \leqslant \text{Min}\ {}_t F(t, u) = F(u) \leqslant F(u_*(s), u) = \int_0^s u_*(\sigma)d\sigma$. ◇

Corollary 1.2.3 (Of Proposition 1.2.1) *Let u, v be two integrable functions. Then, we have the following equivalence:*

$$[A]\begin{cases} \displaystyle\int_0^s u_*(\sigma)d\sigma \leqslant \int_0^s v_*(\sigma)d\sigma, \ \forall s \in \overline{\Omega_*}, \\ \displaystyle\int_\Omega u dx = \int_\Omega v dx, \end{cases}$$

$$\iff [B]\begin{cases} \forall \Phi : \mathbb{R} \to \mathbb{R} \text{ convex and Lipschitzian}, \\ \displaystyle\int_\Omega \Phi(u)dx \leqslant \int_\Omega \Phi(v)dx. \end{cases}$$

Proof Let us show that [B] implies [A] . Indeed, for all $t \in \mathbb{R}$, $\Phi(\sigma) = (\sigma - t)_+$ is Lipschitz convex. Consequently,

$$\int_\Omega (u - t)_+ dx \leqslant \int_\Omega (v - t)_+ dx.$$

By the previous proposition we conclude that

$$\int_0^s u_*(\sigma)d\sigma \leqslant \int_0^s v_*(\sigma)\ \forall s \in [0, |\Omega|].$$

Considering $\Phi(t) = -t$, we deduce $\int_\Omega u(x)dx \geqslant \int_\Omega v(x)dx$, hence [A]. Conversely if Φ is convex and Lipschitzian then Φ' is a bounded increasing function. By convexity

$$\big(v_*(\sigma) - u_*(\sigma)\big)\Phi'\big(u_*(\sigma)\big) \leqslant \Phi\big(v_*(\sigma)\big) - \Phi\big(u_*(\sigma)\big) \qquad \forall \sigma \in \Omega_*.$$

Hence $\int_{\Omega_*} (v_* - u_*)\,\Phi'(u_*)d\sigma \leqslant \int_\Omega \Phi(v)dx - \int_\Omega \Phi(u)dx$. Moreover,

$$\int_{\Omega_*} (v_* - u_*)\,\Phi'(u_*)d\sigma = \int_{\Omega_*} \frac{d}{d\sigma}\int_0^\sigma (v_* - u_*)\,dt\Phi'(u_*)d\sigma$$

$$= -\int_{\Omega_*}\left(\int_0^\sigma (v_* - u_*)\,dt\right)d\Phi'(u_*)$$

(because $\int_0^{|\Omega|}(v_* - u_*)\,dt = 0$). Since $-d\Phi'(u_*) \geqslant 0$ and $\int_0^\sigma (v_* - u_*)\,dt \geqslant 0$, we deduce

$$\int_\Omega \Phi(v)dx - \int_\Omega \Phi(u)dx \geqslant \int_{\Omega_*} (v_* - u_*)\,\Phi'(u_*)d\sigma \geqslant 0\,.$$

◇

Theorem 1.2.1 (2nd Hardy–Littlewood Inequality) *Let $f \in L^p(\Omega)$, $g \in L^{p'}(\Omega)$, $\frac{1}{p} + \frac{1}{p'} = 1$, $1 \leqslant p \leqslant +\infty$. Then*

$$\int_\Omega f(x)g(x)dx \leqslant \int_{\Omega_*} f_*(\sigma)g_*(\sigma)d\sigma\,.$$

We need the following lemma:

Lemma 1.2.3 (Fubini) *Let $f \in L^p(\Omega)$, $f \geqslant a > -\infty$, $g \in L^{p'}(\Omega)$. Then*

$$\int_\Omega f(x)g(x)dx = a\int_\Omega g(x)dx + \int_a^{+\infty} dt \int_{f>t} g(x)dx.$$

Proof Since $f(x) - a = \int_a^{f(x)} dt = \int_a^{+\infty} \chi_{]a,f(x)[}(t)dt$, by the classical Fubini–Tonelli Theorem we have

$$\int_\Omega (f(x) - a)\, g(x)dx = \int_a^{+\infty} dt \int_\Omega g(x)\chi_{]a,f(x)[}(t)dx$$

$$= \int_a^{+\infty} dt \int_\Omega g(x)\chi_{\{f>t\}}(x)dx = \int_a^{+\infty} dt \int_{\{f>t\}} g(x)dx.$$

◇

Proof of the Theorem Start with the case where $f \in L^\infty(\Omega)$, $a = \operatorname{ess\,inf}_\Omega f > -\infty$. By equimeasurability and the first Hardy–Littlewood lemma:

$$\int_\Omega f(x)g(x)dx \leqslant a \int_{\Omega_*} g_* d\sigma + \int_a^{+\infty} dt \int_{f_*>t} g_*(\sigma)d\sigma = \int_{\Omega_*} f_* g_*.$$

If $f \in L^p(\Omega)$, then $T_k(f) \in L^\infty(\Omega)$, $\quad T_k(f)_* = T_k(f_*)$ and

$$\int_\Omega T_k(f)g(x)dx \leqslant \int_\Omega T_k(f_*)g_* d\sigma.$$

When $k \to +\infty$, by the dominated convergence theorem, we obtain

$$\int_\Omega fgdx \leqslant \int_{\Omega_*} f_* g_* d\sigma.$$

◇

1.3 Study of the Continuity of $u \to u_*$

Let's start with some remarks that are consequences of the results we saw above.

Proposition 1.3.1 *If u, v are in $L^\infty(\Omega)$ then*

$$|u_*(\sigma) - v_*(\sigma)| \leqslant |u - v|_\infty \qquad \forall \sigma \in \Omega_*.$$

Proof First note that $(u+c)_* = u_* + c$ ($c =$ constant) (by choosing $\psi(t) = t + c$ in Lemma 1.1.1). Since $|u(x) - v(x)| \leqslant |u - v|_\infty$,

$$v(x) - |u - v|_\infty \leqslant u(x) \leqslant v(x) + |u - v|_\infty \ a.e.$$

Hence

$$v_*(\sigma) - |u - v|_\infty \leqslant u_*(\sigma) \leqslant v_*(\sigma) + |u - v|_\infty \qquad \forall \sigma \, .$$

◇

Proposition 1.3.2 *Let $u,\ v \in L^2(\Omega)$, then*

$$|u_* - v_*|_{L^2(\Omega_*)} \leqslant |u - v|_{L^2(\Omega)} \, .$$

Proof $|u_* - v_*|^2_{L^2(\Omega_*)} = \int_{\Omega_*} u_*^2 d\sigma + \int_{\Omega*} v_*^2 d\sigma - 2\int_{\Omega*} u_* v_* d\sigma$. As $\int_{\Omega_*} u_* v_* \geqslant \int_\Omega uv$ by équimeasurability we have:

$$|u_* - v_*|^2_{L^2(\Omega_*)} \leqslant \int_\Omega u^2 dx + \int_\Omega v^2 dx - 2\int_\Omega uv dx = |u - v|^2_{L^2} \, .$$

◇

Proposition 1.3.3 *Let $u,\ v \in L^1(\Omega)$, then $|u_* - v_*|_{L^1} \leqslant |u - v|_{L^1}$.*

Proof Following the ideas of Crandall and Tartar [48], we have:

$$\min(u, v)(x) = \frac{u(x) + v(x) - |u(x) - v(x)|}{2} \leqslant (u(x) \textit{ and } v(x)) \, .$$

Hence

$$\min(u, v)_* \leqslant \min(u_*, v_*) = \frac{u_* + v_* - |u_* - v_*|}{2} \, .$$

By equimeasurability, we have:

$$\int_{\Omega} [(u+v) - |u-v|]\,(x)dx \leqslant \int_{\Omega_*} [(u_* + v_*) - |u_* - v_*|]\,d\sigma \,.$$

Thus we have: $\displaystyle\int_{\Omega} |u-v|\,(x)dx \geqslant \int_{\Omega_*} |u_* - v_*|\,d\sigma \,.$ ◇

Corollary 1.3.1 *For all $p \in [1, +\infty]$, the mapping $u \in L^p(\Omega) \to u_* \in L^p(\Omega_*)$ is (strongly) continuous.*

We will even show that it is a contraction.

Theorem 1.3.1 (Contraction Property) *Let ρ be a convex function of $\mathbb{R}$ in $\mathbb{R}$, and u and v two measurable functions such that $v \in L^\infty(\Omega)$. Then*

$$\int_{O_*} \rho\left((u+v)_* - u_*\right) d\sigma \leqslant \int_{\Omega} \rho(v)dx. \tag{1.1}$$

In particular, if $u \in L^\infty(\Omega)$ then:

$$\int_{\Omega_*} \rho\left(v_* - u_*\right) d\sigma \leqslant \int_{\Omega} \rho(v-u)dx. \tag{1.2}$$

Corollary 1.3.2 *We further assume that ρ satisfies the following growth condition: $\exists \alpha > 0,\ \beta > 0$: s.t.*

$$\forall t \in \mathbb{R}, \quad |\rho(t)| \leqslant \alpha\,|t|^p + \beta \qquad for\ 1 \leqslant p < +\infty.$$

So $\forall\,(u, v) \in L^p(\Omega) \times L^p(\Omega)$ we have:

$$\int_{\Omega_*} \rho(u_* - v_*)d\sigma \leqslant \int_{\Omega} \rho(u-v)dx.$$

In particular, the map $u \in L^p(\Omega) \to u_* \in L^p(\Omega_*)$ is a contraction.

Proof of the Theorem

1st case: $u \in L^\infty(\Omega),\ \ \rho \in C^2(\mathbb{R})$.

We set $\gamma = \min\left(-|u|_\infty, -|u+v|_\infty\right)$. Then a.e. in x

$$-\int_\gamma^{u(x)} \int_\gamma^{u(x)+v(x)} \rho''(t-s)dtds = -\int_\gamma^{u(x)} \Big[\rho'\left(u(x)+v(x)-s\right) - \rho'(\gamma - s)\Big]ds$$

$$= -\rho\big(\gamma - u(x)\big) + \rho(0) + \rho\big(v(x)\big) - \rho\big(u(x)+v(x)-\gamma\big).$$

Thus

$$\rho\left(v(x)\right) = -\int_\gamma^{u(x)} \int_\gamma^{u(x)+v(x)} \rho''(t-s)dtds + \rho\big(\gamma - u(x)\big) - \rho(0) + \rho\big(u(x)+v(x)-\gamma\big).$$

Repeating the same argument, we have the identity:

$$\begin{aligned}\rho(\alpha{-}\beta) &\\ = -&\int_\gamma^{+\infty}\int_\gamma^{+\infty} \rho''(t-s)\, H(\beta - s)\, H(\alpha - t)dsdt \\ &+\rho(\gamma-\beta) - \rho(0) + \rho(\alpha-\gamma)\end{aligned}$$

if α and $\beta \geqslant \gamma$, where H is the Heaviside function

$$H(t) = \begin{cases} 0 & \text{if } t < 0 \\ 1 & \text{otherwise.}\end{cases}$$

We integrate the function $\rho(v)$ over Ω and use equimeasurability:

$$\int_\Omega \rho\left(v(x)\right)dx$$

$$= -\int_\gamma^{+\infty}\int_\gamma^{+\infty} \rho''(t-s) \int_\Omega H\left(u(x)-s\right)\cdot H\left(u(x)+v(x)-t\right)dxdsdt$$

$$+\int_{\Omega_*} \rho\left(\gamma - u_*(s)\right)ds - \rho(0)\,|\Omega| + \int_{\Omega_*} \rho\left[(u+v)_*(s) - \gamma\right]ds. \tag{1.3}$$

By the Hardy–Littlewood inequality, we have:

$$\int_\Omega H(u(x)-s)\,H\,((u+v)(x)-t)\,dx \leqslant \int_{\Omega_*} H(u-s)_* H\,((u+v)-t)_*\,d\sigma. \qquad (1.4)$$

But $\sigma \to H(\sigma - t)$ is increasing, thus $H\,(u(t)-s)_* = H(u_* - s)$, $H\,((u+v)-t)_* = H\,((u+v)_* - t)$. Since ρ is convex, i.e. $-\rho'' \leqslant 0$, we obtain from the relations (1.3), (1.4) the following equality:

$$\int_\Omega \rho\,(v(x))\,dx \geqslant -\int_\gamma^{+\infty}\int_\gamma^{+\infty} \rho''(t-s) \int_{\Omega_*} H\,(u_*(\sigma)-s) H((u+v)_*(\sigma)-t)\,d\sigma\,dt\,ds$$

$$+\int_{\Omega_*} \rho\,(\gamma - u_*(\sigma))\,d\sigma + \int_{\Omega_*} \rho\,((u+v)_*(\sigma)-\gamma)\,d\sigma - \rho(0)\,|\Omega|\,. \qquad (1.5)$$

A calculation similar to the one done previously (or the previous identity) shows that the right-hand side of the relation (1.5) is none other than $\int_{\Omega_*} \rho\,((u+v)_* - u_*)\,(\sigma)d\sigma$:

$$\int_{\Omega_*} \rho\,((u+v)_* - u_*)\,(\sigma)d\sigma \leqslant \int_\Omega \rho(v)(x)dx.$$

2nd case: $\rho \in C^2$, u **measurable.**
Let the function be:

$$T_n(\sigma) = [n - (n - |\sigma|)_+]\,\mathrm{sign}(\sigma)$$

then $T_n(u) \in L^\infty$ and $T_n(u)_* = T_n(u_*)$ a.e. and

$$\int_{\Omega_*} \rho\,(T_n(u+v)_* - T_n(u_*)) \leqslant \int_\Omega \rho\,(T_n(u+v) - T_n(u))\,.$$

By the dominated convergence theorem

$$\int_{\Omega_*} \rho\,\underbrace{[(u+v)_* - u_*]}_{\in L^\infty(\Omega_*)} = \lim_n \int_{\Omega_*} \rho\,(T_n(u+v)_* - T_n(u_*))$$

$$\leqslant \lim_n \int_\Omega \rho\,(T_n(u+v) - T_n(u)) = \int_\Omega \rho(v)dx.$$

3rd case: ρ is convex.
If (θ_j) is a regularizing sequence $\theta_j \geqslant 0$ then $\rho_j = \theta_j * \rho \in C^\infty(\mathbb{R})$ is convex. Since ρ_j converges uniformly to ρ on any compact set, we have the result (here again we used the fact that

$$(u+v)_* - u_* \in L^\infty(\Omega_*)).$$

By replacing v with $v-u$, we have:

$$\int_{\Omega_*} \rho(v_* - u_*)d\sigma \leqslant \int_\Omega \rho(v-u)(x)dx.$$

◇

Proof of the Corollary Let the function:

$$T_k(\sigma) = [k - (k - |\sigma|)_+]\,\mathrm{sign}(\sigma)$$

$$\int_{\Omega_*} \rho\,(T_k(v_*) - T_k(u_*))\,d\sigma \leqslant \int_\Omega \rho\,(T_k(v) - T_k(u))\,dx.$$

We also have,

$$\begin{cases} |\rho\,(T_k(v_*) - T_k(u_*))| \leqslant c\left(|v_*|^p + |u_*|^p\right) + \beta \in L^1(\Omega_*) \\ \lim\limits_{k\to+\infty} \rho\,(T_k(v_*) - T_k(u_*)) = \rho(v_* - u_*)\ a.e. \end{cases}$$

By the dominated convergence theorem:

$$\lim_{k\to+\infty} \int_{\Omega_*} \rho\,(T_k(v_*) - T_k(u_*)) = \int_{\Omega_*} \rho(v_* - u_*)d\sigma,$$

likewise

$$\lim_{k\to+\infty} \int_\Omega \rho\,(T_k(v) - T_k(u)) = \int_\Omega \rho(v-u)d\sigma\ .$$

◇

Example $\rho(t) = |t|^p \quad p \geqslant 1$ is convex.
Here is a direct corollary of Corollary 1.3.2 and Proposition 1.2.1.

Corollary 1.3.3 $\forall (u, v) \in L^1(\Omega) \times L^1(\Omega)$

1. $\forall t \in \mathbb{R} \quad \int_{\Omega_*}(|u_* - v_*| - t)_+ d\sigma \leqslant \int_\Omega (|u - v| - t)_+ dx.$
2. $\forall s \in \overline{\Omega}_* \quad \int_0^s |u_* - v_*|_*(\sigma) d\sigma \leqslant \int_0^s |u - v|_*(\sigma) d\sigma.$

Proof of the Theorem For statement 1., we apply Corollary 1.3.2 with the convex function $\sigma \to (|\sigma| - t)_+$. For part 2., we combine statement 1. with Proposition 1.2.1 ◇

Note The same approach leads to the Lorentz–Shimogaki inequality (see Bennett–Sharphey [21] or Bénilan–Crandall [20]).

1.4 Function Spaces Related to Rearrangement

Let us now define the following function spaces:

$$L^0(\Omega) = \left\{ v : \Omega \to \mathbb{R} \text{ measurable} \right\}$$

$$L^0_+(\Omega) = L^0(\Omega) \cap \{v : \Omega \to \mathbb{R}_+\}.$$

Definition 1.4.1 (norm on $L^0_+(\Omega)$) Let $\rho : L^0_+(\Omega) \to \overline{\mathbb{R}}_+$. We will say that ρ is a norm if it satisfies the following four axioms: $\forall f, g \in L^0_+(\Omega)$:

1. if $0 \leqslant f \leqslant g$ *then* $\rho(f) \leqslant \rho(g)$ (ρ is monotonous)
2. $\rho(\lambda f) = |\lambda| \rho(f), \quad \forall \lambda \in \mathbb{R}$ (ρ is homogeneous)
3. $\rho(f + g) \leqslant \rho(f) + \rho(g)$ (ρ satisfies the triangle inequality)
4. $\rho(f) = 0 \iff f = 0$ (ρ is definite).

Sometimes, we will use only some of the properties 1. to 4., as is the case for the applications $f \to |f|_{p,q}$ defined below.

If ρ is a map defined on $L^0_+(\Omega)$ then we extend it to $L^0(\Omega)$ by setting, for $f \in L^0(\Omega)$, $\rho(f) = \rho(|f|)$.

We will say that ρ is non-trivial if there exists $f_0 > 0$ such that $\rho(f_0) > 0$ and finite.

Definition 1.4.2 Let ρ be a norm on $L^0(\Omega)$. We will say that ρ is an invariant norm under rearrangement (r.i) if satisfying $f_* = g_*$ $(f, g \in L^0(\Omega))$ implies $\rho(f) = \rho(g)$.

Definition 1.4.3 (Fatou Norm) We will say that a norm ρ is a Fatou norm if in addition we have: $0 \leqslant f_n(x) \leqslant f_{n+1}(x) \xrightarrow[n\to+\infty]{} f(x)$ a.e. implies $\rho(f_n) \xrightarrow[n\to+\infty]{} \rho(f)$.

We describe below some spaces which, like the Lebesgue spaces, have Fatou norms that are non-trivial and invariant under rearrangement. These spaces are the most frequently used in the literature.

To introduce Lorentz spaces, let us define for $f \in L^0(\Omega)$, $s \in \Omega_*$

$$|f|_{**}(s) = \frac{1}{s}\int_0^s |f|_*(\sigma)d\sigma$$

and for $1 \leqslant p \leqslant +\infty, \quad 0 < q \leqslant +\infty$

$$|f|_{(p,q)} = \left[\int_{\Omega_*}\left[t^{\frac{1}{p}}|f|_{**}(t)\right]^q \frac{dt}{t}\right]^{\frac{1}{q}} \quad if\ q < +\infty$$

and

$$|f|_{(p,q)} = \sup_{0\leqslant t<|\Omega|} t^{\frac{1}{p}}|f|_{**}(t) \quad if\ q = +\infty\,.$$

Definition 1.4.4 (Lorentz Spaces) Let $1 \leqslant p \leqslant +\infty, \quad 0 < q \leqslant +\infty$. The Lorentz space $L^{p,q}(\Omega)$ is the set

$$\left\{f \in L^0(\Omega), \qquad |f|_{(p,q)} < +\infty\right\}.$$

Lemma 1.4.1 *Let $1 \leqslant p \leqslant +\infty, \quad 1 \leqslant q \leqslant +\infty$. Then $L^{p,q}(\Omega)$ is a normed vector space with $\rho(f) = |f|_{(p,q)}$. Moreover, ρ is a Fatou norm, invariant under rearrangement (and non-trivial).*

Proof of the Theorem Since $\displaystyle\int_0^s |f|_*(t)dt = \text{Max}\left\{\int_E |f|dx, \quad |E| \leqslant s\right\}$, thus $f \to \displaystyle\int_0^s |f|_*(t)dt$ is sublinear, i.e.

$$\int_0^s |f+g|_*(t)dt \leqslant \int_0^s |f|_*(t)dt + \int_0^s |g|_*(t)dt.$$

As $|\lambda f|_* = |\lambda|\,|f|_*\ \forall\,\lambda \in \mathbb{R}$, we have $\displaystyle\int_0^s |\lambda f|_*(t)dt = |\lambda| \int_0^s |f|_*(t)dt$.

If $0 \leqslant f \leqslant g$ then $0 \leqslant f_* \leqslant g_*$, hence $\displaystyle\int_0^s f_*(t)dt \leqslant \int_0^s g_*(t)dt$.

We deduce from these remarks that $f \to |f|_{(p,q)}$ is a norm if $1 \leqslant q \leqslant +\infty$.

This norm is invariant under rearrangement.

First note that if $f_* = g_*$ then $|f|_* = |g|_*$.

Indeed, $\displaystyle\int_{\Omega_*} \Phi(T_k(|f_*|)) = \int_{\Omega_*} \Phi(T_k(|g_*|))$, so $\displaystyle\int_\Omega \Phi(T_k(|f|)) = \int_\Omega \Phi(T_k(|g|))$, i.e. $\displaystyle\int_\Omega \Phi\left(T_k\left(|f|_*\right)\right) = \int_\Omega \Phi\left(T_k\left(|g|_*\right)\right)$ for all convex Lipschitzian. $\Phi : \mathbb{R} \to \mathbb{R}$. Thus,

$$\int_0^s T_k\left(|f|_*\right) = \int_0^s T_k\left(|g|_*\right) \Longrightarrow |f|_* = |g|_* \ .$$

From this, we see that if $f_* = g_*$ then $|f|_{(p,q)} = |g|_{(p,q)}$.

To show that this is a Fatou norm, we know (see Exercise 11.1.6), that: if $0 \leqslant f_n(x) \leqslant f_{n+1}(x) \xrightarrow[n\to+\infty]{} f(x)$ then

$$f_{n*}(\sigma) \leqslant f_{(n+1)*}(\sigma) \underset{n}{\to} f_*(\sigma) \text{ everywhere.}$$

By the Beppo Levi theorem,

$$\int_0^t f_{n*}(\sigma)d\sigma \underset{n}{\to} \int_0^t f_*(\sigma)d\sigma \quad \forall\, t \in \overline{\Omega_*},$$

which leads to $q < +\infty$:

$$\left[t^{\frac{1}{p}}\,|f_n|_{**}(t)\right]^q \leqslant \left[t^{\frac{1}{p}}\,|f_{n+1}|_{**}(t)\right]^q \xrightarrow[n\to+\infty]{} \left[t^{\frac{1}{p}}\,|f|_{**}(t)\right]^q .$$

Again by the Beppo Levi theorem we deduce:

$$|f_n|_{(p,q)} \xrightarrow[n\to+\infty]{} |f|_{(p,q)} \ .$$

If $q = +\infty$, we consider a s.t. $0 < a < \rho(f)$.

By definition of the essential supremum, there exists $t_a \in]0, |\Omega|[$ s.t.

$$0 < t_a^{1/p} |f|_{**}(t_a) \leqslant \rho(f).$$

By the Beppo Levi theorem, we have

$$\lim_{n\to+\infty} \int_0^{t_a} f_{n*}(s)ds = \int_0^{t_a} f_*(s)ds.$$

Thus, from rank n_a we have, $\forall n \geqslant n_a$,

$$a < t_a^{1/p} |f_n|_{**}(t_a) \leqslant \rho(f_n) \leqslant \rho(f).$$

When $n \to +\infty$, we then have: $a \leqslant \lim_n \rho(f_n) \leqslant \rho(f)$. By making a tend towards $\rho(f)$, we deduce $\rho(f) = \lim_n \rho(f_n)$. ◇

We can replace the previous norm by the equivalent quantity:

$$|f|_{p,q} = \left[\int_{\Omega_*} \left[t^{\frac{1}{p}} |f|_*(t)\right]^q \frac{dt}{t}\right]^{\frac{1}{q}} \quad if\ q < +\infty$$

and

$$|f|_{p,+\infty} = \sup_{0<t<|\Omega|} \left[t^{\frac{1}{p}} |f|_*(t)\right] \quad if\ q = +\infty\,.$$

The map $f \to |f|_{p,q}$ is not in general a norm (unless $1 \leqslant q \leqslant p$, see [21] and Lemma 4.2.1) but nevertheless, it is monotone, definite and homogeneous in the sense of Definition 1.4.1. If $p = q$, by equimeasurability we recover the classical Lebesgue spaces.

Hardy's inequality will allow us to show part of the equivalence:

Lemma 1.4.2 (Hardy) *Let* $1 \leqslant p < +\infty$, $r > 0$, $f \geqslant 0$ *be measurable on* $[0, +\infty[$. *Then*

$$\int_0^{+\infty} F(x)^p x^{p-r-1} dx \leqslant \left(\frac{p}{r}\right)^p \int_0^{+\infty} f(t)^p t^{p-r-1} dt$$

where

$$F(x) = \frac{1}{x}\int_0^x f(t)dt,\ x > 0\,.$$

Proof of the Theorem Applying Hölder's inequality with $d\mu(t) = t^{\frac{r}{p}-1}dt$,

$$\left(\int_0^x f(t)dt\right)^p = \left(\int_0^x f(t)t^{1-\frac{r}{p}}\, t^{\frac{r}{p}-1}dt\right)^p$$

$$\leqslant \left(\frac{p}{r}\right)^{p-1} x^{r(1-\frac{1}{p})} \int_0^x [f(t)]^p\, t^{p-r-1+\frac{r}{p}}dt.$$

From this we have:

$$\int_0^{+\infty}\left(\int_0^x f(t)dt\right)^p x^{-r-1}dx \leqslant \left(\frac{p}{r}\right)^{p-1}\int_0^{+\infty} x^{-1-\frac{r}{p}} \int_0^x [f(t)]^p\, t^{p-r-1+\frac{r}{p}}dt. \tag{1.6}$$

By applying Fubini's theorem

$$\int_0^{+\infty} x^{-1-\frac{r}{p}} \int_0^x [f(t)]^p\, t^{p-r-1+\frac{r}{p}}dt$$

$$= \int_0^{+\infty} [f(t)]^p\, t^{p-r-1+\frac{r}{p}} \left(\int_t^{+\infty} x^{-1-\frac{r}{p}}dx\right) dt$$

$$= \frac{p}{r}\int_0^{+\infty} t^p\, [f(t)]^p\, t^{-r-1}dt. \tag{1.7}$$

By combining the relations (1.6) and (1.7), we have the result. ◇

Using Hardy's lemma, we deduce:

Corollary 1.4.1 *Let $f \in L^{p,q}(\Omega)$, $1 < p < +\infty$, $1 \leqslant q \leqslant +\infty$. Then*

$$|f|_{(p,q)} \leqslant \frac{p}{p-1}|f|_{p,q}\,.$$

Proof of the Theorem If $q < +\infty$, we choose $r = q\left(1 - \dfrac{1}{p}\right)$ and we replace p by q in the previous lemma. After extending $|f|_*$ by zero ($|f| \geqslant 0$), we have:

$$\int_{\Omega_*} |f|_{**}(t)^q t^{q-r-1}dt \leqslant \left(\frac{q}{r}\right)^q \int_{\Omega_*} |f|_*(t)^q t^{q-r-1}dt,$$

hence the result.

If $q = +\infty$ we write

$$|f|_{(p,+\infty)} = \sup_t \left[t^{\frac{1}{p}-1} \int_0^t \sigma^{-\frac{1}{p}} \sigma^{\frac{1}{p}} |f|_*(\sigma) d\sigma \right]$$

$$\leqslant \sup_t \left[t^{\frac{1}{p}-1} \int_0^t \sigma^{-\frac{1}{p}} d\sigma \right] |f|_{p,+\infty} = \frac{p}{p-1} |f|_{p,+\infty} .$$

◇

Definition 1.4.5 (Banach Function Space) We always assume that $|\Omega| < +\infty$ (for simplicity). The Fatou norms ρ on $L^0(\Omega)$ satisfying $\rho(\chi_\Omega) < +\infty$, with $\displaystyle\int_\Omega |f|(x)dx \leqslant C(\Omega)\rho(f), \ \forall f \in L^0(\Omega)$, are associated with spaces called Banach Function Spaces, defined by:

$$L(\Omega, \rho) = \left\{ g \in L^0(\Omega) : \rho(g) < +\infty \right\}.$$

Proposition 1.4.1 *$L(\Omega, \rho)$ is a Banach space for the natural norm*

$$||g|| = \rho(g).$$

Related measure-theoretic theorems, such as the Beppo Levi theorem, Fatou's lemma, and Lebesgue's dominated convergence theorem can be found in the books of Federer [71], Hewitt–Stromberg [98], Brézis [29] and J.E. Rakotoson-J.M. Rakotoson [137].

Here are some recent examples of such spaces.

The first example is the space G-Gamma($G\Gamma$), which was introduced in 2005 by Fiorenza–Rakotoson [84]. Its study has been widely developed by several authors, including Gogatishvili et al. [95], M.R. Formica and L. Zitouni et al. [89].

To simplify, we assume that the measure of Ω is 1. We let w be a positive measurable function on Ω and p and m be two real numbers in $[1, +\infty[$. We further assume that

$$\int_0^t w(s) s^{\frac{m}{p}} ds + \int_t^1 w(s) ds < +\infty \quad \forall t \in]0, 1[.$$

We then define
$G\Gamma(p, m, w) =$

$$\left\{ f \in L^0(\Omega) : ||f||_{G\Gamma(p,m,w)} = \left[\int_0^1 w(t) \left(\int_0^t |f|_*^p(s) ds \right)^{\frac{m}{p}} \right]^{\frac{1}{m}} \text{ is finite} \right\}$$

Proposition 1.4.2

1. *The map* $f \in L^0(\Omega) \to ||f||_{G\Gamma(p,m,w)}$ *is a Fatou norm, invariant under rearrangement.*
2. *The space* $G\Gamma(p, m, w)$ *is a Banach function space equipped with the previous norm*

Before proving this last proposition, let us note that for appropriate choices of p, m and w, we find the previous spaces. In particular:

$$G\Gamma\left(1, q, t^{\frac{q}{p}-q-1}\right) = L^{p,q}(\Omega)$$

$1 \leqslant p < +\infty$, $1 \leqslant q < +\infty$, $G(1, 1, t) = L(\text{Log}\, L)$ (Zygmund space.)

If $w(t) = t^{-1}(1 - \text{Log}\, t)^{\frac{\theta}{p-1}}$, $\theta > 0$, $1 < p < +\infty$ the space is denoted $L^{(p,\theta}(\Omega)$ and is called the small Lebesgue space. This space was introduced by A. Fiorenza to characterize the dual and associated space of the following Banach function space:

$$L^{p),\theta}(\Omega) = \left\{ f \in L^0(\Omega) : \sup_{0<\varepsilon<p-1} \left[\varepsilon^\theta \int_\Omega |f|^{p-\varepsilon} dx \right]^{\frac{1}{p-\varepsilon}} < +\infty \right\}.$$

This last space was introduced by Iwaniec and Sbordone [101], who named it the large or grand Lebesgue space. We will see later other generalizations of this space.

Proof of Proposition 1.4.2

1. Let us first show that it is a norm. By the Hardy–Littlewood equality (see Corollary 1.2.1), we have for $f \in L^0(\Omega)$, $t \in [0, 1]$

$$\left(\int_0^t |f|_*^p(s) ds \right)^{\frac{1}{p}} = \sup \left\{ ||f||_{L^p(E)} : |E| = t \right\}.$$

Consequently, we have for $g \in L^0(\Omega)$

$$\left(\int_0^t |f+g|_*^p(s)ds\right)^{\frac{1}{p}} \leqslant \left(\int_0^t |f|_*^p(s)ds\right)^{\frac{1}{p}} + \left(\int_0^t |g|_*^p(s)ds\right)^{\frac{1}{p}}.$$

From this inequality, we apply the triangle inequality in the space $L^m\Big(]0,1[;w\Big)$ to obtain:

$$||f+g||_{G\Gamma(p,m,w)} \leqslant ||f||_{G\Gamma(p,m,w)} + ||g||_{G\Gamma(p,m,w)}.$$

The homogeneity and the fact that the map is definite follow directly from its definition (because $|\lambda f|_* = \lambda |f|_*$).

As for monotonicity, we know that if $0 \leqslant f \leqslant g$ then $0 \leqslant f_* \leqslant g_*$ everywhere, and consequently

$$||f||_{G\Gamma(p,m,w)} \leqslant ||g||_{G\Gamma(p,m,w)}.$$

This norm is invariant under rearrangement because we know that if $f_* = g_*$ then $|f|_* = |g|_*$, which implies the result.

To show the second statement, let us consider $E \subset \Omega$. Then by hypothesis

$$||\chi_E||_{G\Gamma(p,m,w)} \leqslant \left(\int_0^{|E|} w(t)t^{\frac{m}{p}}dt\right)^{\frac{1}{p}} < +\infty.$$

Let $f \in L^0(\Omega)$, $f \geqslant 0$. Let us show that, for any measurable E in Ω,

$$\int_E f(x)dx \leqslant \frac{|E|}{\left(\int_0^{|E|} w(t)t^{\frac{m}{p}}dt\right)^{\frac{1}{m}}} ||f||_{G\Gamma(p,m,w)}. \tag{1.8}$$

According to Hölder's inequality, we have:

$$\int_E f(x)dx \leqslant |E|^{1-\frac{1}{p}} \left(\int_E f(x)^p dx\right)^{\frac{1}{p}}. \tag{1.9}$$

According to the previous Hardy–Littlewood inequality (Corollary 1.2.1) $|E|^{1-\frac{1}{p}} \left(\int_E f(x)^p dx\right)^{\frac{1}{p}}$

$$\leqslant \frac{|E|\left(\frac{1}{|E|}\int_0^{|E|} f_*^p(s)ds\right)^{\frac{1}{p}} \left(\int_0^{|E|} w(t)t^{\frac{m}{p}}dt\right)^{\frac{1}{m}}}{\left(\int_0^{|E|} w(t)t^{\frac{m}{p}}dt\right)^{\frac{1}{m}}}$$

Since the map $t \to \frac{1}{t}\int_0^t f_*^p(s)ds$ is decreasing, we deduce $|E|^{1-\frac{1}{p}}\left(\int_E f(x)^p dx\right)^{\frac{1}{p}}$

$$\leqslant \frac{|E|\left[\int_0^{|E|} w(t)t^{\frac{m}{p}}\left(\frac{1}{t}\int_0^t f_*^p(s)ds\right)^{\frac{m}{p}} dt\right]^{\frac{1}{m}}}{\left(\int_0^{|E|} w(t)t^{\frac{m}{p}}dt\right)^{\frac{1}{m}}}$$

which leads to the relation 1.8. This completes the proof that $G\Gamma(p, m, w)$ provided with the previous norm is a Banach function space.

Over the last two decades, many authors have been interested Lebesgue and Sobolev spaces with variable exponents. We will not describe these spaces, but we refer the reader to two complementary works on this subject, one by Diening et al. [66] and the other by Cruz-Uribe and Fiorenza [50].

We have studied monotone rearrangement for Lebesgue spaces with variable exponents (see Fiorenza–Rakotoson [85] and [86]). Some of these results will be given in Chap. 10 . ◇

1.5 Ryff's Theorem and Consequences

In this last section, we will give some general properties that will be useful later.

Definition 1.5.1 A map $\sigma : \overline{\Omega} \to \overline{\Omega}_* = [0, |\Omega|]$ is said to be a measure-preserving if:

$$\forall\, E \subset \overline{\Omega}_* \text{ measurable,} \quad \left|\sigma^{-1}(E)\right| = |E|\,.$$

We then have the following theorem due to Ryff:

Theorem 1.5.1 (Ryff's Theorem) *Let Ω be a set of finite measure, $f \in L^0(\Omega)$. Then there exists a measure-preserving map $\sigma : \overline{\Omega} \to \overline{\Omega}_*$ such that:*

$$f_* \circ \sigma = f, \text{ almost everywhere in } \Omega.$$

Note If f is stepless, i.e. without a plateau, so that $\text{meas}\big(P(f)\big) = 0$, then we can take $\sigma(x) = |f > f(x)|$, for $x \in \Omega$.

As a consequence of this theorem we have:

Proposition 1.5.1 *Let f and g be two functions of $L^0_+(\Omega)$. Then there exists $g_1 \geqslant 0$ equimeasurable with g (i.e. $g_{1*} = g_*$) such that*

$$\int_\Omega f g_1 dx = \int_{\Omega_*} f_* g_* d\sigma.$$

Proof of Proposition 1.4.2 By Ryff's theorem, there exists preserving-measure σ such that $f = f_* \circ \sigma$. Let $g_1 = g_* \circ \sigma$. As $g_{1*} = g_*$, since σ preserves measures, we verify, $\forall t \in \mathbb{R}$, $|g_1 > t| = |g > t|$. For the same reason, we have:

$$\int_\Omega f g_1 dx = \int_\Omega (f_* g_*)\big(\sigma(x)\big) dx = \int_{\Omega_*} f_* g_* dt.$$

$\diamondsuit$

As a direct consequence of this proposition, we have:

Proposition 1.5.2 *Let f and g be two elements of $L^0_+(\Omega)$. Then,*

$$\int_{\Omega_*} f_* g_* dt = \text{Max}\left\{\int_\Omega f \overline{g} dx,\ g_* = \overline{g}_*\right\}.$$

Remark If $(f, g) \in L^p(\Omega) \times L^{p'}(\Omega)$, $\dfrac{1}{p} + \dfrac{1}{p'} = 1$, we can prove Proposition 1.5.2 by a density argument (see Exercise 11.1.12).

An application of Proposition 1.5.2 concerns associate norms:

Definition 1.5.2 (Of an Associate Norm) Let ρ be a non-trivial norm on $L^0(\Omega)$. The map $\rho' : L^0(\Omega) \to \mathbb{R}_+$ given by:

$$\rho'(f) = \sup\left\{\int_\Omega |f\psi|\, dx,\ \rho(\psi) \leqslant 1\right\}.$$

is called the associate norm of ρ.

We verify that ρ' is a Fatou norm.

Proposition 1.5.3 *Let ρ be a norm invariant under rearrangement. Then,*

$$\rho'(f) = \sup\left\{\int_{\Omega_*} |f|_* \, |g|_* \, dt, \quad \rho(g) \leqslant 1\right\},$$

$$\rho''(f) = \sup\left\{\int_{\Omega_*} |f|_* \, |g|_* \, dt, \quad \rho'(g) \leqslant 1\right\}.$$

In particular, the norm ρ' is invariant under rearrangement.

Proof of Proposition 1.4.2 By definition of ρ' and the Hardy–Littlewood inequality we have :

$$\rho'(f) \leqslant \sup\left\{\int_{\Omega_*} |f|_* \, |g|_* \, dt, \quad \rho(g) \leqslant 1\right\}. \qquad (E1)$$

By the previous Proposition 1.5.2, and by the fact that ρ is invariant under rearrangement, we have:

$$\begin{aligned} \int_{\Omega_*} |f|_* \, |g|_* \, dt &= \sup\left\{\int_{\Omega} |f| \, |\overline{g}| \, dx, \quad |\overline{g}|_* = |g|_*\right\} \\ &\leqslant \sup\left\{\int_{\Omega} |f| \, |\overline{g}| \, dx, \quad \rho(\overline{g}) \leqslant 1\right\} \\ &= \rho'(f). \qquad (E2) \end{aligned}$$

By combining the previous inequalities (E1) and (E2), we deduce the first equality.

The second equality follows from the first since ρ' is then invariant under rearrangement. ◇

Remarks We have the following Lorentz–Luxemburg theorem (see [21, 39]):

$$\text{if } \rho \text{ is a Fatou norm then } \rho'' = \rho.$$

1.6 Construction of the Monotone Rearrangement of a Function u in Dimension 1

Since the map $u \in L^1(a,b) \to u_* \in L^1(0,b-a)$ is strongly continuous, it suffices to construct the monotone rearrangement of a function u in dimension 1 in the framework of simple functions.

Let

$$u(x) = \sum_{j=1}^{n} u_j \chi_{E_j}(x), \quad x \in [a,b].$$

We sort the values of u in strict descending order, i.e. $t_1 > t_2 > \ldots > t_p$. Each value t_j is associated with the set $\{u = t_j\} = F_j$.

By setting:

$$a_0 = 0,\ a_1 = |F_1|,\ a_2 = |F_1|+|F_2|, \ldots, a_p = |F_1|+|F_2|+\ldots+|F_p|=b-a$$

$$t_1 = \text{Max}\ _{x\in[a,b]}u(x), \qquad t_p = \text{Min}\ _{x\in[a,b]}u(x),$$

we deduce:

$$u_*(s) = \begin{cases} t_1 & 0 \leqslant s < a_1 \\ t_2 & a_1 \leqslant s < a_2 \\ \vdots & \\ t_p & a_{p-1} \leqslant s \leqslant a_p = b-a. \end{cases}$$

The associated distribution function $m_u(t)$ can be constructed simultaneously since

$$m_u(t) = \begin{cases} b-a = a_p & if\ t < t_p, \\ a_{p-1} & if\ t_p \leqslant t < t_{p-1}, \\ \vdots & \\ a_1 & if\ t_2 \leqslant t < t_1, \\ 0 & if\ t \geqslant t_1. \end{cases}$$

This description can be extended to functions in dimension $\geqslant 2$. The choice here is motivated by clarity and ease of calculation in dimension 1.

The algorithm can be interpreted as follows: if we call "block u_j" the "rectangle" $E_j \times \{u_j\}$, we first gather all the rectangles of the same height u_j. We thus obtain new "blocks" $[a_{j-1}, a_j] \times \{t_j\}$ (Fig. 1.1).

Then we put on the left the highest block $[a_0, a_1] \times \{t_1\}$ and so on in descending order.

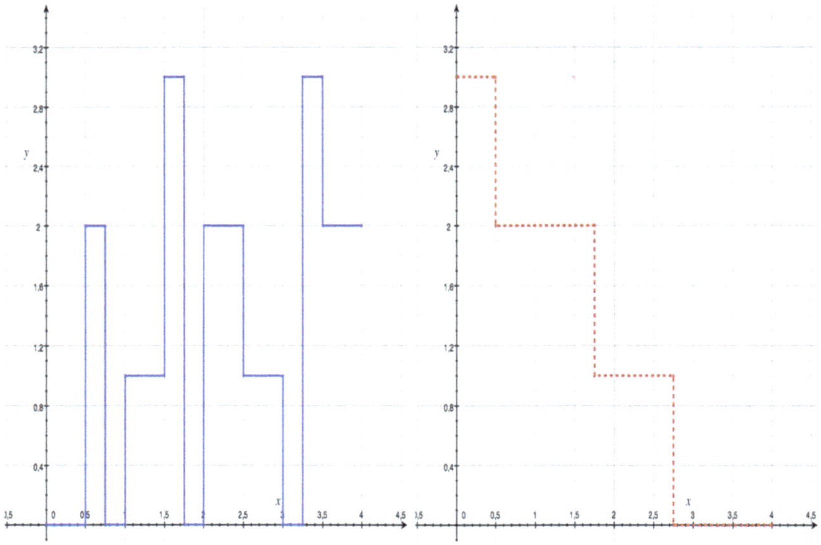

Fig. 1.1 Rearrangement of a staircase function

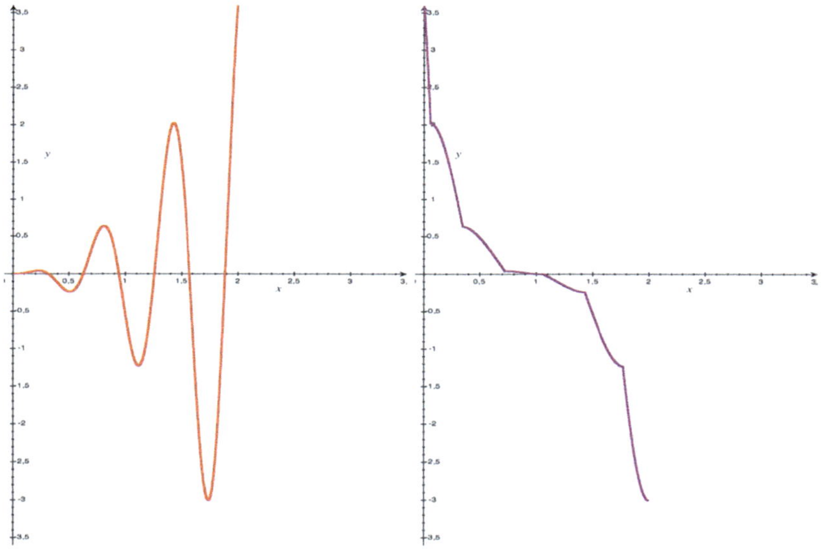

Fig. 1.2 Rearrangement of $x^2 \sin(10x)$

The same algorithm allows us to trace the rearrangement of a function u that is not necessarily simple (see Figs. 1.2 and 1.3).

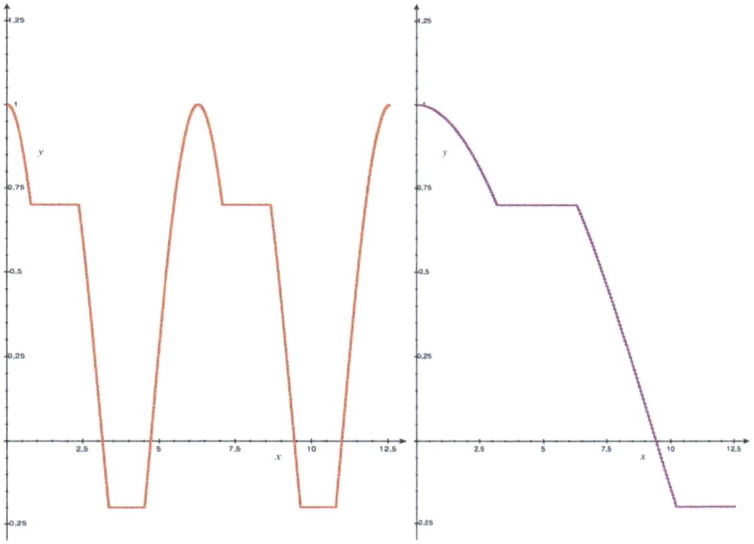

Fig. 1.3 Rearrangement of $\max(\cos x, \min(0.7, \max(\sin x - 0.2)))$

Pre-bibliographic Notes

Ryff's theorem is proven in the book by Chong and Rice [39].

The results of this chapter are often classical, nevertheless some proofs can be found in recent works, like the books by Ziemer [171] and Jacqueline Mossino [115], articles of Alvino et al. [10] and Chiti [38], and the thesis of B. Simon [156].

An exposition of Lorentz spaces is given in the article by Richard Hunt [99].

Other norms are proposed as an exercise (see Exercise 11.2.6).

The list of authors who have worked on G-Gamma spaces is not exhaustive. One of the most recent works is that of Feo et al. [72]. See also [2, 3].

Chapter 2
Relative Rearrangement

In this chapter, we will focus on the directional derivative of the rearrangement, i.e. the map

$$u \in L^p(\Omega),\ (1 \leqslant p \leqslant +\infty) \mapsto u_* \in L^p(\Omega_*)\ .$$

The use of directional derivatives is natural for optimization problems, when we want to characterize an optimal solution. A well-known example is the following: For $f \in L^\infty(\Omega)$,

$$\text{Min}\left\{\frac{1}{2}\int_\Omega |\nabla v|^2 - \int_\Omega f v = J(v),\ v \in H_0^1(\Omega)\right\} = \frac{1}{2}\int_\Omega |\nabla u|^2 - \int_\Omega f\, u = J(u).$$

The optimal solution u then satisfies

$$\lim_{\lambda \to 0,\ \lambda > 0} \frac{J(u+\lambda v) - J(u)}{\lambda} \geqslant 0.$$

It is the derivative of J at the point u in the direction v, hence

$$\int_\Omega \nabla u \cdot \nabla v\, dx - \int_\Omega f v = 0 \quad \forall v \Longleftrightarrow \begin{cases} -\Delta u = f & \text{in } \Omega \\ u = 0 & \text{on } \partial\Omega \end{cases}.$$

If instead of J, we consider the functional:

$$\widehat{J}(v) = J(v) + \int_{\Omega_*} f_*^2(\sigma) v_*^2(\sigma) d\sigma\ ,$$

J. M. Rakotoson, *Relative Rearrangement*, Lecture Notes in Mathematics 2376,
https://doi.org/10.1007/978-3-032-02228-8_2

then there is an optimal solution $\overline{u}$ satisfying

$$\widehat{J}(\overline{u}) = \inf\left\{\widehat{J}(v),\ v \in H_0^1(\Omega)\right\}.$$

What should be the equation given by $\lim\limits_{\lambda\to 0} \dfrac{\widehat{J}(\overline{u}+\lambda v) - \widehat{J}(\overline{u})}{\lambda} \geqslant 0$?.

Another motivation for the study of relative rearrangement is provided by the problem introduced by J.I. Dìaz [57], modeling some aspects of nuclear fusion. This problem is multi-constrained and thanks to relative rearrangement, it can be formulated in the form of an equilibrium equation

$$-\Delta u = G(u).$$

It is stated as follows (see [60, 61] for more details):

Find a pair $(u, F) \in H^1(\Omega) \times W_{loc}^{1,1}(\mathbb{R})$ such that

$$-\Delta u = aF(u) + F(u)F'(u) + bp'(u) \text{ in } \Omega,\quad u - \gamma \in H_0^1(\Omega),$$

$$\int_{\{u>t\}} \left[F(u)F'(u) + bp'(u)\right] dx = j(t, ||u_+||_{L^\infty(\Omega)}),\ \forall t \in \left[\inf_\Omega u, \sup_\Omega u\right],$$

where p, γ, a, b, j are given data.

We show that this problem is equivalent to:

$$\begin{aligned} -\Delta u(x) &= a(x)\mathcal{F}_u(x) + p'(u(x))[b(x) - b_{*u}(|u > u(x)|)] \\ &\quad + j_t'(u_+(x), u_{+*}(0))u'_{+*}(|u > u(x)|) \qquad \text{in } \Omega, \end{aligned}$$

where

$$\mathcal{F}_u(x) = \left[F_v^2 - 2\int_{|u>0|}^{|u>u_+(x)|} [p(u_*)]'(s)]b_{*u}(s)ds \right.$$
$$\left. +2\int_{|u>0|}^{|u>u_+(x)|} j_t'(u_{+*}(s), u_{+*}(0))(u'_{+*}(s))^2 ds\right]_+^{\frac{1}{2}},$$

and

$$b_{*u} = \lim_{\lambda\to 0} \frac{(u+\lambda b)_* - u_*}{\lambda}.$$

The problem is solved in the above references (see also [63, 126, 127]).

Open Problem *We currently do not know how to directly calculate the point limit of*

$$\frac{(u+\lambda v)_* - u_*}{\lambda}(\sigma) \text{ when } \lambda \searrow 0, (\sigma \in \Omega_*).$$

Nevertheless, an answer can be given (see Chap. 7 or the article by Rakotoson–Seoane [138]) if u is regular or simple and $v \in L^p(\Omega)$.

In the absence of direct calculation, the idea of J. Mossino and R. Temam [117] was to consider the directional derivative of $u \xrightarrow{F(s,\cdot)} \int_0^s u_*(\sigma)d\sigma$ for $u \in L^1(\Omega)$, $s \in \overline{\Omega_*}$. Their method is totally different from the one presented here.

2.1 Calculation of a Directional Derivative: Relative Rearrangement

Consider two integrable functions u and v and define for $s \in \overline{\Omega_*}$

$$w(s) = \int_{\{u>u_*(s)\}} v(x)dx + \int_0^{s-|u>u_*(s)|} \left(v|_{\{u=u_*(s)\}}\right)_* (\sigma)d\sigma$$

where $v|_{\{u=u_*(s)\}}$ denotes the restriction of v to $\{u = u_*(s)\}$ and $\left(v|_{\{u=u_*(s)\}}\right)_*$ its decreasing rearrangement.

Theorem 2.1.1 (Directional Derivative) *For all $s \in \overline{\Omega_*}$, for u and v in $L^1(\Omega)$,*

$$\lim_{\lambda\to 0,\ \lambda>0} \frac{F(s,u+\lambda v) - F(s,u)}{\lambda} = w(s).$$

Corollary 2.1.1 (Of the Directional Derivative Theorem) *Let us assume that $v \in L^p(\Omega)$, $1 \leqslant p \leqslant +\infty$. Then,*

1. $w \in W^{1,p}(\Omega_*)$
2. $\dfrac{(u+\lambda v)_* - u_*}{\lambda} \underset{\lambda\to 0}{\rightharpoonup} \dfrac{dw}{ds} \begin{cases} \text{in } L^p(\Omega_*)\text{-weak} & \text{if } 1 \leqslant p < +\infty, \\ \text{in } L^\infty(\Omega_*)\text{-weak-}* & \text{otherwise.} \end{cases}$

Some Preliminary Results
Our proof will be based on some results of convex analysis. This is due to the fact that:

Lemma 2.1.1 *The map* $u \in L^1(\Omega) \xrightarrow{F(s,\cdot)} \int_0^s u_*(\sigma)d\sigma$ *is convex and continuous for the strong topology,* $\forall s \in \overline{\Omega_*}$.

Proof Let $s \in \overline{\Omega_*}$. We have seen (see Corollary 1.2.1 of the Hardy–Littlewood Lemma) that for all $\lambda > 0$, u and v integrable, we have:

$$\int_0^s (\lambda u + v)_*(\sigma)d\sigma \leqslant \int_0^s (\lambda u)_*(\sigma)d\sigma + \int_0^s v_*(\sigma)d\sigma$$
$$= \lambda \int_0^s u_*(\sigma)d\sigma + \int_0^s v_*(\sigma)d\sigma\ .$$

Hence the convexity of $F(s,\cdot)$. Thanks to the property of contraction, we deduce:

$$|F(s,u) - F(s,v)| \leqslant \int_\Omega |u - v|\, dx\ .$$

◇

Definition 2.1.1 (Polar or Conjugate Function) The polar function of $F(s,\cdot)$ is the function defined on $L^1(\Omega)' = L^\infty(\Omega)$ given by

$$F^*(s,q) = \sup_{v \in L^1(\Omega)} \left\{ \int_\Omega qv\, dx - F(s,v) \right\}, \qquad \text{for } q \in L^\infty(\Omega)\ .$$

Definition 2.1.2 (Subdifferential) Let $u \in L^1(\Omega)$, $s \in \overline{\Omega_*}$. The subdifferential of $F(s,\cdot)$ at point u is the set:

$$\partial F(s,u) = \left\{ q \in L^\infty(\Omega) : \int_\Omega q(v-u) \leqslant F(s,v) - F(s,u), \quad \forall v \in L^1(\Omega) \right\}.$$

Proposition 2.1.1

$$\partial F(s,u) = \left\{ q \in L^\infty(\Omega),\ F^*(s,q) + F(s,u) = \int_\Omega qu\,dx \right\}.$$

Proof For $q \in L^\infty(\Omega)$, we have

$$\int_\Omega q(v-u) \leqslant F(s,v) - F(s,u), \quad \forall v \in L^1(\Omega)$$

$$\iff \int_\Omega qv\,dx - F(s,v) \leqslant \int_\Omega qu\,dx - F(s,u), \quad \forall v \in L^1(\Omega)$$

$$\iff F^*(s,q) \leqslant \int_\Omega qu\,dx - F(s,u) \leqslant F^*(s,q)$$

$$\iff F^*(s,q) + F(s,u) = \int_\Omega qu\,dx.$$

Hence the result. ◇

Proposition 2.1.2 *Let u and v be in $L^1(\Omega)$.*

$$F'(s,u;v) = \lim_{\lambda\to 0, \lambda>0} \frac{F(s,u+\lambda v) - F(s,u)}{\lambda}$$
$$= \operatorname{Max}\left\{ \int_\Omega qv\,dx, \quad q \in \partial F(s,u) \right\}.$$

Proof If $q \in \partial F(s,u)$, then $\forall \lambda > 0$,

$$\lambda \int_\Omega qv\,dx \leqslant F(s,u+\lambda v) - F(s,u). \text{Hence}$$

$$\int_\Omega qv\,dx \leqslant F'(s,u;v) \Longrightarrow \operatorname{Max}\left\{ \int_\Omega qv,\ q \in \partial F(s,u) \right\} \leqslant F'(s,u;v)\,.$$

To show the converse, let us set, for u and s fixed,

$$\varphi_\lambda(v) = \frac{F(s,u+\lambda v) - F(s,u)}{\lambda}.$$

The function $\lambda \to F(s, u+\lambda v)$ is convex, then the map $\lambda \to \varphi_\lambda(v)$ is increasing on $\mathbb{R}_+$. By continuity,

$$F'(s, u; v) = \lim_{\lambda \to 0,\ \lambda > 0} \varphi_\lambda(v) = \operatorname{Inf}_{\lambda > 0} \varphi_\lambda(v).$$

Let's calculate the polar function of $F'(s, u; \cdot)$:
i.e. $q \in L^\infty(\Omega) = L^1(\Omega)'$

$$\begin{aligned}
\Big[F'(s, u; \cdot)\Big]^*(q) &= \sup_{v \in L^1(\Omega)} \left\{ \int_\Omega q v\, dx - F'(s, u; v) \right\} \\
&= \sup_{v \in L^1(\Omega)} \left\{ \int_\Omega q v\, dx + \sup_{\lambda > 0} \Big(- \varphi_\lambda(v) \Big) \right\} \\
&= \sup_{v \in L^1(\Omega)} \sup_{\lambda > 0} \left\{ \int_\Omega q v\, dx - \varphi_\lambda(v) \right\} \\
&= \sup_{\lambda > 0} \sup_{v \in L^1(\Omega)} \left\{ \int_\Omega q v\, dx - \varphi_\lambda(v) \right\} \\
&= \sup_{\lambda > 0} \varphi_\lambda^*(q),
\end{aligned}$$

moreover

$$\begin{aligned}
\varphi_\lambda^*(q) &= \sup_{v \in L^1(\Omega)} \left\{ \int_\Omega q v\, dx - \frac{F(s, u+\lambda v)}{\lambda} \right\} + \frac{F(s, u)}{\lambda} \\
&= \sup_{w \in L^1(\Omega)} \left\{ \int_\Omega q \frac{w-u}{\lambda} dx - \frac{F(s, w)}{\lambda} \right\} + \frac{F(s, u)}{\lambda} \\
&= \frac{1}{\lambda} \left(F^*(s, q) + F(s, u) - \int_\Omega q u\, dx \right).
\end{aligned}$$

Since

$$F^*(s, q) + F(s, u) - \int_\Omega q u\, dx \geqslant 0 \quad \forall q \in L^\infty(\Omega),\ \forall u \in L^1(\Omega)$$

we have

$$\Big[F'(s, u; \cdot)\Big]^*(q) = \sup_{\lambda > 0} \varphi_\lambda^*(q) = \begin{cases} 0 & \text{if } q \in \partial F(s, u) \\ +\infty & \text{otherwise.} \end{cases}$$

Consequently, the polar of $\left[F'(s,u;\cdot)\right]^*$ is

$$\left[F'(s,u;\cdot)\right]^{**}(v) = \sup_{q\in L^\infty(\Omega)}\left\{\int_\Omega qv\,dx - \left[F'(s,u;\cdot)\right]^*(q)\right\}$$
$$= \sup_{q\in\partial F(s,u)}\left(\int_\Omega qv\,dx\right).$$

But the map $v \to F'(s,u;v)$ is convex, and continuous because

$$\left|F'(s,u;v) - F'(s,u;\overline{v})\right| \leqslant |v-\overline{v}|_{L^1(\Omega)} ,$$

(and therefore also finite everywhere). As a result (see H. Brezis, [29]) we deduce:

$$\left[F'(s,u;\cdot)\right]^{**} = F'(s,u;\cdot)\,.$$

Hence

$$F'(s,u;v) = \sup_{q\in\partial F(s,u)}\left(\int_\Omega qv\,dx\right).$$

◇

The next step is to compute $\partial F(s,u)$ and this supremum.

Lemma 2.1.2 *For s fixed in* $\overline{\Omega_*}$, $u\in L^1(\Omega)$, *let us define the functional*

$$G_s(v) = \int_\Omega (v-u_*(s))_+dx + su_*(s),$$

$v\in L^1(\Omega)$. *Then*

$$\partial F(s,u) \subset \partial G_s(u)\,.$$

Proof Let $q\in\partial F(s,u)$, and let s and u be fixed. We have seen previously (see Proposition 1.2.1 expression $\displaystyle\int_0^s u_*(\sigma)d\sigma$, in Chap. 1) that $F(s,v)\leqslant G_s(v)$, $F(s,u)=G_s(u)$. So $\forall v\in L^1(\Omega)$, we have

$$\int_\Omega q(v-u)dx \leqslant F(s,v)-F(s,u) \leqslant G_s(v)-F(s,u)\,.$$

Hence $\displaystyle\int_\Omega q(v-u) \leqslant G_s(v)-G_s(u) : q\in\partial G_s(u)$. ◇

Lemma 2.1.3 *For s and u fixed, we have:*

$$\partial G_s(u) = \Big\{ q \in L^\infty(\Omega) : 0 \leqslant q \leqslant 1,\ q = \chi_{\{u>u_*(s)\}} + z$$

$$\text{with support } z \subset \{u = u_*(s)\} \Big\} .$$

Proof Let $q \in \partial G_s(u)$. Then $\forall v \in L^1(\Omega)$

$$\int_\Omega q(v-u)dx \leqslant \int_\Omega (v-u_*(s))_+ dx - \int_\Omega (u-u_*(s))_+ dx .$$

Let $\varphi \in L^\infty(\Omega)$ and $\lambda > 0$, then by choosing $v = u + \lambda\varphi\chi_{\{u\neq u_*(s)\}}$, we deduce that

$$\int_{\{u\neq u_*(s)\}} q\varphi\, dx = \int_{\{u\neq u_*(s)\}} \varphi\, dx \qquad \forall\varphi \in L^\infty(\Omega).$$

Indeed, if we define:

$$f_\lambda(x) = \frac{\big(u + \lambda\varphi\chi_{u\neq u_*(s)} - u_*(s)\big)_+(x) - \big(u - u_*(s)\big)_+(x)}{\lambda}$$

then when $\lambda \to 0$, $f_\lambda(x)$ tends to $\varphi\chi_{\{u>u_*(s)\}}$, from which, by the dominated convergence theorem, we have

$$\int_{\{u\neq u_*(s)\}} \varphi q\, dx \leqslant \lim_{\lambda\to 0} \int_\Omega f_\lambda(x) dx = \int_{\{u>u_*(s)\}} \varphi\, dx \quad ,$$

which provides the equality. Thus $q(x) = \chi_{\{u>u_*(s)\}}$ if $u \neq u_*(s)$. Similarly, if we choose $v = u + \varphi\chi_{\{u=u_*(s)\}}$, $\varphi \in L^\infty(\Omega)$, then:

$$\int_{\{u=u_*(s)\}} \varphi q\, dx \leqslant \int_{\{u=u_*(s)\}} \varphi_+\, dx .$$

If $\varphi \geqslant 0$ this leads to $q \leqslant 1$ and if we choose $\varphi \leqslant 0$ then we have

$$\int_{\{u=u_*(s)\}} \varphi q \, dx \leqslant 0 : q \geqslant 0.$$

We conclude:

$$q(x) = \chi_{\{u>u_*(s)\}}(x) + z(x), \quad 0 \leqslant z \leqslant 1, \quad \operatorname{supp} z \subset \{u = u_*(s)\} \ .$$

Conversely, if q is of this form, then

$$\begin{aligned} \int_\Omega q(v-u) &= \int_\Omega q(v - u_*(s))dx - \int_{\{u>u_*(s)\}} (u - u_*(s))dx \\ &\leqslant \int_\Omega (v - u_*(s))_+ dx - \int_\Omega (u - u_*(s))_+ dx : q \in \partial G_s(u) \ . \end{aligned}$$

◇

To calculate $\partial F(s, u)$, recall that

$$\partial F(s,u) = \left\{ q \in L^\infty(\Omega) : F^*(s,q) + F(s,u) = \int_\Omega qu \, dx \right\} .$$

We need to calculate the polar function $F^*(s, q)$.

Lemma 2.1.4 *Let $q \in \partial G_s(u)$. Then,*

$$F^*(s,q) = \begin{cases} 0 & \text{if } \int_\Omega q(x)dx = s \\ +\infty & \text{else} \end{cases} .$$

Proof

$$\begin{aligned} F^*(s,q) &= \sup_{v \in L^1(\Omega)} \left\{ \int_\Omega qv \, dx - \min_t \left\{ ts + \int_\Omega (v-t)_+ dx \right\} \right\} \\ &= \sup_{v \in L^1(\Omega)} \sup_t \left\{ \int_\Omega qv \, dx - ts - \int_\Omega (v-t)_+ dx \right\} \\ &= \sup_t \left\{ -ts + \sup_{v \in L^1(\Omega)} \left\{ \int_\Omega qv \, dx - \int_\Omega (v-t)_+ dx \right\} \right\} . \end{aligned}$$

Moreover,

$$\int_\Omega qv - \int_\Omega (v-t)_+ dx = \int_\Omega q(v-t)dx + t\int_\Omega q\,dx - \int_\Omega (v-t)_+ dx$$
$$= \int_\Omega (q-1)(v-t)_+ dx - \int_\Omega q(v-t)_- dx + t\int_\Omega q dx\,.$$

Hence

$$F^*(s,q) = \sup_t \left\{ t\left(\int_\Omega q\,dx - s\right)\right.$$
$$\left. + \sup_{v\in L^1(\Omega)} \left\{\int_\Omega (q-1)(v-t)_+ dx - \int_\Omega q(v-t)_- dx\right\}\right\}.$$

Since $q \in \partial G_s(u)$, we have $0 \leqslant q \leqslant 1$. Consequently,

$$\int_\Omega (q-1)(v-t)_+ dx - \int_\Omega q(v-t)_- dx \leqslant 0\,.$$

By taking $v = t$, we see that:

$$\sup_{v\in L^1(\Omega)} \left\{\int_\Omega (q-1)(v-t)_+ dx - \int_\Omega q(v-t)_- dx\right\} = 0\,.$$

Hence

$$F^*(s,q) = \sup_t \left\{ t\left(\int_\Omega q\,dx - s\right)\right\} = \begin{cases} 0 & \text{if } \int_\Omega q(x)dx = s \\ +\infty & \text{otherwise} \end{cases}.$$

◇

We thus deduce:

Corollary 2.1.2

$$\partial F(s,u) = \left\{q \in \partial G_s(u) : F(s,u) = \int_\Omega qu\,dx, \quad \int_\Omega q(x)dx = s\right\} \Longleftrightarrow$$

$$\partial F(s,u) = \left\{q \in L^\infty(\Omega),\ q = \chi_{\{u>u_*(s)\}} + z,\ \textit{with}\ \mathrm{supp} z \subset \{u = u_*(s)\}\right.$$
$$\left.\textit{and}\ \int_0^s u_*(\sigma)d\sigma = \int_\Omega qu\,dx,\ \int_\Omega q(x)dx = s\right\}.$$

Corollary 2.1.3 (Of Corollary 2.1.2) *For all* $s \in \overline{\Omega_*}$

$$\partial F(s,u) = \Big\{ q = \chi_{\{u>u_*(s)\}} + z,\ 0 \leqslant q \leqslant 1,\ \text{supp}\, z \subset \{u = u_*(s)\}$$

$$\text{and} \int_{\{u=u_*(s)\}} z(x)\,dx = s - |u > u_*(s)| \Big\}.$$

Proof Let's take a closer look at $\partial F(s,u)$

$$\int_0^s u_*(\sigma)d\sigma = \int_\Omega qu\,dx \iff \int_0^s u_*(\sigma)d\sigma - \int_0^{|u_*>u_*(s)|} u_*(\sigma)d\sigma = \left(\int_{\{u=u_*(s)\}} z\,dx \right) u_*(s)$$

$$\iff u_*(s)\big(s - |u > u_*(s)|\big) = u_*(s) \left(\int_{\{u=u_*(s)\}} z\,dx \right).$$

If $u_*(s) \neq 0$ then we have:

$$\int_{\{u=u_*(s)\}} z\,dx = s - |u > u_*(s)| \iff \int_0^s u_*(\sigma)d\sigma = \int_\Omega qu\,dx\,,$$

hence we deduce that $\displaystyle\int_\Omega q(x)dx = s$.

If $u_*(s) = 0$ then by equimeasurability we have

$$\int_0^s u_*(\sigma)d\sigma = \int_\Omega qu\,dx \text{ and } \int_\Omega q(x)dx = s \iff \int_{\{u=0\}} z(x)\,dx = s - |u > 0|\,.$$

◇

Theorem 2.1.2 (Expression of $F'(s,u;v)$) *Let u and v be in $L^1(\Omega)$, $s \in \overline{\Omega_*}$. Then,*

$$F'(s,u;v) = \int_{\{u>u_*(s)\}} v\,dx + \int_0^{s-|u>u_*(s)|} \left(v|_{\{u=u_*(s)\}}\right)_*(\sigma)d\sigma .$$

Proof We know that

$$F'(s,u;v) = \text{Max}\left\{\int_\Omega qv\,dx,\ q \in \partial F(s,u)\right\} \text{ with } q = \chi_{\{u>u_*(s)\}} + z.$$

We have: $F'(s,u;v) = \displaystyle\int_{\{u>u_*(s)\}} v\,dx \ +$

$$\text{Max}\left\{\int_\Omega zv\,dx\,,\ 0\leqslant z \leqslant 1,\ \text{supp}\,z\subset\{u = u_*(s)\},\ \int_{\{u=u_*(s)\}} z(x)dx = s - |u > u_*(s)|\right\}$$

$$= \int_{\{u>u_*(s)\}} v\,dx + \text{Max}\left\{\int_{\{u=u_*(s)\}} zv\,dx\ ,\ 0\leqslant z\leqslant 1,\ \int_{\{u=u_*(s)\}} z(x)dx = s - |u > u_*(s)|\right\}.$$

By the corollaries of the Hardy–Littlewood inequalities, we have:

$$\text{Max}\left\{\int_{\{u=u_*(s)\}} zv\,dx\ ,\ 0\leqslant z\leqslant 1,\ \int_{\{u=u_*(s)\}} z(x)dx = s - |u > u_*(s)|\right\}$$

$$= \int_0^{s-|u>u_*(s)|} \left(v|_{\{u=u_*(s)\}}\right)_* d\sigma .$$

Hence the result. ◇

Theorem 2.1.3 *Let* $u \in L^1(\Omega)$, $v \in L^p(\Omega)$, $1 \leqslant p \leqslant +\infty$. *Then*

1. $\frac{d}{ds}\left(\frac{F(\cdot, u+\lambda v) - F(\cdot, u)}{\lambda}\right) \underset{\lambda \searrow 0}{\rightharpoonup} \frac{d}{ds} F'(\cdot, u; v)$
   in $L^p(\Omega_*)$*-weak if* $1 \leqslant p < +\infty$; *in* $L^\infty(\Omega_*)$*-weak-* if* $p = +\infty$.
2. *In particular,* $F'(\cdot, u; v) \in W^{1,p}(\Omega_*)$.

Definition 2.1.3 (Of Relative Rearrangement) Under the same conditions as the theorem above, the relative rearrangement of v with respect to u is the function

$$v_{*u} = \frac{d}{ds} F'(\cdot, u; v) \in L^p(\Omega_*).$$

Proof of Theorem 2.1.3 Let $\frac{F(s, u+\lambda v) - F(s, u)}{\lambda} = w_\lambda(s)$, $\lambda > 0$. Using the contraction property, we have: $\forall s \in \overline{\Omega_*}$

(1) $|w_\lambda(s)| \leqslant |v|_{L^1(\Omega)}$, $\left|\frac{dw_\lambda}{ds}\right|_{L^p(\Omega_*)} \leqslant |v|_{L^p(\Omega)}$.

(2) Since $\lim_{\lambda \to 0} w_\lambda(s) = F'(s, u; v)$ we have $\forall \varphi \in L^\infty(\Omega_*)$,

$$\int_{\Omega_*} \varphi(s) F'(s, u; v) ds = \lim_{\lambda \to 0} \int_{\Omega_*} \varphi(s) w_\lambda(s) ds$$

$\left(\text{so } \frac{dw_\lambda}{ds} \rightharpoonup \frac{d}{ds} F'(\cdot, u; v). \text{ in } \mathcal{D}'(\Omega_*) \right)$.

Consequently, if $1 < p \leqslant +\infty$ then according to the relations (1) and (2) we have:

(3) $\left|F'(s, u; v)\right| \leqslant |v|_{L^1(\Omega)}$.

(4) $\forall \varphi \in C_c^1(\Omega_*)$, we have $\left|\int_\Omega \varphi'(s) F'(s, u; v) ds\right| \leqslant |v|_{L^p(\Omega)} \cdot |\varphi|_{L^{p'}(\Omega_*)}$ (where $\varphi' = \frac{d\varphi}{ds}$).

Therefore $F'(\cdot, u; v) \in W^{1,p}(\Omega_*)$. Since $\dfrac{dw_\lambda}{ds}$ remains within a bound of $L^p(\Omega_*)$ and $\displaystyle\lim_{\lambda\to 0}\int_\Omega \varphi\frac{dw_\lambda}{ds} = -\int_\Omega \varphi'(s)F'(s, u; v)ds$, we deduce

$$\frac{dw_\lambda}{ds} \underset{\lambda\to 0}{\rightharpoonup} \frac{d}{ds}F'(\cdot, u; v)$$

in $\begin{cases} L^p(\Omega_*)\text{-weak} & \text{if } 1 < p < +\infty \\ L^\infty(\Omega_*)\text{-weak-*} & \text{if } p = +\infty. \end{cases}$

For the case $p = 1$, we will use the Dunford–Pettis criterion, i.e. $\forall\varepsilon > 0$, $\exists\delta(\varepsilon) > 0$ s.t. if $|A| \leqslant \delta(\varepsilon)$ then

$$\int_A \left|\frac{dw_\lambda}{ds}\right| ds \leqslant \varepsilon,\ \forall\lambda > 0\,.$$

Indeed, this property ensures that the fact that $\dfrac{dw_\lambda}{ds}$ remains within a bound of $L^1(\Omega_*)$ entails that $\dfrac{dw_\lambda}{ds}$ converges to $\dfrac{d}{ds}F'(\cdot, u; v)$ in $L^1(\Omega_*)-$ weak (using (2)).

To show the Dunford–Pettis property, let us fix $\varepsilon > 0$ and $v_\varepsilon \in L^\infty(\Omega)$ s.t. $|v - v_\varepsilon|_{L^1} \leqslant \dfrac{\varepsilon}{2}$ and $\delta_\varepsilon > 0$ s.t. $2|v_\varepsilon|_\infty \leqslant \dfrac{\varepsilon}{\delta_\varepsilon}$. Let's put $w_{\lambda,\varepsilon}(s) = \dfrac{F(s, u + \lambda v_\varepsilon) - F(s, u)}{\lambda}$ and let A be a measurable subset of Ω_*.

If $|A| \leqslant \delta_\varepsilon$ then we have:

$$\begin{aligned}\int_A \left|\frac{dw_\lambda}{ds}\right| &\leqslant \int_A \left|\frac{dw_\lambda}{ds} - \frac{dw_{\lambda,\varepsilon}}{ds}\right| + \int_A \left|\frac{dw_{\lambda,\varepsilon}}{ds}\right| \\ &= \int_A \left|\frac{(u + \lambda v_\varepsilon)_* - (u + \lambda v)_*}{\lambda}\right| + \int_A \left|\frac{(u + \lambda v_\varepsilon)_* - u_*}{\lambda}\right| \\ &\leqslant |v_\varepsilon - v|_{L^1} + |v_\varepsilon|_{L^\infty}|A| \leqslant \frac{\varepsilon}{2} + \frac{\varepsilon}{2} = \varepsilon.\end{aligned}$$

◇

2.2 Immediate Properties of Relative Rearrangement

Proposition 2.2.1 *Let* $u \in L^1(\Omega)$ *and* $v \in L^p(\Omega)$, $1 \leqslant p \leqslant +\infty$.

(a) $|v_{*u}|_{L^p(\Omega_*)} \leqslant |v|_{L^p(\Omega)}$.
(b) If $v_1 \leqslant v_2$ $\ v_i \in L^p(\Omega)$ *then* $v_{1*u} \leqslant v_{2*u}$.

(continued)

Proposition 2.2.1 (continued)
(c) The map $v \in L^p(\Omega) \to v_{*u} \in L^p(\Omega_*)$ *is a contraction.*
(d) For all Lipschitzian and convex $\Phi : \mathbb{R} \to \mathbb{R}$ *we have:*

$$\int_{\Omega_*} \Phi(v_{1*u} - v_{2*u}) d\sigma \leqslant \int_{\Omega} \Phi(v_1 - v_2) dx\,.$$

Proof (a) To show that $|v_{*u}|_p \leqslant |v|_p$. We have

$$|v_{*u}|_p = \sup\left\{\int_{\Omega_*} v_{*u}\varphi d\sigma,\ \varphi \in L^{p'}(\Omega_*),\ |\varphi|_{L^{p'}} \leqslant 1\right\}\ \frac{1}{p} + \frac{1}{p'} = 1.$$

But by Hölder's inequality, we conclude that:

$$\int_{\Omega_*} v_{*u}\varphi d\sigma = \lim_{\lambda \to 0} \int_{\Omega_*} \frac{dw_\lambda}{d\sigma}\varphi d\sigma \leqslant |\varphi|_{p'} \cdot |v|_p$$

$\left(\text{according to the previous lemma, } \left|\frac{dw_\lambda}{d\sigma}\right|_p \leqslant |v|_p\right)$. Hence the result.

(b) Let v_1 and v_2 be two elements of $L^1(\Omega)$, $u \in L^1(\Omega)$. So, $\forall \lambda > 0$, $(u + \lambda v_1)_* \leqslant (u + \lambda v_2)_*$ if $v_1 \leqslant v_2$ a.e. Thus $\forall \varphi \in \mathcal{D}(\Omega_*)$, $\varphi \geqslant 0$, we have:

$$\int_{\Omega_*} \frac{(u + \lambda v_1)_* - u_*}{\lambda}\varphi \leqslant \int_{\Omega_*} \frac{(u + \lambda v_2)_* - u_*}{\lambda}\varphi$$

and when $\lambda \searrow 0$, we deduce

$$\int_{\Omega *} v_{1*u}\varphi \leqslant \int_{\Omega_*} v_{2*u}\varphi \Longrightarrow v_{1*u} \leqslant v_{2*u}\ a.e.$$

(c) Reasoning as in (a) and (b), if v_1 and v_2 are in $L^p(\Omega)$ with $1 \leqslant p \leqslant +\infty$, $u \in L^1(\Omega)$, then, $\forall \varphi \in L^{p'}(\Omega_*)$ the contraction property and Holder's inequality imply that

$$\int_{\Omega_*} \left[\frac{(u + \lambda v_1)_* - (u + \lambda v_2)_*}{\lambda}\right] \varphi d\sigma \leqslant |\varphi|_{p'}\, |v_1 - v_2|_p\,.$$

By making λ tend to zero, we have:

$$\int_{\Omega_*} (v_{1*u} - v_{2*u})\varphi d\sigma \leqslant |\varphi|_{p'}\, |v_1 - v_2|_p \qquad \forall \varphi \in L^{p'}(\Omega),$$

which implies: $|v_{1*u} - v_{2*u}|_p \leqslant |v_1 - v_2|_p$. This proves that the map $v \in L^p(\Omega) \to v_{*u} \in L^p(\Omega_*)$ is a contraction.

(d) Since the mapping $v \to v_{*u}$ is continuous, it suffices to show that for all convex, Lipschitzian $v \in L^\infty(\Omega)$, for all $\Phi : \mathbb{R} \to \mathbb{R}$ we have:

$$\int_{\Omega_*} \Phi(v_{*u})d\sigma \leqslant \int_\Omega \Phi(v)dx \qquad \forall u \in L^1(\Omega)\,.$$

Indeed, if we denote by $v^k = T_k(v)$ the truncation of v, then $v^k \to v$ if $v \in L^p(\Omega)$, $1 \leqslant p < +\infty$. Then, as we have:

$$\left|\Phi(v^k_{*u})(\sigma) - \Phi(v_{*u})(\sigma)\right| \leqslant \left|\Phi'\right|_\infty \left|v^k_{*u}(\sigma) - v_{*u}(\sigma)\right|\,,$$

we deduce that:

$$\lim_k \int_{\Omega_*} \Phi(v^k_{*u})(\sigma) = \int_{\Omega_*} \Phi(v_{*u})d\sigma.$$

Since

$$\lim_k \int_\Omega \Phi(v^k) = \int_\Omega \Phi(v),$$

the inequality $\int_{\Omega_*} \Phi(v^k_{*u})d\sigma \leqslant \int_\Omega \Phi(v^k)$ would entail $\int_{\Omega_*} \Phi(v_{*u}) \leqslant \int_\Omega \Phi(v)$.

To prove the inequality in the case of a bounded function, we introduce for $v_i \in L^\infty(\Omega)$, $i = 1, 2$, $v = v_1 - v_2$ and the convex set:

$$\mathcal{K}(v) = \left\{h \in L^2(\Omega_*),\quad \int_{\Omega_*} \Phi(h)d\sigma \leqslant \int_\Omega \Phi(v)dx\right\}$$

(where $\Phi : \mathbb{R} \to \mathbb{R}$ is convex Lipschitzian). The set $\mathcal{K}(v)$ is closed in the strong topology of $L^2(\Omega_*)$. Consequently it is weakly closed. For $u \in L^1(\Omega)$,

$$\frac{dw_\lambda}{ds} = \frac{(u + \lambda v_1)_* - (u + \lambda v_2)_*}{\lambda} \text{ satisfies}$$

$$\int_{\Omega_*} \Phi\left(\frac{dw_\lambda}{d\sigma}\right)d\sigma \leqslant \int_\Omega \Phi(v)dx$$

(see the contraction theorems) and

$$\frac{dw_\lambda}{d\sigma} \underset{\lambda\to 0}{\rightharpoonup} v_{1*u} - v_{2*u} \text{ in } L^2(\Omega_*)\text{-weak.}$$

So $\int_{\Omega_*} \Phi(v_{1*u} - v_{2*u}) \leqslant \int_\Omega \Phi(v)$ because $v_{1*u} - v_{2*u} \in \mathcal{K}(v)$. ◇

Note Other extensions are possible for the contraction, for example

Proposition 2.2.2 *If $\psi : \mathbb{R}_+ \to \mathbb{R}_+$ is convex, increasing, and v_1 and v_2 are in $L^\infty(\Omega)$, then*

$$\int_{\Omega_*} \psi\left(|v_{1*u} - v_{2*u}|\right) d\sigma \leqslant \int_\Omega \psi\left(|v_1 - v_2|\right) dx, \quad \forall u \in L^1(\Omega) .$$

Corollary 2.2.1 (Of Proposition 2.2.1) *If $v \in L^p(\Omega)$, $1 \leqslant p \leqslant +\infty$, $u \in L^1(\Omega)$*

1. $\forall s \in \overline{\Omega_*}$, $\quad \int_0^s |v_{*u}|_*(\sigma)d\sigma \leqslant \int_0^s |v|_*(\sigma)d\sigma$.
2. *If $v \in L^{p,q}(\Omega)$ then $v_{*u} \in L^{p,q}(\Omega_*)$ and*

$$|v_{*u}|_{(p,q)} \leqslant |v|_{(p,q)} .$$

Proof This is a consequence of the equivalences we showed before and of the definition of $L^{p,q}(\Omega)$. Indeed the function $\sigma \to (|\sigma| - t)_+$ is convex, Lipschitzian from $\mathbb{R}$ to $\mathbb{R}$. Consequently according to Proposition 2.2.1,

$$\int_{\Omega_*} (|v_{*u}| - t)_+ ds \leqslant \int_\Omega (|v| - t)_+ dx, \quad \forall t \in \mathbb{R}.$$

We deduce, by the properties given in Chap. 1 (see Proposition 1.2.1), that

$$\forall s : \int_0^s |v_{*u}|_*(\sigma)d\sigma \leqslant \int_0^s |v|_*(\sigma)d\sigma.$$

Assertion 1 implies 2. We know that if $1 \leqslant q \leqslant p < +\infty$ then the map $f \to |f|_{p,q}$ is a norm.

We can then, in this case, show that $|v_{*u}|_{p,q} \leqslant |v|_{p,q}$. The proof is included in the following general result, which, by the Corollary 1.3.3, shows that the rearrangement mapping is a contraction of $L(\Omega, \rho)$ into $L(\Omega_*, \rho)$ if ρ is invariant under rearrangement, $L(\Omega, \rho)$ being a Banach function space. By Luxemburg's theorem [21, 39], rearrangement invariance will imply the existence of a norm $\overline{\rho} : L^0(\Omega_*) \to [0, +\infty]$ s.t. $\rho(f) = \overline{\rho}(|f|_*)$. By abuse of notation, we will denote $\overline{\rho}$ by ρ.

Theorem 2.2.1 (Lorentz–Luxemburg) *Let ρ be a non-trivial norm on $L^0(\Omega_*)$. We assume that the norm ρ is a Fatou norm, invariant under rearrangement. Let f, g be in $L^0(\Omega_*)$.*

If $f \geqslant 0$, $g \geqslant 0$ satisfy $\forall s \geqslant 0$, $\int_0^s f_*(\sigma)d\sigma \leqslant \int_0^s g_*(\sigma)d\sigma$, *then*

$$\rho(f) \leqslant \rho(g)\,.$$

We will admit this theorem. Here are some elements of its proof.

We recall on $L^0(\Omega_*)$ the so-called associate norm:

$$\rho'(\varphi) = \sup\left\{\int_{\Omega_*} |\psi\varphi|\, d\sigma : \rho(\psi) \leqslant 1\right\}.$$

Then

$$\int_{\Omega_*} |f\varphi|\, d\sigma \leqslant \rho'(\varphi)\cdot\rho(f) \quad if\ f \in L^0(\Omega_*),\ \rho'(\varphi) < +\infty. \ \diamond$$

We deduce the following lemma from Proposition 1.5.3 and the Lorentz–Luxemburg theorem concerning the associate norm of ρ'.

Lemma 2.2.1 *If ρ is a Fatou norm then*

$$\rho(f) = \sup\left\{\int_{\Omega_*} |f\varphi|\, d\sigma : \rho'(\varphi) \leqslant 1\right\}$$

and if ρ is invariant under rearrangement

$$f \geqslant 0,\ \rho(f) = \sup\left\{\int_{\Omega_*} f_*\, |\varphi|_*\, d\sigma : \rho'(\varphi) \leqslant 1\right\}.$$

We can then show

Lemma 2.2.2 *If* $f \geqslant 0,\ g \geqslant 0,\ \int_0^s f_*(\sigma)d\sigma \leqslant \int_0^s g_*(\sigma)d\sigma\ \forall s \in \Omega_*$, *then* $\forall \varphi \in L^0(\Omega_*)$, *we have:*

$$\int_{\Omega_*} f_* \left|\varphi\right|_* d\sigma \leqslant \int_{\Omega_*} g_* \left|\varphi\right|_* d\sigma \ .$$

Proof If $\left|\varphi\right|_* (|\Omega|) = 0$ then

$$\int_{\Omega_*} f_* \left|\varphi\right|_* = -\int_{\Omega_*} \left(\int_0^s f_*(t)dt\right) d\left|\varphi\right|_*$$

$$\leqslant -\int_{\Omega_*} \left(\int_0^s g_*(t)dt\right) d\left|\varphi\right|_* = \int_{\Omega_*} g_* \left|\varphi\right|_* d\sigma \ .$$

If $\left|\varphi\right|_* (|\Omega|) \neq 0$, then we consider

$$\varphi_n(\sigma) = \begin{cases} \left|\varphi\right|(\sigma) & \text{if } 0 \leqslant \sigma < |\Omega| - \dfrac{1}{n+1} \\ 0 & \text{otherwise} \end{cases} .$$

Consequently $0 \leqslant \varphi_n \leqslant \varphi_{n+1} \to |\varphi|$ a.e. and $\varphi_{n*}(\sigma) = 0$, if $\sigma > |\Omega| - \dfrac{1}{n+1}$.

As $\int_{\Omega_*} f_*\varphi_{n*}(\sigma)d\sigma \leqslant \int_{\Omega_*} g_*(\sigma)\varphi_{n*}(\sigma)d\sigma$, the result follows from the Beppo Levi theorem. ◇

Thus, we deduce Theorem 2.2.1,

$$\rho(f) = \sup\left\{\int_{\Omega_*} f_* \left|\varphi\right|_*,\ \rho'(\varphi) \leqslant 1\right\} \leqslant \rho(g)$$

$$= \sup\left\{\int_{\Omega_*} g_* \left|\varphi\right|_*,\ \rho'(\varphi) \leqslant 1\right\} . \diamond$$

As an application,

$$\rho(f) = |f|_{p,q} = \left[\int_{\Omega_*} \left[t^{\frac{1}{p}} \left|f\right|_* (t)\right]^q \frac{dt}{t}\right]^{\frac{1}{q}} \quad 1 \leqslant q \leqslant p < +\infty$$

is a norm on $L^0(\Omega_*)$ and $L^{p,q} = \left\{f \in L^0(\Omega_*) : \rho(f) < +\infty\right\}$. ρ is Fatou norm and invariant under rearrangement.

We have shown that

$$\int_0^s |v_{*u}|_*(\sigma)d\sigma \leqslant \int_0^s |v|_*(\sigma)d\sigma \quad \forall s \in \overline{\Omega_*}$$

then

$$\rho(v_{*u}) \leqslant \rho(|v|_*) = \left[\int_{\Omega_*}\left[t^{\frac{1}{p}}|v|_*(t)\right]^q \frac{dt}{t}\right]^{\frac{1}{q}},$$

or

$$|v_{*u}|_{L^{p,q}(\Omega_*)} \leqslant |v|_{L^{p,q}(\Omega)} .$$

◇

More generally, under the assumptions of Theorem 2.2.1,

$$\rho(|v_{*u}|) \leqslant \rho(|v|_*).$$

In particular, if u is without a plateau, we find the contraction property

$$\rho(|v_{1*v} - v_{2*u}|) \leqslant \rho(|v_1 - v_2|),$$

ρ being a Fatou norm invariant under rearrangement.

2.3 Mean Operators of the First Kind

The mean operators will allow us to go from a definite integral on Ω_* to a definite integral on Ω.

Definition 2.3.1 Let $u : \Omega \to \mathbb{R}$ be measurable. P, P_* are respectively the set of levels of u and u_*: $P = \bigcup_{i\in D} P_i$, $\quad P_* = \bigcup_{i\in D} P_i^*$, $|P_i| = |\{u = t_i\}| = |P_i^*| = |\{u_* = t_i\}|$.

For almost all $x \in \Omega$, we define:

$$\beta_u(x) = \underline{\beta}(u)(x) = |u > u(x)| = m_u\big(u(x)\big)$$

(continued)

Definition 2.3.1 (continued)

$$\overline{\beta}_u(x) = \overline{\beta}(u)(x) = |u \geqslant u(x)|$$

$$\delta(u)(x) = \left[\underline{\beta}(u)(x),\ \overline{\beta}(u)(x)\right].$$

If $x \in P_i$, we write $s_i' = \underline{\beta}(u)(x),\ s_i'' = \overline{\beta}(u)(x)$, so $P_i^* = [s_i', s_i'')$.

Lemma 2.3.1 *Let P be the set of levels of u, P_* that of u_* and N be a measurable set of Ω_*. We have*

$$\beta_u^{-1}(N) - P = \{x \in \Omega \backslash P : \beta(u)(x) \in N\} \text{ with } \beta = \underline{\beta} \text{ or } \overline{\beta}\,.$$

Then if $|N| = 0$, $\left|\beta_u^{-1}(N) - P\right| = 0$.

Proof Let $\varepsilon > 0$. Then there exists an open $\mathcal{O}_\varepsilon$ such that $N \subset \mathcal{O}_\varepsilon \subset \Omega_*$ with $|\mathcal{O}_\varepsilon| \leqslant \varepsilon$, so

$$\beta_u^{-1}(N) - P \subset \beta_u^{-1}(\mathcal{O}_\varepsilon) - P\,.$$

It is therefore sufficient to show that

$$\left|\beta_u^{-1}(\mathcal{O}_\varepsilon) - P\right| \leqslant \varepsilon\,.$$

$$\left|\beta_u^{-1}(\mathcal{O}_\varepsilon) - P\right| = \int_\Omega \chi_{\mathcal{O}_\varepsilon}\left(\beta(u)(x)\right) \cdot \chi_{\mathbb{R} - \bigcup\limits_{i \in D} t_i}\big(u(x)\big)dx.$$

The function $t \to \chi_{\mathcal{O}_\varepsilon}(m_u(t)) \cdot \chi_{\mathbb{R} - \bigcup\limits_{i \in \mathbb{R}} t_i}(t)$ is a positive Borel function so, by equimeasurability, we have:

$$\left|\beta_u^{-1}(\mathcal{O}_\varepsilon) - P\right| = \int_{\Omega_*} \chi_{\mathcal{O}_\varepsilon}\Big(\beta_{u_*}(u_*((s))\Big) \cdot \chi_{\mathbb{R} - \bigcup\limits_{i \in \mathbb{R}} t_i}\big(u_*(s)\big)ds$$

because $m_u\big(t\big) = m_{u_*}\big(t\big)$. But if $s \notin P_*$ then $m_{u_*}\big(u_*(s)\big) = s$, and we have

$$\Big|\beta_u^{-1}(\mathcal{O}_\varepsilon) - P\Big| = \int_{\Omega_*} \chi_{\mathcal{O}_\varepsilon}(s) \cdot \chi_{\mathbb{R}-\bigcup\limits_{i\in\mathbb{R}} t_i}\big(u_*(s)\big)ds \leqslant |\mathcal{O}_\varepsilon| \leqslant \varepsilon\ . \quad \diamond$$

Definition 2.3.2 Let $u : \Omega \to \mathbb{R}$ be measurable, $P(u) = P$ be the set of levels of u and P_* be that of u_*, $g : \Omega_* \to \mathbb{R}$ measurable (defined almost everywhere) and positive, or $g \in L^1(\Omega_*)$.

We define $M_u(g) : \Omega \to \overline{\mathbb{R}}$ by:

$$M_u(g)(x) = \begin{cases} g\Big(\beta_u(x)\Big) & \text{if } x \in \Omega\backslash P \\ \dfrac{1}{|P_i|}\displaystyle\int_{s_i'}^{s_i''} g(s)ds & \text{if } x \in P_i\ . \end{cases}$$

$M_u(g)$ is well defined almost everywhere in Ω because, by Lemma 2.3.1, we have

$$g = g'\ a.e. \implies M_u(g)(x) = M_u(g')(x)\ a.e.$$

Lemma 2.3.2

(i) If $g : \Omega_* \to \mathbb{R}$ *measurable and positive, or if* g *is integrable, then*

$$\int_{\Omega_*} g(\sigma)d\sigma = \int_\Omega M_u(g)dx\ .$$

(ii) $\forall p \in [1, +\infty]$, $\quad M_u \in \mathcal{L}\Big(L^p(\Omega_*),\ L^p(\Omega)\Big)$ *and* $\|M_u\| = 1$.

Proof It suffices to prove, for $g \geqslant 0$ measurable,

$$\int_{\Omega_*\backslash P_*} g(\sigma)d\sigma = \int_{\Omega\backslash P} M_u(g)dx$$

and

$$\int_{P_i^*} g(\sigma) = \int_{P_i} M_u(g)dx.$$

Let $\overline{g}$ be a Borel function s.t. $g = \overline{g}$ a.e. So, we have:

$$\int\limits_{\Omega_*\setminus P_*} g(\sigma)\,d\sigma = \int\limits_{\Omega_*\setminus P_*} \overline{g} = \int_{\Omega_*} \chi_{\mathbb{R}-\bigcup\limits_{i\in D} t_i}(u_*(s))\overline{g}\Big(m_{u_*}(u_*(s))\Big)ds$$

$$= \int_{\Omega} \chi_{\mathbb{R}-\bigcup\limits_{i\in D} t_i}(u(x))\overline{g}\Big(m_u(u(x))\Big)dx = \int\limits_{\Omega\setminus P} \overline{g}\Big(\beta_u(x)\Big)dx = \int\limits_{\Omega\setminus P} g\Big(\beta_u(x)\Big)dx\,.$$

Moreover, we have:

$$\int_{P_i^*} g = \int_{s_i''}^{s_i''} g(\sigma) = \int_{P_i} M_u(g)dx.$$

This completes the proof of the lemma.

Let us note that: $|M_u(g)| \leqslant M_u(|g|)$. Hence if $g \in L^1(\Omega_*)$ then $M_u(g) \in L^1(\Omega)$, and:

$$\int_{\Omega} M_u(g) = \int_{\Omega} M_u(g_+) - \int_{\Omega} M_u(g_-) = \int_{\Omega_*} g(\sigma)d\sigma\;.$$

For $1 \leqslant p < +\infty$, we have almost everywhere:

$$|M_u(g)(x)|^p \leqslant \big[M_u(|g|)\big]^p(x) \leqslant M_u\left(|g|^p\right)(x),$$

hence

$$\left.\begin{array}{l} \displaystyle\int_{\Omega} |M_u(g)|^p \leqslant \int_{\Omega} M_u\Big(|g|^p\Big) = \int_{\Omega_*} |g|^p\,(\sigma)d\sigma \\[2ex] \qquad\qquad for \quad g = c^{te}, one \quad has \quad M_u g = g \end{array}\right\} \Longrightarrow \|M_u\| = 1$$

if $p = \infty \quad |M_u(g)(x)| \leqslant |g|_\infty : |M_u(g)|_\infty \leqslant |g|_\infty\,.$ ◇

Definition 2.3.3 (Mean Operators of the First Kind) The map $M_u : L^p(\Omega_*) \to L^p(\Omega)$ is called the mean operator of the first kind.

2.4 Mean Operators of the Second Kind

Let u, v be two real measurable functions. We can define $M_{u,v}$ in the following way:

$$P(u) = \text{ set of levels of } u = \bigcup_{i \in D} P_i(u)$$

$v_i = v|_{P_i(u)}$, $g : \Omega_* \to \mathbb{R}$ measurable $\geqslant 0$ or integrable. Then $M_{u,v}(g) : \Omega \to \overline{\mathbb{R}}$ is given by:

$$M_{u,v}(g)(x) = \begin{cases} g\big(\beta_u(x)\big) & \text{if } x \in \Omega \backslash P, \\ M_{v_i}(h_i)(x) & \text{if } x \in P_i, \end{cases}$$

$h_i : \big[0, s_i'' - s_i' = |P_i|\big] \to \mathbb{R} : h_i(s) = g(s_i' + s)$.
M_{v_i} is defined as M_u (Ω replaced by P_i).
We check

if $g = g'$ a.e. in Ω_* then $M_{u,v}(g) = M_{u,v}(g')\ a.e.$
if $g \geqslant 0$ then $\displaystyle\int_{\Omega_*} g d\sigma = \int_\Omega M_{u,v}(g)$
if g is integrable then $\displaystyle\int_{\Omega_*} g = \int_\Omega M_{u,v}(g)$
if $v = c^{te} then \quad M_{u,v} = M_u$.

$M_{u,v} \in \mathcal{L}\Big(L^p(\Omega_*), L^p(\Omega)\Big), \quad 1 \leqslant p \leqslant +\infty, \quad \|M_{u,v}\| = 1.$
Note that:

$$\int P_i M_{u,v}(g)dx = \int P_i\, M_{v_i} h_i =_{lemma\ on\ M_{v_i}} \int_0^{s_i''-s_i'} h_i = \int_{s_i'}^{s_i''} g = \int_{P_i} M_u(g)\,.$$

Definition 2.4.1 (Mean Operators of the Second Kind)
$M_{u,v}$ is called the mean operator of the second kind.

Lemma 2.4.1 *Let $g \in L^1(\Omega_*)$ and g_n be a sequence of $L^1(\Omega_*)$ such that*

$$\begin{cases} g_n(\sigma) \underset{n}{\to} g(\sigma) \ a.e. \\ g_n \to g \text{ in } L^1(\Omega_*). \end{cases}$$

Then

$$M_u(g_n)(x) \to M_u(g)(x) \ a.e.$$

and

$$M_{u,v}(g_n)(x) \to M_{u,v}(g)(x) \ a.e.$$

in Ω.

Proof $N = \Big\{\sigma \in \Omega_* : g_n(\sigma)$ does not tend not to $g(\sigma)$ or $g(\sigma)$ does not exist$\Big\}$ then $|N| = 0$. Therefore, $\Big\{x \in \Omega \backslash P : \beta_u(x) \in N\Big\}$ is of measure zero. So,

* $g_n\Big(\beta_u(x)\Big) \underset{n}{\to} g\Big(\beta_u(x)\Big)$ a.e. $x \in \Omega \backslash P$.

  As $\displaystyle\int_{s_i'}^{s_i''} g_n(\sigma)d\sigma \underset{n}{\to} \int_{s_i'}^{s_i''} g(\sigma)d\sigma$, $M_u(g_n)(x) \to M_u(g)(x)$.
* $h_{i,n}(s) = g_n(s_i' + s) \underset{n}{\to} g(s_i' + s)$ a.e. in P_i^* then

$$M_{v_i}(h_{i,n})(x) \to M_{v_i}(h_i)(x) \text{ a.e. in } P_i.$$

Since D is at most countable, we deduce:

$$M_{u,v}(g_n)(x) \to M_{u,v}(g)(x) \ a.e.$$

◇

2.5 Integral Formulas for a Function of Two Variables and Consequences

Let $F : \Omega_* \times \mathbb{R} \to \mathbb{R}$ be a function defined a.e. in Ω_* and everywhere in $\mathbb{R}$.

We assume that F is Carathéodory, that is to say

$$a.e.\ s : t \in \mathbb{R} \mapsto F(s,t) \text{ is continuous}$$

and

$$\forall t, \qquad s \in \Omega_* \mapsto F(s,t) \text{ is measurable.}$$

We define $\mathcal{M}F$ (resp for $M_u F$, $M_{u,v} F$) by:

$$\big(\mathcal{M}F\big)(x,t) = \Big(\mathcal{M}F(\cdot,t)\Big)(x)\,.$$

Lemma 2.5.1

(i) $\mathcal{M}\big(F \circ u_*\big) = (\mathcal{M}F) \circ u,$

(ii) If $F \geqslant 0$ $\displaystyle\int_{\Omega_*} F \circ u_* = \int_\Omega (\mathcal{M}F) \circ u\, dx,$

(iii) If $F \circ u_* \in L^p(\Omega_*)$, $1 \leqslant p \leqslant +\infty$ *then* $\big(\mathcal{M}F\big) \circ u \in L^p(\Omega)$*:*

$$\Big|\Big(\mathcal{M}F\Big) \circ u\Big|_{L^p(\Omega)} \leqslant |F \circ u_*|_{L^p(\Omega_*)}\,.$$

Proof In the case where $\mathcal{M} = M_u$, if $u(x) = t_i$, $i \in D$ then:

$$\bullet \Big(\mathcal{M}F\Big) \circ u(x) = \frac{1}{|P_i|}\int_{s_i'}^{s_i''} F\big(s, u(x)\big)ds = \frac{1}{|P_i|}\int_{s_i'}^{s_i''} F\big(s, t_i\big)ds$$

$$= \frac{1}{|P_i|}\int_{s_i'}^{s_i''} F\big(s, u_*(s)\big) = \Big(\mathcal{M}F\Big) \circ u_*(x)\,.$$

For a.e. $x \in \Omega \backslash P$ we have:

$$\bullet \Big(M_u F\Big) \circ u(x) = F\big(\beta_u(x), u(x)\big) = F\Big(\beta_u(x), u_*\big(\beta_u(x)\big)\Big) = M_u\big(F \circ u_*\big)(x).$$

(i) and (ii) follow from this. The proof is identical for $M_{u,v}$. ◇

Lemma 2.5.2 *Let* $v \in L^p(\Omega)$, $1 \leqslant p \leqslant +\infty$, $u \in L^1(\Omega)$. *Let* $w = w(u,v)$ *be the function defined by Theorem 2.1.1 (directional derivative). Let* $g \in L^q(\Omega_*)$, $\dfrac{1}{q} + \dfrac{1}{p} = 1$, *then:*

$$\int_{\Omega_*} g v_{*u} = \int_{\Omega_*} g \cdot \frac{dw}{ds} = \int_\Omega M_{u,v}(g) \cdot v\, dx\,.$$

Proof If $1 < p \leqslant +\infty$, as the map $g \in L^q(\Omega_*) \to M_{u,v}(g) \in L^q(\Omega)$ is continuous it suffices to show the equality for a dense subspace: $g \in \mathcal{D}(\Omega_*)$. We have

$$w(s) = \int_\Omega v(x)\chi_{]u_*(s),+\infty[}\big(u(x)\big)dx + \sum_{i\in D} \chi_{P_i^*}(s) \int_0^{s-s_i'} v_{i*}d\sigma \,.$$

We write:

$$\int_{\Omega_*} g \cdot \frac{dw}{ds} = -\int_{\Omega_*} \frac{dg}{ds} w = A + \sum_{i\in D} B_i$$

with $A = -\int_{\Omega_*} \frac{dg}{ds} \int_\Omega \chi_{]u_*(s),+\infty[}\big(u(x)\big)v(x)dx$ and $B_i = -\int_{s_i'}^{s_i''} \frac{dg}{ds} \cdot \left[\int_0^{s-s_i'} v_{i*}d\sigma\right] ds$.

By Fubini's theorem

$$A = -\int_\Omega v(x) \int_{\Omega_*} \frac{dg}{ds} \cdot \chi_{]-\infty,u(x)[}\big(u_*(s)\big)ds$$

$= -\int_\Omega v(x) \int_{\overline{\beta_u}(x)}^{|\Omega|} \frac{dg}{ds} ds = \int_\Omega v(x) g\Big(\overline{\beta_u}(x)\Big)$, since we have

$$\{u_*(\cdot) < u(x)\} = \Big(\overline{\beta_u}(x), |\Omega|\Big] \text{ and } \{u_* \geqslant u(x)\} = \Big[0, \overline{\beta_u}(x)\Big) \,.$$

For B_i we integrate by parts:

$$B_i = -g(s_i'') \cdot \int_0^{s_i''-s_i'} v_{i_*}(\sigma)d\sigma + \int_{s_i'}^{s_i''} g(s)v_{i_*}(s - s_i')ds \,.$$

The first term is: $= -\int_{P_i^*} g(s_i'') \cdot v_{i*}(\sigma) = -\int_{P_i} g(s_i'')v_i(x)dx \,.$

As, $s_i'' = \overline{\beta_u}(x)$, this last integral is written:

$$- g(s_i'') \cdot \int_0^{s_i''-s_i'} v_{i_*}(\sigma) d\sigma = - \int_{P_i} g(\overline{\beta_u}(x)) v_i(x) dx .$$

As for the second term,

$$\int_{s_i'}^{s_i''} g(s) v_{i_*}(s - s_i')\, ds = \int_0^{s_i''-s_i'} g(\sigma + s_i') \cdot v_{i_*}(\sigma) d\sigma = \int_{P_i^*} (F \circ v_{i*})(s) ds$$

(with $F(s,t) = g(s + s_i') \cdot t$)

$$= \int_{P_i} (M_{v_i} F) \circ v_i dx = \int_{P_i} M_{v_i} \left(g\left(\cdot + s_i'\right)\right) v dx = \int_{P_i} M_{u,v}(g) v\, dx.$$

Hence, by combining the two terms, we obtain:

$$B_i = - \int_{P_i} g(\overline{\beta_u}(x)) v(x) dx + \int_{P_i} M_{u,v}(g) v(x) dx ,$$

$$\int_{\Omega_*} g \frac{dw}{ds} = \int_{\Omega \backslash P} g\left(\overline{\beta_u}(x)\right) v\, dx + \int_P M_{u,v}(g) v\, dx = \int_\Omega M_{u,v}(g) v\, dx .$$

Case where $p = 1$. For $g \in L^\infty(\Omega_*)$, we consider a sequence $g_n \in \mathcal{D}(\Omega_*)$ s.t.

$$\begin{cases} g_n \to g \text{ in } L^1(\Omega_*) \\ |g_n|_\infty \leqslant |g|_\infty \\ g_n \to g \; a.e. \end{cases} .$$

Then $M_{u,v}(g_n)(x) \to M_{u,v}(g)(x)$ a.e. in Ω (by the Lemma 2.4.1). So if $v \in L^1(\Omega)$,

$$\left| M_{u,v}(g_n) v(x) \right| \leqslant |g_n|_\infty \cdot |v(x)| \leqslant |g|_\infty |v(x)| \in L^1(\Omega)$$

and by the dominated convergence theorem, we have:

$$\int_\Omega M_{u,v}(g_n)v\,dx \to \int_\Omega M_{u,v}(g)v\,dx.$$

From the same reason,

$$\int_\Omega g_n \cdot v_{*u} d\sigma \to \int_\Omega g(\sigma) \cdot v_{*u}(\sigma)d\sigma,$$

hence $\int_\Omega M_{u,v}(g)v = \int_{\Omega_*} g(\sigma)v_{*u}(\sigma)d\sigma$. In particular, for $g(s) = F \circ u_*(s)$

$$\int_{\Omega_*} (F \circ u_*)\, v_{*u} = \int_\Omega \left[\left(M_{u,v}F\right) \circ u\right] v dx.$$

◇

2.6 Applications of Mean Operators

The applications of these formulas are numerous, here are some examples.

Proposition 2.6.1 *[weak continuity of $u \to v_{*u}$] Let u and v be two elements of $L^1(\Omega)$. We assume that u is without a plateau i.e. $mes\left(P(u)\right) = 0$, u_n a sequence of $L^1(\Omega)$ s.t. $u_n(x) \xrightarrow[n\to+\infty]{} u(x)$ a.e. in Ω.*

Then $v_{*u_n} \underset{n\to+\infty}{\rightharpoonup} v_{*u}$ in $L^1(\Omega_*)$-weak.

Proof For $x \in \Omega$, $\underline{\beta}(u_n)(x) = |u_n > u_n(x)|$, $\overline{\beta}(u_n)(x) = |u_n \geqslant u_n(x)|$, $\delta(u_n)(x) = \left[\underline{\beta}(u_n)(x), \overline{\beta}(u_n)(x)\right]$. Let us then note, for $g \in \mathcal{D}(\Omega_*)$

$$\text{Min}_{\,\sigma\in\delta(u_n)(x)} g(\sigma) \leqslant M_{u_{n,v}}(g)(x) \leqslant \text{Max}_{\,\sigma\in\delta(u_n)(x)} g(\sigma)\,.$$

We have,

$$|u > u(x)| = \liminf_n \underline{\beta}(u_n)(x) \leqslant \limsup_n \overline{\beta}(u_n)(x) \leqslant |u \geqslant u(x)|\,.$$

If $mes\Big(P(u)\Big) = 0$ then $\lim_n \delta(u_n)(x) = \underline{\beta}(u)(x)$. So $\lim_n M_{u_{n,v}}(g)(x) = g\Big(|u > u(x)|\Big)$. By the dominated convergence theorem and the previous formula, we have

$$\lim_n \int_{\Omega_*} g(\sigma) v_{*u_n}(\sigma) d\sigma = \lim_n \int_\Omega M_{u_n,v}(g) v\, dx = \int_{\Omega_*} g(\sigma) v_{*u}(\sigma) d\sigma \ .$$

Thus, $v_{*u_n} \rightharpoonup v_{*u}$ in $\mathcal{D}'(\Omega_*)$. But $\left|v_{*u_n}\right|_{L^1} \leqslant |v|_{L^1}$ and for all measurable set $E \subset \Omega_*$

$$\int_E \left|v_{*u_n}\right| \leqslant \int_0^{|E|} |v|_* \ .$$

Thus, v_{*u_n} satisfies the Dunford–Pettis criterion, therefore

$$v_{*u_n} \rightharpoonup v_{*u} in L^1(\Omega_*)\text{-weak.}$$

The assumption $mes\big(P(u)\big) = 0$ is useful. We can construct a function $u \in L^1(\Omega)$ and a sequence $u_n \in L^1(\Omega)$ such that v_{*u_n} does not tend to v_{*u} ($\forall\, v \in L^1(\Omega)$) in $\mathcal{D}'(\Omega_*)$, and $\text{mes}\big(P(u)\big) \neq 0$ (see Chap. 7). ◇

In the case where u has at least one plateau, we have the following result generalizing the previous one:

Theorem 2.6.1 *Let $v \in L^1(\Omega)$ and u_n be a sequence of $L^1\Omega)$ such that $u_n(x) \to u(x)$ a.e. in Ω, $u \in L^1(\Omega)$. Let $\theta \in Ł^1(\Omega_*)$ s.t. $\theta(\sigma) = 0$ if $\sigma \in P(u_*)$. Then,*

$$\theta v_{*u_n} \rightharpoonup \theta v_{*u} \; in \; L^1(\Omega_*)\text{-weak.}$$

Proof By density, it suffices to consider $\theta \in C(\overline{\Omega}_*)$. As above, if $E \subset \Omega_*$ is measurable then

$$\int_E \left|\theta v_{*u_n}\right| dt \leqslant |\theta|_\infty \int_0^{|E|} |v|_* \, dt \ .$$

It is then sufficient to show $\forall\, g \in C(\overline{\Omega}_*)$

$$\lim_n \int_{\Omega_*} g\theta v_{*u_n} = \int_{\Omega_n} g\theta v_{*u} dt.$$

By the average formula, we have:

$$\int_{\Omega_*} g\theta v_{*u_n} dt = \int_{\Omega_*} M_{u_n v}(g\theta)(x) v(x) dx \ .$$

Since $\theta(\sigma) = 0$ if $\sigma \in P(u_*)$, we deduce that

$$\lim_n M_{u_n v}(g\theta)(x) = (g\theta)\big(|u > u(x)|\big).$$

Note that $|u > u(x)| \in P(u_*)$ if $x \in P(u)$.

By the dominated convergence theorem, and again the mean formula, we deduce the result. ◇

Theorem 2.6.2 *Let u and v be in $L^1(\Omega)$ and let $\theta \in L^1(\Omega_*)$ s.t. $\theta(\sigma) = 0$ if $\sigma \in P(u_*)$. Then,*

$$\theta \left(v \chi_{\Omega \backslash P(u)}\right)_{*u} = \theta \chi_{\Omega_* \backslash P(u_*)} v_{*u} \ .$$

Proof If $g \in C(\overline{\Omega}_*)$ then,

$$\int_{\Omega_*} g\theta \left(v \chi_{\Omega \backslash P(u)}\right)_{*u} dt = \int_{\Omega \backslash P(u)} (\theta g)\big(|u > u(x)|\big) v \, dx$$

and

$$\int_{\Omega_*} g\theta \left(\chi_{\Omega_* \backslash P(u_*)}\right) v_{*u} dt = \int_{\Omega} M_{u,v} \left(g\theta \chi_{\Omega_* \backslash P(u_*)}\right) v \, dx$$

$$= \int_{\Omega \backslash P(u)} (g\theta)\big(|u > u(x)|\big) v \, dx$$

(because $\chi_{\Omega_* \backslash P(u_*)}\big(|u > u(x)|\big) = 1$ if $x \in \Omega \backslash P(u)$). ◇

Here are some recent results showing the use of mean operators and a directional derivative calculation of the map

$$u \to \int_0^s |u_*(t)|^p dy, \qquad s \in \Omega_* .$$

2.7 Some Additional Results on Relative Rearrangement

Lemma 2.7.1 *Let* $u \in L^1_+(\Omega),\ v \in L^\infty(\Omega)$. *Then*

$$(uv)_{*u}(s) = u_*(s)v_{*u}(s),$$

for almost all $s \in \Omega_* =]0, |(\Omega_*)[|$.

Proof Since the map $v \in L^\infty(\Omega) \to (uv)_{*u} \in L^1(\Omega)$ is continuous, we can assume that v has no plateau (the set of functions v without plateau (i.e. $|v = t| = 0\,\forall\, t \in \mathbb{R}$, is dense in $L^\infty(\Omega)$, see Exercise 11.1.3). Using Lemma 2.5.2 on the mean operator, we have for all $\varphi \in L^\infty(\Omega_*)$

$$\int_0^{|\Omega|} (uv)_{*u}(s)\varphi(s)ds = \int_\Omega M_{u,uv}(\varphi)uvdx.$$

By the definition of $M_{u,uv}$, we have

$$u(x)M_{u,uv}(\varphi)(x) = \begin{cases} u(x)\varphi\big(|u > u(x)|\big) & if\ x \in \Omega\backslash P(u), \\ 0 & if\ u(x) = 0, \\ u(x)\varphi\big(|u > u(x)| + \theta(x)\big) & otherwise, \end{cases}$$

where $\theta(x) = \Big|\big\{y : u(y) = u(x),\, u(x)v(y) > u(x)v(x)\big\}\Big|$.

In this case, as long as $u(x) > 0$, $\theta(x) = \Big|\big\{y : v\big|_{\{u=u(x)\}}(y) > v(x)\big\}\Big|$. As a result, $u(x)M_{u,v}(\varphi)(x) = u(x)M_{u,uv}(\varphi)(x)$ a.e. x and

$$\int_{\Omega_*} (uv)_{*u}(s)\varphi(s)ds = \int_\Omega M_{u,v}(\varphi)uvdx. \tag{2.1}$$

On the other hand,

$$\int_{\Omega_*} u_*(s)v_{*u}(s)\varphi(s)ds = \int_\Omega M_{u,v}(u_*\varphi)vdx, \tag{2.2}$$

and by the definition of $M_{u,v}$

$$M_{u,v}(u_*\varphi)(x) = \begin{cases} u_*\big(|u > u(x)|\big)\varphi\big(|u > u(x)|\big) & if\ x \in \Omega\backslash P(u), \\ \varphi\big(|u > u(x)| + \theta(x)\big)u(x) & otherwise, \end{cases}$$

where $\theta(x) = \Big|\big\{y : \big(v\big|_{\{u=u(x)\}}\big)(y) > v(x)\big\}\Big|$.

(Since $u_*\big(|u > u(x)| + \theta(x)\big) = u(x)$, $|u > u(x)| + \theta(x)$ belongs to the interval $\big[|u > u(x)|; |u \geqslant u(x)|\big)$, where $u_* = u(x)$.)

For $x \in \Omega\backslash P(u)$, $u_*\big(|u > u(x)|\big) = u(x)$. So,

$$M_{u,v}(u_*\varphi)(x) = u(x)M_{u,v}(\varphi)(x) \text{ a.e.} \tag{2.3}$$

The relations (2.1), (2.2), (2.3) imply $\forall\varphi \in L^\infty(\Omega_*)$ $\displaystyle\int_{\Omega_*} u_*(s)v_{*u}(s)\varphi(s)ds = \int_\Omega M_{u,v}(\varphi)uv\,ds = \int_{\Omega_*}(uv)_{*u}(s)\varphi(s)ds$

$$: u_*v_{*u} = (uv)_{*u}.$$

◇

Corollary 2.7.1 (Of Lemma 2.7.1) *If $u \in L^p_+(\Omega)$, $v \in L^p(\Omega)$, $1 < p < +\infty$, then*

$$(u^{p-1}v)_{*u} = u_*^{p-1}v_{*u} \; a.e \; in \; \Omega_*.$$

Proof Since

$$\{u^{p-1} > u_*^{p-1}(s)\} = \{u > u_*(s)\}\text{and}\{u^{p-1} = u_*^{p-1}(s)\} = \{u = u_*(s)\},$$

the definition of relative rearrangement implies

$$v_{*(u^{p-1})} = v_{*u}.$$

By Lemma 2.7.1, for $v \in L^\infty(\Omega)$,

$$(u^{p-1}v)_{*u} = (u^{p-1}v)_{*(u^{p-1})} = u_*^{p-1}v_{*(u^{p-1})} = u_*^{p-1}v_{*u}.$$

By the contraction property of relative rearrangement, we deduce for all $v_n \in L^\infty(\Omega)$, $v \in L^p(\Omega)$

$$\left|(u^{p-1}v_n)_{*u} - (u^{p-1}v)_{*u}\right|_1 \leqslant |u|_p^{p-1}|v_n - v|_p\,,$$

and

$$|u_*^{p-1}(v_{n*u} - v_{*u})|_1 \leqslant |u|_p^{p-1}|v_{n*u} - v_{*u}|_p \leqslant |u|_p^{p-1}|v_n - v|_p.$$

The result follows by the density of $L^\infty(\Omega)$ in $L^p(\Omega)$. ◇

We propose to calculate the directional derivative of

$$u \in L^p(\Omega) \to \int_0^t |u|_*^p(\sigma)d\sigma, \ \forall t > 0, \ 1 < p < +\infty.$$

Theorem 2.7.1 *If $u \in L^p(\Omega)$, $v \in L^p(\Omega)$, $1 < p < +\infty$, then*

$$\frac{|u + \lambda v|_*^p - |u|_*^p}{\lambda} \underset{\lambda \searrow 0}{\rightharpoonup} p\Big(|u|^{p-2}uv\Big)_{*|u|} \text{ in } L^1(\Omega_*)\text{-weak}.$$

If $p \geqslant 2$, we then have

$$\Big(|u|^{p-2}uv\Big)_{*|u|} = |u|_*^{p-2}(uv)_{*|u|}.$$

Proof Let us set

$$g_{1\lambda}(x) = \frac{|u + \lambda v|^p(x) - |u|^p(x)}{\lambda} - pv(x)|u(x)|^{p-2}u(x), \ x \in \Omega.$$

Since

$$\frac{d}{dt}|t|^p = p|t|^{p-2}t,$$

we deduce that for almost all $x \in \Omega$

$$g_{1\lambda}(x) \xrightarrow[\lambda \to 0]{} 0.$$

Moreover, for $0 < \lambda < 1$ we have

$$|g_{1\lambda}(x)| \leqslant c_p\Big[|u(x)| + |v(x)|\Big]^p = h_1(x) \in L^1(\Omega).$$

Consequently, Lebesgue's dominated convergence theorem leads to

$$g_{1\lambda} \to 0 \; in \; L^1(\Omega)\text{-strongly}.$$

By writing

$$|u+\lambda v|^p(x) = |u|^p(x) + \lambda p v(x)|u(x)|^{p-2}u(x) + \lambda g_{1\lambda}(x),$$

we deduce from the contraction property that $\forall \varphi \in L^\infty(\Omega_*)$

$$\int_0^{|\Omega|} \frac{|u+\lambda v|_*^p - \big(|u|^p + \lambda p v|u|^{p-2}u\big)_*}{\lambda}\varphi = O\Big(|g_{1\lambda}|_{L^1(\Omega)}\Big) \xrightarrow[\lambda\to 0]{} 0. \tag{2.4}$$

Using Theorem 2.1.3:

$$\lim_{\lambda\searrow 0}\int_0^{|\Omega|}\varphi(s)\frac{\big(|u|^p+\lambda p v|u|^{p-2}u\big)_* - |u|_*^p}{\lambda}ds = p\int_0^{|\Omega|}\varphi(s)\big(|u|^{p-2}uv\big)_{*|u|^p}ds.$$

Since the map $t \in \mathbb{R}_+ \to t^p$ is strictly increasing, we have

$$\big(|u|^{p-2}uv\big)_{*|u|^p} = \big(|u|^{p-2}uv\big)_{*|u|}.$$

Therefore,

$$\lim_{\lambda\searrow 0}\int_0^{|\Omega|}\varphi(s)\frac{|u+\lambda v|_*^p - |v|_*^p}{\lambda}(s)ds = p\int_0^{|\Omega|}\varphi(s)\big(|u|^{p-2}uv\big)_{*|u|}ds.$$

If $p \geqslant 2$, we then check

$$\big(|u|^{p-2}uv\big)_{*|u|} = \big(|u|^{p-2}uv\big)_{*|u|^{p-2}} = |u|_*^{p-2}(uv)_{*|u|}.$$

◇

Corollary 2.7.2 (Case where $u \geqslant 0$**)** *If* $u \in L^p_+(\Omega)$, $v \in L^p(\Omega)$, $1 < p < +\infty$, *then*

$$\frac{|u+\lambda v|_*^p - u_*^p}{\lambda} \underset{\lambda\searrow 0}{\rightharpoonup} p u_*^{p-1} v_{*u} \; in \; L^1(\Omega_*)\text{-weak}.$$

Proof We combine Theorem 2.7.1 and Corollary 2.7.1 of Lemma 2.7.1 to obtain the result. ◇

2.8 Construction of a Relative Rearrangement of a Function v with Respect to a Step Function u

In this section, we will give an example of a plot of a relative rearrangement of a function v with respect to a simple function u. As the map $v \in L^1(a,b) \to v_{*u} \in L^1(0, b-a)$ is a contraction, we can restrict ourselves to simple functions v to illustrate the plot.

Let $u : [a,b] \to \mathbb{R},\ u(x) = \sum_{j=1}^{n} u_j \chi_{E_j}(x),\ x \in [a,b]$. We start by sorting the values u_j in strictly decreasing order $t_1 > \cdots > t_p$.

We define $F_j = \{u = t_j\}$ and $a_0 = 0,\ a_1 = |F_1|,\ a_2 = |F_1| + |F_2|, \ldots, a_p = b - a = |F_1| + \ldots + |F_p|$ (so $a_j - a_{j-1} = |F_j| = |\{u = t_j\}|$).

We define v_j to be the restriction of v to F_j. Applying the algorithm for calculating the monotone rearrangement of v_j, we obtain v_{j*} (which is also a simple function when v is a simple function) (Fig. 2.1).

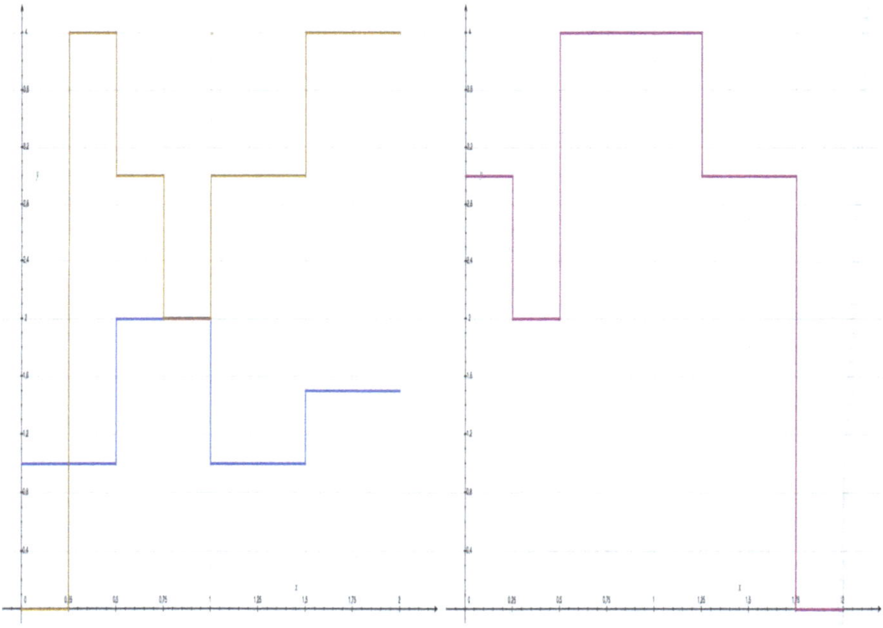

Fig. 2.1 Relative rearrangement of two simple functions

We obtain v_{*u} by translating the rearrangements as follows:

$$v_{*u}(s) = \begin{cases} v_{1*}(s) & \text{if } 0 \leqslant s < a_1 \\ v_{2*}(s - a_1) & \text{if } a_1 \leqslant s < a_2 \\ & \text{(translated from } v_{2*} \text{ to } a_1 \\ & \text{taking a length } |u = t_1|) \\ \vdots & \\ v_{p*}(s - a_{p-1}) & \text{if } a_{p-1} \leqslant s < a_p = b - a \\ & \text{(translated from } v_{p*} \text{ to } [a_{p-1}, a_p]) \end{cases}$$

Note
For other plots of the relative rearrangement when v and u are piecewise affine, for example, and u satisfying $|\{x : u'(x) = 0\}| = 0$, we refer to the article by Rakotoson and Seoane [138].

Here is an explicit example of the relative rearrangement of continuous functions: let

$$I =]0, \frac{5}{2}[= I_*, \ v(x) = 3x, \ u(x) = \begin{cases} x & 0 \leqslant x \leqslant 1 \\ \frac{1}{2}x + \frac{1}{2} & 1 < x \leqslant 2 \\ \frac{7}{2} - x & 2 < x \leqslant \frac{5}{2} \end{cases}$$

then

$$v_{*u}(s) = \begin{cases} 6 - s & 0 \leqslant s < \frac{3}{2} \\ 3(\frac{5}{2} - s) & \frac{3}{2} \leqslant s \leqslant \frac{5}{2} \end{cases}$$

As the map $u \to b_{*u}$ is not generally strongly continuous, we cannot construct b_{*u} from the simple functions u.

If $u \in W^{1,1}(\Omega)$ s.t. $\left|\left\{x : \nabla u(x) = 0\right\}\right| = 0$ and $v \in L^1(\Omega)$ then from Federer's formula (cf., p. 214), applied to $w(s)$, we deduce after differentiation, that for almost all $\sigma \in \Omega_*, \ N \geqslant 2$:

$$v_{*u}(s) = \frac{1}{a(s)} \int_{\left\{u=u_*(s)\right\}} \frac{v(x) d H_{N-1}(x)}{|\nabla u|(x)} \quad \text{where } a(s) = \int_{\left\{u=u_*(s)\right\}} \frac{d H_{N-1}}{|\nabla u|(x)}.$$

The previous calculation gives a weak meaning to v_{*u} for $u \in L^1(\Omega)$. An analogous expression can be shown for $N = 1$.

This calculation is justified in the articles by Rakotoson [123] and Díaz–Rakotoson [61]. It is left to the readers.

Here is an example where v is continuous and u is simple (see Figs. 2.2 and 2.3):

$$v(x) = x \quad 0 \leqslant x \leqslant 3,$$

and $u(x) = \begin{cases} 2 & 0 \leqslant x < 1, \\ 1 & 1 \leqslant x < 2, \\ 3 & 2 \leqslant x \leqslant 3. \end{cases}$ So $v_{*u}(s) = \begin{cases} 3 - s & 0 \leqslant s < 1, \\ 2 - s & 1 \leqslant s < 2, \\ 4 - s & 2 \leqslant s \leqslant 3. \end{cases}$

For the following figure, on the left side we have the plots of u, u_*, v, v_* with

$$u(x) = \begin{cases} 2 & 0 < x < 1 \\ 1 & 1 < x < 2 \\ 2 & 2 < x < 3 \\ 0 & 3 < x < 4 \\ 1 & 4 < x < 5 \\ 2 & 5 < x < 2\pi \end{cases}, \qquad v(x) = \sin(x)$$

and on the right side the relative rearrangement of v_{*u} and the monotonic rearrangement $\{v_{*u}\}_*$. We see that $\{v_{*u}\}_* = v_*$. This property will be demonstrated in Chap. 7.

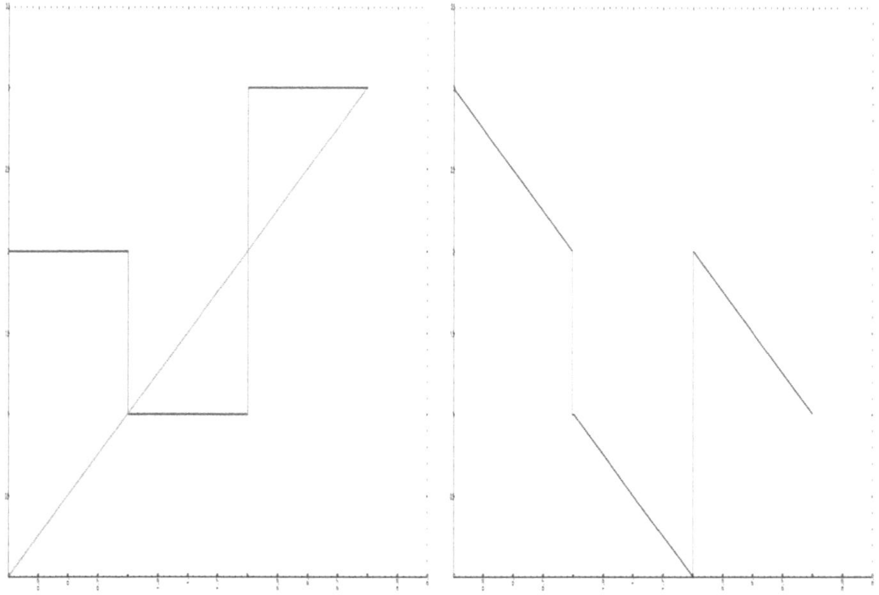

Fig. 2.2 Relative rearrangement of the continuous function $v(x) = x$ and a step function

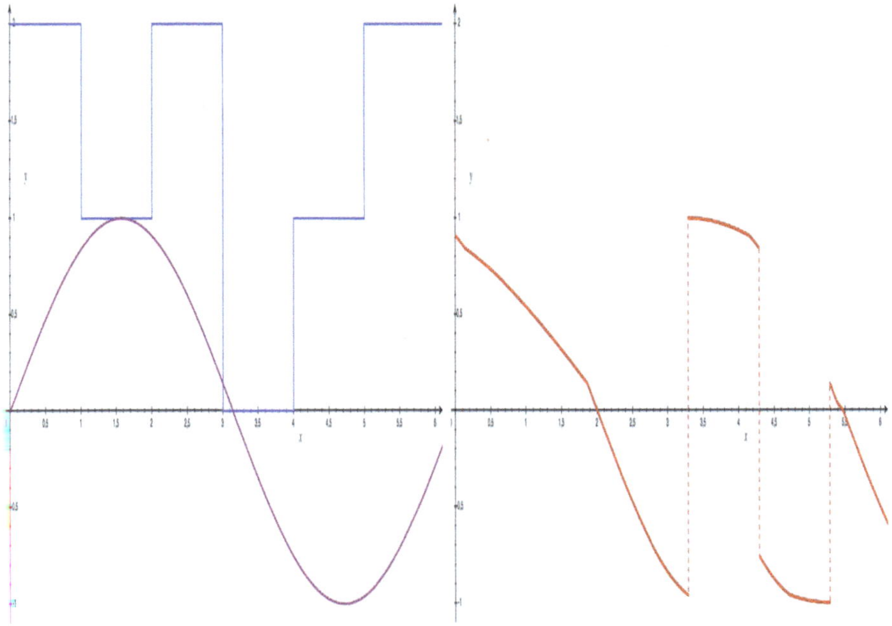

Fig. 2.3 Relative rearrangement of the continuous function $v(x) = \sin(x)$ and a step function

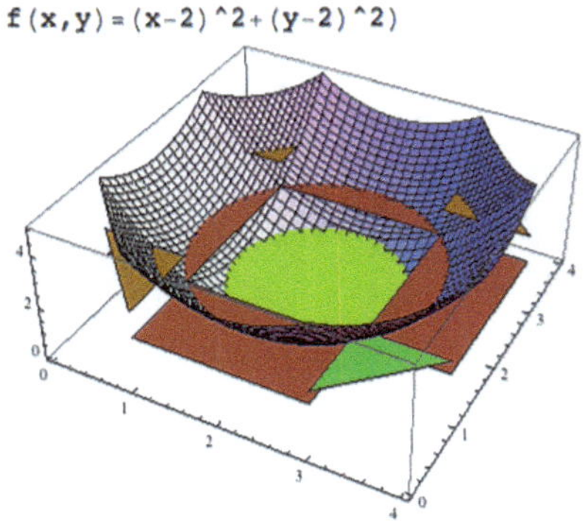

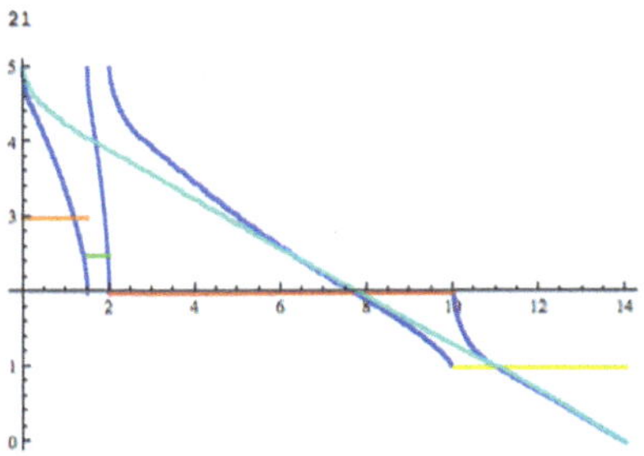

Pre-bibliographical Notes
All the results of this chapter (except the Lorentz–Luxemburg Theorem 2.2.1, which can be found in the book by Chong and Rice) are stated in the book by J. Mossino, the thesis by B. Simon, or in the articles by Mossino–Temam, the author and/or his collaborators.

The last plot was made by Catherine Rakotoson, using the software Mathematica™.

Chapter 3
Polyà–Szegö Inequalities and Regularity of Monotone Rearrangement

3.1 Introduction

One of the equalities that we will frequently use to estimate u_* pointwise is the following: $u_*(s) - u_*(t) = \displaystyle\int_t^s u'_*(\sigma)d\sigma$. But such an equality is true only if u_* is absolutely continuous on $[t, s]$. Thus in this chapter, we will show how the relative rearrangement will allow us to answer the following natural question: if u is regular (say $C^k(\overline{\Omega})$), what can we say about the regularity of u_*? Through this study, we will deduce inequalities that will link the gradient of u and the derivative of u_*. These inequalities will give the classical Polyà–Szegö inequalities but also various other extensions.

First, let us make some preliminary remarks. The nature of the domain must intervene in the general case.

Indeed if $\Omega = \Omega_1 \cup \Omega_2$, $\quad \Omega_i$ open disjoint ($\overline{\Omega}_1 \cap \overline{\Omega}_2 = \emptyset$), then the characteristic function of the set $\overline{\Omega}_1$: $u(x) = \chi_{\overline{\Omega}_1}(x)$ is a continuous function on Ω, and $u_* : [0, |\Omega|] \to \mathbb{R}$ is given by

$u_*(s) = \begin{cases} 1 & if\ s \in [0, |\Omega_1|[\\ 0 & otherwise \end{cases}$. So $u_* \notin C(\overline{\Omega}_*)$. We will then show that $u_* \in C(\overline{\Omega}_*) \iff \overline{u(\Omega)}$ is connected, if $u \in C(\Omega) \cap L^\infty(\Omega)$.

Then, to show that even if the function u indefinitely differentiable or analytic, its rearrangement is not necessarily C^1. Here is an example on $[-1, 2]$. Let us consider $u(x) = x^2$, then, for $s \in [0, 3]$

$$u_*(s) = \begin{cases} (2-s)^2 & if\ 0 \leqslant s < 1 \\ \left(\dfrac{3-s}{2}\right)^2 & if\ 1 \leqslant s \leqslant 3 \end{cases}.$$

We note that $u_* \notin C^1[0, 3]$.

J. M. Rakotoson, *Relative Rearrangement*, Lecture Notes in Mathematics 2376,
https://doi.org/10.1007/978-3-032-02228-8_3

We will nevertheless show that if $u \in C^{0,1}(\overline{\Omega})$, where Ω is regular connected or $u = 0$ on $\partial\Omega$, then $u_* \in C^{0,1}_{loc}(\Omega_*)$.

Let's start by studying the continuity of u_*. From now on, Ω will always be an open, bounded set (for simplicity).

3.2 Continuity $s \to u_*(s)$

Theorem 3.2.1 (Continuity of u_*) *Let $u \in C(\Omega) \cap L^\infty(\Omega)$. The following conditions are equivalent:*

(i) $u_* \in C(\overline{\Omega_*})$,
(ii) $\overline{u(\Omega)}$ *is connected.*

Proof $(ii) \Longrightarrow (i)$ Since $\overline{u(\Omega)}$ is connected, we have:

$$\overline{u(\Omega)} = \left[\alpha = \text{ess inf}_{\Omega} u,\ \beta = \text{ess sup}_{\Omega} u\right] \subset \mathbb{R}\,.$$

We always have $u_*(0) = \lim\limits_{s \searrow 0} u_*(s)$, $u_*(|\Omega|) = \lim\limits_{s \nearrow |\Omega|} u_*(s)$. Now, $s \in \Omega_*$, and since u_* is monotone, it has a limit on the left and on the right. Furthermore, we know that u_* is right continuous. Consequently,

$$u_*(s_-) = \lim_{h \to 0\ h>0} u_*(s-h),\ u_*(s) = \lim_{h \to 0\ h>0} u_*(s+h)\,.$$

If u_* were discontinuous at the point s then $|u_*(s) < u_* < u_*(s_-)| = 0$. Or by equimeasurability, $|u_*(s) < u < u_*(s_-)| = 0$.

Since $\left\{u_*(s) < u < u_*(s_-)\right\}$ is contained in $]\alpha, \beta[$, there exists $x \in \Omega : u_*(s) < u(x) < u_*(s_-)$ and by continuity, this set $\left\{u_*(s) < u < u_*(s_-)\right\}$ contains a ball centered at the point x. Consequently $|u_*(s) < u < u_*(s_-)| > 0$ (hence the contradiction).

$(i) \Longrightarrow (ii)$ We argue by contradiction. If $\overline{u(\Omega)}$ is not connected then there would exist two closed non-empty sets F_1 and F_2 such that $F_1 \cap F_2 = \emptyset$ and $F_1 \cup F_2 = \overline{u(\Omega)}$. $\overline{u(\Omega)}$ being a compact set, F_1 and F_2 are also compact. Let $t_1 \in F_1$ and $t_2 \in F_2$. We can assume $t_1 < t_2$. We set $c = \sup F_1 \cap [t_1, t_2]$. By compactness, we have $c \in F_1 \cap [t_1, t_2]$. Let $d = \inf F_2 \cap [c, t_2]$, likewise $d \in F_2 \cap [c, t_2]$. We have $c < d$ because $F_1 \cap F_2 = \emptyset$ and $]c, d[\cap\big(F_1 \cup F_2\big) = \emptyset$. But as $\overline{u(\Omega)} = F_1 \cup F_2$, we have $\left\{x \in \Omega : c < u(x) < d\right\} = \emptyset$. Thus $|c < u_* < d| = 0$. But $\text{ess inf}_{\Omega} u \leqslant c < d \leqslant \text{ess sup}_{\Omega} u$, so if u_* were continuous on $\overline{\Omega_*}$, there would exist $s \in \Omega_*$ s.t. $u_*(s) \in]c, d[$ and $h > 0 : u_*\Big([s-h, s+h]\Big) \subset]c, d[$. Then $|c < u_* < d| \geqslant 2h > 0$. This contradicts the above result. ◇

3.3 Fleming–Rishel Formula

To analyze the derivative of u_* we need some lemmas. First of all, we introduce the notion of perimeter in the sense of De Giorgi, which extends the classical notion of perimeter for smooth sets.

Definition 3.3.1 (Perimeter of a Measurable Set)

- Let E be a measurable set of Ω. We call the perimeter of E following Ω the number
$$P_\Omega(E) = \sup\left\{\int_E \operatorname{div}(\vec{\Phi})dx, \quad \vec{\Phi} \in C_c^\infty(\Omega)^N, \left|\vec{\Phi}\right|_\infty \leqslant 1\right\}.$$
- We call the perimeter of E following $\mathbb{R}^N$ the number
$$P_{\mathbb{R}^N}(E) = \sup\left\{\int_E \operatorname{div}(\vec{\Phi})dx, \ \vec{\Phi} \in C_c^\infty(\mathbb{R}^N)^N, \ |\vec{\Phi}|_\infty \leqslant 1\right\}.$$

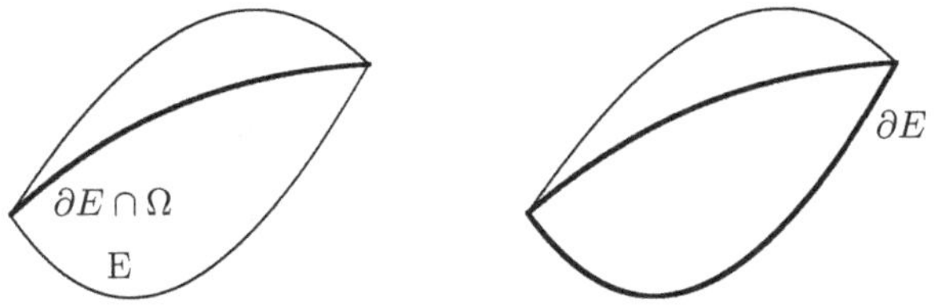

Proposition 3.3.1

1. $P_\Omega(\Omega) = 0 = P_\Omega(E)$ *if* $|E| = 0$.
2. $P_\Omega(E) \leqslant P_{\mathbb{R}^N}(E)$, *for all measurable* E *in* Ω. *If* $E \subset \overline{\Omega_1} \subset \Omega$, Ω_1 *relatively compact open in* Ω, *we have an equality.*
3. *Let* $\{E_n\}$ *be a sequence of measurable sets of* Ω *such that* $\chi_{E_n}(x) \to \chi_E(x)$ *a.e. where* E *measurable* $\subset \Omega$.
   Then $P_\Omega(E) \leqslant \liminf_n P_\Omega(E_n)$ *(lower semi-continuity).*

Proof

1. Since $\overrightarrow{\Phi} \in C_c^\infty(\Omega)^N$, $\Phi = (\Phi_1, \dots, \Phi_N)$,
   $\displaystyle\int_\Omega \operatorname{div} \overrightarrow{\Phi}\, dx = \int_{\partial\Omega'} \Phi \cdot \overrightarrow{n}\, dH_{N-1}(x) = 0$, thanks to Green's formula with support $\overrightarrow{\Phi} \subset \Omega' \subset\subset \Omega$, Ω' class C^1, $\overrightarrow{n}(x)$ the exterior normal at the point $x \in \partial\Omega'$.
2. If $\overrightarrow{\Phi} \in C_c^\infty(\Omega)^N$, then the extension by zero in $\mathbb{R}^N$ is an element of $C_c^\infty(\mathbb{R}^N)^N$ hence the inequality.
   If $E \subset \overline{\Omega_1} \subset \Omega$, consider $\psi \in C_c^\infty(\Omega)$ s.t. $\psi = 1$ over $\overline{\Omega_1}$.
   $\forall\, \overrightarrow{\Phi} \in C_c^\infty(\mathbb{R}^N)^N$, we have

$$\int_E \operatorname{div}(\psi \overrightarrow{\Phi}) dx = \int_E \operatorname{div}(\overrightarrow{\Phi}) dx, \quad \psi \overrightarrow{\Phi} \in C_c^\infty(\Omega)^N .$$

   Hence $P_{\mathbb{R}^N}(E) \leqslant P_\Omega(E) : P_\Omega(E) = P_{R^N}(E)$.
   Here is an example of a pair (E, Ω_1) satisfying $E \subset \overline{\Omega_1} \subset \Omega$. Let E : $\operatorname{dist}(E, \partial\Omega) > 0$ then we can take

$$\Omega_1 = \left\{x \in \Omega : \operatorname{dist}(x, \partial\Omega) > \frac{1}{2}\operatorname{dist}(E, \partial\Omega)\right\} .$$

3. Let $\overrightarrow{\Phi} \in C_c^\infty(\Omega)^N$ then by the dominated convergence theorem, we have:

$$\lim_n \int_{E_n} \operatorname{div}(\overrightarrow{\Phi}) dx = \int_E \operatorname{div}(\overrightarrow{\Phi}) dx .$$

   As $\displaystyle\int_{E_n} \operatorname{div}(\overrightarrow{\Phi}) dx \leqslant P_\Omega(E_n)$ we have $\displaystyle\int_E \operatorname{div}(\overrightarrow{\Phi}) dx \leqslant \liminf_n P_\Omega(E_n)$ or $P_\Omega(E) \leqslant \liminf_n P_\Omega(E_n)$. ◇

The perimeter thus defined is a good generalization because, we have the following proposition:

Proposition 3.3.2 *Let U be a bounded open subset of Ω such that $\partial U \cap \Omega$ is of class C^1. Then, for $N \geqslant 2$,*

$$P_\Omega(U) = \int_{\partial U \cap \Omega} dH_{N-1}(x) .$$

Proof Let $\Phi \in C_c^\infty(\Omega)^N$. Since $\Phi(x) = 0$ in a neighborhood of $\partial\Omega$, then by Green's formula,

$$\int_U \operatorname{div}(\Phi)dx = \int\limits_{\partial U \cap \Omega} \Phi \cdot \vec{n}\, dH_{N-1}(x) \leqslant |\Phi|_\infty \cdot \int\limits_{\partial U \cap \Omega} dH_{N-1}(x)\,.$$

From this inequality, we conclude that $P_\Omega(U) \leqslant \int\limits_{\partial U \cap \Omega} dH_{N-1}(x)$, because $\left|\vec{n}\right| = 1$.

Conversely, consider the vector field $\vec{n}(x), x \in \partial U$. The map $x \in \partial U \cap \Omega \to \vec{n}(x)$ ($\left|\vec{n}\right| = 1$, unitary exterior normal) is of class C^1. We can extend it to $\mathbb{R}^N$ in a field $\vec{V}(x)$ such that $\vec{V} \in C^1$ and $\left|\vec{V}(x)\right| \leqslant 1, \forall x \in \mathbb{R}^N$.

First note that since C_c^∞ is dense in C_c^1, we also have

$$\begin{aligned}
P_\Omega(U) &= \sup\left\{\int_U \operatorname{div}(\Phi)dx,\ \Phi \in C_c^1(\Omega)^N,\ |\Phi|_\infty \leqslant 1\right\} \\
&\geqslant \sup\left\{\int_U \operatorname{div}(\psi \vec{V})dx,\ \forall \psi \in C_c^1(\Omega),\ |\psi|_\infty \leqslant 1\right\} \\
&= \sup\left\{\int\limits_{\partial U \cap \Omega} \psi \vec{V} \cdot \vec{n}\, dH_{N-1}(x),\ \forall \psi \in C_c^1(\Omega),\ |\psi|_\infty \leqslant 1\right\}.
\end{aligned}$$

As $\vec{V} \cdot \vec{n} = 1$ on $\partial U \cap \Omega$,

$$P_\Omega(U) \geqslant \sup\left\{\int\limits_{\partial U \cap \Omega} \psi dH_{N-1}(x)dx,\ \psi \in C_c^1(\Omega),\ |\psi|_\infty \leqslant 1\right\}.$$

Hence $P_\Omega(U) \geqslant \int\limits_{\partial U \cap \Omega} dH_{N-1}(x)$. (Note that there exists a sequence $\psi_j \in C_c^\infty(\Omega)$ s.t. $\left|\psi_j(x)\right| \leqslant 1$ and $\psi_j(x) \xrightarrow[j \to +\infty]{} 1\ \forall x \in \Omega$.) ◇

Theorem 3.3.1 (Fleming–Rishel Formula) *Let $u \in W^{1,1}(\Omega)$. Then*

$$\int_\Omega |\nabla u|\, dx = \int_{-\infty}^{+\infty} P_\Omega\left(\{u > t\}\right) dt\ .$$

Proof Let $\overrightarrow{V}(x) = \begin{cases} \dfrac{\nabla u}{|\nabla u|}(x) & if\ \nabla u(x) \neq 0 \\ 0 & otherwise \end{cases}$. Then $|V(x)| \leqslant 1$. By regularization, there exists a sequence $\overrightarrow{\Phi}_j$ of $C_c^\infty(\Omega)^N$ such that $\left|\overrightarrow{\Phi}_j(x)\right| \leqslant 1$ for all $x \in \Omega$ and $\Phi_j(x) \xrightarrow[j\to+\infty]{} V(x)$ a.e. in Ω; we then have

$$\int_\Omega |\nabla u|\,(x)dx = \lim_{j\to+\infty} \int_\Omega \nabla u \cdot \Phi_j dx\ .$$

By Green's formula, and Fubini's theorem (see Lemma 1.2.3)

$$\int_\Omega \nabla u \cdot \Phi_j dx = -\int_\Omega u \operatorname{div}(\Phi_j) dx = -\int_{-\infty}^{+\infty} dt \int_{\{u>t\}} \operatorname{div}(\Phi_j) dx$$

(note that $\int_\Omega \operatorname{div}(\Phi_j)dx = 0$).

Since $-\int_{-\infty}^{+\infty} dt \int_{\{u>t\}} \operatorname{div}(\Phi_j)dx \leqslant \int_{-\infty}^{+\infty} P_\Omega(u > t)dt$, we deduce

$$\int_\Omega |\nabla u|\, dx = \lim_j \int_\Omega \nabla u \cdot \Phi_j(x)dx \leqslant \int_{-\infty}^{+\infty} P_\Omega(u > t)dt\ .$$

For the inverse inequality, start with the case where $u \in C^\infty(\Omega) \cap W^{1,1}(\Omega)$. Consider for $\varepsilon > 0$, $\psi_\varepsilon \in \mathcal{D}(\Omega)$ s.t. $0 \leqslant \psi_\varepsilon \leqslant \psi_\delta \leqslant 1$ if $\delta < \varepsilon$ and $\psi_\varepsilon(x) \xrightarrow[\varepsilon\to 0]{} 1$ for all $x \in \Omega$. Then by the dominated convergence theorem,

$$\lim_{\varepsilon\to 0} \int_\Omega \psi_\varepsilon \frac{|\nabla u|^2}{\sqrt{\varepsilon + |\nabla u|^2}} dx = \int_\Omega |\nabla u|\, dx\ .$$

By Green's formula, then Fubini's theorem, we have:

$$\int_\Omega \psi_\varepsilon \frac{|\nabla u|^2}{\sqrt{\varepsilon+|\nabla u|^2}}dx = -\int_\Omega u \operatorname{div}\left(\psi_\varepsilon \frac{\nabla u}{\sqrt{\varepsilon+|\nabla u|^2}}\right)dx$$

$$= -\int_{-\infty}^{+\infty} dt \int_{\{u>t\}} \operatorname{div}\left(\psi_\varepsilon \frac{\nabla u}{\sqrt{\varepsilon+|\nabla u|^2}}\right)dx$$

(note $\int_\Omega \operatorname{div}(\psi_\varepsilon U_\varepsilon)\,dx = 0$ where $U_\varepsilon = \dfrac{\nabla u}{\sqrt{\varepsilon+|\nabla u|^2}}$).

Since $u \in C^\infty(\Omega)$, by Sard's theorem, for almost any $t \in \mathbb{R}$, $u^{-1}(t)$ is a submanifold of dimension $N-1$ of class C^∞ on which $\nabla u \neq 0$. Therefore for almost all t, the sets $\{u > t\}$ are open with boundary C^∞ and the exterior normal at a point x on the boundary is $\vec{n}(x) = -\dfrac{\nabla u}{|\nabla u|}(x)$.

Using Green's formula, we deduce:

$$-\int_{-\infty}^{+\infty} dt \int_{u>t} \operatorname{div}(\psi_\varepsilon U_\varepsilon)dx = \int_{-\infty}^{+\infty} dt \int_{\partial\{u>t\}} \psi_\varepsilon\, |U_\varepsilon|\, dH_{N-1}(x)\,.$$

When $\varepsilon \to 0$, this last integral converges to:

$$\int_{-\infty}^{+\infty} dt \int_{u^{-1}(t)} dH_{N-1} = \int_{-\infty}^{+\infty} P_\Omega(u>t)dt\ .$$

Combining these last relations, we have:

$$\int_\Omega |\nabla u|\,dx = \int_{-\infty}^{+\infty} P_\Omega(u>t)dt\ .$$

If $u \in W^{1,1}(\Omega)$ then there exists a sequence $u_n \in C^\infty(\Omega) \cap W^{1,1}(\Omega)$ s.t. $u_n(x) \xrightarrow[n\to+\infty]{} u(x)$ a.e. and in $W^{1,1}(\Omega)$-strong. Then for almost all $t \in \mathbb{R}$, we have $\chi_{\{u_n>t\}}(x) \to \chi_{\{u>t\}}(x)$ a.e. in x. By the lower semicontinuity $\liminf_n P_\Omega(u_n > t) \geqslant P_\Omega(u>t)$ and by Fatou's lemma, we have:

$$\int_{-\infty}^{+\infty} P_\Omega(u>t)dt \leqslant \liminf_n \int_{-\infty}^{+\infty} P_\Omega(u_n>t)dt = \liminf_n \int_\Omega |\nabla u_n|\,dx$$

$$= \int_\Omega |\nabla u|\,dx.$$

◇

3.4 Absolute Local Continuity of Monotone Rearrangement

We will admit the following lemma, whose proof can be found in the books of Federer [71] and V. Maz'ja [112], or the article of L. Ljusternik [109]:

Lemma 3.4.1 (Isoperimetric and Relative Isoperimetric Inequality)

1. *For any measurable subset E of $\mathbb{R}^N$, of finite measure, we have: $P_{R^N}(E) \geqslant N\alpha_N^{\frac{1}{N}} |E|^{1-\frac{1}{N}}$ where α_N denotes the measure of the unit ball of $\mathbb{R}^N$. We have equality if and only if E is a ball of $\mathbb{R}^N$. This inequality is called the isoperimetric inequality in $\mathbb{R}^N$.*
2. *We assume that Ω satisfies the property of the inner cone (for example Ω is Lipschitzian) and that it is connected.*

   Then there exists a constant $Q = Q(\Omega, N) > 0$ such that for any measurable subset E of Ω, we have:

$$\min\left(|E|^{1-\frac{1}{N}}, |\Omega\backslash E|^{1-\frac{1}{N}}\right) \leqslant QP_\Omega(E)$$

(this is the so-called relative isoperimetric inequality).

Note

1. If Ω is a ball then $Q(\Omega, N) = \dfrac{1}{\alpha_{N-1}}\left(\dfrac{\alpha_N}{2}\right)^{1-\frac{1}{N}}$ (where α_m is the measure of the unit ball of $\mathbb{R}^m$).
2. If Ω is a rectangle with sides $a, b, a \geqslant b > 0$ then $Q(\Omega, 2) = a^{1/2}(2b)^{-1/2}$, (so $\Omega \subset \mathbb{R}^2$).
3. If Ω is a triangle of $\mathbb{R}^2$ whose smallest angle is ω then $Q(\Omega, 2) = (2\omega)^{-1/2}$.

These last two statements are due to Cianchi [43]. We will now apply these inequalities and the Fleming–Rishel lemma to the search for the regularity of the monotone rearrangement.

Theorem 3.4.1 (Fundamental (of Regularity)) *Let $u \in W^{1,1}(\Omega)$. We assume that there exists a continuous function*

$k : \Omega_* \to]0, +\infty[$ s.t. $P_\Omega(u > t) \geqslant k(|u > t|)$ for almost all $t \in]\text{ess inf}_\Omega u, \text{ess sup}_\Omega u[$. Then, $u_* \in W^{1,1}_{loc}(\Omega_*)$. Moreover, u satisfies the following pointwise inequality

$$-k(s)\frac{du_*}{ds}(s) \leqslant |\nabla u|_{*u}(s)\ a.e.$$

Proof Let $0 < \varepsilon < |\Omega|$ and $I(\varepsilon) =]\varepsilon, |\Omega| - \varepsilon[$. If $u \in W^{1,1}(\Omega)$ then for all $s \in \Omega_*$, $\big(u - u_*(s)\big)_+ \in W^{1,1}(\Omega)$ and according to the Fleming–Rishel formula, we deduce

$$\int\limits_{u>u_*(s)} |\nabla u|\,dx = \int_{\mathbb{R}} P_\Omega\Big(\big(u - u_*(s)\big)_+ > t\Big)dt$$

$$= \int_0^{+\infty} P_\Omega\Big(u > t + u_*(s)\Big)dt = \int\limits_{u_*(s)}^{+\infty} P_\Omega\big(u > \sigma\big)d\sigma\ . \tag{3.1}$$

Let $c = \dfrac{1}{\text{Min}\ \big\{k(\sigma),\ \sigma \in I(\varepsilon)\big\}} \in]0, +\infty[$. Let $\eta > 0$ then there exists $\delta > 0$ such that $c \displaystyle\int_0^\delta |\nabla u|_*(\sigma)d\sigma \leqslant \eta$. Let us then considerm pairwise disjoint intervals $]a_i, b_i[\subset I(\varepsilon)$, $i = 1, \ldots, m$, such that $\displaystyle\sum_{i=1}^m (b_i - a_i) \leqslant \delta$. According to the relationship (3.1), we deduce

$$\int\limits_{u_*(b_i)<u<u_*(a_i)} |\nabla u|\,dx \quad = \int\limits_{u_*(b_i)}^{u_*(a_i)} P_\Omega(u > t) \geqslant \int\limits_{u_*(b_i)}^{u_*(a_i)} k(|u > t|)dt\,. \tag{3.2}$$

But for almost all t s.t. $u_*(b_i) \leqslant t \leqslant u_*(a_i)$, we have

$$a_i \leqslant |u \geqslant u_*(a_i)| \leqslant |u \geqslant t| = |u > t| \leqslant |u > u_*(b_i)| \leqslant b_i\ .$$

Hence

$$k\big(|u > t|\big) \geqslant \text{Min}\ \Big\{k(\sigma),\ a_i \leqslant \sigma \leqslant b_i\Big\} \geqslant \text{Min}\ \Big\{k(\sigma) : \sigma \in I(\varepsilon)\Big\} \tag{3.3}$$

and by definition of relative rearrangement:

$$\int\limits_{u_*(b_i)<u<u_*(a_i)} |\nabla u|\,dx \quad = \int\limits_{a_i}^{b_i} |\nabla u|_{*u}(\sigma)d\sigma\,. \tag{3.4}$$

By combining the relations (3.2) with (3.4), we end up with $\forall\, i = 1, \dots, m$:

$$\int_{a_i}^{b_i} |\nabla u|_{*u}(\sigma)d\sigma \geqslant \text{Min}\left\{k(\sigma),\ \sigma \in I(\varepsilon)\right\} |u_*(a_i) - u_*(b_i)|\,. \tag{3.5}$$

Hence

$$c\int_{\bigcup_i]a_i,b_i[} |\nabla u|_{*u}(\sigma)d\sigma \geqslant \sum_{i=1}^{m}\Big[u_*(a_i) - u_*(b_i)\Big]. \tag{3.6}$$

But by Corollary 2.2.1, we know:

$$\int_{\bigcup_i]a_i,b_i[} |\nabla u|_{*u} \leqslant \int_0^{|\bigcup_i]a_i,b_i[|} \Big(|\nabla u|_{*u}\Big)_*(\sigma)d\sigma \leqslant \int_0^{\delta} |\nabla u|_*(\sigma)d\sigma, \tag{3.7}$$

hence

$$\sum_{i=1}^{m}\Big[u_*(a_i) - u_*(b_i)\Big] \leqslant c\int_0^{\delta} |\nabla u|_*(\sigma)d\sigma \leqslant \eta\,, \tag{3.8}$$

which proves that u_* is absolutely continuous, like $u_* \in L^1(\Omega)$:

$$u_* \in W^{1,1}_{loc}(\Omega_*).$$

To obtain the point inequality, we take (3.2) from (3.6) with
$a_i = s, b_i = s + h, h > 0,$

$$\frac{1}{h}\int_s^{s+h} |\nabla u|_{*u}(\sigma)d\sigma \geqslant \text{Min}\left\{k(\sigma),\ s \leqslant \sigma \leqslant s+h\right\} \cdot \frac{u_*(s) - u_*(s+h)}{h}. \tag{3.9}$$

Since $u_* \in W^{1,1}_{loc}(\Omega_*)$, we have $\lim\limits_{h\to 0}\dfrac{u_*(s) - u_*(s+h)}{h} = -\dfrac{du_*}{ds}(s)$ for almost all s and as $|\nabla u|_{*u} \in L^1(\Omega_*)$, it follows that

$$\lim_{h\to 0}\frac{1}{h}\int_s^{s+h} |\nabla u|_{*u}(\sigma)d\sigma = |\nabla u|_{*u}(s) \text{ for almost all } s\,.$$

From these facts, the relation (3.9) leads to:

$$-k(s)\cdot\frac{du_*}{ds}(s)\leqslant|\nabla u|_{*u}(s)\text{ a.e. in } s\,.$$

◇

Corollary 3.4.1 *Let $u\in W_0^{1,p}(\Omega),\ u\geqslant 0,\ 1\leqslant p\leqslant+\infty$. So*

$$u_*\in W^{1,p}(\varepsilon,|\Omega|)\ \forall\,\varepsilon\in]0,|\Omega|[.$$

Moreover one has $k(\sigma)=N\alpha_N^{\frac{1}{N}}\sigma^{1-\frac{1}{N}},\sigma\in\Omega_*$ and we have the following Polyà–Szegö inequalities

$$N\alpha_N^{\frac{1}{N}}\left|s^{1-\frac{1}{N}}\frac{du_*}{ds}\right|_{L^p(\Omega_*)}\leqslant\left||\nabla u|_{*u}\right|_{L^p(\Omega_*)}\leqslant|\nabla u|_{L^p(\Omega)}\,. \tag{3.10}$$

Proof If $u\in W_0^{1,p}(\Omega),\ u\geqslant 0$ then $\forall t>0\ \{u>t\}\cap\partial\Omega$ is of H_{N-1}-measure zero. Then for all $t>0,\quad P_\Omega(u>t)=P_{\mathbb{R}^N}(u>t)\geqslant N\alpha_N^{\frac{1}{N}}\,|u>t|^{1-\frac{1}{N}}$. We apply the fundamental Theorem 3.4.1 with $k(\sigma)=N\alpha_N^{\frac{1}{N}}\sigma^{1-\frac{1}{N}}$. We first deduce the inequalities (3.10).

Since $u_*\in W_{loc}^{1,1}(\Omega_*)$ and $\forall\,\varepsilon\in]0,|\Omega|[$:

$$\left|\frac{du_*}{ds}\right|_{L^p(\varepsilon,|\Omega|)}\leqslant c_\varepsilon\,|\nabla u|_{L^p(\Omega)}<+\infty\,,$$

from the fact that $u_*\in L^p(\Omega_*)$ we deduce that $u_*\in W^{1,p}(\varepsilon,|\Omega|)$. ◇

Corollary 3.4.2 *We assume that Ω is a connected open set satisfying the property of the interior cone (for example $C^{0,1}$). We let $k(s)=\dfrac{1}{Q}\min\left(s^{1-\frac{1}{N}},(|\Omega|-s)^{1-\frac{1}{N}}\right)$. So, if $1\leqslant p\leqslant+\infty$,*

$$\forall\,u\in W^{1,p}(\Omega),\quad u_*\in W_{loc}^{1,p}(\Omega_*)\,.$$

In addition, we have the Polyà–Szegö inequalities:

$$\left|k\cdot\frac{du_*}{ds}\right|_{L^p(\Omega_*)}\leqslant\left||\nabla u|_{*u}\right|_{L^p(\Omega_*)}\leqslant|\nabla u|_{L^p(\Omega)}\,. \tag{3.11}$$

Proof Let $u \in W^{1,p}(\Omega)$. Then according to the relative isoperimetric inequality: $P_\Omega(u > t) \geqslant k(|u > t|)\ \forall\ , t \in \mathbb{R}$. So

$$u_* \in W^{1,1}_{loc}(\Omega_*) \text{ and } -k(s)\frac{du_*}{ds} \leqslant |\nabla u|_{*u}(s) \text{ a.e.},$$

from where

$$\left|k(s)\frac{du_*}{ds}\right|_{L^p(\Omega_*)} \leqslant \left| |\nabla u|_{*u} \right|_{L^p(\Omega_*)} \leqslant |\nabla u|_{L^p(\Omega)}. \tag{3.12}$$

So, $\forall\, \varepsilon > 0$, $\left|\dfrac{du_*}{ds}\right|_{L^p(\varepsilon, |\Omega|-\varepsilon)} \leqslant c_\varepsilon |\nabla u|_{L^p(\Omega)} \Longrightarrow u_* \in W^{1,p}_{loc}(\Omega_*)$. ◇

Corollary 3.4.3 *Let $u \in W_0^{1,p}(\Omega)$, $1 \leqslant p \leqslant +\infty$. We define for $s \in \overline{\Omega}_*$*

$$k(s) = N\alpha_N^{\frac{1}{N}} \operatorname{Min}\left(s^{1-\frac{1}{N}}, (|\Omega| - s)^{1-\frac{1}{N}}\right).$$

Then $\forall t \neq 0$ $P_\Omega(u > t) \geqslant k(|u > t|)$, and we have the Polyà–Szegö inequalities:

$$\left|k \cdot \frac{du_*}{ds}\right|_{L^p(\Omega_*} \leqslant \left| |\nabla u|_{*u} \right|_{L^p(\Omega_*)} \leqslant |\nabla u|_{L^p(\Omega)}.$$

Proof Let $t \neq 0$. If $t > 0$ then $P_\Omega(u > t) = P_{\mathbb{R}^N}(u > t) \geqslant N\alpha_N^{\frac{1}{N}} |u > t|^{1-\frac{1}{N}}$ and if $t < 0$ then $P_\Omega(u > t) = P_\Omega(u \leqslant t) = P_{\mathbb{R}^N}(u \leqslant t) \geqslant N\alpha_N^{\frac{1}{N}} |u \leqslant t|^{1-\frac{1}{N}}$.

Hence,

$$P_\Omega(u > t) \geqslant N\alpha_N^{\frac{1}{N}} \min\left(|u > t|^{1-\frac{1}{N}},\ (|\Omega| - |u > t|)^{1-\frac{1}{N}}\right).$$

We apply the fundamental Theorem 3.4.1 and follow the same reasoning as for Corollaries 3.4.1 and 3.4.2. ◇

We can give a global regularity for u_* if u is sufficiently smooth.

Theorem 3.4.2 *Let Ω be a connected bounded open set satisfying the property of the inner cone (e.g. Lipschitzian). Let $Q = Q(\Omega, N)$ be the relative isoperimetric constant associated with Ω.*

Then, if $u \in W^{1,p}(\Omega)$, $p > N$, we have $u_* \in W^{1,q}(\Omega_*)$, $1 \leqslant q < q_0$ with $q_0 = \dfrac{1}{1 + \frac{1}{p} - \frac{1}{N}}$.

(continued)

Theorem 3.4.2 (continued)
In addition, we have: $\left|\frac{du_*}{ds}\right|_q \leqslant Qc(N,p)\,|\Omega|^{\beta(N,p)}\,\left||\nabla u|_{*u}\right|_p$, *with* $\beta(N,p) = \frac{1}{N} + \frac{1}{q} - \frac{1}{p} - 1,$

$$c(N,p) = 4\left(1 - \frac{pq}{N'(p-q)}\right)^{\nu} \quad \nu = \frac{1}{p} - \frac{1}{q}.$$

Proof It suffices to show that $u'_* \in L^q(\Omega_*)$ (because $u_* \in W^{1,1}_{loc}(\Omega_*)$). By writing:

$$\int_{\Omega_*} \left|\frac{du_*}{d\sigma}\right|^q d\sigma = \int_{\Omega_*} \overline{k}(\sigma)^{-q} \left|\overline{k}(\sigma)\frac{du_*}{d\sigma}\right|^q d\sigma$$

where $\overline{k}(\sigma) = \min\left(\sigma^{1-\frac{1}{N}}, (|\Omega| - \sigma)^{1-\frac{1}{N}}\right)$, we then apply Hölder's inequality to this last inequality to obtain:

$$\left[\int_{\Omega_*} \left|\frac{du_*}{d\sigma}\right|^q d\sigma\right]^{\frac{1}{q}} \leqslant \left[\int_{\Omega_*} \overline{k}(\sigma)^{-\frac{qp}{p-q}} d\sigma\right]^{\frac{(p-q)}{pq}} \left|\overline{k}\frac{du_*}{d\sigma}\right|_p$$

$$\leqslant Qc(N,p)\,|\Omega|^{\beta(N,p)}\,\left||\nabla u|_{*u}\right|_p .$$

◇

We have the same result if $u \in W^{1,p}_0(\Omega),\ p > N, and\ \Omega$ un open bounded.

In reality, the $W^{1,1}_{loc}(\Omega_*)$-regularity of u_* is true for any connected open set but in this case we no longer have an estimate for $\left|u'_*\right|_{L^1(\varepsilon,|\Omega|-\varepsilon)}$, for $\varepsilon > 0$. We have the following theorem

Theorem 3.4.3 *Let Ω be any open connected set. If $u \in W^{1,1}_{loc}(\Omega)$ then*

$$u_* \in W^{1,1}_{loc}(\Omega_*).$$

The proof of this theorem requires the following lemmas.

Lemma 3.4.2 *Let* $u : \Omega \to \mathbb{R}$ *be measurable,* $B \subset \Omega_*$ *s.t.* $|B| = 0$ *and* $\forall s \in B$, $\dfrac{du_*}{ds}(s) \in \mathbb{R}$. *Then,*

$$|u_*(B)| = 0\,.$$

Proof of Lemma 3.4.2 Let $k \in \mathbb{N}$ and let us define for $n \in \mathbb{N}^*$, $B_k = \Big\{s \in B : |u_*'(s)| \leqslant k\Big\}$ and $B_{n,k} = \Big\{s \in B_k : |u_*(s) - u_*(s')| \leqslant (k+1)\,|s - s'|\,, |s - s'| \leqslant \dfrac{1}{n}\Big\}$.

Then, $B_k = \bigcup\limits_{n \geqslant 1} B_{n,k}$ and $|B_k| = 0$. For $n \geqslant 1$ (fixed) and $\delta \in]0, \dfrac{1}{n}[$, there exists a family of pairwise disjoint intervals $(J_{n,p})_{p \geqslant 1}$ such that $B_{n,k} \subset \bigcup\limits_{p \geqslant 1} J_{n,p}$ and $\sum\limits_{p \geqslant 1} \text{meas}(J_{n,p}) \leqslant \delta$. Let $b_p = |u_*(B_{n,k} \cap J_{n,p})|$. Since $u_*(B_{n,k}) = \bigcup\limits_{p=1}^{+\infty} u_*(B_{n,k} \cap J_{n,p})$ we have $|u_*(B_{n,k})| \leqslant \sum\limits_{p=1}^{+\infty} b_p$ and $b_p \leqslant (k+1)\text{meas}(J_{n,p})$ (in fact if $J_{n,p} = [b, a]$ then we have $u_*\Big(B_{n,k} \cap J_{n,p}\Big) \subset [u_*(a), u_*(b)]$ and $b_p \leqslant u_*(b) - u_*(a) \leqslant (k+1)\text{meas}(J_{n,p})$).

We deduce $|u_*(B_{n,k})| \leqslant (k+1)\delta$: $|u_*(B_{n,k})| = 0$ $\forall n, k$ which implies $|u_*(B_k)| = 0 : |u_*(B)| = 0$. ◇

Lemma 3.4.3 *Let* $v : \Omega \to \mathbb{R}$ *s.t.* $v_* \in W^{1,1}_{loc}(\Omega_*)$. *So, the set*

$$I'(v) = \Big\{t \in I(v) = \Big]\text{ess inf}_{\,\Omega} v, \text{ess sup}_{\,\Omega} v\Big[,\ m_v'(t) = 0\Big\}$$

has zero measure. Here, $m_v(t) = |v > t|$.

Proof Consider the set

$$E_{nd} = \Big\{s \in \Omega_*,\ v_*'(s) \text{ does not exist or is not finite}\Big\}.$$

Then $|E_{nd}| = 0$ so $\gamma_j = \left|v_*\left(E_{nd} \cap \left[\dfrac{1}{j}, |\Omega| - \dfrac{1}{j}\right]\right)\right| = 0, \quad \forall j \geqslant j_\Omega.$

Hence $|v_*(E_{nd})| \leqslant \sum_{j \geqslant j_\Omega} \gamma_j = 0$. Since $v_*\big(m_v(t)\big) = t, \quad \forall t \in I(v)$, we have

$I'(v) \subset \Big\{t \in I(v), \quad m_v(t) \in E_{nd}\Big\} \subset v_*(E_{nd})$. Hence $\big|I'(v)\big| \leqslant |v_*(E_{nd})| = 0$. ◇

Lemma 3.4.4 *If $v : \Omega \to \mathbb{R}$ measurable s.t. the set*

$$I'(v) = \Big\{t \in I(v) =]\text{ess inf}_{\,\Omega} v, \text{ess sup}_{\,\Omega} v[,\ m_v'(t) = 0\Big\}$$

is of zero measure and $v_* \in C(\Omega_*)$, then, $v_* \in W^{1,1}_{loc}(\Omega_*)$.

Proof It suffices to show that for all $\varepsilon > 0$, if $E \subset [\varepsilon, |\Omega| - \varepsilon]$ with $|E| = 0$ then $|v_*(E)| = 0$.

Let us set

$$D_v = \Big\{t \in I(v), |v = t| > 0\Big\},$$

$$I''(v) = \Big\{t \in I(v), \text{ s.t. } m_v'(t) \text{ does not exist or is not finished}\Big\}.$$

As $v_*\big(m_v(t)\big) = t \ \forall t \in I(v)$, for all $t \in I_3(v) = I(v)\backslash\Big(D_v \cup I'(v) \cup I''(v)\Big)$ we can apply the chain rule for compound derivatives $v_*'\big(m_v(t)\big) = \dfrac{1}{m_v'(t)} \in \mathbb{R}$. As $\big|D_v \cup I'(v) \cup I''(v)\big| = 0$,

$$\text{for almost all } t \in I(v),\ \ v_*'\big(m(t)\big) \in \mathbb{R}.$$

Let $E_d = \Big\{s \in E : v_*'(s) \text{ is finite}\Big\}$, then

$$\Big\{t \in I_3(v) : \exists s \in E, t = v_*(s)\Big\} \subset \Big\{t \in I_3(v) : m_v(t) \in E_d\Big\} \subset v_*(E_d).$$

Hence $|v_*(E)| = \Big|\Big\{t \in I_3(v), \exists s \in E, t = v_*(s)\Big\}\Big| \leqslant |v_*(E_d)|$. By applying Lemma 3.4.2, we deduce $|v_*(E_d)| = 0$. ◇

Proof of Theorem 3.4.3. Let $\big(\Omega_j\big)_{j \geqslant 0}$ be a sequence of connected bounded open sets with Lipschitzian boundary s.t. $\overline{\Omega}_j \subset \Omega_{j+1} \subset \bigcup_{p \geqslant 0} \Omega_p = \Omega$. Let $u_j = u|_{\Omega_j}$ be the restriction to Ω_j of $u \in W^{1,1}_{loc}(\Omega)$.

Then $u_j \in W^{1,1}(\Omega_j)$ and $u_{j^*} \in W^{1,1}_{loc}(\Omega_{j*})$. Moreover, $u_{j^*}(0) \leqslant (u_{j+1})_*(0) \xrightarrow[j\to+\infty]{} u_*(0)$, $u_{j^*}(|\Omega_j|) \to u_*(|\Omega|)$.

Thus, if $u_*(|\Omega|) < a < b < u_*(0)$, then for j large,

$$|a < u < b| = |a < u_* < b| \geqslant |a < u_j < b| > 0$$

(because $u_{j^*}(|\Omega_j|) < a < b < u_{j^*}(0)$), which proves that necessarily $u_* \in C(\Omega_*)$. If $I_j(u) =]u_{j^*}(|\Omega_j|), u_{j^*}(0)[$ then for almost all $t \in I_j(u)$, $\forall h > 0$,

$$0 \leqslant \frac{m_j(t) - m_j(t+h)}{h} \leqslant \frac{m(t) - m(t+h)}{h} \tag{3.13}$$

where $m_j(t) = |u_j > t|$, $m(t) = |u > t|$.

Which implies $0 \leqslant -m'_j(t) \leqslant -m'(t)$ for almost all $t \in I_j(u)$.

Hence $$\{t \in I_j(u) : m'(t) = 0\} \subset \{t \in I_j(u) : m'_j(t) = 0\}.$$

As $u_{j*} \in W^{1,1}_{loc}(\Omega_{j*})$ so $\left|\{t \in I_j(u) : m'_j(t)=0\}\right| = 0$ according to Lemma 3.4.3. Hence

$$\left|\{t \in I(u) : m'(t) = 0\}\right| \leqslant \sum_j \left|\{t \in I_j(u) : m'_j(t)\}\right| = 0.$$

We now apply Lemma 3.4.4 to conclude that $u_* \in W^{1,1}_{loc}(\Omega_*)$. ◇

Note We can use a decomposition of Ω into cubes, but to show that u_* is continuous, we proceed differently (see Exercise 11.1.25, Chap. 10, or the articles cited).

3.5 Spherical Rearrangements and Classical Polyà–Szegö Inequalities

Definition 3.5.1 Let $u : \Omega \to \mathbb{R}$ be measurable and $\underset{\sim}{\Omega}$ the ball centered at the origin and of the same measure as Ω. We define $\underset{\sim}{u} : \underset{\sim}{\Omega} \to \mathbb{R}$ by $\underset{\sim}{u}(x) = u_*(\alpha_N |x|^N)$, where $|x|$ *isthe* Euclidean norm of $x \in \underset{\sim}{\Omega}$.

$\underset{\sim}{u}$ is called the <u>spherical rearrangement of u</u>.

Property 3.5.1 (Immediate)

1. $\underset{\sim}{u}$ is a radial function, which decreases along the radius, i.e. if $|x| = r_1 < r_2 = |y|$, (x, y) in $\underset{\sim}{\Omega}$ then $\underset{\sim}{u}(x) \geqslant \underset{\sim}{u}(y)$.
2. u and $\underset{\sim}{u}$ are equimeasurable.
3. $u \in L^p(\Omega) \to \underset{\sim}{u} \in L^p(\underset{\sim}{\Omega})$ is a contraction for $1 \leqslant p \leqslant +\infty$.

Proof 1. follows from the definition of $\underset{\sim}{u}$. As for 2., we have for all $t \in \mathbb{R}$, by change of variables,

$$\left|\underset{\sim}{u} > t\right| = \int_{\underset{\sim}{\Omega}} \chi_{]t,+\infty[}\big(\underset{\sim}{u}(x)\big)dx = \int_0^{\text{Radius of } \underset{\sim}{\Omega}} \chi_{]t,+\infty[}\big(u_*(\alpha_N r^N)\big) N\alpha_N r^{N-1} dr$$

$$= \int_{\Omega_*} \chi_{]t,+\infty[}\big(u_*(s)\big)ds = |u_* > t| = |u > t| \,.$$

Statement 3. follows from the contraction property of u_*. ◇

Let us now return to the regularity of $\underset{\sim}{u}$ and the Polyà–Szegö inequalities.

Proposition 3.5.1 (Classical Polyà–Szegö Inequalities) *Let* $u \in W_0^{1,p}(\Omega)$, $u \geqslant 0$, $1 \leqslant p \leqslant +\infty$. *Then*

$$\underset{\sim}{u} \in W_0^{1,p}(\underset{\sim}{\Omega}) \text{ and } \left|\nabla \underset{\sim}{u}\right|_{L^p(\underset{\sim}{\Omega})} \leqslant \big|\,|\nabla u|_{*u}\big|_{L^p(\Omega_*)} \leqslant |\nabla u|_{L^p(\Omega)} \,.$$

Proof Start with the case where $u \in C_c^\infty(\Omega)$, $u \geqslant 0$.

Then $\forall \varepsilon > 0$ (small), $\underset{\sim}{u} \in W^{1,\infty}\left(\underset{\sim}{\Omega} \backslash B(0,\varepsilon)\right)$ where $B(0,\varepsilon)$ denotes the ball centered at the origin of radius ε. Indeed, $u_* \in W^{1,\infty}(\alpha_N \varepsilon^N, |\Omega|)$. Thus, if $(x, y) \in \left(\underset{\sim}{\Omega} \backslash B(0,\varepsilon)\right)^2$, then

$$\left|\underset{\sim}{u}(x) - \underset{\sim}{u}(y)\right| \leqslant |u_*'|_{L^\infty(\alpha_N\varepsilon^N, |\Omega|)}\, \alpha_N \big|\, |x|^N - |y|^N \,\big| \leqslant c_N^\varepsilon \,|x - y| \,.$$

With equimeasurability, this inequality leads to the fact that $\underset{\sim}{u}$ is Lipschitzian on $\underset{\sim}{\Omega}\backslash B(0,\varepsilon)=\Omega_{\varepsilon}$.

Again, by changing variables, $1 \leqslant p < +\infty$

$$\int_{\Omega_\varepsilon} |\nabla \underset{\sim}{u}|^p dx = (N\alpha_N^{\frac{1}{N}})^p \int_{\alpha_N \varepsilon^N}^{|\Omega|} \left[s^{1-\frac{1}{N}} \left| \frac{du_*}{ds} \right| \right]^p dx \leqslant \left| \, |\nabla u|_{*u} \right|^p_{L^p(\Omega_*)} \leqslant |\nabla u|^p_{L^p(\Omega)} \,. \tag{3.14}$$

Hence when $\varepsilon \to 0$, $\forall\, p \in [1,+\infty[$

$$\left| \nabla \underset{\sim}{u} \right|_{L^p(\underset{\sim}{\Omega})} \leqslant \left| \, |\nabla u|_{*u} \right|_{L^p(\Omega_*)} \leqslant |\nabla u|_{L^p(\Omega)} \,. \tag{3.15}$$

In the relation (3.14) or (3.15), we conclude that $\underset{\sim}{u} \in W^{1,\infty}(\underset{\sim}{\Omega})$ when $p \to +\infty$. As $\underset{\sim}{u}(x) = \inf\limits_{\overline{\Omega}} u = 0$ for $x \in \partial\underset{\sim}{\Omega}$ we have $\underset{\sim}{u} \in W_0^{1,\infty}(\underset{\sim}{\Omega})$.

If $u \in W_0^{1,p}(\Omega)$, $u \geqslant 0$, then, there exists $u_n \in \mathcal{D}(\Omega)$, $u_n \geqslant 0$ s.t. $u_n \to u$ strongly in $W^{1,p}(\Omega)$.

Since $\left| \nabla \underset{\sim}{u}_n \right|_{L^p(\underset{\sim}{\Omega})} \leqslant |\nabla u_n|_{L^p}(\Omega) \leqslant c^{te}$ and $\left| \underset{\sim}{u}_n - \underset{\sim}{u} \right|_{L^p(\underset{\sim}{\Omega})} = |u_{n*} - u_*|_{L^p(\Omega_*)} \xrightarrow[n\to+\infty]{} 0$, we conclude, for $p > 1$, that $\underset{\sim}{u}_n$ converges to $\underset{\sim}{u}$ in $W^{1,p}(\underset{\sim}{\Omega})$-weak: $\underset{\sim}{u} \in W^{1,p}(\underset{\sim}{\Omega})$.

Hence $\left| \nabla \underset{\sim}{u} \right|_{L^p(\underset{\sim}{\Omega})} \leqslant \left| \, |\nabla u|_{*u} \right|_{L^p(\Omega_*)} \leqslant |\nabla u|_{L^p(\Omega)}$ (using a change of variables as in relation (3.14)).

If $p = 1$, let us verify that $(\nabla \underset{\sim}{u}_n)_{n \geqslant 0}$ satisfies the Dunford-Pettis conditions.

Since $\left| \nabla \underset{\sim}{u}_n \right|(x) = N\alpha_N\, |x|^{N-1} \left| u'_{n*}(\alpha_n\, |x|^N) \right|$, we have the desired result.

Let us introduce some general results.

Lemma 3.5.1 *Let* $u \in W_0^{1,1}(\Omega)$*,* $u \geqslant 0$*. Let:* $v(\sigma) = \left|\nabla \underset{\sim}{u}\right|_*(\sigma)$*,* $z(\sigma) = N\alpha_N^{\frac{1}{N}}\sigma^{1-\frac{1}{N}}\left|u'_*(\sigma)\right|$*, for* $\sigma \in \overline{\Omega}_*$*. Then* $v = z_*$*.*

Proof For $t \in \mathbb{R}$, using the expression of $\left|\nabla \underset{\sim}{u}\right|$, we have:

$$|v > t| = \int_{\Omega_*} \chi_{]t,+\infty[}\big(v(\sigma)\big)d\sigma = \int_{\underset{\sim}{\Omega}} \chi_{]t,+\infty[}\Big(z\big(\alpha_N |x|^N\big)\Big)dx$$

(this by equimeasurability). Indeed with a change of variables in this last integral, we conclude:

$$|v > t| = \int_{\Omega_*} \chi_{]t,+\infty[}\big(z(s)\big)ds = |z_* > t| : v = z_* \,.$$

Lemma 3.5.2 (See also Chap. 4) *Under the conditions of Lemma 3.5.1,* $\forall s \in \overline{\Omega}_*$*, we have the pointwise Polyà–Szegö inequalities:*

$$\int_0^s \left|\nabla \underset{\sim}{u}\right|_*(\sigma)d\sigma \leqslant \int_0^s \left(|\nabla u|_{*u}\right)_*(\sigma)d\sigma \leqslant \int_0^s |\nabla u|_*(\sigma)d\sigma \,.$$

Proof By Corollary 3.4.1 of the fundamental (regularity) Theorem 3.4.1, we have a.e. in σ, $z(\sigma) \leqslant |\nabla u|_{*u}(\sigma)$. Hence $v(\sigma) = z_*(\sigma) \leqslant \left(|\nabla u|_{*u}\right)_*(\sigma)$. Thus,

$$\int_0^s \left|\nabla \underset{\sim}{u}\right|_*(\sigma)d\sigma \leqslant \int_0^s \left(|\nabla u|_{*u}\right)_*(\sigma)d\sigma \leqslant \int_0^s |\nabla u|_*(\sigma)d\sigma.$$

(see the properties of the relative rearrangement of Chap. 2).

Corollary 3.5.1 *The sequence* $(\nabla \underset{\sim}{u}_n)_n$ *satisfies the Dunford–Pettis condition.*

In particular, $\nabla \underset{\sim}{u}_n$ *tends to* $\nabla \underset{\sim}{u}$ *in* $L^1(\underset{\sim}{\Omega})$*-weak and* $\underset{\sim}{u} \in W_0^{1,1}(\underset{\sim}{\Omega})$*.*

Proof Let $\eta > 0$. There exists an integer n_0 s.t.

$$\forall n \geqslant n_0,\ |\nabla(u_n - u)|_{L^1} \leqslant \frac{1}{2}\eta..$$

Let $\delta > 0$ s.t.

$$\int_0^{\delta} |\nabla u|_* (\sigma) d\sigma \leqslant \frac{1}{2}\eta, \quad \int_0^{\delta} \left|\nabla u_j\right|_* (\sigma) d\sigma \leqslant \frac{1}{2}\eta \qquad j = 0, \ldots, n_0 .$$

Then, by the Hardy–Littlewood theorem, and Lemma 3.5.2, $\forall\, E$ measurable in $\underset{\sim}{\Omega}$ satisfying $|E| \leqslant \delta$,

$$\int_E \left|\nabla \underset{\sim}{u_n}\right| dx \leqslant \int_0^{|E|} \left|\nabla \underset{\sim}{u_n}\right|_* d\sigma \leqslant \int_0^{|E|} |\nabla u_n|_* \, d\sigma .$$

If $n \leqslant n_0$ then $\displaystyle\int_E \left|\nabla \underset{\sim}{u_n}\right| dx \leqslant \eta$.

If $n \geqslant n_0$ then $\displaystyle\int_0^{|E|} \left|\nabla \underset{\sim}{u_n}\right|_* d\sigma \leqslant |\nabla(u_n - u)|_{L^1} + \int_0^{|E|} |\nabla u|_* \, d\sigma \leqslant \eta$ (by the contraction property).

By combining these relations, we have: $\displaystyle\int_E \left|\nabla \underset{\sim}{u_n}\right| dx \leqslant \eta$. We conclude as the case $p > 1$.

Finally, let us conclude this section with the following general theorem:

Theorem 3.5.1 (Extension of Polyà–Szegö Inequalities) *Let ρ be a Fatou norm invariant by rearrangement and non-trivial on $L^0(\Omega_*)$. Then, for all $u \in W_0^{1,1}(\Omega)$, $u \geqslant 0$ we have the inequality:*

$$\rho\left(\left|\nabla \underset{\sim}{u}\right|_*\right) \leqslant \rho\left(|\nabla u|_{*u}\right) \leqslant \rho\left(|\nabla u|_*\right) .$$

Proof This is a consequence of Lemma 3.5.2 and the Lorentz–Luxemburg theorem 2.2.1. ◇

Remark If for all $u \in W^{1,1}(\Omega)$ we have a $\left|k \cdot \frac{du_*}{ds}\right|_{L^1(\Omega_*)} \leqslant |\nabla u|_{L^1(\Omega)}$ with $k(0) = k(|\Omega|) = 0$, $k \in C^1_{loc}(\Omega_*)$, then for any measurable set E in Ω,

$$P_\Omega(E) \geqslant k(|E|) .$$

Indeed, if $E \subset \Omega$ measurable then, there exists a sequence $u_m \in C^\infty(\Omega) \cap W^{1,1}(\Omega)$ s.t.:

(i) $\lim_{m\to+\infty} \int_\Omega |\nabla u_m|\, dx = P_\Omega(E)$,
(ii) $\lim_{m\to+\infty} |u_m - \chi_E|_{L^1} = 0$.

Such a result is proven in the book by V. Maz'ja [112].

Let $\varphi \in \mathcal{D}(\Omega_*)$ s.t. $|\varphi|_\infty \leqslant 1$ and $\varphi(|E|) = 1$. Assuming that $0 < |E| < |\Omega|$, we then have $\int_{\Omega_*} (k\varphi)' u_{m*} d\sigma \leqslant |\varphi|_\infty\, |\nabla u_m|_{L^1}$ but $u_{m*} \to \chi_{[0,|E|]}$. Hence

$$k(|E|) = \lim_m \int_{\Omega_*} (k\varphi)' u_{m*} \leqslant \lim_m |\nabla u_m|_{L^1} = P_\Omega(E).$$

If $|E| = 0$ or $|E| = |\Omega|$ then $k(|E|) = P_\Omega(E) = 0$.

3.6 Pointwise Polyà–Szegö Inequalities for the α-Rearrangement

We can define rearrangements similar to the spherical case, called α-rearrangements. We will need some preliminaries.

Definition 3.6.1 (And Notations) Let Σ_α be an open cone of $\mathbb{R}^N$, $N \geqslant 2$, with vertex the origin and solid angle $\alpha \in]0, \alpha_{N-1}[$, where α_{N-1} is the measure of the unit ball of $\mathbb{R}^{N-1}$ (i.e. also the $(N-1)$ Hausdorff measure of the unit sphere S^{N-1}). Let A_α be a subset of S^{N-1} of measure α i.e. $H_{N-1}(A_\alpha) = \alpha$, then we have precisely,

$$\Sigma_\alpha = \left\{\lambda x,\ x \in A_\alpha,\ \lambda \in]0, +\infty[\right\} .$$

($A_\alpha = \Sigma_\alpha \cap S^{N-1}$). We assume $\partial\Sigma_\alpha$ is Lipschitzian.

We denote by $\Sigma(\alpha, R)$ the open sector of $\mathbb{R}^N$ of solid angle α and radius R, that is to say, $\Sigma(\alpha, R) = \Sigma_\alpha \cap B(0, R)$ where $B(0, R)$ is the ball of radius R centered at the origin.

We will denote by σ_N the measure of the unit sector $\Sigma(\alpha, 1)$ i.e.

$$\text{meas}\{\Sigma(\alpha, 1)\} = \sigma_N.$$

Example for $N = 2$, $\sigma_2 = \dfrac{\alpha}{2}$.

In the plane $\mathbb{R}^2$, we consider the polar coordinates. For $\alpha \in]0, \pi[$, we have:

$$\Sigma_\alpha = \left\{(\rho, \theta) \in \mathbb{R}^2 : 0 < \theta < \alpha,\ \rho > 0\right\}$$

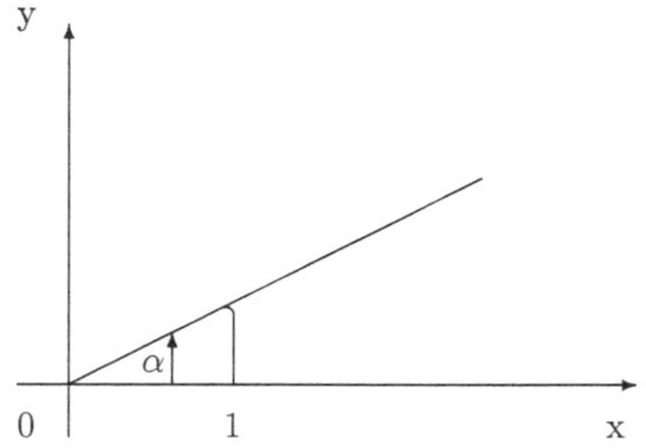

We have the following theorem.

Theorem 3.6.1 (Isoperimetric Inequality for a Cone) *If Σ_α is a convex cone of $\mathbb{R}^N$ then $P_{\Sigma_\alpha}(E) \geqslant N\sigma_N^{\frac{1}{N}} |E|^{1-\frac{1}{N}}$, for any measurable set $E \subset \Sigma_\alpha$ of finite measure. If $\partial\Sigma_\alpha \backslash \{0\}$ is regular then we have equality if and only if E is a convex sector $\Sigma(\alpha, R)$ homothetic to Σ_α.*

This theorem will be admitted. For more details one can consult the article by Lions et al. [107].

For an open bounded connected Lipschitzian Ω, we consider $\Gamma_0 \subset \partial\Omega$, $H_{N-1}(\Gamma_0) > 0$ and $\Gamma_1 = \partial\Omega\backslash\Gamma_0$. We define the following relative isoperimetric constant:

$$Q(\Gamma_1, \Omega) = \sup_E \frac{|E|^{1-\frac{1}{N}}}{P_\Omega(E)}$$

where the supremum is taken over all measurable sets E of Ω such that $\partial E \cap \Gamma_0$ does not contain any subset of positive H_{N-1} measure. We admit the following propositions:

Proposition 3.6.1 (Example of Explicit Calculation) *Let* $\alpha \in]0, \alpha_{N-1}[$ *and* $\Sigma(\alpha, R)$ *be the convex sector defined previously. We define* $\widetilde{\Gamma}_0 = \Big\{x \in \partial\Sigma(\alpha, R),\ |x| = R\Big\}$ *and* $\widetilde{\Gamma}_1 = \partial\Sigma(\alpha, R)\backslash\widetilde{\Gamma}_0$. *Then*

$$Q\Big(\widetilde{\Gamma}_1, \Sigma(\alpha, R)\Big) = \frac{1}{N\sigma_N^{\frac{1}{N}}}\ with\ \sigma_N = H_N\Big(\Sigma(\alpha, 1)\Big).$$

Proposition 3.6.2 (Finiteness $Q(\Gamma_1, \Omega)$) *We have* $Q(\Gamma_1, \Omega) < +\infty$. *Moreover, there exists a unit sector* $\Sigma(\alpha, 1)$ *such that* $Q(\Gamma_1, \Omega) = \dfrac{1}{N\sigma_N^{\frac{1}{N}}}$.

We will define the α-rearrangement:

Definition 3.6.2 (α-rearrangement) Let Ω be a connected bounded Lipschitzian open set, with isoperimetric constant $\dfrac{1}{N\sigma_N^{\frac{1}{N}}}$. We call the α-rearrangement of a measurable function $u : \Omega \to \mathbb{R}$, the function $C_\alpha u : \sum(\alpha, R) \to \mathbb{R}$, with $H_N\big(\sum(\alpha, R)\big) = \sigma_N R^N = H_N(\Omega)$ defined by

$$C_\alpha u(x) = u_*\left(\sigma_N |x|^N\right).$$

The function $C_\alpha u$ has the same properties as $\underset{\sim}{u}$. The proofs are identical to those of $\underset{\sim}{u}$.

Property 3.6.1

1. $|u|_{L^q(\Omega)} = |C_\alpha u|_{L^q(\sum(\alpha,R))} \qquad 1 \leqslant q \leqslant +\infty.$
2. If $u \in W^{1,p}_{\Gamma_0}(\Omega) = \Big\{v \in W^{1,p}(\Omega) : v = 0\ sur\ \Gamma_0\Big\}$, $1 \leqslant p \leqslant +\infty$, then for $u \geqslant 0$

(continued)

Property 3.6.1 (continued)

$$-u'_*(s) \leqslant \frac{s^{\frac{1}{N}-1}}{N\sigma_N^{\frac{1}{N}}} |\nabla u|_{*u}(s)$$

and

$$|\nabla C_\alpha(u)|_{L^p(\sum(\alpha,R))} \leqslant |\nabla u|_{L^p(\Omega)} .$$

We have the same as the spherical rearrangement of the pointwise Polyà–Szegö type inequalities.

Theorem 3.6.2 *Let $u \in W^{1,1}_{\Gamma_0}(\Omega)$, $u \geqslant 0$. Then $\forall\, s \in \overline{\Omega}_*$, we have*

$$\int_0^s |\nabla C_\alpha(u)|_*(\sigma)d\sigma \leqslant \int_0^s (|\nabla u|_{*u})_*(\sigma)d\sigma \leqslant \int_0^s |\nabla u|_*(\sigma)d\sigma .$$

Lemma 3.6.1 *Let $u \in W^{1,1}_{\Gamma_0}(\Omega)$, $u \geqslant 0$.*
Let $w(s) = -N\sigma_N^{\frac{1}{N}} s^{1-\frac{1}{N}} u'_*(s)$ for $s \in \Omega_*$. Then, $\forall\, \sigma \in \Omega_*$

$$|\nabla C_\alpha(u)|_*(\sigma) = w_*(\sigma) \leqslant (|\nabla u|_{*u})_*(\sigma).$$

Proof of the Lemma Since $C_\alpha u(x) = u_*(\sigma_N |x|^N)$, we have a.e. in x

$$|\nabla C_\alpha(u)|(x) = -N\sigma_N |x|^{N-1} u'_*(\sigma_N |x|^N) = w(\sigma_N |x|^N) .$$

For all $t \in \mathbb{R}$, we deduce after changing variables

$$\text{measure}\Big\{x : |\nabla C_\alpha(u)|(x) > t\Big\} = \int_{\sum(\alpha,R)} \chi_{]t,+\infty[}(w(\sigma_N |x|^N)dx$$

$$= \int_{\Omega_*} \chi_{]t,+\infty[}\big(w_*(s)\big)ds : |\nabla C_\alpha(u)|_* = w_* .$$

By the pointwise inequalities (see Property 3.6.1), we have

$+\, w(s) \leqslant |\nabla u|_{*u}(s)$, a.e. in $s \in \Omega_*$ from which $w_* \leqslant \Big(|\nabla u|_{*u}\Big)_*$. ◇

Proof of Theorem 3.6.2 From the previous lemma, we deduce by combining with the properties of the relative rearrangement:

$$\int_0^s |\nabla C_\alpha(u)|_*(\sigma)d\sigma \leqslant \int_0^s (|\nabla u|_{*u})_*(\sigma)d\sigma \leqslant \int_0^s |\nabla u|_*(\sigma)d\sigma\ .$$

◇

Pre-bibliographical Notes

The proofs given in Sects. 3.1 and 3.2 are taken from the thesis of B. Simon [156]. Fundamental Theorem 3.4.1 is a combination of the results of Rakotoson–Temam [144, 146] and Rakotoson [130]. Corollaries 3.4.1 and 3.4.2 are improved variants of the Polyà–Szegö inequalities. All the rest of this section is due to the author. The theorems of Sects. 3.3 and 3.4 are improved versions of existing results, see the book by J. Mossino [115].

α-rearrangements were introduced by Lions et al. [107], the pointwise inequalities relating the α-rearrangement and the relative rearrangement are due to the author.

Chapter 4
Pointwise Inequalities and Sobolev Inclusions

We saw in the previous chapter that if Ω is a connected bounded open set $u \in W^{1,1}_{loc}(\Omega)$, then its decreasing rearrangement u_* is in $W^{1,1}_{loc}(\Omega_*)$. It turns out that when we know a little about the regularity of the boundary of Ω then we can relate the derivative u'_* of u_* with the relative rearrangement of $|\nabla u|$ with respect to u when $u \in W^{1,1}(\Omega)$, which leads us to an inequality of the form

$$-\frac{du_*}{ds}(s) \leqslant K(s;\Omega)\cdot|\nabla u|_{*u}(s) \quad a.e\ in\ \Omega_*. \tag{4.1}$$

The function $K(\cdot;\Omega)$ depends essentially on the domain Ω. In the articles [145], [146], we used the domain class $\mathcal{J}_\alpha$, $\alpha \geqslant \dfrac{1}{N'} = \dfrac{N-1}{N}$ introduced by V. Maz'ja [112]. When $\Omega \in \mathcal{J}_\alpha$, there exists a constant $Q_\alpha(\Omega) > 0$ s.t for all measurable $E \subset \Omega$

$$P_\Omega(E) \geqslant Q_\alpha(\Omega)^{-1}\min\Big(|E|^\alpha;\ \big(|\Omega| - |E|^\alpha\big)\Big). \tag{4.2}$$

Considering Theorem 3.4.1 for this domain class, we have

$$-\frac{du_*}{ds} \leqslant Q_\alpha(\Omega)\max\Big(s^\alpha;\ (|\Omega|-s)^\alpha\Big)\cdot|\nabla u|_{*u}(s). \tag{4.3}$$

The inequalities of type (4.1) (therefore (4.3)) lead to a simple method to obtain Sobolev-type inequalities for spaces invariant under rearrangement, such as $L^p(\Omega)$.

Better still, the results obtained fill some gaps in classical inclusions. Indeed, classical results on Sobolev inclusions tell us that $W_0^{1,N}(\Omega)$ is contained in $L^q(\Omega)$ for all finite q, and that if $p > N$ then $W_0^{1,p}(\Omega)$ is contained (also continuously) in the Hölderian spaces $C^{0,1-\frac{N}{p}}(\overline{\Omega})$. We note that there are "missing" intermediate

J. M. Rakotoson, *Relative Rearrangement*, Lecture Notes in Mathematics 2376,
https://doi.org/10.1007/978-3-032-02228-8_4

spaces $X(\Omega)$ satisfying:

$W_0^{1,p}(\Omega) \underset{\neq}{\hookrightarrow} X(\Omega) \underset{\neq}{\hookrightarrow} W_0^{1,N}(\Omega)$ and $X(\Omega) \subset C(\overline{\Omega})$. E.M. Stein showed that the Lorentz–Sobolev space

$$W^1L^{N,1}(\Omega) = \left\{ f \in L^1(\Omega) : |\nabla f| \in L^{N,1}(\Omega) \right\}$$

is contained in the set of continuous functions $C(\Omega)$. To do this, using an integral representation of f, he shows that for any cube $Q(h)$ of side $|h|$, for any x and $x+h$ in $Q(h)$, we have: $|f(x+h) - f(x)| \leqslant c|\nabla f \chi_{Q(h)}|_{L^{N,1}}$.

The method that we propose in this chapter will lead us to re-prove this result but at the same time, to specify the constant, we will prove in particular that $W^1L^{N,1}(\Omega) \subset C(\Omega)$ and $\forall\, u \in W^1L^{N,1}(\Omega)$, $\forall\, x \in \Omega$ s.t. $B(x,r) \subset \Omega$, we have:

$$\underset{B(x,r)}{\operatorname{osc}}\ u \leqslant \frac{\alpha_N^{1-\frac{1}{N}}}{\alpha_{N-1}} \int_0^{\alpha_N r^N} t^{\frac{1}{N}} |\nabla u|_* \frac{dt}{t}\ . \tag{4.4}$$

This inequality follows from a more general inequality on $W^{1,1}(\Omega)$ stating that if $\overline{u}$ is the restriction of u to $B(x,r)$ then

$$\underset{B(x,r)}{\operatorname{osc}}\ u \leqslant \frac{\alpha_N^{1-\frac{1}{N}}}{\alpha_{N-1}} \int_0^{\alpha_N r^N} s^{\frac{1}{N}-1} \Big(|\nabla \overline{u}|_{*\overline{u}} \Big)_* (s) ds\ . \tag{4.5}$$

When the open Ω is Lipschitzian, we will have;

$$W^1L^{N,1}(\Omega) \hookrightarrow C(\overline{\Omega}). \tag{4.6}$$

We define $W^1L(\Omega,\rho) = \left\{ v \in L^1(\Omega),\ \rho(|\nabla v|) < +\infty \right\}$ when $L(\Omega,\rho)$ is a Banach function space.

We will show that the method could be applied to different inclusions concerning Sobolev spaces $W^1L(\Omega,\rho)$ associated with Banach function spaces $L(\Omega,\rho)$ which is invariant under rearrangement. Among other things, if we consider the spaces $bmo_r(\Omega)$ (bounded mean oscillation) equipped with the norm

$$|v|_{bmo} = \sup_{B\subset\Omega} \left[|B|^{-1} \int_B |v(x) - v_B| dx \right] + |v|_1$$

where $v_B = |B|^{-1} \displaystyle\int_B v(y)dy$, B being any ball in Ω, then

$$W^1L^{N,\infty}(\Omega) \hookrightarrow bmo_r(\Omega). \tag{4.7}$$

Furthermore,

$$W^1L^N(\Omega) \hookrightarrow vmo_r(\Omega) \tag{4.8}$$

($vmo_r(\Omega)$ is a subspace of bmo_r, vmo is short for vanishing mean oscillation).

If we consider the set $W^1L^{N)}(\Omega)$, $L^{N)}(\Omega)$ being the large Lebesgue space of exponent N, we will have

$$W^1L^{N)}(\Omega) \hookrightarrow L_{exp}(\Omega).$$

The space $L_{exp}(\Omega)$ will be defined later (see Theorem 4.4.1).

Note that we have the following inclusions:

$$L^{(N}(\Omega) \subset L^{N,1}(\Omega) \subset L^N(\Omega) \subset L^{N,\infty}(\Omega) \subset L^{N)}(\Omega),$$

$$C(\overline{\Omega}) \subset vmo_r(\Omega) \subset bmo_r(\Omega) \subset L_{exp}(\Omega).$$

These spaces are called "limit" spaces.

The previous injections are continuous and the inclusion constants are different. For example, for

$$W^1L^{(N}(\Omega) = \left\{u \in L^1(\Omega) : |\nabla u|_{(N} < +\infty\right\}, \quad N \geqslant 2$$

we have

$$\underset{B(x,r)}{\operatorname{osc}} u \leqslant \frac{\alpha_N^{1-\frac{1}{N}}}{\alpha_{N-1}} \left(\frac{|\Omega| N}{N-1}\right)^{1-\frac{1}{N}} \left|\nabla u \cdot \chi_{B(x,r)}\right|_{(N}.$$

We will find other classical Poincaré–Sobolev inequalities but with the additional estimate of the inclusion constant. In particular

$$W^1L^p(\Omega) = W^{1,p}(\Omega) \hookrightarrow C^{0,\alpha}(\overline{\Omega}), \quad \alpha = 1 - \frac{p}{N},$$

$$W^1L^{p,q}(\Omega) \subset L^{p^*,q}(\Omega) \text{ for } 1 \leqslant p < N, \quad 1 \leqslant q \leqslant +\infty, \quad p^* = \frac{Np}{N-p}.$$

All estimates will first be obtained using the relative rearrangement of the gradient $|\nabla u|_{*u}$.

4.1 Pointwise Inequalities for Relative Rearrangement

Definition 4.1.1 (PSR Property) We will say that a subset V of $W^{1,1}(\Omega)$ satisfies the Poincaré–Sobolev inequalities for relative rearrangement (denoted PSR) if

(i) $u_* \in W^{1,1}_{loc}(\Omega_*)$, $\forall u \in V$
(ii) There exists a measurable function $K(\cdot, \Omega, V) : \Omega_* \to [0, +\infty[$ s.t.:

$$-u'_*(s) \leqslant K(s, \Omega, V) \cdot |\nabla u|_{*u}(s), \text{ a.e. all } s \text{ and } \forall u \in V.$$

Before giving general properties of such sets V, we first give some examples V (consequence of Theorem 3.4.1).

Theorem 4.1.1 (Existence of a Set V Satisfying the Property PSR)

1. *The set $V = W_0^{1,1}(\Omega) \cap L^0_+(\Omega)$ satisfies the PSR property. In addition, we can choose $K(s, \Omega, V) = \dfrac{s^{\frac{1}{N}-1}}{N\alpha_N^{\frac{1}{N}}}$.*
2. *The set $V = W_0^{1,1}(\Omega)$ satisfies the PSR property with*

$$K(s, \Omega, V) = \frac{\max\left(s^{\frac{1}{N}-1}, (|\Omega| - s)^{\frac{1}{N}-1}\right)}{N\alpha_N^{\frac{1}{N}}}.$$

3. *If Ω is a connected open set with a Lipschitzian boundary then $V = W^{1,1}(\Omega)$ satisfies the PSR property. Moreover, there exists a constant $Q(\Omega, N) = Q > 0$ (which is a relative isoperimetric constant) such that*

$$K(s, \Omega, V) = Q \max\left(s^{\frac{1}{N}-1}, (|\Omega| - s)^{\frac{1}{N}-1}\right).$$

4. *If Ω is a connected Lipschitzian open set then $V = W^{1,1}_{\Gamma_0}(\Omega) \cap L^0_+(\Omega)$ satisfies the PSR property with $K(s, \Omega, V) = \dfrac{s^{\frac{1}{N}-1}}{N\sigma_N^{\frac{1}{N}}}$ with $\sigma_N = \left|\sum(\alpha, 1)\right|$.*

Proof This is a consequence of the Fundamental Theorem 3.4.1 and its Corollaries 3.4.1, 3.4.2, 3.4.3, Property 3.6.1.

Corollary 4.1.1 *Let $\Omega = B(R)$ be a ball of radius $R > 0$. Then $V = W^{1,1}(B(R))$ satisfies the PSR property and*

$$K(s, B(R), V) = \frac{1}{\alpha_{N-1}}\left(\frac{\alpha_N}{2}\right)^{1-\frac{1}{N}} \text{Max}\left(s^{\frac{1}{N}-1}, \left(\alpha_N R^N - s\right)^{\frac{1}{N}-1}\right)$$

where α_m denotes the measure of the unit ball of $\mathbb{R}^m$.

Proof We know that in the case of a ball the best relative isoperimetric constant is $Q = \dfrac{1}{\alpha_{N-1}}\left(\dfrac{\alpha_N}{2}\right)^{1-\frac{1}{N}}$ (see [112]), hence the result. ◇

Here are some general theorems for sets V satisfying the PSR property.

Proposition 4.1.1 *Let $V \subset W^{1,1}(\Omega)$ satisfy the PSR property.*
For all $u \in V \cap W^{1,q}(\Omega)$, $1 \leqslant q < +\infty$, we also have:

$$-u_*'(s) \leqslant K(s, \Omega, V) \cdot \left[\left(|\nabla u|^q\right)_{*u}\right]^{\frac{1}{q}}(s) \quad a.e \; in \; s\,.$$

Proof For $h > 0$, we have $|u_*(s+h) < u < u_*(s)| \leqslant h$, and by Hölder's inequality

$$\int_s^{s+h} |\nabla u|_{*u}\, d\sigma = \int_{u_*(s+h)<u<u_*(s)} |\nabla u|\, dx \quad \leqslant h^{1-\frac{1}{q}} \left(\int_s^{s+h} \left(|\nabla u|^q\right)_{*u}(\sigma) d\sigma \right)^{\frac{1}{q}}.$$

Hence, for almost all $s \in \Omega_*$ we have:

$$|\nabla u|_{*u}(s) = \lim_{h \to 0} \frac{1}{h} \int_s^{s+h} |\nabla u|_{*u}(\sigma) d\sigma \leqslant \left[\left(|\nabla u|^q\right)_{*u}(s)\right]^{\frac{1}{q}}. \quad ◇$$

Proposition 4.1.2 *Let V satisfy the PSR property. Let $K(s) = K(s, \Omega, V)$ for $0 < \sigma < t$, $\operatorname{ess}\sup_{\sigma<s<t} [sK(s)] = g(t, \sigma)$. Then*

$$u_*(\sigma) - u_*(t) \leqslant g(t, \sigma) \left(\frac{t}{\sigma}\right) |\nabla u|_{**}(t)$$

where $|\nabla u|_{**}(t) = \frac{1}{t} \int_0^t |\nabla u|_*(\tau) d\tau.$

Proof Indeed, since $u_* \in W^{1,1}_{loc}(\Omega_*)$, we have

$$u_*(\sigma) - u_*(t) = -\int_\sigma^t u'_*(s) ds \leqslant \int_\sigma^t K(s, \Omega, V) |\nabla u|_{*u}(s) ds$$

$$\leqslant g(t, \sigma) \int_\sigma^t \frac{1}{s} |\nabla u|_{*u}(s) ds$$

$$\leqslant g(t, \sigma) \left(\frac{t}{\sigma}\right) \frac{1}{t} \int_0^t |\nabla u|_{*u}(s) ds \leqslant g(t, \sigma) \left(\frac{t}{\sigma}\right) |\nabla u|_{**}(t) .$$

◇

It can happen that instead of using u_* we need u_{**}. In this case we can replace u'_* by u'_{**}. We then have:

Theorem 4.1.2 *Let V be a subset of $W^{1,1}(\Omega)$ satisfying the PSR property. So, $\forall u \in V$*

$$-u'_{**}(s) \leqslant \widetilde{K}(s) \left[|\nabla u|_{*u}\right]_{**}(s) \ a.e.\ s \in \Omega_*.$$

In particular, we have:

$$-u'_{**}(s) \leqslant \widetilde{K}(s) |\nabla u|_{**}(s) \ a.e.\ s \text{ where } \widetilde{K}(s) = \frac{1}{s} \operatorname{ess}\sup_{0\leqslant t\leqslant s} [tK(t)] .$$

Proof By integration by parts, for $s \in \Omega_*$, $u \in V$, we have

$$\frac{1}{s} \int_0^s [u_*(t) - u_*(s)] dt = \frac{1}{s} \int_0^s t \left|u'_*(t)\right| dt. \tag{4.9}$$

By the PSR and Hardy–Littlewood property, we deduce

$$u_{**}(s)-u_*(s) \leqslant \frac{1}{s}\int_0^s tK(t)\,|\nabla u|_{*u}(t)dt \leqslant \widetilde{K}(s)\int_0^s \Big(|\nabla u|_{*u}\Big)_*(t)dt.$$

As $-\dfrac{d}{ds}u_{**}(s) = \dfrac{1}{s}\left[u_{**}(s) - u_*(s)\right]$, we deduce

$$-u'_{**}(s) \leqslant \widetilde{K}(s)\Big(|\nabla u|_{*u}\Big)_{**}(s) \leqslant \widetilde{K}(s)\,|\nabla u|_{**}(s).$$

◇

For example if $V = W^{1,1}_{0+}(\Omega)$ then $\widetilde{K}(s) = K(s) = \dfrac{s^{\frac{1}{N}-1}}{N\alpha_N^{\frac{1}{N}}}$.

4.2 General Type Inclusions: Applications to Lorentz Spaces

To study Poincaré-Sobolev type inclusions associated to a norm ρ let us recall:

Proposition 4.2.1 *If $L(\Omega, \rho)$ is a Banach function space invariant under rearrangement then the associate space $L(\Omega, \rho')$, where ρ' is the associate norm of ρ, is also a Banach function space invariant under rearrangement and*

$$\int_{\Omega_*} |f|_*|g|_*d\sigma \leqslant \rho(f)\rho'(g) \quad \forall f \in L(\Omega, \rho), \quad \forall g \in L(\Omega, \rho').$$

From now on, we will consider a norm ρ defined on both $L^0(\Omega)$ and $L^0(\Omega_*)$. If ρ is invariant under rearrangement, we will write $\rho(f) = \rho(|f|_*)$.

Theorem 4.2.1 (Inclusion in $L^\infty(\Omega)$)
Let V satisfy the PSR property. The set

$$\Big\{u \in V,\ s.t.\ \rho\left(|\nabla u|_{*u}\right) < +\infty\Big\}$$

is included in $L^\infty(\Omega)$ if $\rho'\big(K(\cdot, \Omega, V)\big)$ is finite. Moreover,

$$\operatorname*{osc}_{\Omega} u \leqslant \rho'\Big(K(\cdot, \Omega, V)\Big)\rho\Big(|\nabla u|_{*u}\Big).$$

Proof For all $\sigma, s \in \Omega_*$, we have $|u_*(\sigma) - u_*(s)| \leqslant \int_{\Omega_*} |\nabla u|_{*u}(t) K(t) dt$, where $K(t) = K(t, \Omega, V)$. By definition of ρ'

$$\int_{\Omega_*} K(t)\, |\nabla u|_{*u}(t) \leqslant \rho'(K) \cdot \rho\big(\, |\nabla u|_{*u}\big)\,,$$

from which we get

$$\operatorname*{osc}_{\Omega} u \leqslant \rho'(K) \rho(|\nabla u|_{*u}) < +\infty \; if \;\; \rho'(K) < +\infty$$

and

$$\rho(|\nabla u|_{*u}) < +\infty.$$

◇

We can replace the previous set by another more common set if ρ has more properties.

Corollary 4.2.1 *Let ρ be a Fatou norm invariant under rearrangement on $L^0(\Omega_*)$. Then, for V satisfying the PSR property,*

$$\Big\{u \in V : \rho(|\nabla u|) < +\infty\Big\} \subset L^\infty(\Omega) \; if \;\; \rho'(K) < +\infty$$

where $K(\cdot) = K(\cdot, \Omega, V)$.

Proof We know that for $u \in W^{1,1}(\Omega)$, we always have,

$$\int_0^s (|\nabla u|_{*u})_*(\sigma) d\sigma \leqslant \int_0^s |\nabla u|_*(\sigma) d\sigma, \;\; \forall s \in \Omega_*\,.$$

Then $\rho(|\nabla u|_{*u}) \leqslant \rho(|\nabla u|_*)$. Therefore we have:

$$\Big\{u \in V : \rho(|\nabla u|_*) < +\infty\Big\} \subset \Big\{u \in V : \rho(|\nabla u|_{*u}) < +\infty\Big\}$$

and

$$\rho(|\nabla u|_*) = \rho(|\nabla u|).$$

◇

The most common application of Corollary 4.2.1 is the case where $V = W_0^{1,1}(\Omega)$.

In this case we have seen that

$$K(s) = (N\alpha_N^{\frac{1}{N}})^{-1}\mathrm{Max}\left(s^{-\frac{1}{N'}}; (|\Omega| - s)^{-\frac{1}{N'}}\right).$$

We then calculate K_*,

$$K_*(s) = 2^{\frac{1}{N'}}\left(N\alpha_N^{\frac{1}{N}}\right)^{-1}s^{-\frac{1}{N'}}.$$

(Note K has no plateau and is continuous.) Thus the condition $\rho'(K) < +\infty$ is equivalent to $\rho'(t^{-\frac{1}{N'}}) < +\infty$. Moreover if $u \in W^1L(\Omega, \rho)$ and $\rho'(t^{-\frac{1}{N'}}) < +\infty$ then

$$\operatorname*{osc}_{\Omega} u = u_*(0) - u_*(|\Omega|) \leqslant \int_{\Omega_*} K_*(t)(|\nabla u|_{*u})_*(t)$$

(using the Hardy–Littlewood inequality.) As a result we obtain:

$$\operatorname*{osc}_{\Omega} u \leqslant \frac{2^{\frac{1}{N'}}}{N\, a_N^{\frac{1}{N}}}\int_0^{|\Omega|} t^{\frac{1}{N}}(|\nabla u|_{*u})_*\frac{dt}{t} \leqslant \frac{2^{\frac{1}{N'}}}{N\alpha_N^{\frac{1}{N}}}\int_0^{|\Omega|} t^{\frac{1}{N}}|\nabla u|_*(t)\frac{dt}{t} \tag{4.10}$$

(according to Theorem 3.5.1). Using the definition of ρ' and the fact that ρ is rearrangement invariant

$$\operatorname*{osc}_{\Omega} u \leqslant \frac{2^{\frac{1}{N'}}}{N\alpha_N^{\frac{1}{N}}}|\nabla u|_{N,1} \leqslant \frac{2^{\frac{1}{N'}}}{N\alpha_N^{\frac{1}{N}}}\rho(t^{-\frac{1}{N'}})\rho(|\nabla u|) < +\infty. \tag{4.11}$$

Hence the following proposition.

Proposition 4.2.2 *If $\rho'(t^{-\frac{1}{N'}}) < +\infty$ and ρ is a Fatou norm invariant under rearrangement, we necessarily have $W_0^1L(\Omega, \rho) \subset W_0^1L^{N,1}(\Omega)$, which implies*

$$L(\Omega, \rho) \subset L^{N,1}(\Omega).$$

This result was given in [45] and furthermore they proved that if we have a continuous injection $W^1L(\Omega, \rho) \hookrightarrow L^\infty(\Omega)$ then necessarily $\rho'(t^{-\frac{1}{N'}}) < +\infty$. We will not give the proof of this converse. We therefore have ultimately

Proposition 4.2.3 *Let ρ be a Fatou norm invariant under rearrangement and* $W_0^1L(\Omega,\rho)=\left\{v\in W_0^{1,1}(\Omega):\rho(|\nabla v|)<+\infty\right\}$. *Then*

$$W_0^1L(\Omega,\rho)\hookrightarrow L^\infty(\Omega) \text{ if and only if } \rho'(t^{-\frac{1}{N'}})<+\infty.$$

For this reason we return to the space $L^{N,1}(\Omega)$ and we will show that its associate is $L^{N',\infty}(\Omega)$.

Lemma 4.2.1 *The map* $f\in L^0(\Omega_*)\xrightarrow{\rho_N}\int_{\Omega_*}t^{\frac{1}{N}}|f|_*(t)\frac{dt}{t}$ *is a Fatou norm, invariant under rearrangement, and non-trivial.*

Proof We easily have the fact if $\rho_N(f)=0$ then $|f|_*(t)=0\ \forall t$ so $f\equiv 0$ and $\rho_N(\lambda f)=|\lambda|\rho_N(f)$. As for the triangle inequality, since the function $t\in\Omega_*\to t^{\frac{1}{N}-1}$ is decreasing then according to Proposition 1.5.2, we have

$$\int_{\Omega_*}|f|_*(t)t^{\frac{1}{N}-1}dt=\text{Max}\left\{\int_\Omega|f|g\,dx:g_*(t)=t^{\frac{1}{N}-1}\right\}.$$

Thus

$$\int_{\Omega_*}\Big(|f_1|+|f_2|\Big)_*(t)t^{\frac{1}{N}-1}dt\leqslant\int_{\Omega_*}|f_1|_*(t)t^{\frac{1}{N}-1}dt+\int_{\Omega_*}|f_2|_*(t)t^{\frac{1}{N}-1}dt.$$

Hence

$$\rho_N(f_1+f_2)\leqslant\int_N\Big(|f_1|+|f_2|\Big)_*(t)t^{\frac{1}{N}-1}dt\leqslant\rho_N(f_1)+\rho_N(f_2).$$

If $f\in L^0_+(\Omega_*)$ $0\leqslant f_n\leqslant f_{n+1}\to f$ almost everywhere, then as $0\leqslant f_{n*}\leqslant(f_{n+1})_*\to f_*$, by the Beppo Levi theorem we have $\rho_N(f_n)\to\rho_N(f)$. Consequently $f\to\rho_N(f)$ is a Fatou norm invariant under rearrangement on $L^0_+(\Omega)$, therefore on $L^0(\Omega)$. It suffices to take $f(t)=1$, $\rho_N(1)=N|\Omega|^{\frac{1}{N}}$, so it is non-trivial. ◇

Lemma 4.2.2 *The associate norm of* ρ_N *of Lemma 4.2.1 satisfies*

$$\rho'_N(f) \leqslant \sup_{0 \leqslant t \leqslant |\Omega|} \left[t^{\frac{1}{N'}} |f|_*(t) \right] = |f|_{N',\infty} \text{ where } \frac{1}{N'} = 1 - \frac{1}{N}.$$

Proof If $\rho_N(g) \leqslant 1$ then $\int_\Omega |fg| \leqslant \int_{\Omega_*} |f|_* |g|_*$ (according to the Hardy–Littlewood inequality) hence

$$\int_\Omega |fg|\, dx \leqslant \sup_t \left[t^{\frac{1}{N'}} |f|_*(t) \right] \cdot \int_{\Omega_*} t^{-\frac{1}{N'}} |g|_*(t) dt \leqslant \sup_t \left[t^{\frac{1}{N'}} |f|_*(t) \right].$$

Hence $\rho'_N(f) \leqslant \sup_t \left[t^{\frac{1}{N'}} |f|_*(t) \right]$. ◇

This inequality is sufficient in our context, nevertheless since ρ_N is invariant under rearrangement, we actually have the equality.

Lemma 4.2.3

$$\frac{1}{N} |f|_{N',\infty} \leqslant \rho'_N(f), \leqslant |f|_{N',\infty} \qquad \forall f \in L^0(\Omega).$$

Proof Since ρ_N is invariant under rearrangement, we deduce:

$$\rho'_N(f) = \sup \left\{ \int_{\Omega_*} |f|_* |g|_* dt,\ \rho_N(g) \leqslant 1 \right\}.$$

Let $t \in \Omega_*$ (fixed) and $E \subset \Omega_*$ such that $|E| = t$.

Consider the function $g(\sigma) = \dfrac{1}{N t^{\frac{1}{N}}} \chi_E(\sigma)$, $\sigma \in \Omega_*$. So $\rho_N(g) = 1$ because $g_*(\sigma) = \dfrac{1}{N t^{\frac{1}{N}}} \chi_{[0,t)}(\sigma)$, and we deduce

$$\rho'_N(f) \geqslant \int_{\Omega_*} |f|_* g_*(\sigma) d\sigma = \frac{t^{\frac{1}{N'}}}{N} \left(\frac{1}{t} \int_0^t |f|_*(\sigma) d\sigma \right).$$

But $|f|_*(t) \leqslant \frac{1}{t}\int_0^t |f|_*(\sigma)d\sigma$. Therefore, $\forall t \in \Omega_*$

$$\rho'_N(f) \geqslant \frac{1}{N}\left[t^{\frac{1}{N'}}|f|_*(t)\right]. \qquad \diamond$$

Recall that $f \to |f|_{N',\infty}$ is not a norm but Corollary 1.4.1 tells us that

$$|f|_{(N',\infty)} \leqslant N|f|_{N',\infty}. \tag{4.12}$$

Moreover, since $|f|_*(t) \leqslant |f|_{**}(t) \quad \forall t$ we have

$$|f|_{N',\infty} \leqslant |f|_{(N',\infty)}. \tag{4.13}$$

Thus, using Lemma 4.2.3, and relations (4.12), (4.13), we deduce that ρ'_N is equivalent to $|\cdot|_{(N',\infty)}$, from which we get the following proposition:

Proposition 4.2.4 *The associate space to the Lorentz space* $\left(L^{N,1}(\Omega, |\cdot|_{N,1}\right)$ *is, modulo a norm equivalence the space* $\left(L^{N',\infty}(\Omega), |\cdot|_{(N',\infty)}\right)$.

Remarks

- The space $L^{p,\infty}(\Omega)$ is also called the Marcinkiewicz space, because it coincides with $\left\{f \in L^0(\Omega) : \sup_{t>0}\Big||f| > t\Big|t^p < +\infty\right\}$
- We also have that the space $L^{p,q}(\Omega)$, $1 < p < +\infty$, $1 \leqslant q \leqslant +\infty$ is (modulo a norm equivalence) $L^{p',q'}(\Omega)$ with $\frac{1}{p} + \frac{1}{p'} = \frac{1}{q} + \frac{1}{q'} = 1$.

Lemma 4.2.4

(a) Let $K_1 = \frac{t^{\frac{1}{N}-1}}{(N\alpha_N^{\frac{1}{N}})}$ *then* $\rho'_{N,1}(K_1) = \left(N\alpha_N^{\frac{1}{N}}\right)^{-1}$.

(b) Let $K_2(t) = \mathrm{Max}\left(t^{\frac{1}{N}-1}, (|\Omega| - t)^{\frac{1}{N}-1}\right)$ *then*

$$\rho'_{N,1}(K_2) = 2^{1-\frac{1}{N}}.$$

Proof

(a) We have $K_{1*} = K_1$. Hence $\sup_\sigma \left[\sigma^{\frac{1}{N'}} K_{1*}(\sigma)\right] = \left(N\alpha_N^{\frac{1}{N}}\right)^{-1}$.

(b) We have $K_{2*}(\sigma) = 2^{\frac{1}{N'}}\sigma^{-\frac{1}{N'}}$. Hence $\sup_\sigma \left[\sigma^{\frac{1}{N'}} K_{2*}(\sigma)\right] = 2^{\frac{1}{N'}}$.

In both cases, we apply Lemma 4.2.3. ◇

Corollary 4.2.2 (Of Lemma 4.2.4)

(a) Let $W_0^1 L^{N,1}(\Omega) = \left\{u \in W_0^{1,1}(\Omega),\ |\nabla u| \in L^{N,1}(\Omega)\right\}$.
Then, $W_0^1 L^{N,1}(\Omega) \subset L^\infty(\Omega)$ *and* $\forall u \in W_0^1 L^{N,1}(\Omega)$, $|u|_\infty \leqslant \dfrac{1}{N\alpha_N^{\frac{1}{N}}} |\nabla u|_{L^{N,1}(\Omega)}$.

(b) Let $r > 0$ *and* $B(r)$ *be a ball of radius* r. *Let us consider*
$W^1 L^{N,1}(B(r)) = \left\{u \in L^1(B(r)) : |\nabla u| \in L^{N,1}(B(r))\right\}$. *Then,*

$$\operatorname*{osc}_{B(r)} u \leqslant \frac{\alpha_N^{1-\frac{1}{N}}}{\alpha_{N-1}} \int_0^{\alpha_N r^N} t^{\frac{1}{N}} |\nabla u|_*(t)\frac{dt}{t} = \frac{\alpha_N^{1-\frac{1}{N}}}{\alpha_{N-1}} |\nabla u|_{L^{N,1}(B(r))}\,.$$

Proof

(a) If $u \in W_0^1 L^{N,1}(\Omega)$ then $v = |u| \in W_0^{1,1}(\Omega)$, $|u|_*(|\Omega|) = 0$. Using ρ_N, we have $\rho_N(|\nabla v|_*) = |\nabla u|_{N,1}$, and using Theorem 4.2.1, we have for all s

$$v_*(s) \leqslant \rho_N'(K_1)\rho_N(|\nabla v|_*) \leqslant \frac{1}{N\alpha_N^{\frac{1}{N}}} |\nabla u|_{N,1}\,.$$

(b) Since $X(B(r)) = W^1 L^{N,1}(B(r))$ satisfies the PSR property with

$$K\Big(s, B(r), X\big(B(r)\big)\Big) = \frac{1}{\alpha_{N-1}}\left(\frac{\alpha_N}{2}\right)^{1-\frac{1}{N}} K_2(s)\,,$$

we have, by Theorem 4.2.1,

$$\operatorname*{osc}_{B(r)} u \leqslant \rho_N'(K)\rho_N(|\nabla u|_*) \leqslant \frac{\alpha_N^{1-\frac{1}{N}}}{\alpha_{N-1}} |\nabla u|_{L^{N,1}(B(r))}\,. \diamond$$

Corollary 4.2.3 *Let* $W^1L^{N,1}(\Omega) = \Big\{u \in L^1(\Omega) : |\nabla u| \in L^{N,1}(\Omega)\Big\}$. *Then* $W^1L^{N,1}(\Omega) \subset C(\Omega)$ *and* $\forall\, x \in \Omega$, *and* $r > 0$ *s.t.* $B(x,r) \subset \Omega$,

$$\operatorname*{osc}_{B(x,r)} u \leqslant \frac{\alpha_N^{1-\frac{1}{N}}}{\alpha_{N-1}} |\nabla u|_{L^{N,1}(B(r))} \,.$$

4.3 General Inclusion Index for Functions with Zero Trace

The pointwise inequalities will allow us to give simple estimates of constants such as: $\inf\limits_{u\in W} \dfrac{|\nabla u|_p}{|u|_q}$, $1 \leqslant q \leqslant p^*$ or more generally if ρ and ρ_0 are two norms, how to estimate

$$\inf_{u\in W} \frac{\rho(|\nabla u|)}{\rho_0(|u|)},$$

just using three "simple" arguments:

(a) $\forall\, s, \sigma \in \Omega_*,\ u_*(\sigma) - u_*(s) = -\displaystyle\int_\sigma^s u_*'(t)dt$ if $u_* \in W^{1,1}_{loc}(\Omega_*)$.

(b) Equimeasurability or invariance by rearrangement.

(c) Poincaré–Sobolev pointwise inequalities for relative rearrangement with Hölder's inequality between a norm and its associated standard.

Here is a general theorem that illustrates this.

Theorem 4.3.1 *Let* $L(\Omega,\rho)$ *and* $L(\Omega,\rho_0)$ *be two spaces of Banach functions invariant under rearrangement. We define*

$$W_0^1L(\Omega,\rho) = \Big\{v \in W_0^{1,1}(\Omega) : \rho(|\nabla v|) < +\infty\Big\}.$$

For $s \in \Omega_*$, *we set* $b(s) = \rho'(K\chi_{]s,|\Omega|[})$, *where*

(continued)

Theorem 4.3.1 (continued)

$$K(t) = \left(N\alpha_N^{\frac{1}{N}}\right)^{-1} t^{-\frac{1}{N'}} \text{ with } t \in \Omega_*.$$

Then

1. $\forall\, v \in W_0^1 L(\Omega, \rho)$, *by setting* $u = |v|$ *we have*

$$\rho_0(u) \leqslant \rho_0(b) \cdot \rho(|\nabla u|).$$

2. *In particular,* $W_0^1 L(\Omega, \rho) \hookrightarrow L(\Omega, \rho_0)$ *if* $\rho_0(b) < +\infty$.

Proof As $v \in W_0^{1,1}(\Omega)$, we have $u = |v| \in W_0^{1,1}(\Omega)$ and $u_*(\Omega|) = 0$. Thus,

$$u_*(s) =- \int_s^{|\Omega|} u_*'(\sigma)d\sigma \leqslant (N\alpha_N^{\frac{1}{N}})^{-1} \int_{\Omega_*} t^{-\frac{1}{N'}} \chi_{]s,|\Omega|[}(t)|\nabla u|_{*u}(t)dt$$

$$\leqslant \rho'\left(K\chi_{]s,|\Omega[}\right) \cdot \rho(|\nabla u|_{*u}).$$

Since ρ is invariant under rearrangement

$$u_*(s) \leqslant \rho'\left(K\chi_{]s,|\Omega|[}\right)\rho\left(|\nabla u|_*\right) = \rho'\left(K\chi_{]s,|\Omega|[}\right)\rho\left(|\nabla u|\right)$$

hence

$$\rho_0(u) \leqslant \rho_0(b) \cdot \rho(|\nabla u|).$$

We deduce statement 2. ◇

Definition 4.3.1 $\rho_0(b)$ is called the inclusion index associated with $W_0^{1,1}(\Omega, \rho)$ and $L(\Omega, \rho_0)$.

We will extend these results to functions that are not zero-trace, but first let us give some examples of applications, starting with the inclusions most used in the literature.

4.4 Poincaré-Sobolev–Lorentz Inclusions

4.4.1 *Case of Zero-Trace Functions*

Consider the case where $V = \left\{u \in W_0^{1,1}(\Omega),\ u \geqslant 0\right\}$. Then we know that $K(t) = \frac{t^{\frac{1}{N}-1}}{N\alpha_N^{\frac{1}{N}}}$. Let us choose $\rho(f) = \left(\int_{\Omega_*} |f|^p\right)^{\frac{1}{p}}$, $1 < p < +\infty$, then $W^1(\Omega,\rho) = W_0^{1,p}(\Omega) \cap L_+^0(\Omega)$.

In this case $\rho'(f) = \left(\int_{\Omega_*} |f|^{p'}\right)^{\frac{1}{p'}}$. Thus, for $s \in \Omega_*$:

$$\rho'(b(s)) = \left(\int_s^{|\Omega|} t^{-\frac{p'}{N'}}\right)^{\frac{1}{p'}} \frac{1}{N\alpha_N^{\frac{1}{N}}} = \begin{cases} \dfrac{1}{N\alpha_N^{\frac{1}{N}}}\left(+\operatorname{Log}\dfrac{|\Omega|}{s}\right)^{\frac{1}{N'}} & if\ p = N \\ \dfrac{1}{N\alpha_N^{\frac{1}{N}}} \cdot \left[\dfrac{|\Omega|^{1-\frac{p'}{N'}} - s^{1-\frac{p'}{N'}}}{N'-p'} N'\right]^{\frac{1}{p'}} & otherwise \end{cases}$$

Note We can also consider the case $p = 1$, $\rho'\big(b(s)\big) = \dfrac{s^{\frac{1}{N}-1}}{N\alpha_N^{\frac{1}{N}}}$.

Theorem 4.4.1 *Let $L(\Omega, \rho_0)$ be a Banach function space invariant under rearrangement s.t. $\rho_0\left(+\left(\operatorname{Log}\frac{|\Omega|}{s}\right)^{\frac{1}{N'}}\right) < +\infty$. Then $W_0^{1,N}(\Omega) \hookrightarrow L(\Omega, \rho_0)$.*

In particular $W_0^{1,N}(\Omega) \hookrightarrow L_{exp}^{\frac{1}{N'}}(\Omega)$

where $L_{exp}^{\alpha}(\Omega) = \left\{v \in L^0(\Omega) : \sup\limits_{0<s<|\Omega|} \dfrac{|v|_{**}(s)}{\left(1+\operatorname{Log}\dfrac{|\Omega|}{s}\right)^{\alpha}} < +\infty\right\}$ for $\alpha \geqslant 0$.

Proof According to the expression of $\rho'(b(s))$ and Theorem 4.3.1, we deduce

$$W_0^{1,N}(\Omega \hookrightarrow L(\Omega, \rho_0).$$

In the case where $\rho_0(v) = \sup_{0<s<|\Omega|} \dfrac{|v|_{**}(s)}{\left(1+\operatorname{Log}\dfrac{|\Omega|}{s}\right)^{\frac{1}{N'}}}$ we have

$\rho_0\left(\left(\operatorname{Log}\dfrac{|\Omega|}{s}\right)^{\frac{1}{N'}}\right) < +\infty$, hence we have inclusion. □

Note that we obtain as a bonus that, if $v \in W_0^{1,N}(\Omega),\ u = |v|$, then

$$\sup_{0<s<|\Omega} \frac{u_{**}(s)}{\left(1+\log\dfrac{|\Omega|}{s}\right)^{\frac{1}{N'}}} \leqslant \frac{2}{N\alpha_N^{\frac{1}{N}}}\left|\,|\nabla u|_{*u}\right|_N \leqslant \frac{2}{N\alpha_N^{\frac{1}{N}}}|\nabla v|_N.$$

As for all $r \in [1, +\infty[$, we have

$$\left(\int_{\Omega_*} u_*^r(t)dt\right)^{\frac{1}{r}} \leqslant \left[\int_{\Omega_*}\left(1-\operatorname{Log}\frac{s}{|\Omega|}\right)^{\frac{r}{N'}} ds\right]^{\frac{1}{r}} |v|_{L_{exp}^{\frac{1}{N}'}} < +\infty,$$

we then deduce:

Theorem 4.4.2 (Sobolev Inclusion for $p = N$) *For all $r < \infty$, we have $W_0^{1,N}(\Omega) \subset L^r(\Omega)$,*

$$|u|_{L^r(\Omega)} \leqslant \frac{|\Omega|^{\frac{1}{r}}}{N\alpha_N^{\frac{1}{N}}}\left(\int_0^1 (-\operatorname{Log} t)^{\frac{r}{N'}}\, dt\right)^{\frac{1}{r}} |\,|\nabla u|_{*u}\,|_N$$

and

$$\int_0^1 (-\operatorname{Log} t)^{\frac{r}{N'}}\, dt = \int_0^{+\infty} \sigma^{r(1-\frac{1}{N})} exp^{-\sigma} d\sigma$$

$$= \Gamma\left(1+\frac{r}{N'}\right) \widetilde{\ _{r\to+\infty}} r(N')^{\frac{1}{N'}} exp^{-\frac{1}{N'}}.$$

The inclusion of Theorem 4.4.1 is equivalent to the precise estimate of Theorem 4.4.2, as shown by the following corollary:

Corollary 4.4.1 (Trudinger's Inequality) *We have* $\forall\, \underline{\lambda} > \dfrac{1}{N\alpha_N^{\frac{1}{N}}}$

$$\int_\Omega exp\left(\frac{|u(x)|}{\underline{\lambda}\,|\nabla u|_N}\right)^{\frac{N}{N-1}} \leqslant \frac{|\Omega|}{1-\left(\dfrac{1}{\underline{\lambda} N\alpha_N^{\frac{1}{N}}}\right)^{\frac{N}{N-1}}}.$$

Proof Idea For integer k, we choose $r = k\frac{N}{N-1}$ in Theorem 4.4.2

$$\int_\Omega |u(x)|^{k\frac{N}{N-1}}\,dx \leqslant \frac{|\Omega|}{(N\alpha_N^{\frac{1}{N}})^{k\frac{N}{N-1}}}\left(\int_0^{+\infty} \sigma^k exp^{-\sigma} d\sigma\right)|\nabla u|_N^{k\frac{N}{N-1}}\ .$$

Knowing that $exp(t) = \sum_{k=0}^{+\infty}\frac{t^k}{k!}$, starting from the previous inequality, by multiplying by $\dfrac{1}{\underline{\lambda}^{k\frac{N}{N-1}}}$, we deduce the inequality after summation. ◇

Note The optimization of the condition on $\underline{\lambda}$ (Trudinger's inequality) was proven by E. Lieb [105].

Theorem 4.4.3 (Inclusion in $C^{0,\alpha}$ (Case $p > N$)) *Let* $u \in W_0^{1,p}(\Omega)$, $p > N$. *Then* $u \in C^{0,\alpha}(\overline{\Omega})$ *with* $\alpha = 1 - \dfrac{N}{p}$. *Moreover,*

(i) $|u|_\infty \leqslant \dfrac{1}{N\alpha_N^{\frac{1}{N}}}\left[\dfrac{N(p-1)}{p-N}\right]^{1-\frac{1}{p}} |\Omega|^{\frac{1}{N}-\frac{1}{p}}\,|\,|\nabla u|_{*u}\,|_{L^p(\Omega_*)}$,

(ii) $\forall r > 0,\ \forall x \in \Omega$ *s.t.* $B(x,r) = B(r) \subset \Omega$, *we have:*

$$\underset{B(x,r)}{\text{osc}}\ u \leqslant \frac{2^{\frac{1}{N}-\frac{1}{p}}}{\alpha_{N-1}}\left(\alpha_N\frac{N(p-1)}{p-N}\right)^{1-\frac{1}{p}} |\nabla u|_{L^p(B(x,r))}\cdot r^{1-\frac{N}{p}}\ .$$

Proof

(i) We choose $\rho_0(f) = |f|_\infty$. As $b(s)=\dfrac{1}{N\alpha_N^{\frac{1}{N}}}\left[\dfrac{N(p-1)}{p-N}\right]^{1-\frac{1}{p}}\left[|\Omega|^{1-\frac{p'}{N'}} - s^{1-\frac{p'}{N'}}\right]$, for $p > N,\ s \in \Omega_*$, we deduce that

$$\rho_0(b) = \frac{1}{N\alpha_N^{\frac{1}{N}}}\left[\frac{N(p-1)}{p-N}\right]^{1-\frac{1}{p}} |\Omega|^{\frac{1}{N}-\frac{1}{p}}.$$

We conclude by the general estimate Theorem 4.3.1, knowing that

$$|u| \in W_0^{1,p}(\Omega)\ if\ u \in W_0^{1,p}(\Omega).$$

(ii) Let $\overline{u}$ be the restriction of u to $B(x,r)$. Then $\overline{u} \in W^{1,1}(B(r))$. By the pointwise Poincaré–Sobolev inequality (see Corollary 4.1.1 of Theorem 4.1.1), we deduce

$$\underset{B(x,r)}{\text{osc}}\ u = \overline{u}_*(\alpha_N r^N) - \overline{u}_*(0)$$

$$\leqslant \frac{1}{\alpha_{N-1}}\left(\frac{\alpha_N}{2}\right)^{1-\frac{1}{N}} \int_0^{\alpha_N r^N} \text{Max}\left(t^{\frac{1}{N}-1}, \left(\alpha_N r^N - t\right)^{\frac{1}{N}-1}\right) |\nabla \overline{u}|_{*\overline{u}}(t)dt.$$

Applying Hölder's inequality to this last integral and knowing

$$\left| |\nabla \overline{u}|_{*\overline{u}} \right|_{L^p(0,\alpha_N r^N)} \leqslant |\nabla \overline{u}|_{L^p(B(r))} = |\nabla u|_{L^p(B(r))},$$

we deduce $\text{osc}_{B(x,r)}\, u \leqslant \frac{1}{\alpha_{N-1}}\left(\frac{\alpha_N}{2}\right)^{1-\frac{1}{N}} \cdot \left(2\displaystyle\int_0^{\alpha_N r^N} t^{(\frac{1}{N}-1)p'}\right)^{\frac{1}{p'}} |\nabla u|_{L^p(B(r))}$

$$= \frac{2^{\frac{1}{N}-\frac{1}{p}}}{\alpha_{N-1}}\left(\alpha_N \frac{N(p-1)}{p-N}\right)^{1-\frac{1}{p}} |\nabla u|_{L^p(B(r))}. \qquad \diamond$$

Note that the zero extension of u in $\mathbb{R}^N$ is in $W^{1,p}(\mathbb{R}^N) \subset C^{0,\alpha}(\overline{\Omega})$.

Theorem 4.4.4 (Inclusion for $p < N$) *If $1 \leqslant p < N$, we have: $W_0^{1,p}(\Omega) \subset_{>} L^{p^*}(\Omega)$ with $\frac{1}{p^*} = \frac{1}{p} - \frac{1}{N}$. Moreover, we have:*

(i) $|u|_{L^{p^*}(\Omega)} \leqslant \frac{1}{N\alpha_N^{\frac{1}{N}}} B^{\frac{1}{p^*}} | |\nabla u|_{*u} |_{L^p(\Omega_*)} if\ 1 < p < N,$

$$B = \left\{ \frac{\Gamma\left(\frac{pp^*}{p^*-p}\right)}{\Gamma\left(\frac{p^*}{p^*-p}\right)\Gamma\left(\frac{p(p^*-1)}{p^*-p}\right)} \right\}^{-1+\frac{p}{p^*}} \left[p^*\left(1-\frac{1}{p}\right)\right]^{p^*-\frac{p^*}{N}}.$$

Here $\Gamma(.)$ is the well-known function $\Gamma(x+1) = \int_0^{+\infty} t^x e^{-t} dt.$

(ii) $|u|_{L^{\frac{N}{N-1}}} \leqslant \frac{1}{N\alpha_N^{\frac{1}{N}}} |\nabla u|_{L^1(\Omega)} \qquad if\ p = 1.$

Proof Let $u \in W_0^{1,p}(\Omega)$, then we know that $|u| \in W_0^{1,p}(\Omega)$, so we can assume that $u \geqslant 0$. We have: $u_*(s) = \int_s^{|\Omega|} \left|u_*'(t)\right| dt$.

Then

$$\int_\Omega u(x)^{p^*} dx = \int_{\Omega_*} u_*(s)^{p^*} ds = \int_0^{|\Omega|} ds \left[\int_s^{|\Omega|} \left|u_*'(t)\right| dt \right]^{p^*}.$$

Applying Bliss' inequality (see below), we deduce

$$|u|_{p^*}^{p^*} = \int_0^{|\Omega|} ds \left[\int_s^{|\Omega|} \left|u_*'(t)\right| dt \right]^{p^*} \leqslant B \left[\int_0^{|\Omega|} \left|u_*'(t) t^{\frac{N-1}{N}}\right|^p dt \right]^{\frac{p^*}{p}}.$$

By the PSR inequality, we know that in this case,

$$\left|u_*'(t) t^{\frac{N-1}{N}}\right| \leqslant \frac{1}{N\alpha_N^{\frac{1}{N}}} |\nabla u|_{*u}(t).$$

Hence by combining the two inequalities:

$$|u|_{p^*} \leqslant \frac{B^{\frac{1}{p^*}}}{N\alpha_N^{\frac{1}{N}}} \left||\nabla u|_{*u}\right|_{L^p(\Omega_*)} \leqslant \frac{B^{\frac{1}{p^*}}}{N\alpha_N^{\frac{1}{N}}} |\nabla u|_{L^p(\Omega)} \ .$$

As for (ii), since $\overline{\mathcal{D}(\Omega)_+} = W_{0+}^{1,1}(\Omega)$ (positive functions), we can assume that $u \in \mathcal{D}(\Omega),\ u \geqslant 0$. Moreover, if we consider the spherical rearrangement $\underset{\sim}{u} \in W^{1,\infty}(\underset{\sim}{\Omega})$, $\underset{\sim}{u} \geqslant 0$, then we have:

$$\int_{\underset{\sim}{\Omega}} \left|\nabla \underset{\sim}{u}(x)\right| dx = N\alpha_N^{\frac{1}{N}} \int_{\underset{\sim}{\Omega}} s^{1-\frac{1}{N}} \left|u_*'(s)\right| ds \leqslant \int_{\Omega_*} |\nabla u|_{*u} (\sigma) d\sigma$$

$$= \int_{\Omega} |\nabla u| \, dx \ .$$

By equimeasurability, it suffices to show that:

$$\left(\int_{\Omega} \left(\underset{\sim}{u}(x)\right)^{\frac{N}{N-1}} dx\right)^{1-\frac{1}{N}} \leqslant \frac{1}{N\alpha_N^{\frac{1}{N}}} \left|\nabla \underset{\sim}{u}\right|_{L^1} \ .$$

Let $T_t, t > 0$, be the truncation at level t, i.e. $T_t(\sigma) = \min(\sigma, t)$ if $\sigma \geqslant 0$. Following the idea of W. Ziemer [171] let us set

$$f(t) = \left(\int_{\Omega} T_t(\underset{\sim}{u})^{\frac{N}{N-1}} dx\right)^{1-\frac{1}{N}} .$$

Then f is absolutely continuous because

$$\left|f(t) - f(t')\right| \leqslant \left(\int_{\Omega} \left|T_t(\underset{\sim}{u}) - T_{t'}(\underset{\sim}{u})\right|^{\frac{N}{N-1}} dx\right)^{1-\frac{1}{N}} \leqslant \left|t - t'\right| \left|\underset{\sim}{u} > \min(t, t')\right|^{1-\frac{1}{N}} ,$$

hence if $t' = t + h$,

$$\left|f(t) - f(t')\right| \leqslant h |\underset{\sim}{u} > t|^{1-\frac{1}{N}} \ for \ h > 0.$$

This implies that $\left|f'(t)\right| \leqslant |\underset{\sim}{u} > t|^{1-\frac{1}{N}}$ almost all t,

$$|\underset{\sim}{u}|_{L^{\frac{N}{N-1}}} \leqslant \left| \int_0^{+\infty} f'(t)dt \right| \leqslant \int_0^{+\infty} |\underset{\sim}{u} > t|^{1-\frac{1}{N}} dt\,.$$

Since $P_{\underset{\sim}{\Omega}}(\underset{\sim}{u} > t) = N\alpha_N^{\frac{1}{N}} |\underset{\sim}{u} > t|^{1-\frac{1}{N}}$ for almost all t, we deduce using the Fleming–Rishel formula

$$|u|_{L^{\frac{N}{N-1}}} \leqslant \frac{1}{N\alpha_N^{\frac{1}{N}}} \int_0^{+\infty} P_{\Omega}(\underset{\sim}{u} > t)dt \leqslant \frac{1}{N\alpha_N^{\frac{1}{N}}} \int_{\underset{\sim}{\Omega}} |\nabla \underset{\sim}{u}|dx\,.$$

Hence the inequality. ◇

Note Using $\rho_0(b) = |b|_q$, we can show in this case ($1 < p < N$) that $\rho_0(b) < +\infty\ \forall q < p^*$. We unfortunately cannot treat the limiting case $q = p^*$ with the above index method. Hence the use of Bliss lemma.

Lemma 4.4.1 (Bliss) *Let $\varphi \geqslant 0$ and $q > p > 1$. Then*

$$\int_0^{+\infty} \left(\int_t^{+\infty} \varphi(s)ds \right)^q dt \leqslant B \left[\int_0^{+\infty} \varphi(t)^p t^{-1+p+\frac{p}{q}} \right]^{\frac{q}{p}}$$

where

$$B = \left\{ \frac{\Gamma\left(\frac{pq}{q-p}\right)}{\Gamma\left(\frac{q}{q-p}\right)\Gamma\left(\frac{p(q-1)}{q-p}\right)} \right\}^{\frac{q}{p}-1} \times \left[q\left(1 - \frac{1}{p}\right)\right]^{q-\frac{q}{p}}.$$

4.4.2 Some Sobolev-Type Inclusions Related to GΓ Spaces, to the Grand Lebesgue Space and to the BMO Space

We have seen that the expression of the $G\Gamma$ spaces introduced in Chap. 1 covers a certain number of Banach function spaces. We can associate a Sobolev space with them and deduce associated Sobolev–Poincaré-type inclusions. We start with a general result

Theorem 4.4.5 *Let* $v \in W_0^{1,1}(\Omega)$, $v \geqslant 0$, $1 \leqslant p < N$, $p_s = \dfrac{Np}{N-p}$. *Then, for all* $t \in \Omega_*$,

$$\left(\int_0^t v_*(\sigma)^{p_s} d\sigma\right)^{\frac{1}{p_s}} \leqslant \frac{(BLC)^{\frac{1}{p_s}}}{N\alpha_N^{\frac{1}{N}}} \left(\int_0^t \left[|\nabla v|_{*v}(\sigma)\right]_*^p d\sigma\right)^{\frac{1}{p}} + t^{\frac{1}{p_s}} v_{**}(t)$$

with the Bliss constant

$$BLC = \begin{cases} 1 \; if \; p = 1, \text{otherwise we have} \\ \left[\dfrac{\Gamma\left(\dfrac{p \cdot p_s}{p_s - p}\right)}{\Gamma\left(\dfrac{p_s}{p_s - p}\right)\Gamma\left(\dfrac{p \cdot (p_s - 1)}{p_s - p}\right)}\right]^{\frac{p_s}{p} - 1} \left[p_s\left(1 - \frac{1}{p}\right)\right]^{\frac{p_s}{p'}} \end{cases}$$

Proof Consider $\underset{\sim}{\Omega}$ the ball having the same measure as Ω centered at the origin and the radial function $\underline{v}(x) = v_*(\alpha_n |x|^N)$, $x \in \underset{\sim}{\Omega}$.

We have $v \in W_0^{1,1}(\Omega)$, $v \geqslant 0$. Then the Polyà–Szegö inequality (see Chap. 3) shows that: $\underline{v} \in W_0^{1,1}(\underset{\sim}{\Omega})$.

For $t \in [0, |\Omega|]$, consider $y_t \in \underset{\sim}{\Omega}$ such that $\alpha_n |y_t|^N = t$.

Let us set $r_t = |y_t|$. By the equimeasurability, we have
$||v_*||_{L^{p_s}(0,t)}$

$$= ||\underline{v}||_{L^{p_s}(B(0;r_t))} \leqslant ||\underline{v} - \underline{v}(y_t)||_{L^{p_s}(B(0,r_t))} + \underline{v}(y_t)|B(0, r_t)|^{\frac{1}{p_s}}. \tag{4.14}$$

By Sobolev's inequality, we have $\underline{v} - \underline{v}(y_t) \in W_0^{1,p}(B(0, r_t))$ (see [95])

$$||\underline{v} - \underline{v}(y_t)||_{L^{p_s}(B(0,r_t))} \leqslant \frac{(\text{Bliss})^{\frac{1}{p_s}}}{N\alpha_N^{\frac{1}{N}}} ||\nabla \underline{v}||_{L^p(B(0,r_t))}. \tag{4.15}$$

From (4.14) and (4.15) with $0 \leqslant \underline{v}(y_t) \leqslant \frac{1}{t}\int_0^t v_*(\sigma)d\sigma$ we deduce

$$||v_*||_{L^{p_s}(0,t)} \leqslant \frac{(\text{Bliss})^{\frac{1}{p_s}}}{N\alpha_N^{\frac{1}{N}}} \left(\int_0^t |\nabla \underline{v}|_*^p(\sigma)d\sigma\right)^{\frac{1}{p}} + t^{\frac{1}{p_s}} v_{**}(t). \tag{4.16}$$

But we have $|\nabla \underline{v}|_*(\sigma) = \left(- N\alpha_N^{\frac{1}{N}} s^{1-\frac{1}{N}} v_*'(s)\right)_*(\sigma)$ and by the Sobolev inequality for the relative rearrangement, we have

$$\left(- N\alpha_N^{\frac{1}{N}} s^{1-\frac{1}{N}} v_*'(s)\right) \leqslant |\nabla v|_{*v}(s). \tag{4.17}$$

So thanks to the relations (4.16), (4.17) we have

$$||v_*||_{L^{p_s}(0,t)} \leqslant \frac{(\text{Bliss})^{\frac{1}{p_s}}}{N\alpha_N^{\frac{1}{N}}} \left(\int_0^t \left[|\nabla v|_{*v}\right]_*^p(\sigma)d\sigma\right)^{\frac{1}{p}} + t^{\frac{1}{p_s}} v_{**}(t). \tag{4.18}$$

◇

Let us recall the following lemma, the proof of which follows from the results of Chap. 1 Corollary 1.2.1 and from the contraction property of relative rearrangement (see Proposition 2.2.2).

Lemma 4.4.2 *For all t,*

$$\int_0^t \left[\,|\nabla v|_{*v}\right]_*^p(\sigma)d\sigma \leqslant \int_0^t |\nabla v|_*^p(\sigma)d\sigma \text{ while } v \in W^1L^p(\Omega).$$

With the help of Lemma 4.4.2, we directly obtain

Corollary 4.4.2 (Of Theorem 4.4.5) *For each* $v \in W^1G\Gamma(p,m,w)$, $1 \leqslant p < N$, $p_s = \dfrac{Np}{N-p}$, $m < +\infty$

$$||v||_{G\Gamma(p_s,m,w)} \leqslant \frac{(Bliss)^{\frac{1}{p_s}}}{N\alpha_N^{\frac{1}{N}}} ||\nabla v||_{G\Gamma(p,m,w)} + \left[\int_0^1 w(t) t^{\frac{m}{p_s}} |v|_{**}(t)^m dt\right]^{\frac{1}{m}}. \tag{4.19}$$

In particular, for $N \geqslant 2$

$$W^1G\Gamma\left(1, 1, t^{-1}(1-\operatorname{Log} t)^{-\frac{1}{N'}}\right) \subset_{>} L^{(N'}(\Omega). \tag{4.20}$$

We have $L(\mathrm{Log}L)(\Omega) \underset{\neq}{\subset_{>}} G\Gamma\left(1, 1, t^{-1}(1-\mathrm{Log}t)^{-\frac{1}{N'}}\right) \underset{\neq}{\subset_{>}} L^1(\Omega)$. *More generally, for* $\theta > 0$ *we have*

$$|v|_{(N',\theta} \leqslant \frac{1}{N\alpha_N^{\frac{1}{N}}} \int_0^1 \frac{|\nabla v|_{**}(t)dt}{(1+|\operatorname{Log} t|)^{1-\frac{\theta}{N}}} + \int_0^1 \frac{t^{-\frac{1}{N}}|v|_{**}(t)}{(1+|\operatorname{Log} t|)^{1-\frac{\theta}{N}}} dt. \tag{4.21}$$

Proof We have $|v| \in W^1G\Gamma(p,m,w)$ if $v \in W^1G\Gamma(p,m,w)$, $1 \leqslant p < N$, $p_s = \dfrac{Np}{N-p}$. We apply the relation (4.18) with $|v|$ and using Lemma 4.4.2 we directly obtain the result (4.19). For the two last equations (4.20) and (4.21), we apply the preceding equation (4.19), with the weight $w(t) = t^{-1}(1+|\operatorname{Log} t|)^{\frac{\theta}{N}-1}$, $p = m = 1$ then $p_s = N'$ and the Bliss constant is equal to 1. ◇

Similar embeddings can be obtained with $m = +\infty$.

The previous theorems like Theorem 4.3.1 are especially suited for inclusions in spaces of Banach functions invariant under rearrangement. Nevertheless, the previous rearrangement technique can be adapted to obtain inclusions in normed spaces without being Banach function spaces. As we did for inclusions in Hölderian spaces, let's start with injections in bounded mean oscillation (BMO) spaces

Theorem 4.4.6 *We have the following injection*

$$W^1L^{N,\infty}(\Omega) \subset_{>} bmo_r(\Omega)$$

(continued)

Theorem 4.4.6 (continued)
where $bmo_r(\Omega) = \left\{v \in L^1(\Omega) : \sup_{B_{ball} \subset \Omega} |B|^{-1} \int_B |v(x) - v_B| dx < +\infty\right\}$ *equipped with the norm*

$$|v|_{bmo} = \sup_{B_{ball} \subset \Omega} |B|^{-1} \int_B |v(x) - v_B| dx + |v|_1 \text{ and } v_B = |B|^{-1} \int_B v(y) dy.$$

The proof requires the following lemma due to [46] and [45]:

Lemma 4.4.3 *Let* $v \in W^{1,1}(\Omega)$, *B be a ball contained in* Ω, *u the restriction of v to B. Let a = measure of* $B = |B|$. *Then*

$$\int_B |v(x) - v_B| dx \leqslant \frac{2}{a} \int_0^a t(a-t) \left|\frac{du_*}{dt}\right| dt < +\infty.$$

Proof Let us first note that by equimeasurability, we have

$$\int_B |v(x) - v_B| dx = \int_B |u(x) - v_B| dx = \int_0^a |u_*(s) - v_B| ds,$$

and that $u_* \in W^{1,1}_{loc}(0, a)$ because B is convex and $u \in W^{1,1}(B)$. By Corollary 4.1.1, we have

$$t u_*'(t) \in L^1\left(0, \frac{a}{2}\right).$$

For s and ε fixed s.t. $0 < s < a - \varepsilon < a$, we will prove that

$$|u_*(s) - v_B| \leqslant \int_0^{a-\varepsilon} \left| \chi_{]s, a-\varepsilon[} - \frac{t}{a - \varepsilon} \right| |u_*'(t)| dt + o(1)_{\varepsilon \to 0}. \tag{4.22}$$

To do this, note that $\varepsilon |u_*(\varepsilon)| \xrightarrow{\varepsilon \to 0} 0$. Indeed, if $u_+ > 0$, then for $\varepsilon > 0$ (small) we have:

$$0 \leqslant \varepsilon \cdot u_*(\varepsilon) \leqslant \int_0^\varepsilon u_*(\sigma) d\sigma \underset{\varepsilon \to 0}{\longrightarrow} 0.$$

Otherwise $u \leqslant 0$ in B then

$$0 \leqslant -u_*(\varepsilon) \leqslant \frac{1}{a-\varepsilon}|u|_{L^1}, \ 0 < \varepsilon < a.$$

Hence

$$0 \leqslant \varepsilon|u_*(\varepsilon)| \leqslant \frac{\varepsilon}{a-\varepsilon}|u|_{L^1} \xrightarrow[\varepsilon\to 0]{} 0.$$

Consequently,

$$v_B = \frac{1}{a}\int_0^a u_*(\sigma)d\sigma = \frac{1}{a-\varepsilon}\int_\varepsilon^{a-\varepsilon} u_*(t)dt + o(1) \text{ when } \varepsilon \to 0.$$

By integration by parts, we have, knowing that $tu_*'(t) \in L^1\left(0, \frac{a}{2}\right)$,

$$v_B = \frac{1}{a-\varepsilon}\int_\varepsilon^{a-\varepsilon}(t)'u_*(t)dt + o(1) = \frac{1}{a-\varepsilon}\int_0^{a-\varepsilon} t|u_*'(t)|dt + u_*(a-\varepsilon) + o(1)$$

and likewise, we have

$$u_*(s) = -\int_s^{a-\varepsilon} u_*'(\sigma)d\sigma + u_*(a-\varepsilon).$$

So, by taking the difference

$$u_*(s) - v_B = \int_0^{a-\varepsilon}\left(\chi_{]s,a-\varepsilon[}(t) - \frac{t}{a-\varepsilon}\right)|u_*'(t)|dt + o(1)_{\varepsilon\to 0}$$

we obtain the relation (4.22).

Integrating the relation (4.22) with respect to s between 0 and $a-\varepsilon$ and using Fubini's theorem,

$$\begin{aligned}\int_0^{a-\varepsilon}|u_*(s) - v_B|ds &\leqslant \int_0^{a-\varepsilon}|u_*'(t)|\left[\int_0^{a-\varepsilon}\left|\chi_{(0,t)}(s) - \frac{t}{a-\varepsilon}\right|ds\right]dt + o(1)\\ &\leqslant 2\int_0^{a-\varepsilon}\frac{t(a-\varepsilon-t)}{a-\varepsilon}|u_*'(t)|dt + o(1)\\ &\leqslant 2\int_0^a \frac{t(a-t)}{a-\varepsilon}|u_*'(t)|dt + o(1).\end{aligned} \tag{4.23}$$

But according to Corollary 4.1.1, we have

$$t(a-t)|u_*'(t)| \leqslant \frac{a}{\alpha_{N-1}}\left(\frac{\alpha_N}{2}\right)^{1-\frac{1}{N}} \operatorname{Max}\left(t^{\frac{1}{N}}; (a-t)^{\frac{1}{N}}\right)|\nabla u|_{*u}(t). \tag{4.24}$$

Therefore, the majorant function is in $L^1(0, a)$ because

$$\int_0^a |\nabla u|_{*u} dt = \int_B |\nabla u| dx = \int_B |\nabla v| dx.$$

Consequently, we let ε tend towards zero in (4.23) to get

$$\int_0^a |u_*(s) - v_B| ds \leqslant 2 \int_0^a \frac{t(a-t)}{a} |u_*'(t)| dt,$$

which proves the lemma. ◇

Lemma 4.4.4 *Under the same hypotheses as Lemma 4.4.3, we have*

$$|B|^{-1} \int_B |v(x) - v_B| dx \leqslant c \int_0^a Max\left(t^{\frac{1}{N}}; (a-t)^{\frac{1}{N}}\right) |\nabla u|_{*u}(t) dt$$

where

$$u = v\Big|_B, \ v_B = |B|^{-1} \int_B v(y) dy, \ a = |B|, c = \frac{2^{\frac{1}{N}} \alpha_N^{1-\frac{1}{N}}}{a \cdot \alpha_{N-1}}.$$

Proof This lemma follows from Eq. (4.24) and Lemma 4.4.3. ◇

Proof of Theorem 4.4.6 Let $H(t) = Max\left(t^{\frac{1}{N}}; (a-t)^{\frac{1}{N}}\right) \leqslant a^{\frac{1}{N}}$ for $t \in [0, a]$. Since $L^{N',1}$ is the associate space with $L^{N,\infty}$ then, we have

$$\int_0^a H(t) |\nabla u|_{*u}(t) dt \leqslant |H|_{L^{N',1}(0,a)} \Big| |\nabla u|_{*u} \Big|_{L^{N,\infty}(0,a)}$$

$$\leqslant N' a \, |\nabla v|_{L^{N,\infty}(\Omega)}.$$

Hence

$$\boxed{\sup_{B \subset \Omega} |B|^{-1} \int_B |v(x) - v_B| dx \leqslant \frac{2^{\frac{1}{N}} \alpha_N^{1-\frac{1}{N}}}{\alpha_{N-1}} N' |\nabla v|_{L^{N,\infty}(\Omega)} < +\infty.}$$

◇

Lemma 4.4.4 allows us to give a sufficient condition for

$$W^1L(\Omega, \rho) \hookrightarrow bmo_r(\Omega) \text{ when } \rho \text{is invariant under rearrangement.}$$

We deduce from Lemma 4.4.4:

$$\boxed{\sup_{B\subset\Omega} |B|^{-1}\int_B |v(x) - v_B|dx \leqslant \frac{2^{\frac{1}{N}}\alpha_N^{1-\frac{1}{N}}}{\alpha_{N-1}} \sup_{a>0}\left[\frac{1}{a}\rho'\big(H\chi_{(0,a)}\big)\right]\rho(|\nabla v|).}$$

Therefore the condition is that

$$\sup_{a>0}\frac{1}{a}\rho'(H\chi_{(0,a)}) < +\infty \text{ where } H(t) = Max\left(t^{\frac{1}{N}};\,(a-t)^{\frac{1}{N}}\right).$$

Recall that the set $vmo_r(\Omega)$ is the subspace of $bmo_r(\Omega)$ defined by

$$f \in vmo_r(\Omega) \text{ if and only if } f \in L^1(\Omega)$$

and

$$\lim_{R\to 0}\sup_{|B|\leqslant R} |B|^{-1}\int_B |v(x) - v_B|dx = 0.$$

Using Lemma 4.4.4, we have

Lemma 4.4.5 *If* $v \in W^{1,N}(\Omega) = W^1L^N(\Omega)$, $N \geqslant 2$, *then*

$$\lim_{R\to 0}\sup_{|B|\leqslant R} |B|^{-1}\int_B |v(x) - v_B|dx = 0.$$

Proof Since $H_*(s) = \left(a - \frac{s}{2}\right)^{\frac{1}{N}}$,
$s \in (0, a)$ if $H(t) = Max\left(t^{\frac{1}{N}};\,(a-t)^{\frac{1}{N}}\right)$, by Hölder's inequality, one has:

$$\frac{1}{a}\int_0^a H(t)|\nabla u|_{*u}(t)dt \leqslant \frac{1}{a}|H|_{L^{N'}(0,a)}|\nabla v|_{L^N(B)}$$

where $|B| = a$, $\frac{1}{a}|H|_{L^{N'}(0,a)} \leqslant c_N = (N-1)^{\frac{1}{N'}}$ and $|\nabla v|_{L^N(B)} \leqslant \left(\int_0^a |\nabla v|_*^N(t)dt\right)^{\frac{1}{N}}$, from which we obtain the result according to Lemma 4.4.4

$$\lim_{R\to 0}\sup_{|B|\leqslant R}|B|^{-1}\int_B|v(x)-v_B|dx\leqslant c_N\lim_{R\to 0}\left(\int_0^R|\nabla v(t)|_*^N dt\right)^{\frac{1}{N}}=0.$$

Here is another inclusion related to large Lebesgue–Sobolev spaces:

Theorem 4.4.7 *Let Ω be a bounded space of $\mathbb{R}^N$. Then for each $\theta>0$*

$$W_0^1L^{N),\theta}(\Omega)\hookrightarrow L_{exp}^{\frac{N-1+\theta}{N}}(\Omega).$$

Furthermore,

$$||v||_{L_{exp}^{\frac{N-1+\theta}{N}}(\Omega)}\leqslant||v||_{L^1(\Omega)}+\frac{1}{N\alpha_N^{\frac{1}{N}}}||\nabla v||_{L^{N),\theta}(\Omega)}.$$

Here, we use the following equivalence (see [68, 81])

$$||v||_{N),\theta}\approx\sup_{0<t<1}(1-\operatorname{Log}t)^{-\frac{\theta}{N}}\left(\int_t^1|v|_{**}^N(s)ds\right)^{\frac{1}{N}}\doteq||v||_{L^{N),\theta}(\Omega)}.$$

Proof Let $v\in W_0^1L^{N),\theta}(\Omega)$. Since $|v|\in W_0^1L^{N),\theta}(\Omega)$, we assume that $v\geqslant 0$ ($v\in W_0^1L^{N),\theta}(\Omega)$). We may also suppose that $|\Omega|=1$. By Sobolev's inequality we have following Theorem 4.1.2 with $\widetilde{K}(s)=K(s)$:

$$-v_{**}'(s)\leqslant\frac{s^{\frac{1}{N}-1}}{N\alpha_N^{\frac{1}{N}}}|\nabla v|_{**}(s). \tag{4.25}$$

Integrating this relation between t and 1 and multiplying by $(1-\operatorname{Log}t)^{-\frac{\beta}{N}}$ with $\beta=N-1+\theta$, we get

$$(1-\operatorname{Log}t)^{-\frac{\beta}{N}}v_{**}(t)$$

$$\leqslant(1-\operatorname{Log}t)^{-\frac{\beta}{N}}v_{**}(1)+\frac{(1-\operatorname{Log}t)^{-\frac{\beta}{N}}}{N\alpha^{\frac{1}{N}}}\int_t^1 s^{\frac{1}{N}-1}|\nabla v|_{**}(s)ds.$$

By applying Holder's inequality to the last integral and taking the supremum as a function of t

$$||v||_{L^{\frac{\beta}{N}}_{exp}(\Omega)} \leqslant ||v||_{L^1(\Omega)} + \frac{1}{N\alpha_N^{\frac{1}{N}}}||\nabla v||_{L^{N),\theta}(\Omega)}. \qquad \diamond$$

4.4.3 Case of Functions Which Vanish on a Part of the Boundary

For this class of functions, we use instead the notion of α-rearrangement seen in Sect. 3.6 of Chap. 3. We then assume that Ω is a connected Lipschitzian open set, $\Gamma_0 \subset \partial\Omega$, $H_{N-1}(\Gamma_0) > 0$, $\Gamma_1 = \partial\Omega\backslash\Gamma_0$.

Note Since we can write $Q(\Gamma_1, \Omega) = \dfrac{1}{N\sigma_N^{\frac{1}{N}}}$, we see that the case of functions with partially zero trace is the same as the case of functions with zero trace. It suffices to replace the α_N-measure of the unit ball by the measure σ_N-measure of a unit sector $\Sigma(\alpha, 1)$. We then easily deduce:

Theorem 4.4.8 (Inclusion of $W^{1,p}_{\Gamma_0}(\Omega)$)

For $1 \leqslant p \leqslant +\infty$, *let* $W^{1,p}_{\Gamma_0}(\Omega) = W^{1,p}(\Omega) \cap W^{1,1}_{\Gamma_0}(\Omega)$.

1. *If* $p > N$ *then,* $W^{1,p}_{\Gamma_0}(\Omega) \subset C^{0,\alpha}(\overline{\Omega})$, $\alpha = 1 - \dfrac{N}{p}$ *and* $\forall\, u \in W^{1,p}_{\Gamma_0}(\Omega)$

$$|u|_\infty \leqslant \frac{1}{N\sigma_N^{\frac{1}{N}}}\left[\frac{N(p-1)}{p-N}\right]^{1-\frac{1}{p}} |\Omega|^{\frac{1}{N}-\frac{1}{p}}\, |\,|\nabla u|\,|_{*u}|_{L^p(\Omega_*)}.$$

The oscillation of u *in* $B(x, r)$ *is the same as in Theorem 4.4.3.*

2. *If* $p = N$, $W^{1,N}_{\Gamma_0}(\Omega) \subset L^r(\Omega)\ \forall\, r \in [1, +\infty[$ *and*

$$|u|_r \leqslant \frac{|\Omega|^{\frac{1}{r}}}{N\sigma_N^{\frac{1}{N}}}\left(\int_0^{+\infty} \sigma^{r(1-\frac{1}{N})} exp^{-\sigma} d\sigma\right)^{\frac{1}{r}} |\,|\nabla u|\,|_{*u}|_N$$

$$W^{1,N}_{\Gamma_0}(\Omega) \hookrightarrow L^{\frac{1}{N'}}_{exp}(\Omega).$$

3. *If* $1 \leqslant p < N$, $W^{1,p}_{\Gamma_0}(\Omega) \subset L^{p^*}(\Omega)$, $\dfrac{1}{p^*} = \dfrac{1}{p} - \dfrac{1}{N}$ *we have:*

(continued)

Theorem 4.4.8 (continued)

(i) $|u|_{p^*} \leqslant \frac{B^{\frac{1}{p^*}}}{N\sigma_N^{\frac{1}{N}}} | |\nabla u| |_{*u}|_p$ *if* $1 < p < N$ *with* B *the same constant as Theorem 4.4.4 (Bliss' constant).*

(ii) $|u|_{\frac{N}{N-1}} \leqslant \dfrac{1}{N\sigma_N^{\frac{1}{N}}} |\nabla u|_1$ *if* $p = 1$.

N.B. To obtain the regularity at the boundary we will use implicitly an extension operator $P : W^{1,p}(\Omega) \to W^{1,p}(\mathbb{R}^N)$ for Ω open Lipschitzian (see Maz'ja V. [112]).

4.4.4 General Inclusion Index of Functions with Arbitrary Trace in $W^{1-\frac{1}{p},p}(\partial\Omega)$

Sobolev inclusions for spaces $W^{1,p}(\Omega)$ are analogous to those presented above when Ω is an open bounded set with Lipschitzian boundary. The estimates in this case are different, i.e. the constants are different and we no longer use the gradient but the norm of $W^{1,p}(\Omega)$.

One of the great interests of this section is the explicit calculation of inclusion constants because the section on functions with zero trace can be used to prove the same inclusions since there is an extension operator $P : W^{1,p}(\Omega) \to W_0^{1,p}(\Omega_0)$, Ω relatively compact in Ω_0.

Here is a general estimation theorem in this case:

Theorem 4.4.9 *For* $s \in \Omega_*$, *let* $I(s) = \left[\min\left(s, \frac{|\Omega|}{2}\right), \max\left(s, \frac{|\Omega|}{2}\right)\right]$.

Let V *be a subset of* $W^{1,1}(\Omega)$ *satisfying the PSR property, associated with the function* $K = K(\cdot, \Omega, V)$. *Let* ρ *be a non-trivial norm on* $L^0(\Omega_*)$. *Let* $b(s) = \rho'(\chi_{I(s)}K)$ *and*

$$V^1(\Omega, \rho) = \left\{v \in V \ s.t.\ \rho\big(|\nabla v|_{*v}\big) < +\infty\right\}.$$

Then for any definite, homogeneous and monotonous map ρ_0 *on* $L^0(\Omega_*)$, *satisfying* $\rho_0(1) < +\infty$, *if* $\rho_0(b) < +\infty$ *we have*

$$V^1(\Omega, \rho) \subset L(\Omega, \rho_0) = \left\{v \in L^0(\Omega) : \rho_0(v_*) < +\infty\right\}.$$

(continued)

Theorem 4.4.9 (continued)
In addition, we have: $\forall u \in V^1(\Omega, \rho)$

$$\rho_0\left(u_* - u_*\left(\tfrac{|\Omega|}{2}\right)\right) \leqslant \rho_0(b)\rho(|\nabla u|_{*u}).$$

Definition 4.4.1 $\rho_0(b)$ is called the inclusion index associated with $V^1(\Omega, \rho)$ and $L(\Omega, \rho_0)$.

Notes

1. If ρ_0 satisfies $\rho_0(f + \ constant) \leqslant \rho_0(f) + constant\ \rho_0(1)$, the last inequality implies:

$$\rho_0(u_*) \leqslant \rho_0(b)\rho(|\nabla u|_{*u}) + \left|u_*\left(\frac{|\Omega|}{2}\right)\right| \rho_0(1) < +\infty\,.$$

Indeed $|u_*| \leqslant \left|u_* - u_*\left(\frac{|\Omega|}{2}\right)\right| + \left|u_*\left(\frac{|\Omega|}{2}\right)\right|$, which implies that

$$\rho_0(u_*) \leqslant \rho_0\left(u_* - u_*\left(\frac{|\Omega|}{2}\right)\right) + \left|u_*\left(\frac{|\Omega|}{2}\right)\right| \rho_0(1).$$

2. If ρ is a rearrangement-invariant Fatou norm then under the conditions of the theorem we have: $W^1L(\Omega, \rho) \subset V^1(\Omega, \rho)$
because $\forall u \in W^1L(\Omega, \rho),\ \rho(|\nabla u|_{*u}) \leqslant \rho(|\nabla u|_*) < +\infty$.

Proof of Theorem 4.4.9 Let $u \in V^1(\Omega, \rho)$. Then $\forall s \in \Omega_*$,

$$\left|u_*(s) - u_*\left(\frac{|\Omega|}{2}\right)\right| \leqslant \int_{\Omega_*} \chi_{I(s)}(\sigma) K(\sigma) \left|\nabla u\right|_{*u}(\sigma) d\sigma,$$

from which we have:

$$\left|u_*(s) - u_*\left(\frac{|\Omega|}{2}\right)\right| \leqslant \rho'(\chi_{I(s)} K) \cdot \rho(|\nabla u|_{*u}).$$

If ρ_0 is homogeneous and monotonous, we deduce:

$$\rho_0\left(u_* - u_*\left(\frac{|\Omega|}{2}\right)\right) \leqslant \rho_0(b)\rho(|\nabla u|_{*u})$$

where $b(s) = \rho'(\chi_{I(s)}K)$ and ρ' is the norm associated to ρ. ◇

Corollary 4.4.3 (Of Theorem 4.4.9) *Under the conditions of the previous theorem, if ρ is a Fatou norm invariant under rearrangement and $\rho_0(f+\lambda) \leqslant \rho_0(f) + \lambda\rho_0(1)$, $\forall\lambda \in \mathbb{R}_+$, $f \in L^0(\Omega_*)$ then*

$$\rho_0(u_*) \leqslant \rho_0(b)\rho(|\nabla u|) + \frac{2}{|\Omega|}\rho_0(1)\,|u|_1 \ ,$$

$$\forall u \in W^1L(\Omega, \rho) = \Big\{v \in V : \rho(|\nabla v|) < +\infty\Big\} .$$

Proof of the Corollary Since ρ is a Fatou norm invariant under rearrangement we deduce that (see remarks above):

$$\rho_0(u_*) \leqslant \rho_0(b)\rho(|\nabla u|_*) + \left|u_*\left(\frac{|\Omega|}{2}\right)\right|\rho_0(1) \ . \tag{4.26}$$

But since u_* is decreasing we have:

$$\frac{2}{|\Omega|}\int_{\frac{|\Omega|}{2}}^{|\Omega|} u_*(\sigma)d\sigma \ \leqslant u_*\left(\frac{|\Omega|}{2}\right) \leqslant \frac{2}{|\Omega|}\int_0^{\frac{|\Omega|}{2}} u_*(\sigma)d\sigma \ ,$$

from which we get

$$\left|u_*\left(\frac{|\Omega|}{2}\right)\right| \leqslant \frac{2}{|\Omega|}\,|u|_{L^1} \ . \tag{4.27}$$

Combining the two relations (4.26) and (4.27), we deduce the inequality. ◇

Using the previous theorem and Corollary 4.4.3 we then have

Theorem 4.4.10 *Let $L(\Omega, \rho)$ and $L(\Omega, \rho_0)$ two spaces of Banach functions invariant under rearrangement. Consider,*

$$W^1L(\Omega, \rho) = \Big\{v \in W^{1,1}(\Omega_*) : \rho(\nabla v|) < +\infty\Big\},$$

(continued)

Theorem 4.4.10 (continued)
and

$$\overline{K}(t) = \max\left(t^{-\frac{1}{N'}}; \left(|\Omega| - t\right)^{-\frac{1}{N'}}\right).$$

We assume that Ω is a connected Lipschitzian bounded open set.

For $(t, s) \in \Omega_*^2$, let $b(s; t) = \chi_{I(s)}\overline{K}(t) = b(s)(t)$, $\overline{b}(s) = \rho'\big(b(s)_*\big)$). Then

$$W^1 L(\Omega, \rho) \hookrightarrow L(\Omega, \rho_0) \; if \; \rho_0(\overline{b}) < +\infty.$$

Note that for $V = W^{1,1}(\Omega)$, Ω connected, $K = Q\overline{K}$ hence Theorem 4.4.10 follows from Theorem 4.4.6 and its corollary.

We need to calculate $(\chi_{I(s)}\overline{K})_*$ for s fixed. Here is the computation of $b(s)$

Lemma 4.4.6 (Calculation of $(\chi_{I(s)}K)_*$) *Let $s \in \Omega_*$. Then for all $\sigma \in \overline{\Omega}_*$,*

$$(\chi_{I(s)}K)_*(\sigma) = Q\begin{cases}(s+\sigma)^{\frac{1}{N}-1}\chi_{[0,\frac{|\Omega|}{2}-s]}(\sigma) & if\ 0 < s \leqslant \dfrac{|\Omega|}{2},\\ (|\Omega| - s + \sigma)^{\frac{1}{N}-1}\chi_{[0,s-\frac{|\Omega|}{2}]} & if\ \dfrac{|\Omega|}{2} < s \leqslant |\Omega|.\end{cases}$$

Proof <u>If $0 < s \leqslant$</u>$\dfrac{|\Omega|}{2}$ then $\chi_{I(s)}K \doteq \dfrac{u}{Q}$ looks like

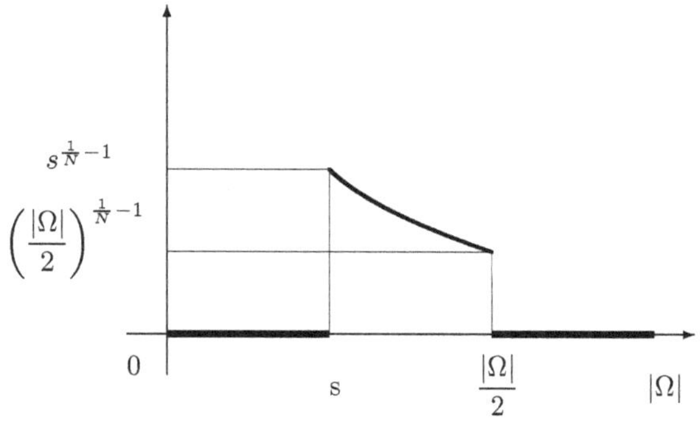

$$\text{i.e. } u(\sigma) = \begin{cases}Q\sigma^{\frac{1}{N}-1} & if\ s \leqslant \sigma \leqslant \dfrac{|\Omega|}{2}\\ 0 & elsewhere.\end{cases}$$

To obtain the decreasing rearrangement of v it is sufficient to translate by a length s to the left. This yields $(\chi_{[s,\frac{|\Omega|}{2}]}K)_*(\sigma) = Q \begin{cases} (\sigma+s)^{\frac{1}{N}-1} & if \quad 0 \leqslant \sigma \leqslant \dfrac{|\Omega|}{2} - s \\ 0 & otherwise. \end{cases}$

<u>If $\frac{|\Omega|}{2} < s \leqslant |\Omega|$</u> then $v = \chi_{[\frac{|\Omega|}{2},s]}K$ looks like:

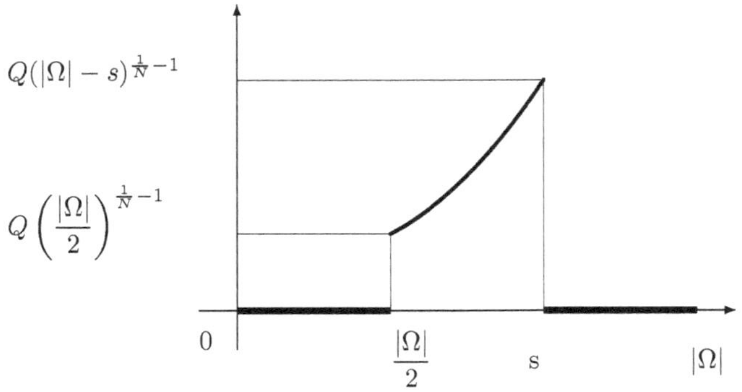

i.e. $v(\sigma) = \begin{cases} Q(|\Omega| - \sigma)^{\frac{1}{N}-1} & if \ \dfrac{|\Omega|}{2} < \sigma \leqslant s \\ 0 & otherwise. \end{cases}$

$\forall t \in \mathbb{R}_+$, we have

$$|v > t| = \begin{cases} 0 & if \ t \geqslant Q\,(|\Omega| - s)^{\frac{1}{N}-1} \\ s - \dfrac{|\Omega|}{2} & if \ t \leqslant Q\left(\dfrac{|\Omega|}{2}\right)^{\frac{1}{N}-1} \\ s - |\Omega| + \left(\dfrac{t}{Q}\right)^{-\frac{N}{N-1}} & otherwise. \end{cases}$$

Hence

$$v_*(\sigma) = \begin{cases} 0 & if \ \sigma \geqslant s - \dfrac{|\Omega|}{2} \\ Q(|\Omega| - s + \sigma)^{\frac{1}{N}-1} & if \ 0 \leqslant \sigma < s - \dfrac{|\Omega|}{2}. \end{cases}$$

◇

For a change, we will consider in this part $\rho(f) = |f|_{(p,q)}$ the norm in the Lorentz space $L^{p,q}(\Omega)$. We will assume that Ω is a connected Lipschitzian bounded open set whose relative isoperimetric constant is denoted $Q > 0$. We have:

Lemma 4.4.7 *Let* $1 < p < +\infty,\ 1 \leqslant q \leqslant +\infty$. *Then, the norm associated with* ρ *is equivalent to* $|\cdot|_{(p',q')}$, $\frac{1}{p}+\frac{1}{p'}=1,\ \frac{1}{q}+\frac{1}{q'}=1$. *Moreover, we have*

$$\rho'(g) \leqslant |g|_{(p',q')}, \qquad \forall g \in L^0(\Omega).$$

Proof We will simply show the last inequality, which is sufficient in our context:

$$\int_{\Omega_*} |fg|\,dx \leqslant \int_{\Omega_*} |f|_*(t)\,|g|_*(t)dt \leqslant \int_{\Omega_*} |f|_{**}(t)\,|g|_{**}(t)dt$$

$$= \int_{\Omega_*} \left[t^{\frac{1}{p}}\,|f|_{**}(t)\right]\left[t^{\frac{1}{p'}}\,|g|_{**}(t)\right]\frac{dt}{t} \leqslant |f|_{(p,q)}\,|g|_{(p',q')}$$

(by Hölder's inequality).

By taking the supremum over all functions $f \in L^0(\Omega_*) such that |f|_{(p,q)} \leqslant 1$, we have the inequality $\rho'(g) \leqslant |g|_{(p',q')}\ \forall g \in L^0(\Omega_*)$. ◇

Let's start with the case where $p \geqslant N,\ q = 1$.

Lemma 4.4.8 (Expression of $b(s)$) *Let* $p \geqslant N,\ q = 1$ *and let, for* $s \in \Omega_*$, $b_1(s) = \left|\chi_{I(s)}K\right|_{p',+\infty}$. *So,*

$$b_1(s) = Q\left|s - \frac{|\Omega|}{2}\right|^{\frac{1}{p'}}\left(\frac{2}{|\Omega|}\right)^{\frac{1}{N'}}, \quad \frac{1}{p}+\frac{1}{p'}=1,\ N' = \frac{N}{N-1},$$

and

$$b_1(s) \leqslant b(s) = \left|\chi_{I(s)}K\right|_{(p',+\infty)} \leqslant p b_1(s) \qquad \forall s \in \overline{\Omega}_*.$$

Proof Since we have $|v|_{p',+\infty} \leqslant |v|_{(p',+\infty)} \leqslant p\,|v|_{p',+\infty}$, let us compute $\left|\chi_{I(s)}K\right|_{p',+\infty}$. Due to the symmetry of the expressions of the rearrangement in the previous Lemma (4.4.6), it is sufficient to make the case where $0 < s \leqslant \frac{|\Omega|}{2}$,

$$b_1(s) = \left|\chi_{I(s)}K\right|_{p',+\infty} = \sup_t\left[t^{\frac{1}{p'}}\left(\chi_{I(s)}K\right)_*(t)\right].$$

Or by taking up the expression of the previous lemma

$$b_1(s) = Q \sup_{0 \leqslant t \leqslant \frac{|\Omega|}{2} - s} t^{\frac{1}{p'}} (s+t)^{-\frac{1}{N'}} = Q \left[\sup_{0 \leqslant t \leqslant \frac{|\Omega|}{2} - s} \frac{t^{\frac{N'}{p'}}}{s+t} \right]^{\frac{1}{N'}} .$$

For $p \geqslant N$, the function $t \to \dfrac{t^{\frac{N'}{p'}}}{s+t}$ is increasing, from which we get

$$b_1(s) = Q \left(\frac{2}{|\Omega|} \right)^{\frac{1}{N'}} \cdot \left(\frac{|\Omega|}{2} - s \right)^{\frac{1}{p'}}, \quad 0 < s \leqslant \frac{|\Omega|}{2} .$$

By symmetry, we have for $\dfrac{|\Omega|}{2} < s < |\Omega|$, $b_1(s) = Q \left(\frac{2}{|\Omega|} \right)^{\frac{1}{N'}} \cdot \left(s - \frac{|\Omega|}{2} \right)^{\frac{1}{p'}}$. ◇

As we stated previously, the inclusions are the same for zero trace functions. Here is a first example:

Theorem 4.4.11 (Inclusions of Lorentz–Sobolev Spaces : p ⩾ N) *Let $p \geqslant N$, $q = 1$. Let us define $W^1 L^{p,1}(\Omega) = \left\{ v \in L^1(\Omega) : |\nabla v| \in L^{p,1}(\Omega) \right\}$. Then,*

1. *$W^1 L^{p,1}(\Omega) \subset L^\infty(\Omega) \cap C^\alpha(\Omega)$, $\alpha = 1 - \dfrac{N}{p}$.*
2. *$\forall v \in W^1 L^{p,1}(\Omega)$, we have*

$$|v|_\infty \leqslant pQ \left(\frac{|\Omega|}{2} \right)^{\frac{1}{N} - \frac{1}{p}} |\nabla v|_{(p,1)} + \frac{2}{|\Omega|} |v|_1$$

and $\underset{B(x,r)}{\operatorname{osc}} v \leqslant \dfrac{\alpha_N^{1-\frac{1}{p}}}{\alpha_{N-1}} |\nabla v|_{L^{p,1}(B(r))} r^{1-\frac{N}{p}}$ *where $B(x,r) = B(r) \subset \Omega$.*

Proof Let us first note that:

$L^{p,1} \subset L^{N,1}$ and $|f|_{(N,1)} \leqslant |\Omega|^{\frac{1}{N} - \frac{1}{p}} |f|_{(p,1)}$, $\forall f \in L^{p,1}$ $p \geqslant N$. Indeed,

$$|f|_{(N,1)} = \int_{\Omega_*} t^{\frac{1}{N} - \frac{1}{p}} t^{\frac{1}{p}} |f|_{**}(t) \frac{dt}{t} \leqslant |\Omega|^{\frac{1}{N} - \frac{1}{p}} |f|_{(p,1)} .$$

Thus $W^1L^{p,1}(\Omega) \hookrightarrow W^1L^{N,1}(\Omega)$ with a continuous injection. We have seen that: $\forall\, v \in W^1L^{p,1}(\Omega)$,

$$\operatorname*{osc}_{B(r)} v \leqslant \frac{\alpha_N^{1-\frac{1}{N}}}{\alpha_{N-1}} |\nabla v|_{L^{N,1}(B(r))} \leqslant \frac{\alpha_N^{1-\frac{1}{p}} r^{1-\frac{N}{p}}}{\alpha_{N-1}} |\nabla v|_{L^{p,1}(B(r))}$$

(using the previous inequality).

To justify the estimate L^∞, we consider, $\rho_0(v) = |v|_\infty$. Then, by applying the Corollary 4.4.3 of Theorem 4.4.9, we deduce: $\forall\, v \in W^1(\Omega,\ |\cdot|_{p,1})$,

$$|v|_\infty \leqslant |b|_\infty\, |\nabla v|_{(p,1)} + \frac{2}{|\Omega|}|v|_1 \leqslant pQ\left(\frac{|\Omega|}{2}\right)^{\frac{1}{N}-\frac{1}{p}} |\nabla v|_{(p,1)} + \frac{2}{|\Omega|}|v|_1 \,.$$

◇

Note If $p > N$, for all $N < q < p$ we have: $L^p \subsetneq L^{q,1} \subsetneq L^{N,1}$. Indeed, $f \in L^p_+(\Omega)$ we have:

$$\int_{\Omega_*} t^{\frac{1}{q}} f_*(t)\frac{dt}{t} \leqslant \left(\int_{\Omega_*} t^{-\frac{p'}{q'}}\right)^{\frac{1}{p'}} |f_*|_p = c\,|\Omega|^{\frac{1}{p'}-\frac{1}{q'}}\,|f|_p < +\infty$$

(because $q' > p'$ for $p > q$).

Consequently, we have:

$$W^{1,p}(\Omega) \subset W^1L^{q,1}(\Omega) \subset W^1L^{N,1}(\Omega)\,.$$

We then deduce the estimate of the oscillation for $W^{1,p}(\Omega)$.

Corollary 4.4.4 (Of the Inclusion Theorem 4.4.11) *For all $p > N$, we have:*

$$W^{1,p}(\Omega) \hookrightarrow C^{0,\alpha}(\overline{\Omega}), \qquad \alpha = 1 - \frac{p}{N},$$

$$\operatorname*{osc}_{B(r)} v \leqslant \frac{\alpha_N^{1-\frac{1}{p}}}{\alpha_{N-1}} |\nabla v|_{L^p(B(r))} \cdot r^{1-\frac{p}{N}}\,.$$

The regularity at the boundary $\partial\Omega$ follows from the continuous extension operator $P : W^{1,p}(\Omega) \to W^{1,p}(\mathbb{R}^N)$. ◇

We will discuss the estimates of the constants for Lorentz–Sobolev inclusion spaces, starting with the case of zero-trace functions.

Theorem 4.4.12 (Inclusion in Lorentz Spaces) *If* $1 \leqslant p < N,\ 1 \leqslant q < +\infty$ *then* $W_0^1 L^{p,q} \subsetneq L^{p^*,q}(\Omega),\ \dfrac{1}{p^*} = \dfrac{1}{p} - \dfrac{1}{N}$.

Moreover, we have for $v \in W^1L(\Omega,\ |\cdot|_{p,q})$:

$$|v|_{p^*,q} \leqslant \frac{p^*}{N\alpha_N^{\frac{1}{N}}} \big\| |\nabla u|_{*u} \big|_{p,q} \leqslant \frac{p^*}{N\alpha_N^{\frac{1}{N}}} |\nabla v|_{p,q} \ \ if\ 1 \leqslant q \leqslant p.$$

and

$$|v|_{(p^*,q)} \leqslant \frac{p^*}{N\alpha_N^{\frac{1}{N}}} \Big| |\nabla u|_{*u} \Big|_{(p,q)} + (p^*)^{\frac{1}{q}} |\Omega|^{\frac{1}{p^*}-1} |v|_1 \ \ if\ p < q \leqslant +\infty.$$

Proof Let's start by recalling a Hardy lemma whose proof is identical to the one given at the beginning of the chapter: (see Exercise II.I.27 Chap. 11).

Lemma 4.4.9 (Hardy (2nd Version)) *Let* $f\ :\ [0,+\infty[\to\ [0,+\infty[$ *be measurable, and* $r < 0,\ q \geqslant 1$ *be two real numbers. Then,*

$$\int_0^{+\infty} \left(\int_x^{+\infty} f(t)dt \right)^q x^{-r-1} dx \leqslant \left(\frac{q}{|r|} \right)^q \int_0^{+\infty} f(t)^q t^{q-r-1} dt\,.$$

1st **case:** $v \in W_0^1 L^{p,q}$ **s.t.** $v = 0$ **on** $\partial\Omega$

Start with case $q = +\infty$. Let us write for $t \in]0, |\Omega|]$

$$u_{**}(t) = -\int_t^{|\Omega|} u'_{**}(\sigma)d\sigma + \frac{1}{|\Omega|}\int_\Omega u(x)dx.$$

Using Theorem 4.1.2, knowing that $\gamma_0 u = 0$ we have

$$-\,u'_{**}(\sigma) \leqslant \frac{\sigma^{\frac{1}{N}-1}}{N\alpha_N^{\frac{1}{N}}} |\nabla u|_{**}(\sigma).$$

After integration, we get

$$t^{\frac{1}{p^*}} u_{**}(t) \leqslant \frac{p^*}{N\alpha_N^{\frac{1}{N}}} \left(1 - t^{\frac{1}{p^*}} |\Omega|^{-\frac{1}{p^*}}\right) |\nabla u|_{(p,+\infty)} + \frac{t^{\frac{1}{p^*}}}{|\Omega|} |v|_1.$$

Thus

$$|v|_{p^*,+\infty)} \leqslant \frac{p^*}{N\alpha_N^{\frac{1}{N}}} |\nabla v|_{(p,+\infty)} + |\Omega|^{\frac{1}{p^*}-1} |v|_1.$$

Case where $q < +\infty$

Let us set $u = |v|$ (reasoning by density we can assume v is bounded), we have:

$$|v|^q_{p^*,q} = \int_{\Omega_*} \left[t^{\frac{1}{p^*}} u_*(t)\right]^q \frac{dt}{t} = \int_{\Omega_*} t^{\frac{q}{p^*}-1} \left(\int_t^{|\Omega|} \left|u_*'(\sigma)\right| d\sigma\right)^q dt\ .$$

If we extend u_*' by zero, we can apply the previous Hardy lemma to obtain:

$$|v|^q_{p^*,q} \leqslant (p^*)^q \int_{\Omega_*} |u_*'(t)|^q\, t^{q+\frac{q}{p^*}-1} dt = (p^*)^q \int_{\Omega_*} \left[t^{\frac{1}{p}} \left(t^{1-\frac{1}{N}} \left|u_*'(t)\right|\right)\right]^q \frac{dt}{t}.$$

By the pointwise Poincaré–Sobolev inequality:

$$t^{1-\frac{1}{N}} \left|u_*'(t)\right| \leqslant \frac{1}{N\alpha_N^{\frac{1}{N}}} |\nabla u|_{*u}(t) \quad a.e.$$

Hence

$$|v|^q_{p^*,q} \leqslant \left(\frac{1}{N\alpha_N^{\frac{1}{N}}}\right)^q (p^*)^q \int_{\Omega_*} \left[t^{\frac{1}{p}} |\nabla u|_{*u}\right]^q \frac{dt}{t}\ .$$

Let

- if $1 \leqslant q \leqslant p$ using the Polyà–Szegö inequality (Theorem 3.5.1), we have:

$$|v|_{p^*,q} \leqslant \frac{p^*}{N\alpha_N^{\frac{1}{N}}} \left||\nabla| u_{*u}\right|_{p,q} \leqslant \frac{p^*}{N\alpha_N^{\frac{1}{N}}} |\nabla v|_{p,q}\ .$$

- if $p < q < +\infty$ then by integration by parts, we have

$$\int_{\Omega_*} t^{\frac{q}{p^*}-1} u_{**}(t)^q dt = \frac{p^*}{q} |\Omega|^{\frac{q}{p^*}} u_{**}(|\Omega|)^q - p^* \int_{\Omega_*} t^{\frac{q}{p^*}} u_{**}^{q-1}(t) u_{**}'(t) dt.$$

Using Theorem 4.1.2, i.e. $-u'_{**}(t) \leqslant \dfrac{t^{\frac{1}{N}-1}}{N\alpha_N^{\frac{1}{N}}}\left(|\nabla u|_{*u}\right)_{**}$, and Hölder's inequality we deduce:

$$\int_{\Omega_*} t^{\frac{q}{p^*}-1} u_{**}(t)^q dt \leqslant \frac{p^*}{q} |\Omega|^{\frac{q}{p^*}} u_{**}(|\Omega|)^q$$

$$+ \frac{p^*}{N\alpha_N^{\frac{1}{N}}} \left(\int_{\Omega_*} t^{\frac{q}{p^*}-1} u_{**}(t)^q dt\right)^{\frac{1}{q'}} \left(\int_{\Omega_*} t^{\frac{q}{p}-1} (|\nabla u|_{*u})^q_{**}(t) dt\right)^{\frac{1}{q}}.$$

Using Young's inequality: $ab \leqslant \dfrac{1}{q}a^q + \dfrac{1}{q'}b^{q'}$, we deduce the result. ◇

The main idea is to give a method similar to the one given above, which allows us to obtain an estimate of the constants where the contribution of each parameter such as the domain Ω and the different powers $p, q, \ldots$ appear.

The estimates will be a function of the relative isoperimetric constant $Q(\Omega)$.

For this, we will use Theorem 4.1.2 with $V = W^{1,1}(\Omega)$, Ω connected, bounded and Lipschitzian. The theorem with the relative constant implies:

Corollary 4.4.5 (Of Theorem 4.1.2) *If $u \in W^{1,1}(\Omega)$ then*

$$-u'_{**}(s) \leqslant Q(\Omega) s^{\frac{1}{N}-1} |\nabla u|_{**}(s) \ a.a.\ s \in \left]0, \frac{|\Omega|}{2}\right[.$$

Thus, the idea is to reduce to half of the interval $\left]0, \dfrac{|\Omega|}{2}\right[$ and follow the previous method given in the context of a zero trace function.

Here are some illustrations of this case:

Proposition 4.4.1 *If Ω is a connected Lipschitzian bounded open set, then*

1. *If $1 \leqslant p < N$ we have*

$$W^1L^{p,q}(\Omega) \hookrightarrow L^{p^*,q}(\Omega) \text{ with } 1 \leqslant q \leqslant +\infty,\ p^* = \frac{Np}{N-p}.$$

2. *If $p = N$ we have*

$$W^1L^{N,q}(\Omega) \hookrightarrow L_{exp}^{\frac{1}{q'}}(\Omega),\ 1 < q < +\infty.$$

Proof Let $v \in W^1L^{p,q}(\Omega)$, $u = |v|$. Let us start with $1 \leqslant p < N$

Case $q = +\infty$ Since the function $t \in \Omega_* \to u_{**}(t)$ is decreasing, we deduce that

$$\sup_{\frac{|\Omega|}{2} \leqslant t \leqslant |\Omega} t^{\frac{1}{p^*}} u_{**}(t) \leqslant 2^{\frac{1}{p^*}} \sup_{0 \leqslant t \leqslant \frac{|\Omega|}{2}} t^{\frac{1}{p^*}} u_{**}(t) :$$

On $\left]0, \dfrac{|\Omega|}{2}\right[$, we write

$$u_{**}(t) = u_{**}\left(\frac{|\Omega|}{2}\right) - \int_t^{\frac{|\Omega|}{2}} u'_{**}(\sigma)d\sigma \leqslant u_{**}\left(\frac{|\Omega|}{2}\right) + Q(\Omega)\int_t^{\frac{|\Omega|}{2}} \sigma^{\frac{1}{N}-1}|\nabla u|_{**}(\sigma)d\sigma.$$

Hence

$$0 \leqslant u_{**}(t) \leqslant \frac{2}{|\Omega|}|v|_1 + Q(\Omega)\sup_{\sigma}[\sigma^{\frac{1}{p}}|\nabla u|_{**}(\sigma)] \cdot \int_t^{|\Omega|} \sigma^{\frac{1}{N}-\frac{1}{p}-1}d\sigma$$

and

$$\sup_{0 \leqslant t \leqslant \frac{|\Omega|}{2}} t^{\frac{1}{p^*}} u_{**}(t) \leqslant |\Omega|^{\frac{1}{p^*}-1} 2^{1-\frac{1}{p^*}}|v|_1 + p^* Q(\Omega)|\nabla u|_{(p,+\infty)}$$

$$|v|_{(p^*,\infty)} \leqslant 2|\Omega|^{\frac{1}{p^*}-1}|v|_1 + 2^{\frac{1}{p^*}} Q(\Omega)|\nabla u|_{(p,+\infty)}.$$

Case where $q < +\infty$ We have

$$|v|_{(p^*,q)} \leqslant \left[\int_0^{\frac{|\Omega|}{2}} t^{\frac{q}{p^*}}[u_{**}(t)]^q \frac{dt}{t}\right]^{\frac{1}{q}} + \left[\int_{\frac{|\Omega|}{2}}^{|\Omega|} t^{\frac{q}{p^*}}[u_{**}(t)]^q \frac{dt}{t}\right]^{\frac{1}{q}} = I_1 + I_2.$$

We find an upper bound of the last integral using the fact that the map $t \to u_{**}(t)$ is decreasing as

$$I_2 \leqslant |\Omega|^{\frac{1}{p^*}} u_{**}\left(\frac{|\Omega|}{2}\right) \leqslant 2|\Omega|^{\frac{1}{p^*}-1}|v|_1 \tag{4.28}$$

As for the first integral, we use the previous inequality (corollary of Theorem 4.2.1) as before, by first writing

$$I_1 \leqslant \left(\int_0^{|\Omega|} t^{\frac{q}{p^*}} \left|u_{**}(t) - u_{**}\left(\frac{|\Omega|}{2}\right)\right|^q \frac{dt}{t}\right)^{\frac{1}{q}} + u_{**}\left(\frac{|\Omega|}{2}\right)\left(\frac{p*}{q}\right)^{\frac{1}{q}}\left(\frac{|\Omega|}{2}\right)^{\frac{1}{p^*}-\frac{1}{q}}.$$

So

$$I_1 \leqslant I_3 + \left(\frac{|\Omega|}{2}\right)^{\frac{1}{p*}-\frac{1}{q}-1}\left(\frac{p*}{q}\right)^{\frac{1}{q}}|v|_1$$

and

$$I_3^q \leqslant \int_0^{\frac{|\Omega|}{2}} t^{\frac{q}{p*}}\left(\int_t^{\frac{|\Omega|}{2}}\left|u'_{**}(\sigma)\right|d\sigma\right)^q \frac{dt}{t}.$$

Using Hardy's inequality (Lemma 4.4.9) we deduce as above:

$$I_3^q \leqslant (p*)^q \int_0^{\frac{|\Omega|}{2}}\left[t^{\frac{1}{p}}(t^{1-\frac{1}{N}}|u'_{**}(t)|\right]^q\frac{dt}{t} \leqslant (p*)^q Q(\Omega)^q|\nabla u|^q_{(p,q)}.$$

Combining the three previous relations we deduce

$$|v|_{(p^*,q)} \leqslant p^* Q(\Omega)|\nabla v|_{(p,q)} + c_0|v|_1$$

with $c_0 = 2|\Omega|^{\frac{1}{p*}-1} + \left(\frac{|\Omega|}{2}\right)^{\frac{1}{p*}-\frac{1}{q}-1}\left(\frac{p*}{q}\right)^{\frac{1}{q}}$.

For $p = N,\ 1 < q < +\infty$, for $0 \leqslant s < \dfrac{|\Omega|}{2}$, we have

$$0 \leqslant u_{**}(s) \leqslant Q(\Omega)\int_s^{|\Omega|} t^{-\frac{1}{N'}}|\nabla u|_{**}(t)dt + u_{**}\left(\frac{|\Omega|}{2}\right)$$

$$u_{**}(s) \leqslant Q(\Omega)|t^{-\frac{1}{N'}}|_{L^{N',q'}(]s,|\Omega|[)}|\nabla u|_{L^{N,q}} + \frac{2}{|\Omega|}|u|_1.$$

Hence

$$|u|_{L_{exp}^{\frac{1}{q'}}(\Omega)} \leqslant Q(\Omega)|\nabla u|_{L^{N,q}} + \frac{2}{|\Omega|}|u|_{L^1}. \qquad \diamond$$

Other inclusions of the Sobolev type can be obtained by the same method. For example, for $1 < p < N$, we have

$$L^{(p}(\Omega) \hookrightarrow L^{p,1}(\Omega) \hookrightarrow L^p,$$

so that $W^1L^{(p}(\Omega) \hookrightarrow W^1L^{p,1}(\Omega) \hookrightarrow W^1L^p(\Omega) \hookrightarrow L^{p*}(\Omega)$. $\diamond$

Note on Compact Injections in Sobolev Spaces Related to the GΓ Spaces

The question of compact injections is closely linked to the inclusions that we have studied in part. For example, the injection of $W^1L^{(p}(\Omega)$ into $L^{p*}(\Omega)$ is compact for $1 < p < N$, $p* = \dfrac{Np}{N-p}$. Similarly, $W^1L^{1,1}(\Omega) = W^1L(\mathrm{Log}\, L)$ in $L^{N'}(\Omega)$ with $N' = \dfrac{N}{N-1}$,

$N \geqslant 2$ is also compact. These two results are included in the following theorem, which was elaborated in [88] and [89].

Theorem 4.4.13 (Of Compactness) *Let Ω be a bounded open Lipschitz of measure 1, $(p, m) \in]1, +\infty[^2$ and w a weight function on $]0, 1[$ s.t $d = \inf_{t\in]0,1[} \left[t^{\frac{m}{p}} w(t)\right] > 0$ and $w(t) \cdot t^{\frac{m}{p}} \in L^1(]0, 1[)$. Consider the G-gamma space : $G\Gamma(p, m; w)$ defined in Chap. 1 and*

$$W^1G\Gamma(p, m; w) = \left\{v \in W^{1,1}(\Omega) : |\nabla v| \in G\Gamma(p, m; w)\right\}.$$

Then

1. *$W^1G\Gamma(p, m; w) \hookrightarrow L^{p*}(\Omega)$, $p^* = \dfrac{Np}{N-p}$. Moreover:*
2. *the previous injection is compact if and only if $w \notin L^1(]0, 1[)$.*

For the proof we refer to the article [89], where we imposed that $m \leqslant p$ because for the case $m > p$ the results follow from the fact that

$$G\Gamma(p, m; w) \hookrightarrow L^m(\Omega) \text{ because } d > 0$$

and in this case

$$W^1G\Gamma(p, m; w) \hookrightarrow W^{1,m}(\Omega) \hookrightarrow L^{p*}(\Omega) \underset{\neq}{\hookrightarrow} L^{m*}(\Omega).$$

As $p* < m*$, in this case the injection of $W^{1,m}(\Omega)$ into L^{p*} is compact.

Note that, even in this case $(m > p)$ we still have

$$w \notin L^1(]0, 1[) \text{ because } w(t) \geqslant d \cdot t^{-\frac{m}{p}} \notin L^1(]0, 1[)$$

and so the theorem still applies.

In the case where $G\Gamma(p, m; w) = L^{(p}(\Omega)$ (the small Lebesgue space), the parameters $m = 1 < p < N$, and $w(t) = t^{-1}(1 - \log t)^{-\frac{1}{p}} \notin L^1(]0, 1[)$ and even if $G\Gamma(p, m; w) = L^{1,1}(\Omega)$ with

$$p = m = 1, \ w(t) = t^{-1} \notin L^1(]0, 1[), \ \ tw(t) = 1.$$

Hence the results stated above.

But the injection of $W^1L^{m,1}(\Omega)$ into $L^{m*}(\Omega)$ is not compact for $1 < m < N$. ◇

4.5 Interpolation Inequalities for a Normed Space

Poincaré–Sobolev (PSR) inequalities can also be used to obtain interpolation inequalities with, as a bonus, estimates of the constants.

Here are some examples:

Theorem 4.5.1 (Generalized Interpolation of Gagliardo-Nirenberg Type) *Let ρ be a non-trivial norm on $L^0(\Omega_*)$ and let $u \in V^1_{0+}(\Omega, \rho) = \Big\{v \in W^{1,1}_{0+}(\Omega),\ \rho\big(|\nabla v|_{*v}\big) < +\infty\Big\}$.*

For $q \in]0, +\infty[$, we define $|\Omega_q| = \begin{cases} |\Omega| & if\ q \geqslant 1 \\ |u > 0| & if\ 0 < q < 1 \end{cases}$.

For any homogeneous and monotone map ρ_0 on $L^0(\Omega_*)$, we then have:

$$\rho_0(u_*) \leqslant \left(\frac{q}{N\alpha^{\frac{1}{N}}}\right)^{\frac{1}{q}} \rho\left(|\nabla u|_{*u}\right)^{\frac{1}{q}} \rho_0\left[\left(\rho'\left(k_1 u_*^{q-1} \chi_{[s,|\Omega|]}\right)\right)^{\frac{1}{q}}\right]$$

with $k_1(t) = t^{\frac{1}{N}-1}$.

Corollary 4.5.1 *For simplicity, we assume that Ω is a bounded open subset of $\mathbb{R}^N$. If $N' \leqslant p < +\infty$, then $\forall\, u \in W^{1,N}_0(\Omega)$, with $a = \dfrac{N'}{p}$*

$$|u|_p \leqslant \left(\frac{1}{aN\alpha_N^{\frac{1}{N}}}\right)^a \left||\nabla u|_{*|u|}\right|_N^a |u|_{p-N'}^{1-a}$$

(continued)

Corollary 4.5.1 (continued)

$$\leqslant \left(\frac{1}{aN\alpha_N^{\frac{1}{N}}}\right)^a |\nabla u|_N^a \, |u|_{p-N'}^{1-a}.$$

Proof of Theorem 4.5.1 Let $u \in W_0^1 L(\Omega, \rho)$, according to the Poincaré–Sobolev inequality (PSR), we have:

$$-u_*'(s) \leqslant \frac{s^{\frac{1}{N}-1}}{N\alpha_N^{\frac{1}{N}}} |\nabla u|_{*u}(s) \quad a.e.$$

Since $u \geqslant 0$, we deduce after multiplication by u_*^{q-1} that

$$-\frac{1}{q}\frac{d}{ds}u_*^q(s) \leqslant \frac{1}{N\alpha_N^{\frac{1}{N}}} |\nabla u|_{*u}(s) \cdot s^{\frac{1}{N}-1} u_*^{q-1}(s), \quad q > 0.$$

By integrating between s and $|\Omega_q|$ we deduce from this inequality that $\forall s \in \overline{\Omega}_*$

$$u_*^q(s) \leqslant \frac{q}{N\alpha_N^{\frac{1}{N}}} \int_{\Omega_*} |\nabla u|_{*u}(t) t^{\frac{1}{N}-1} u_*(t)^{q-1} \chi_{[s,|\Omega_q|]}(t) dt.$$

Using ρ and ρ', we obtain, $\forall s \in \overline{\Omega}_*$

$$u_*^q(s) \leqslant \frac{q}{N\alpha_N^{\frac{1}{N}}} \rho\left(|\nabla u|_{*u}\right) \rho'\left(k_1 u_*^{q-1} \chi_{[s,|\Omega_q|]}\right)$$

where $k_1(t) = t^{\frac{1}{N}-1}$, $t \in \Omega_*$. Since ρ_0 is homogeneous and monotone on $L^0(\Omega_*)$, we then have:

$$\rho_0(u_*) \leqslant \left(\frac{q}{N\alpha_N^{\frac{1}{N}}}\right)^{\frac{1}{q}} \rho\left(|\nabla u|_{*u}\right)^{\frac{1}{q}} \rho_0\left(\left(\rho'\left(k_1 \chi_{[s,|\Omega_q|]} u_*^{q-1}\right)\right)^{\frac{1}{q}}\right). \diamond$$

Proof of Corollary 4.5.1 It suffices to consider $u \geqslant 0$. We choose $\rho_0 = |\cdot|_p$, $\rho \doteq |\cdot|_N$ and $qN' = p$. Then the inequality of the theorem implies:

$$|u|_p \leqslant \left(\frac{q}{N\alpha_N^{\frac{1}{N}}}\right)^{\frac{1}{q}} \left||\nabla u|_{*u}\right|_N^{\frac{1}{q}} \left(\int_{\Omega_*}\left(\int_s^{|\Omega|} t^{-1}u_*^{p-N'}(t)dt\right)ds\right)^{\frac{1}{p}} \tag{4.29}$$

Let $a = 1 - \frac{p-N'}{p} = \frac{N'}{p} = \frac{1}{q}$. Integrating the last integral by parts(or using Fubini's theorem):

$$\int_{\Omega_*}\left(\int_s^{|\Omega|} t^{-1}u_*^{p-N'}(t)dt\right)ds = \int_{\Omega_*} u_*^{p-N'}(t)dt. \tag{4.30}$$

Combining (4.29) and (4.30), we have:

$$|u|_p \leqslant \left(\frac{1}{aN\alpha_N^{\frac{1}{N}}}\right)^{a} \left||\nabla u|_{*u}\right|_N^{a} |u|_{p-N'}^{1-a}.$$

◇

Remarks

- The previous corollary is often used with $N = 2,\ p = 4,\ p - N' = 2$, so $a = \dfrac{1}{2}$ and

$$|u|_4 \leqslant \left(\frac{1}{\pi}\right)^{\frac{1}{4}} |\nabla u|_2^{\frac{1}{2}} |u|_2^{\frac{1}{2}}$$

- Theorem 4.5.1 can be stated for functions with non-zero trace, i.e. $W^{1,1}(\Omega)$. It suffices to apply the appropriate inequality (PSR).

  From this last interpolation, other interpolations follow. Here is an example

Corollary 4.5.2 (Of Theorem 4.5.1) *Under the same conditions as the previous theorem, if* $\max(N', p - N') \leqslant r \leqslant p < +\infty$ *then:*

$$|u|_p \leqslant c\,|\nabla u|_N^{\frac{r}{p}} |u|_r^{1-\frac{r}{p}}$$

with $c = \left(\dfrac{1}{aN\alpha_N^{\frac{1}{N}}}\right)^{a} C_r(N,\Omega)^{\frac{r}{p}-a} |\Omega|^{(1-a)(1-\frac{p-N'}{r})}$,

$$a = \frac{N'}{p},\ C_r(N,\Omega) = \frac{|\Omega|^{\frac{1}{r}}}{N\alpha_N^{\frac{1}{N}}}\left(\int_0^{+\infty} \sigma^{r(1-\frac{1}{N})}e^{-\sigma}d\sigma\right)^{\frac{1}{r}}.$$

Proof We saw in Theorem 4.4.2 that for $u \not\equiv 0$: $|u|_r \leqslant C_r(N,\Omega)\,|\nabla u|_N$ for $W^{1,N}_{0+}(\Omega)$. We deduce that:

$$|\nabla u|_N^a = |\nabla u|_N^{a-\frac{r}{p}}\,|\nabla u|_N^{\frac{r}{p}} \leqslant C_r(N,\Omega)^{\frac{r}{p}-a}\,|u|_r^{a-\frac{r}{p}}\,|\nabla u|_N^{\frac{r}{p}} \tag{4.31}$$

and by Hölder 's inequality, we know that:

$$|u|_{p-N'}^{1-a} \leqslant |\Omega|^{(1-a)(1-\frac{p-N'}{r})}\,|u|_r^{1-a}\ . \tag{4.32}$$

Combining this with the previous Theorem 4.5.1, we deduce the result. ◇

Pre-bibliographical Notes

The results from this chapter are mostly due to the author (see the references cited at the end of the book) except those of the classical Sobolev inclusions. Nevertheless, the proofs are based on the author's results.

Proposition 4.2.1, Theorem 4.4.6 and Lemma 4.4.5 come from the work of A. Cianchi and L. Pick [45]. Nevertheless, the proofs are slightly different. The results of Proposition 4.2.1 overlap partly with the results previously given in [132]

Inclusions in limit spaces such as L^{α}_{exp} or bmo_r, vmo_r have been proven in [45, 65] among other works.

On the other hand, the results related to $G\Gamma$ spaces are due to the author and his collaborators.

Chapter 5
Formalism of Estimates for Boundary Value Problems

Pointwise estimates using monotone rearrangements in partial differential equations were initiated by G. Talenti [160] (using in part a technique due to V. Maz'ja) where he compared the solution of $-\Delta u = f \in L^2_+(\Omega)$, $u \in H^1_0(\Omega)$ to the radial solution of $-\Delta U = f$, $U \in H^1_0(\underset{\sim}{\Omega})$, where $\underset{\sim}{f}$ is the spherical rearrangement function of f and $\underset{\sim}{\Omega}$ is the ball of the same measure as Ω. He then proved that $u_*(s) \leqslant U_*(s)$, $\forall s$. Since U_* is known explicitly, we deduce the regularities of u in any invariant space by rearrangement. Similar results have been proven by various authors adopting the Maz'ja–Talenti method.

In this chapter, we adopt a slightly different method, but with the same aim of finding an ordinary differential inequality for the rearrangement $|u|_*$ of the solution u and using a test function analogous to Talenti's. We will try to systematically develop a mechanism for a priori estimates by using the various properties of monotone and relative rearrangement. Roughly speaking, the idea is as follows: Consider for example a solution u, (in a set V), of a variational equation of the form:

$$\big(A(u), \varphi\big) = \big\langle T, \varphi\big\rangle, \quad \forall \varphi \in W \ (V \subset W) \, and \, T \in W' \text{ (dual space of } W).$$

For each $s \in \Omega_*$, we construct a test function $\varphi(s)$ so as to transform the previous equation into a pointwise inequality in s. In general, since we must estimate $v = |u|$, this inequality can be written as:

$$\Big[a\,(\nabla v)\Big]_{*v}(s) \leqslant \Phi\big(G(T)_{*v}\big),$$

where $G(T)$ is a function of the data T only and $\Phi : \mathbb{R} \to \mathbb{R}$ is well determined in general. With the assumptions of growth on the operator, if V satisfies the PSR

J. M. Rakotoson, *Relative Rearrangement*, Lecture Notes in Mathematics 2376,
https://doi.org/10.1007/978-3-032-02228-8_5

property then the inequality leads to the differential inequality:

$$-v_*'(s) \leqslant K(s, \Omega, V)\Phi\big(G(T)_{*v}\big)(s)\,. \tag{5.1}$$

From here, an analysis similar to that of Chap. 4 leads to estimates of v in various normed spaces $L(\Omega_*, \rho)$. For example, if ρ and ρ_0 are two distinct norms s.t. $\rho\Big(\Phi\big(G(T)_{*v}\big)\Big) < +\infty$ and if the injection index $\rho_0\big(\rho'(\chi_{I(s)}K)\big)$ (where ρ' is the associate norm of ρ) is finite then $v_* \in L(\Omega_*, \rho_0)$ i.e. $\rho_0(v_*) < +\infty$.

A recent result by A. Cianchi and V. Maz'ja [44] shows us that we can use rearrangement techniques to obtain an estimate on the gradient of the solution for some equations, even nonlinear ones.

By examining their method closely, we can adapt it here based on two fact:

1. We can have a pointwise inequality analogous to that of the relation (5.1) for the relative rearrangement of the gradient of $v = |u|$, u solution of the variational equation , of the preceding form.

$$|\nabla v|_{*v}(s) \leqslant \Big[\big(|\nabla v|^p\big)_{*v}\Big]^{\frac{1}{p}} \leqslant \Phi_0\Big(G(t)_{*v}(s)\Big). \tag{5.2}$$

2. We use the inequality $\forall\,\varphi \geqslant 0, \quad \varphi \in L^\infty(\Omega_*)$

$$\int_{\Omega_*} \varphi^*(t)|\nabla v|_*(t)dt \leqslant \int_{\Omega*} \varphi(t)\Big[|\nabla v|_{*v}(t)\Big]dt \leqslant \int_{\Omega_*} \varphi\Big[|\nabla v|^p)_{*v}\Big]^{\frac{1}{p}} dt \tag{5.3}$$

where $\varphi^*(t) = \varphi_*(|\Omega| - t)$ is the monotonically increasing rearrangement of φ.

The inequality given in statement (2), see Theorem 12 of [128], will be taken up again in Chap. 7 (Theorem 5.2.3), (see also [15]).

To illustrate this, we will show that if $u \in W_0^{1,p}(\Omega)$ solution of the following quasilinear equation

$$-\operatorname{div}\Big(|\nabla u|^{p-2}\nabla u\Big) = f \in L^1(\Omega), \quad \text{where } 2 \leqslant p < N$$

then it necessarily satisfies

$$|\nabla u|_*(s) \sup_{\varphi \in L_+^\infty(\Omega_*)} \left[\frac{\displaystyle\int_0^s (s-t)\varphi(t)dt}{\displaystyle\int_0^s \varphi(t)t^{1-\frac{p'}{N'}}dt}\right]^{\frac{1}{p}} \leqslant \left(\frac{N'}{p'-N'}\right)^{\frac{1}{p}} c_N^{\frac{1}{p-1}} |f|_{L^1}^{\frac{1}{p-1}}$$

with $c_N = \dfrac{1}{N\alpha_N^{\frac{1}{N}}}$.

We will give a lower bound for the previous supremum.

The basic pointwise inequality to obtain this last inequality is

$$\left(|\nabla v|^p\right)_{*v}(s) \leqslant \left(\frac{1}{N\alpha_N^{\frac{1}{N}}}\right)^{p'} s^{-\frac{p'}{N'}} \left(\int_0^s |f|_*(t)dt\right)^{p'}, \ s \in \Omega_*.$$

We will apply Theorem 5.2.3 of Chap. 7. The method then leads to $\forall \varphi \geqslant 0, \quad \varphi \in L^\infty(\Omega_*)$.

$$\int_{\Omega_*} \varphi^*(t)|\nabla v|_*^p(t)dt \leqslant \int_{\Omega_*} \varphi\left(|\nabla v|^p\right)_{*v} dt \leqslant C_N^{p'} \left(\int_{\Omega_*} \varphi(t)t^{-\frac{p'}{N'}}\right) |f|_{L^1}^{p'}.$$

5.1 Some Preliminary Lemmas

We will illustrate this approach with a few examples, varying the boundary conditions, and we will finish with some direct consequences .

Let us start with some general lemmas:

Lemma 5.1.1 (Hölder Inequality for the Relative Rearrangement) *If* $1 < p < +\infty$, $p' = \dfrac{p}{p-1}$, *if* $F_1 \in L^p(\Omega)^N$ *and* $F_2 \in L^{p'}(\Omega)^N$ *then* $\forall u \in L^1(\Omega)$, *we have* $s \in \Omega_*$

$$\left(F_1 \cdot F_2 \chi_{\Omega\backslash P(u)}\right)_{*u}(s) \leqslant \left[|F_1|^p \chi_{\Omega\backslash P(u)}\right]_{*u}^{\frac{1}{p}}(s) \left[|F_2|^{p'} \chi_{\Omega\backslash P(u)}\right]_{*u}^{\frac{1}{p'}},$$

where $\chi_{\Omega\backslash P(u)}$ *denotes the characteristic function of* $\Omega\backslash P(u)$, $P(u)$ *is the set of plateaus of* u, *and* $F_1 \cdot F_2$ *is the scalar product in* $\mathbb{R}^N$, $|F_i|$ *is the Euclidean norm of* F_i, $i = 1, 2$.

Proof By definition of relative rearrangement, the function: $s \to \displaystyle\int_{\{u>u_*(s)\}} F_1 \cdot F_2 \chi_{\Omega\backslash P(u)} dx$

is in $W^{1,1}(\Omega_*)$, so it is almost everywhere differentiable. Thus

$$\left(F_1 \cdot F_2 \chi_{\Omega\backslash P(u)}\right)_{*u}(s) = \lim_{h \searrow 0} \frac{1}{h} \int_{\{u_*(s+h)<u<u_*(s)\}} F_1 \cdot F_2 \chi_{\Omega\backslash P(u)} dx \ .$$

Applying Hölder's inequality to this last integral and making h tend to zero we deduce the lemma. ◇

Lemma 5.1.2 *Let $1 \leqslant p \leqslant +\infty$, and let $v \in L^p(\Omega)$ s.t. $v_* \in W^{1,1}_{loc}(\Omega_*)$. Then $\forall\, g \in L^{p'}(\Omega)$ the function $G(s) = \int_\Omega g\big(v - v_*(s)\big)_+ dx$ is in $W^{1,1}_{loc}(\Omega_*)$. Moreover, we have for almost all s*

$$G'(s) = -\frac{dv_*}{ds}(s) \int_{\{v>v_*(s)\}} g(x)dx = -\frac{dv_*}{ds}(s) \int_0^s g_{*v}(\sigma)d\sigma.$$

Proof The function $H : \mathbb{R} \to \mathbb{R}$ defined by $H(t) = \int_\Omega g(v-t)_+ dx$ is locally Lipschitzian because, $\forall\, t_1,\ t_2$

$$|H(t_1) - H(t_2)| \leqslant \left(\int_\Omega |g|\, dx\right) |t_1 - t_2|\,,\ |H(t)| \leqslant \int_\Omega |gv| + |t| \int_\Omega |g|\ .$$

Consequently, the composite function $G(s) = H\big(v_*(s)\big)$ is in $W^{1,1}_{loc}(\Omega_*)$ as soon as $v_* \in W^{1,1}_{loc}(\Omega_*)$ and obeys the chain rule for deriving a compound function, i.e. $G'(s) = H'\big(v_*(s)\big)\dfrac{dv_*}{ds}(s)$ a.e. in s. A direct calculation of $H'(t) = \lim\limits_{h\searrow 0} \dfrac{H(t+h) - H(t)}{h}$ leads to the expression $H'(t) = -\int_{\{v>t\}} g dx\,.$

5.2 Point Estimates for Quasilinear Equations

In this section, we will consider, for $p \in]1, +\infty[$, an operator commonly called the Leray–Lions operator

$$\widehat{a} : \Omega \times \mathbb{R} \times \mathbb{R}^N \to \mathbb{R}^N \text{ satisfying:}$$

LL1 $\widehat{a}$ is Carathéodory i.e. for almost all x, $(\sigma, \xi) \in \mathbb{R} \times \mathbb{R}^N \to \widehat{a}(x, \sigma, \xi) \in \mathbb{R}^N$ is continuous
and $\forall\, (\sigma, \xi) \in \mathbb{R} \times \mathbb{R}^N$, $(x) \to \widehat{a}(x, \sigma, \xi) \in \mathbb{R}^N$ is measurable.

LL2 (Coercivity), there exists a real $\alpha > 0$ s.t. $a.e.\ x \in \Omega$, $\forall (\sigma, \xi) \in \mathbb{R} \times \mathbb{R}^N$

$$\widehat{a}(x, \sigma, \xi) \cdot \xi \geqslant \alpha\, |\xi|^p \ .$$

LL3 There is a constant $c > 0$ s.t. a.e. $x \in \Omega$, $\forall (\sigma, \xi) \in \mathbb{R} \times \mathbb{R}^N$ we have:

$$|\widehat{a}(x, \sigma, \xi)| \leqslant c\Big(|\xi|^{p-1} + |\sigma|^{p-1} + a_0(x)\Big) \text{ where } a_0 \in L^{p'}(\Omega) \ .$$

Remarks

1. Assumptions (LL1) and (LL3) will not be directly useful. They essentially ensure that the operations we will perform will make sense. For example, that $\forall u \in W^{1,p}(\Omega)$

$$\int_\Omega \widehat{a}\Big(x, u(x), \nabla u(x)\Big) \cdot \nabla\varphi(x)dx \text{ is finite, } \forall \varphi \in W^{1,p}(\Omega) \ .$$

2. We will assume the existence of solutions throughout the chapter.

For $T \in W^{-1,p'}(\Omega) = \Big(W_0^{1,p}(\Omega)\Big)'$, we will denote by $|T|_{-1,p'}$ its (dual) norm and we will use the decomposition $T = -\sum_{i=1}^{N} \frac{\partial f_i}{\partial x_i}$ with $f_i \in L^{p'}(\Omega)$,

$$F(x)=\Big(f_1(x), \dots, f_n(x)\Big),\ |F(x)|=\left[\sum_{i=1}^{N} f_i(x)^2\right]^{\frac{1}{2}} \ with\ |F|_{L^{p'}(\Omega)}=|T|_{-1,p'} \ .$$

5.2.1 *Case of Dirichlet Problems: Basic Pointwise Inequalities*

Case Where $div(F) = T \in W^{-1,p'}(\Omega)$

Lemma 5.2.1 (Pointwise Equation on Ω_*) *Under the previous conditions, for any solution $u \in W_0^{1,p}(\Omega)$ of $-\,div\Big(\widehat{a}(x, u, \nabla u)\Big) = T$ in $\mathcal{D}'(\Omega)$, we have*

$$\Big[\widehat{a}(\nabla u) \cdot \nabla u\Big]_{*v}(s) = \Big[F \cdot \nabla u\Big]_{*v}(s)\ a.e.\ s \in \Omega_*$$

where we set $\widehat{a}(\nabla u)(x) = \widehat{a}\Big(x, u(x), \nabla u(x)\Big)$, $v = |u|$.

Proof Let $u \in W_0^{1,p}(\Omega)$ be a solution of $-div\Big(\widehat{a}(x,u,\nabla u)\Big) = T$, which is equivalent to $\forall \varphi \in W_0^{1,p}(\Omega)$,

$$\int_\Omega \widehat{a}(\nabla u)\cdot \nabla\varphi dx = \int_\Omega F\cdot\nabla\varphi dx\ .$$

For $s \in \Omega_*$, consider the locally Lipschitzian function $\Phi_s : \mathbb{R} \to \mathbb{R}$

$$\text{defined , for } \sigma \in \mathbb{R}, by \Phi_s(\sigma) = \Big(|\sigma| - v_*(s)\Big)_+ \text{sign}\,(\sigma)\ .$$

This function satisfies $\Phi_s'(\sigma) = \begin{cases} 1 & if\ |\sigma| > v_*(s), \\ 0 & else \end{cases}.$

By choosing $\varphi = \Phi_s(u)$ we have $\forall\, s \in \Omega_*$

$$\int\limits_{\{v>v_*(s)\}} \widehat{a}(\nabla u)\cdot\nabla u\,dx = \int\limits_{\{v>v_*(s)\}} F\cdot\nabla u\,dx$$

which is equivalent to

$$\Big[\widehat{a}(\nabla u)\cdot\nabla u\Big]_{*v} = \Big[F\cdot\nabla u\Big]_{*v}. \tag{5.4}$$

◇

We will need the following lemma:

Lemma 5.2.2 *Let $G = |F|^{p'}$, $p < m < +\infty$. Then, we have $\forall\, v \in L^1(\Omega)$*

1. $(G_{*v})_* \leqslant G_{**}$
2. If $F \in L^{\frac{m}{p-1}}(\Omega)$ then

$$(G_{*v})^{\frac{1}{p}} \in L^m(\Omega_*) \quad and \quad |(G_{*v})^{\frac{1}{p}}|_{L^m} \leqslant \left(\frac{m}{m-p}\right)^{\frac{1}{p}} |F|^{\frac{1}{p-1}}_{L^{\frac{m}{p-1}}}.$$

Proof

1. From Corollary 2.2.1. we know that $\forall\sigma \in \Omega_*$,

$$(G_{*v})_*(\sigma) \leqslant (G_{*v})_{**}(\sigma) \leqslant G_{**}.$$

2. We deduce from (1) and the equimeasurability that

$$|(G_{*v})^{\frac{1}{p}}|_{L^m} \leqslant |G_{**}|^{\frac{1}{p}}_{L^{\frac{m}{p}}} \leqslant \left(\frac{m}{m-p}\right)^{\frac{1}{p}} |F|^{\frac{1}{p-1}}_{L^{\frac{m}{p-1}}}.$$

Theorem 5.2.1 *Under the same assumptions as in Lemma 5.2.1, we have, for any solution $u \in W_0^{1,p}(\Omega)$ of $-div\Big(\widehat{a}(x,u,\nabla u)\Big) = T$, a.e. in s*

1. $|\nabla v|_{*v}(s) \leqslant \Big[\big(|\nabla v|^p\big)_{*v}(s)\Big]^{\frac{1}{p}} \leqslant \dfrac{1}{\alpha^{\frac{1}{p-1}}}\Big[|F|^{p'}\Big]^{\frac{1}{p}}_{*v}(s),$
2. $-v_*'(s) \leqslant \dfrac{s^{\frac{1}{N}-1}}{N\alpha_N^{\frac{1}{N}}\alpha^{\frac{1}{p-1}}}\Big[|F|^{p'}\Big]^{\frac{1}{p}}_{*v}(s)$, *where* $v=|u|$, $T=-div(F)$.

Proof Recall that the map $v \to v_{*u}$ is increasing, i.e. if $v_1 \leqslant v_2$ a.e. then $v_{1*u} \leqslant v_{2*u}$ a.e. in s and $(\alpha v)_{*u} = \alpha v_{*u}$ if $\alpha > 0$.

Since $\Big(\widehat{a}(\nabla u)\cdot\nabla u\Big)(x) \geqslant \alpha\,|\nabla v|^p\,(x)$ a.e. in Ω, Lemma 5.2.1 implies, for almost all s:

$$\alpha\Big(|\nabla v|^p\Big)_{*v}(s) \leqslant \Big[F\cdot\nabla u\Big]_{*v}(s) \leqslant \Big[|F|^{p'}\Big]^{\frac{1}{p'}}_{*v}(s)\Big[|\nabla v|^p\Big]^{\frac{1}{p}}_{*v}(s). \tag{5.5}$$

Indeed, it suffices to apply Lemma 5.1.1 (Hölder inequality) and the fact that $|\nabla u| = |\nabla v|,\quad \nabla u(x) = 0$ a.e. on $P(u)$.

By simplifying the inequality (5.5), we deduce:

$$\Big[|\nabla v|^p\Big]^{\frac{1}{p}}_{*v}(s) \leqslant \frac{1}{\alpha^{\frac{p'}{p}}}\Big[|F|^{p'}\Big]^{\frac{1}{p}}_{*v}(s), \quad a.e.\ s. \tag{5.6}$$

Hölder's inequality for relative rearrangement leads to:

$$|\nabla v|_{*v}(s) \leqslant \Big[|\nabla v|^p\Big]^{\frac{1}{p}}_{*v}(s) \quad a.e.\ s. \tag{5.7}$$

Relations (5.6) and (5.7) give statement 1 of the theorem. As for the second relation, since $v \in W_0^{1,p}(\Omega)$ $v \geqslant 0$, we have according to the PSR property associated with this class of functions (see Theorem 4.1.1):

$$-v_*'(s) \leqslant \frac{s^{\frac{1}{N}-1}}{N\alpha_N^{\frac{1}{N}}}|\nabla v|_{*v}(s). \tag{5.8}$$

We combine this last relation with statement 1 to obtain statement 2. ◇

Case Where $\operatorname{div}(F) = T \in L^1(\Omega) \cap W^{-1,p'}$
In this case, the previous equality (see relation (5.4)) becomes

$$\left[\widehat{a}(\nabla u) \cdot \nabla u\right]_{*v}(s) = \frac{d}{ds}\int_\Omega \Phi_s(v(x))T(x)dx. \tag{5.9}$$

Using Lemma 5.1.2, we arrive at

$$\left[\widehat{a}(\nabla u) \cdot \nabla u\right]_{*v}(s) = -v_*'(s)\int_{v>v_*(s)} \operatorname{sign}(u)T(x)dx. \tag{5.10}$$

By the Hardy–Littlewood inequality, this relation (5.10) becomes

$$\left[\widehat{a}(\nabla u) \cdot \nabla u\right]_{*v}(s) \leqslant -v_*'(s)\int_0^s |T|_*(\sigma)d\sigma. \tag{5.11}$$

Hence, combining this with the (PSR) inequality, we obtain the following theorem.

Theorem 5.2.2 *Let $u \in W_0^{1,p}(\Omega)$ be the solution of*

$$-\operatorname{div}\left(\hat{a}(x, u, \nabla u)\right) = T \in L^1(\Omega) \cap W^{-1,p'}(\Omega).$$

Let $c_{N\alpha} = \dfrac{1}{\alpha N \alpha_N^{\frac{1}{N}}}$. *Then for almost all* $s \in \Omega_*$

1. $|\nabla v|_{*v}(s) \leqslant c_{N\alpha}^{\frac{1}{p-1}} s^{-\frac{N-1}{N(p-1)}} \left(\int_0^s |T|_*(s)d\sigma\right)^{\frac{1}{p-1}}$.
2. $\left(|\nabla v|^p\right)_{*v}(s) \leqslant c_{N\alpha}^{p'} s^{-\frac{p(N-1)}{N(p-1)}} \left(\int_0^s |T|_*(\sigma)d\sigma\right)^{p'}$ *where* $v=|u|$.

Note Statement 1. allows us to give the estimates in any Banach functions spaces of $L(\Omega, \rho)$, while statement 2. allows us to give an estimate of the gradient. The growth in s in the neighborhood of zero plays an important role.

5.2.2 *Applications of the Previous Pointwise Inequalities to Estimates of the Solution u in a Banach Function Space*

Corollary 5.2.1 (Of Theorem 5.2.1) *Let ρ be a norm on $L^0(\Omega_*)$. Assume that $\rho\Big(\big(|F|^{p'}\big)_{*v}^{\frac{1}{p}}\Big) < +\infty$ and if $k(\sigma) = \sigma^{\frac{1}{N}-1}$ we define $\overline{b}(s) = \rho'(k\chi_{[s,|\Omega|]})$.*

Let ρ_0 be a homogeneous and monotone map. Assume that the injection index $\rho_0(\overline{b}) < +\infty$. Then, we have:

$$\rho_0(v_*) \leqslant \frac{\rho_0(\overline{b})}{N\alpha_N^{\frac{1}{N}}\alpha^{\frac{1}{p-1}}}\rho\Big(\big(|F|^{p'}\big)_{*v}^{\frac{1}{p}}\Big) < +\infty\,.$$

Proof Since $v \in W_0^{1,p}(\Omega),\ v \geqslant 0$ we have $v_*(|\Omega|) = 0$ and $v_* \in W^{1,p}\big(\varepsilon, |\Omega|\big)$, $\forall\, 0 < \varepsilon < |\Omega|$. Thus $\forall\, s \in \Omega_*$, applying Theorem 5.2.1:

$$v_*(s) = -\int_s^{|\Omega|} v_*'(\sigma)d\sigma \;\leqslant \int_{\Omega_*} \frac{k(\sigma)\Big[\,|F|^{p'}\Big]_{*v}^{\frac{1}{p}}(\sigma)}{N\alpha_N^{\frac{1}{N}}\alpha^{\frac{1}{p-1}}}\chi_{[s,|\Omega|]}(\sigma)d\sigma\,.$$

Using successively ρ and ρ', then ρ_0, we deduce from this last inequality:

$$\rho_0(v_*) \leqslant \frac{\rho_0\Big(\rho'\big(k\chi_{[s,|\Omega|]}\big)\Big)}{N\alpha_N^{\frac{1}{N}}\alpha^{\frac{1}{p-1}}}\cdot\rho\Big(\Big[\,|F|^{p'}\Big]_{*v}^{\frac{1}{p}}\Big) < +\infty\,.$$

◇

From this result, we find theL^∞-regularity when $F \in L^r(\Omega)^N$, and $r > \frac{N}{p-1}$.

Proposition 5.2.1 *Let $T \in W^{-1,r}(\Omega)$ with $r > \max\left(\dfrac{N}{p-1}, \dfrac{p}{p-1}\right)$.Then any solution $u \in W_0^{1,p}(\Omega)$ of $-\,div\Big(\widehat{a}(x,u,\nabla u)\Big) = T$ in $\mathcal{D}'(\Omega)$ is bounded and*

$$|u|_\infty \leqslant \frac{\overline{b}(0)}{N\alpha_N^{\frac{1}{N}}\alpha^{\frac{1}{p-1}}}\,|F|_{L^r(\Omega)}^{\frac{1}{p-1}}\,.$$

Proof We choose $\rho_0(\cdot) = |\cdot|_\infty$, $\rho(\cdot) = |\cdot|_{(p-1)r}$ then

$$\rho'\left(k\chi_{[s,|\Omega|]}\right) = \left[\int_s^{|\Omega|} \sigma^{(\frac{1}{N}-1)(\frac{(p-1)r}{(p-1)r-1})}\right]^{1-\frac{1}{(p-1)r}} = \overline{b}(s)\,.$$

Since $r > \dfrac{N}{p-1}$ we have $\overline{b}(0) < +\infty$, so $\left|\overline{b}\right|_\infty = \overline{b}(0) < +\infty$. Hence, by applying the previous Corollary 5.2.1,

$$|v|_\infty \leqslant \frac{\overline{b}(0)}{N\alpha_N^{\frac{1}{N}}\alpha^{\frac{1}{p-1}}}\left|\left(|F|^{p'}\right)_{*v}^{\frac{1}{p'}}\right|_{L^{(p-1)r}} \leqslant \frac{\overline{b}(0)}{N\alpha_N^{\frac{1}{N}}\alpha^{\frac{1}{p-1}}}|F|_{L^r(\Omega)}^{\frac{1}{p-1}}. \quad \diamond$$

To obtain a better result analogous to that of this property, we can introduce the following set:

$$W_{-1}^{\frac{N}{p},\frac{1}{p}}(\Omega) = \left\{T \in \mathcal{D}'(\Omega) : T = -div\, F,\ |F|^{p'} \in L^{\frac{N}{p},\frac{1}{p}}(\Omega)\right\}.$$

Proposition 5.2.2

(i) $W_{-1}^{\frac{N}{p},\frac{1}{p}}(\Omega) \subset W^{-1,p'}(\Omega)$, $W^{-1,r}(\Omega) \subset W_{-1}^{\frac{N}{p},\frac{1}{p}}(\Omega)$, $r > \dfrac{N}{p-1}$.

(ii) If $T \in W_{-1}^{\frac{N}{p},\frac{1}{p}}(\Omega)$, *then* $u \in L^\infty(\Omega)$ *and* $|u|_\infty \leqslant \dfrac{1}{N\alpha_N^{\frac{1}{N}}\alpha^{\frac{1}{p-1}}}\left||F|^{p'}\right|_{\left(\frac{N}{p},\frac{1}{p}\right)}^{p}$.

Proof of Proposition 5.2.2

(i) To show that $W_{-1}^{\frac{N}{p},\frac{1}{p}}(\Omega) \subset W^{-1,p'}(\Omega)$, consider $T = -div(F)$, $|F|^{p'} \in L^{\frac{N}{p},\frac{1}{p}}(\Omega)$. As, $\forall s \in \Omega_*$,

$$\frac{1}{|\Omega|}\int_\Omega |F|^{p'}dx \leqslant \frac{1}{s}\int_0^s |F|_*^{p'}(\sigma)d\sigma = \left(|F|^{p'}\right)_{**}(s)\,,$$

we obtain

$$\left(\frac{1}{|\Omega|}\int_\Omega |F|^{p'}dx\right)^{\frac{1}{p}}\int_{\Omega_*} s^{\frac{1}{N}-1}ds \leqslant \int_{\Omega_*} s^{\frac{1}{N}}\left(|F|^{p'}\right)_{**}^{\frac{1}{p}}(s)\frac{ds}{s} < +\infty\,,$$

which results in

$$|F|^{p'}_{L^{p'}} \leqslant c \left||F|^{p'}\right|_{(\frac{N}{p},\frac{1}{p})} .$$

To prove $W^{-1,r}(\Omega) \subset W_{-1}^{\frac{N}{p},\frac{1}{p}}(\Omega)$, $r > \max\left(\frac{N}{p-1}, \frac{p}{p-1}\right)$ we apply Hölder's inequality as before

$$\int_{\Omega_*} s^{\frac{1}{N}-1} \left(|F|^{p'}\right)^{\frac{1}{p}}_{**} (s)ds \leqslant \gamma \left|\left(|F|^{p'}\right)_{**}\right|^{\frac{1}{p}}_{L^{\frac{r}{p'}}} .$$

By Hardy's inequality, (see Lemma 1.4.2) we then deduce:

$$\int_{\Omega} s^{\frac{1}{N}-1} \left(|F|^{p'}\right)^{\frac{1}{p}}_{**} (s)ds \leqslant c\, |F|^{\frac{1}{p-1}}_{L^r} < +\infty \ if\ T = -div(F) \in W^{-1,r}(\Omega).$$

(ii) Let $G = |F|^{p'}$, $c_N = \frac{1}{N\alpha_N^{\frac{1}{N}} \alpha^{\frac{1}{p-1}}}$. Then, by the Hardy–Littlewood inequality, we have:

$$|u|_\infty \leqslant c_N \int_0^{|\Omega|} \sigma^{\frac{1}{N}-1}\left(G_{*v}\right)^{\frac{1}{p}}(\sigma)d\sigma \leqslant c_N \int_{\Omega_*} \sigma^{\frac{1}{N}-1}\left(G_{*v}\right)^{\frac{1}{p}}_{*}(\sigma)d\sigma .$$

But at any point $\sigma \in \Omega_*$, we have (cf. Corollary 2.2.1 or Lemma 5.2.2):

$$\left(G_{*v}\right)_*(\sigma) \leqslant \left(G_{*v}\right)_{**}(\sigma) \leqslant G_{**}(\sigma) .$$

So,

$$|u|_\infty \leqslant c_N \int_{\Omega_*} \sigma^{\frac{1}{N}-1} G_{**}(\sigma)^{\frac{1}{p}} d\sigma = c_N \left||F|^{p'}\right|^{\frac{1}{p}}_{(\frac{N}{p},\frac{1}{p})} .$$

◇

We can obtain the regularity of a radial solution decreasing along the radius.

Theorem 5.2.3 (Regularity of the Spherical Rearrangement) *Let $\underset{\sim}{v}$ be the spherical rearrangement of $v = |u|$. Then, $\forall\, \sigma \in \Omega_*$*

$$|\nabla \underset{\sim}{v}|_*(\sigma) \leqslant \frac{1}{\alpha^{\frac{1}{p-1}}} G_{**}(\sigma)^{\frac{1}{p}} \text{ where } G = |F|^{p'} .$$

Proof According to Theorem 5.2.1, we have a.e. in $s \in \Omega_*$,

$$- N\alpha_N^{\frac{1}{N}} s^{1-\frac{1}{N}} v_*'(s) \leqslant \frac{1}{\alpha^{\frac{1}{p-1}}} \left(G_{*v}\right)^{\frac{1}{p}}(s) .$$

Hence, $\forall\, \sigma \in \Omega_*$ (via Lemma 3.5.1 and Corollary 2.2.1),

$$|\nabla \underset{\sim}{v}|_*(\sigma) = \left(- N\alpha_N^{\frac{1}{N}} s^{1-\frac{1}{N}} v_*' \right)_*(\sigma) \leqslant \frac{1}{\alpha^{\frac{1}{p-1}}} \left(G_{*v}\right)_*^{\frac{1}{p}}(\sigma) \leqslant \frac{1}{\alpha^{\frac{1}{p-1}}} G_{**}(\sigma)^{\frac{1}{p}} .$$

◇

Corollary 5.2.2 (Of Theorem 5.2.3) *If* $1 \leqslant r \leqslant +\infty$ *and* $|F|^{p'} \in L^{\frac{r}{p},+\infty}(\Omega)$ *then* $|\nabla \underset{\sim}{v}| \in L^{r,+\infty}(\Omega)$.

Proof It follows immediately from Theorem 5.2.3. ◇

Corollary 5.2.3 (Of Theorem 5.2.3) *We assume that* $T \in L^r(\Omega)$, $r > N$. *Then* $\underset{\sim}{v}$ *is Lipschitzian (in particular* $\nabla \underset{\sim}{v} \in L^\infty(\Omega)^N$*).*

Proof Let $w \in H_0^1(\Omega) : -\Delta w = T$. Then, $w \in W^{2,r}(\Omega)$. Thus $|\nabla w| \in L^\infty(\Omega)$. Since $T = -div(F)$, $F = \nabla w$ we deduce $\left|\nabla \underset{\sim}{v}\right|_*(\sigma) \leqslant \frac{|F|_\infty^{\frac{1}{p-1}}}{\alpha^{\frac{1}{p-1}}}$ $\forall\, \sigma \in \overline{\Omega}_*$. Hence the result, via Proposition 5.2.1, because

$$L^r(\Omega) \subset W^{-1,s} \text{ for some } s > \max\left(\frac{N}{p-1}, \frac{p}{p-1}\right) \;\diamond.$$

We can show the existence of PDE solutions that are radial and decreasing along the radius. Here is an example,

Proposition 5.2.3 *Let* $\Omega = B(R)$ *be the ball of radius* $R > 0$ *centered at the origin. If* $f \in L^\infty(\Omega)$ *is such that* $f(x) = \underset{\sim}{f}(x) \geqslant 0$, *then there exists one and only one solution* $u \in W^{1,\infty}(\Omega) \cap W_0^{1,p}(\Omega)$ *satisfying*

$$- \Delta_p u = -div\left(|\nabla u|^{p-2}\nabla u\right) = f \text{ in } \Omega .$$

Proof We know that there exists $u \in W_0^{1,p}(\Omega)$ satisfying:

$$J(u)=\frac{1}{p}\int_\Omega |\nabla u|^p\,dx-\int_\Omega fu=\inf\left\{\frac{1}{p}\int_\Omega |\nabla v|^p\,dx-\int_\Omega fv,\ v\in W_0^{1,p}(\Omega)\right\}.$$

Furthermore, $u = u_\sim \geqslant 0$. Indeed, $u \in W_0^{1,p}(\Omega)$ satisfies $-\Delta_p u = f$. By the principle of the maximum $u \geqslant 0$. Moreover, the operator $-\Delta_p$ being strictly monotone on $W_0^{1,p}(\Omega)$, we deduce that u is the unique solution of this partial differential equation. As the spherical rearrangement $u_\sim \geqslant 0$, $u_\sim$ belongs to $W_0^{1,p}(\Omega)$.

Moreover, by the Polyà–Szegö inequality and the Hardy–Littlewood inequality, we have:

$$\int_\Omega |\nabla u_\sim|^p dx \leqslant \int_\Omega |\nabla u|^p\,dx \text{ and } \int_\Omega fu \leqslant \int_\Omega f_\sim u_\sim = \int_\Omega fu_\sim.$$

Hence $J(u_\sim) \leqslant J(u)$. Thus $J(u_\sim) = J(u) : u = u_\sim$.

The $W^{1,\infty}$-regularity follows from the previous Corollary 5.2.2. ◇

We will see in the next section how, using rearrangement, we can obtain the regularity of the gradient of the solution to the problem given in Lemma 5.2.1 when $T \in L^1(\Omega) \cap W^{-1,p'}(\Omega)$.

The existence of radial solutions can be shown in other ways (for example by a fixed point theorem).

Combining Bliss' inequality with Theorem 5.2.1, we have:

Theorem 5.2.4

1. *If $T \in W^{-1,\frac{N}{p-1}}(\Omega)$, $1 < p < N$ then any solution $u \in W_0^{1,p}(\Omega)$, $-div\left(\widehat{a}(x,u,\nabla u)\right) = T$ in $\mathcal{D}'(\Omega)$ satisfies:*

   $u \in L_{exp}^{\frac{1}{N'}}(\Omega)$ and

$$\sup_{0<s\leqslant|\Omega|}\frac{|u|_*(s)}{\left(1+\operatorname{Log}\frac{|\Omega|}{s}\right)^{\frac{1}{N'}}}\leqslant\left(\frac{N}{N-p}\right)^{\frac{1}{p}}|F|^{\frac{1}{p-1}}_{L^{\frac{N}{p-1}}}.$$

   In particular $u \in L^q(\Omega)$ $\forall q < +\infty$.

2. *If $T \in W^{-1,r}(\Omega)$, $p' < r < \dfrac{N}{p-1}$ then $u \in L^{m^*}(\Omega)$,*

   with $m^* = \dfrac{(p-1)rN}{N-(p-1)r}$ and $|u|_{m^*} \leqslant B_0|F|^{\frac{1}{p-1}}_{L^r(\Omega)}$.

Proof

1. By Theorem 5.2.1, we have

$$- v_*'(t) \leqslant c_{N\alpha} t^{-\frac{1}{N'}} (G_{*v})^{\frac{1}{p}}(t), \quad t \in \Omega_* \tag{5.12}$$

where $G = |F|^{p'}$. Integrating this relation between $s > 0$ and $|\Omega|$, we obtain

$$v_*(s) \leqslant c_{N\alpha} \int_s^{|\Omega|} t^{-\frac{1}{N'}} (G_{*v})^{\frac{1}{p}}(t)dt. \tag{5.13}$$

Applying Hölder's inequality and Lemma 5.2.2, we deduce

$$v_*(s) \leqslant \left(1 + \mathrm{Log}\,\frac{|\Omega|}{s}\right)^{\frac{1}{N'}} |(G_{*v})^{\frac{1}{p}}|_{L^N(\Omega)},$$

and

$$\sup_{0<s\leqslant|\Omega|} \frac{|u|_*(s)}{\left(1 + \mathrm{Log}\,\dfrac{|\Omega|}{s}\right)^{\frac{1}{N'}}} \leqslant \left(\frac{N}{N-p}\right)^{\frac{1}{p}} |F|^{\frac{1}{p-1}}_{L^{\frac{N}{p-1}}(\Omega)}.$$

2. For statement 2, let $m = (p-1)r$, therefore $p < m < N$, $m^* = \dfrac{Nm}{N-m}$.
Then according to Bliss' inequality (see Lemma 4.4.1)

$$|u|^{m^*}_{L^{m^*}} \leqslant B\left[\int_{\Omega_*} \left|v_*'(t)t^{\frac{N-1}{N}}\right|^m dt\right]^{\frac{m^*}{m}}$$

using the estimate $- v_*'(t)t^{1-\frac{1}{N}} \leqslant c_{N\alpha}\big(G_{*v}\big)^{\frac{1}{p}}(t)$ of Theorem 5.4.2 with $G = |F|^{p'}$, we deduce

$$|v|^{m^*}_{m^*} \leqslant B c^{m^*}_{N\alpha} \left[\int_{\Omega_*} (G_{*v})^{\frac{m}{p}}(t)dt\right]^{\frac{m^*}{m}}$$

$$|v|_{m^*} \leqslant B^{\frac{1}{m^*}} c_{N\alpha} |G_{*v}|^{\frac{1}{p}}_{L^{\frac{m}{p}}} \leqslant B^{\frac{1}{m^*}} c_\Omega |F|^{\frac{1}{p-1}}_{L^{\frac{m}{p-1}}} < +\infty$$

for $F \in L^r(\Omega)$, $\quad r = \dfrac{m}{p-1} < N.$ ◇

5.2.3 A Method for Estimating the Gradient of the Solution $u \in W_0^{1,p}(\Omega)$ of $\operatorname{div}\left(\widehat{a}(x; u; \nabla u)\right) = T \in L^1(\Omega) \cap W^{-1,p'}(\Omega)$

The main theorem of this section is the following

Theorem 5.2.5 *Let $u \in W_0^{1,p}(\Omega)$ with $2 \leqslant p < N$ be a solution of $-\operatorname{div}\left(\widehat{a}(x; u; \nabla u)\right) = T$ with $T \in L^1(\Omega) \cap W^{-1,p'}(\Omega)$, where $\widehat{a}$ satisfies in particular the coercivity hypothesis (LL2). Then, for all $s \in \Omega_*$*

1. $$|\nabla u|_*(s) \sup_{\psi \in L_+^\infty(\Omega_*)} \left[\frac{\displaystyle\int_0^s (s-t)\psi(t)dt}{\displaystyle\int_0^s \psi(t)t^{1-\frac{p'}{N'}}dt}\right]^{\frac{1}{p}} \leqslant \left(\frac{N'}{p'-N'}\right)^{\frac{1}{p}} c_{N\alpha}^{\frac{1}{p-1}} |T|_{L^1(\Omega)}^{\frac{1}{p-1}}.$$
2. *In particular, we have $|\nabla u| \in L^{\frac{N}{N-1}(p-1),\infty}(\Omega)$ and*

$$s^{\frac{(N-1)}{N(p-1)}} |\nabla u|_*(s) \leqslant c'_{N\alpha} |T|_{L^1}^{\frac{1}{p-1}}$$

$$\text{with} \quad c'_{N\alpha} = \frac{2^{\frac{1}{p}}(N')^{\frac{2}{p}}}{(p'-N')^{\frac{1}{p}}(2N'-p')^{\frac{1}{p}}} \cdot c_{N\alpha}^{\frac{1}{p-1}}.$$

Proof Using Theorem 5.2.2 we have

$$\left(|\nabla v|^p\right)_{*v}(s) \leqslant c_{N\alpha}^{p'} s^{-\frac{p'}{N'}} \left(\int_0^s |T|_*(\sigma)d\sigma\right)^{p'}. \tag{5.14}$$

We use the following lemma, whose proof is given in theorem 12 of [128], and taken up again in Chap. 7.

Lemma 5.2.3 *Let $\Phi : \mathbb{R}_+ \to \mathbb{R}_+$, be an increasing concave function and h be a measurable function s.t $\Phi(h) \in L^1(\Omega)$, $v \in L^1(\Omega)$. Then*

$$\int_{\Omega_*} \Phi(h_*(\sigma)(\sigma))\varphi^*(\sigma)d\sigma \leqslant \int_{\Omega_*} \Phi(h_{*v}(t))\varphi(t)dt,$$

(continued)

Lemma 5.2.3 (continued)
$$\forall \varphi \geqslant 0, \quad \varphi \in L^{\infty}(\Omega_*).$$

Here $\varphi^*(s) = \varphi_*(|\Omega| - s)$ is the increasing rearrangement of φ.

Applying this lemma with the affine function $\Phi(t) = t$, we deduce from the relation (5.14) that

$$\int_{\Omega*} \varphi^*(\sigma)|\nabla v|_*^p(\sigma)d\sigma \leqslant c_{N\alpha}^{p'} \left(\int_{\Omega_*} \varphi(s) s^{-\frac{p'}{N'}} ds \right) |T|_{L^1}^{p'}. \tag{5.15}$$

Note that if $\varphi : \Omega_* \to \mathbb{R}_+$ is increasing then $\varphi^* = \varphi$.

Let us then consider $\psi \in L_+^{\infty}(\Omega_*)$, s fixed in Ω_* and the increasing function

$$\varphi(t) = \begin{cases} \displaystyle\int_0^t \psi(\sigma)d\sigma & if\ 0 \leqslant t \leqslant s, \\ \displaystyle\int_0^s \psi(\sigma)d\sigma & if\ s < t \leqslant |\Omega|, \end{cases}$$

then, since $t \to |\nabla v|_*^p(t)$ is decreasing

$$\int_{\Omega_*} \varphi(t)|\nabla v|_*^p(t)dt \geqslant |\nabla v|_*^p(s) \int_0^s \left(\int_0^t \psi(\sigma)d\sigma \right) dt, \tag{5.16}$$

and by integration by parts

$$\int_0^s \left(\int_0^t \psi(\sigma)d\sigma \right) dt = \int_0^s (s-t)\psi(t)dt. \tag{5.17}$$

As for the integral on the right-hand side of (5.15), we have, after integration by parts

$$\int_{\Omega_*} \varphi(t)t^{-\frac{p'}{N'}} dt = \frac{N'}{p'-N'} \int_0^s \psi(t) \left(t^{1-\frac{p'}{N'}} - s^{1-\frac{p'}{N'}} \right) dt$$

$$+ \frac{N'}{p'-N'} \left(\int_0^s \psi(t)dt \right) s^{1-\frac{p'}{N'}} - N'|\Omega|^{1-\frac{p'}{N'}} \int_0^{|\Omega|} \psi(t)dt$$

hence

$$\int_{\Omega_*} \varphi(t)t^{-\frac{p'}{N'}} dt \leqslant \frac{N'}{p'-N'} \int_0^s \psi(t)t^{1-\frac{p'}{N'}} dt. \tag{5.18}$$

Combining the relations (5.15) to (5.18) with inequality 1 of the Theorem, we lower the supremum by choosing $\psi(t) = \chi_{]0,s[}(t)$ thus

$$\sup_{\psi \in L^\infty_+(\Omega_*)} \left[\frac{\int_0^s (s-t)\psi(t)dt}{\int_0^s \psi(t) t^{1-\frac{p'}{N'}} dt} \right]^{\frac{1}{p}} \geqslant \left(1 - \frac{p'}{2N'}\right)^{\frac{1}{p}} s^{\frac{p'}{pN'}}. \tag{5.19}$$

Using this relation (5.19) and inequality 1. of the Theorem, we arrive at inequality 2. with $c'_{N\alpha}$ as given. ◇

5.2.4 Case of Neumann–Dirichlet Problems

Similar results can be obtained for Neumann–Dirichlet problems. If the domain is a connected Lipschitzian bounded open set, and if we denote by Γ_0 a positive measure subset of $\partial\Omega$, $\Gamma_1 = \partial\Omega \backslash \Gamma_0$, then we will set as in Chap. 4, Sect. 4.3,

$\dfrac{1}{N\sigma_N^{\frac{1}{N}}}$ the isoperimetric constant associated with Ω and

$W^{1,p}_{\Gamma_0}(\Omega) = \left\{ v \in W^{1,p}(\Omega) : v = 0 \text{ } on \text{ } \Gamma_0 \right\}$.

Then

$$\frac{1}{N\sigma_N^{\frac{1}{N}}} = \sup_{u \in W^{1,1}_{\Gamma_0}} \frac{|u|_{\frac{N}{N-1}}}{|\nabla u|_1}.$$

Following the same approach as in the previous section we have:

Theorem 5.2.6 *Let* $T = -div(F) \in \left(W^{1,p}_{\Gamma_0}(\Omega)\right)'$ *and* $F \in L^{p'}(\Omega)^N$. *Then for any solution* $u \in W^{1,p}_{\Gamma_0}(\Omega)$ *satisfying*

$$\int_\Omega \widehat{a}(x, u, \nabla u) \cdot \nabla\varphi dx = \int_\Omega F \cdot \nabla\varphi, \quad \forall \varphi \in W^{1,p}_{\Gamma_0}(\Omega)$$

we have for almost every $s \in \Omega_*$:

1. $\left[\widehat{a}(\nabla u) \cdot \nabla u\right]_{*v}(s) = \left[F \cdot \nabla u\right]_{*v}(s)$ *where* $\widehat{a}(\nabla u) = \widehat{a}(x, u, \nabla u)$, $v = |u|$,

(continued)

Theorem 5.2.6 (continued)

2. $|\nabla u|_{*v}(s) \leqslant \dfrac{(G_{*v})^{\frac{1}{p}}(s)}{\alpha^{\frac{1}{p-1}}}$ *where* $G = |F|^{p'}$,
3. $-v'_*(s) \leqslant \widehat{c}_N s^{\frac{1}{N}-1}(G_{*v})^{\frac{1}{p}}(s)$ *where* $\widehat{c}_N = \dfrac{1}{N\sigma_N^{\frac{1}{N}}\alpha^{\frac{1}{p-1}}}$.

Therefore the corollaries of this theorem are identical to those of the Dirichlet problem in Sect. 5.2.1 of it suffices to replace α_N by σ_N.

Corollary 5.2.4 (From Theorem 5.2.6) *Consider a solution* $u \in W^{1,p}_{\Gamma_0}(\Omega)$ *of*

$$\int_\Omega \widehat{a}(\nabla u)\cdot\nabla\varphi dx = \int_\Omega F\cdot\nabla\varphi \;\; \forall\varphi \in W^{1,p}_{\Gamma_0}(\Omega). \qquad \text{Then}$$

1. *if* $F \in L^r(\Omega)^N$, $r > \max\left(\dfrac{N}{p-1}, \dfrac{p}{p-1}\right)$ *then*
$$u \in L^\infty(\Omega)$$
$$|u|_\infty \leqslant \frac{\overline{b}(0)}{N\sigma_N^{\frac{1}{N}}\alpha^{\frac{1}{p-1}}}|F|^{\frac{1}{p-1}}_{L^r(\Omega)},$$
2. *if* $F \in L^{\frac{N}{p-1}}(\Omega)$, $p < N$ *then* $u \in L^{\frac{1}{N'}}_{exp}(\Omega)$ *and in particular* $u \in L^q(\Omega)$ $\forall q < +\infty$, $|u|_q \leqslant c|F|^{\frac{1}{p-1}}_{L^{\frac{N}{p-1}}}$,
3. *if* $F \in L^r(\Omega)$, $p' < r < \dfrac{N}{p-1}$, *then* $u \in L^{m^*}(\Omega)$, $m^* = \dfrac{N(p-1)r}{N-(p-1)r}$ *and* $|u|_{m^*} \leqslant c|F|^{\frac{1}{p-1}}_{L^r}$.

Using α-rearrangement, we then obtain:

Corollary 5.2.5 (From Theorem 5.2.6) *Let $T \in L^r(\Omega)$, $r > N$ and $u \in W^{1,p}_{\Gamma_0}(\Omega)$ be a solution of*

$$\int_\Omega \widehat{a}(\nabla u)\cdot\nabla\varphi = \int_\Omega F\cdot\nabla\varphi, \qquad \varphi \in W^{1,p}_{\Gamma_0}(\Omega),\ T = -div(F),$$

we assume that $u \geqslant 0$. Then $\nabla C_\alpha(u) \in L^\infty(\Omega)^N$. Moreover, we have

$$|\nabla C_\alpha(u)|_*(\sigma) \leqslant \frac{|F|_\infty^{\frac{1}{p-1}}}{\alpha^{\frac{1}{p-1}}}, \ \forall\sigma \in \overline{\Omega}_*\,.$$

Proof We follow exactly the same proof as the Corollary 5.2.3. ◇

Note

- The Neumann–Dirichlet problem is formally equivalent to

$$\begin{cases} -div\big(\widehat{a}(x,u,\nabla u)\big) = T & in\ \Omega, \\ \widehat{a}(\nabla u)\cdot\vec{n} = 0 & on\ \Gamma_1 = \partial\Omega\backslash\Gamma_0, \\ u = 0 & on\ \Gamma_0. \end{cases}$$

- In the following, we will always write $\widehat{a}(\nabla u) = \widehat{a}(x,u,\nabla u)$.
- When $\mathrm{div}\,(F) = T \in L^1(\Omega)\cap W^{-1,p'}(\Omega)$ then the same argument as in the previous section leads to the same type of estimate as in Theorem 5.2.5.

5.3 Case of Homogeneous Neumann Equations

Consider $f \in L^{p'}(\Omega)$ with zero mean. Consider the following formal problem:

$$(\mathcal{P}_N)\begin{cases} u \in W^{1,p}(\Omega), \\ -div\big(\widehat{a}(x,u,\nabla u)\big) = f & in\ \Omega, \\ \widehat{a}(\nabla u)\cdot\vec{n} = 0 & on\ \partial\Omega, \end{cases}$$

where $\vec{n}(x)$ is the exterior unit normal to $\partial\Omega$ at point x and Ω is a connected Lipschitzian open set. The corresponding variational problem is

$$\int_\Omega \widehat{a}(\nabla u)\cdot\nabla\varphi dx = \int_\Omega f\varphi \quad \forall\varphi \in W^{1,p}(\Omega). \qquad (\mathcal{P}_{vN})$$

We then have the following pointwise estimate:

Theorem 5.3.1 *Any variational solution $u \in W^{1,p}(\Omega)$ of $(\mathcal{P}_{vN})$ satisfies for almost all s*

1. $\left(\widehat{a}(\nabla u)\cdot\nabla u\right)_{*v}(s) \leqslant -v_*'(s)\int_0^s |f|_{*v}(\sigma)d\sigma,$
2. $$|\nabla v|_{*v}(s) \leqslant \left[\left(|\nabla v|^p\right)_{*v}(s)\right]^{\frac{1}{p}}$$
$$\leqslant \left[\frac{Q}{\alpha}\max\left(s, |\Omega|-s\right)^{\frac{1}{N}-1}\right]^{\frac{1}{p-1}}\left(\int_0^s |f|_{*v}(\sigma)d\sigma\right)^{\frac{1}{p-1}},$$
3. $$-v_*'(s) \leqslant \left[\frac{Q}{\alpha^{\frac{1}{p}}}\max\left(s, |\Omega|-s\right)^{\frac{1}{N}-1}\right]^{p'}\left(\int_0^s |f|_{*v}(\sigma)d\sigma\right)^{\frac{1}{p-1}}$$

where $v = |u|$ and Q the relative isoperimetric constant associated with Ω.

Proof Consider the function $\varphi = \left(v - v_*(s)\right)_+\mathrm{sign}\,(u) \in W^{1,p}(\Omega)$. Applying Lemma 5.1.2, we deduce from

$$\int_{v>v_*(s)} \widehat{a}(\nabla u)\cdot\nabla u\,dx = \int_\Omega f\mathrm{sign}\,(u)\left(v - v_*(s)\right)_+dx$$

that

$$\left[\widehat{a}(\nabla u)\cdot\nabla u\right]_{*v}(s) = -v_*'(s)\int_0^s \left(f\mathrm{sign}\,(u)\right)_{*v}(\sigma)d\sigma\ .$$

Since $f\mathrm{sign}\,(u) \leqslant |f|$ and $-v_*' \geqslant 0$, we have statement 1.

For statement 2, we use the coercivity of $\widehat{a}$ to deduce

$$\left[\widehat{a}(\nabla u)\cdot\nabla u\right]_{*v}(s) \geqslant \alpha\left[\,|\nabla v|^p\,\right]_{*v}(s),$$

which gives in combination with statement 1 and the PSR property that

$$\left(|\nabla v|^p\right)_{*v}(s) \leqslant \frac{Q}{\alpha}\max\left(s, |\Omega|-s\right)^{\frac{1}{N}-1}|\nabla v|_{*v}(s)\left(\int_0^s |f|_{*v}(\sigma)d\sigma\right).$$

According to Hölder's inequality, we have:

$$\left(|\nabla v|^p\right)_{*v}(s) \geqslant \left(|\nabla v|_{*v}(s)\right)^p.$$

Thus the last two inequalities lead to statement 2, i.e.

$$|\nabla v|_{*v}(s) \leqslant \left[\left(|\nabla v|^p\right)_{*v}(s)\right]^{\frac{1}{p}}$$

$$\leqslant \left[\frac{Q}{\alpha}\max\left(s, |\Omega|-s\right)^{\frac{1}{N}-1}\right]^{\frac{1}{p-1}} \left(\int_0^s |f|_{*v}(\sigma)d\sigma\right)^{\frac{1}{p-1}}.$$

Again, using the PSR inequality statement 2./ leads to 3./. ◇

5.3.1 Estimates in Lorentz Spaces $L^{m,q}(\Omega)$

Corollary 5.3.1 (Of Theorem 5.3.1) *We assume that* $f \in L^{\frac{N}{p},\frac{1}{p-1}}(\Omega)$. *Then*

$$u \in L^\infty(\Omega) \text{ and } |u|_\infty \leqslant \frac{2}{|\Omega|}|u|_1 + \left(\frac{Q}{\alpha^{\frac{1}{p}}}\right)^{p'} |f|^{\frac{1}{p-1}}_{(\frac{N}{p},\frac{1}{p-1})}.$$

Proof We know that $\int_0^s |f|_{*v}(\sigma)d\sigma \leqslant \int_0^s |f|_*(\sigma)d\sigma$. We then have

$$v_*(0) - v_*\left(\frac{|\Omega|}{2}\right) \leqslant \int_0^{\frac{|\Omega|}{2}} \left[\frac{Q}{\alpha^{\frac{1}{p}}}\max\left(s, |\Omega|-s\right)^{\frac{1}{N}-1}\right]^{p'} \left(\int_0^s |f|_* d\sigma\right)^{\frac{1}{p-1}} ds$$

$$= \left(\frac{Q}{\alpha^{\frac{1}{p}}}\right)^{p'} \int_0^{\frac{|\Omega|}{2}} s^{(\frac{1}{N}-1)p'} \left(\int_0^s |f|_*(\sigma)d\sigma\right)^{\frac{1}{p-1}} ds\,.$$

Now we have for the latter integral

$$\int_0^{\frac{|\Omega|}{2}} \left[s^{(\frac{1}{N}-1)p+1} |f|_{**}(s)\right]^{\frac{1}{p-1}} ds = \int_0^{\frac{|\Omega|}{2}} \left[s^{\frac{p}{N}} |f|_{**}(s)\right]^{\frac{1}{p-1}} \frac{ds}{s} \leqslant |f|^{\frac{1}{p-1}}_{(\frac{N}{p},\frac{1}{p-1})}.$$

Combining these last inequalities, we obtain:

$$|u|_\infty = v_*(0) \leqslant$$

$$\leqslant v_*\left(\frac{|\Omega|}{2}\right) + \left(\frac{Q}{\alpha^{\frac{1}{p}}}\right)^{p'} |f|^{\frac{1}{p-1}}_{(\frac{N}{p},\frac{1}{p-1})} \leqslant \frac{2}{|\Omega|}|u|_1 + \left(\frac{Q}{\alpha^{\frac{1}{p}}}\right)^{p'} |f|^{\frac{1}{p-1}}_{(\frac{N}{p},\frac{1}{p-1})}.$$

◇

Note In the case of the homogeneous Dirichlet problem (or Neumann–Dirichlet) problem, we have the same type of result, i.e. if $u \in W_0^{1,p}(\Omega)$ is a solution of $-\,div\big(\widehat{a}(\nabla u)\big) = f$ and if $f \in L^{\frac{N}{p},\frac{1}{p-1}}(\Omega)$ then $u \in L^\infty(\Omega)$ and we have:

$$|u|_\infty \leqslant \left(\frac{1}{N\alpha_N^{\frac{1}{N}}\alpha^{\frac{1}{p}}}\right)^{p'} |f|^{\frac{1}{p-1}}_{(\frac{N}{p},\frac{1}{p-1})}.$$

To treat the case where $f \in L^{m,\frac{q}{p-1}}(\Omega)$, $1 \leqslant m < \dfrac{N}{p}$, let's start with the following observation:

Proposition 5.3.1 *$\forall u \in L^1(\Omega)$, if $v = |u|$ then we have: $\forall r > 0$, $q > 0$*

$$\left(\int_{\frac{|\Omega|}{2}}^{|\Omega|} \sigma^{\frac{q}{r}-1} v_*(\sigma)^q d\sigma\right)^{\frac{1}{q}} \leqslant 2\,|\Omega|^{\frac{1}{r}-1}\left(\frac{r}{q}\right)^{\frac{1}{q}} |u|_1 \left(1-\left(\frac{1}{2}\right)^{\frac{q}{r}}\right)^{\frac{1}{q}}.$$

Proof Since v_* is decreasing, we have:

$$\int_{\frac{|\Omega|}{2}}^{|\Omega|} \sigma^{\frac{q}{r}-1} v_*(\sigma)^q d\sigma \leqslant v_*\left(\frac{|\Omega|}{2}\right)^q \left(\frac{r}{q}\right) |\Omega|^{\frac{q}{r}} \left(1-\left(\frac{1}{2}\right)^{\frac{q}{r}}\right)$$

and

$$v_*\left(\frac{|\Omega|}{2}\right) \leqslant \frac{2}{|\Omega|}\int_0^{\frac{|\Omega|}{2}} v_*(\sigma)d\sigma \leqslant \frac{2}{|\Omega|}\,|u|_1\,.$$

Combining these last two inequalities, we obtain the proposition. ◇

We now have the regularity theorem, for $u \in W^{1,p}(\Omega)$ a solution of $-div\big(\widehat{a}(\nabla u)\big) = f$.

Theorem 5.3.2 *Let* $m \in \left[1, \dfrac{N}{p}\right[$ *and* $f \in L^{m,\frac{q}{p-1}}(\Omega)$ *with* $1 \leqslant q < +\infty$. *Then any solution* $u \in W^{1,p}(\Omega)$ *of*

$$\int_\Omega \widehat{a}(\nabla u)\cdot\nabla\varphi\, dx = \int_\Omega f\varphi\, dx, \quad \forall\varphi \in W^{1,p}(\Omega)$$

is in $u \in L^{r,q}(\Omega)$ *with* $r = \dfrac{mN(p-1)}{N-p}$. *Moreover,*

$$|u|_{r,q} \leqslant 4\,|\Omega|^{\frac{1}{r}-1}\left(\frac{r}{q}\right)^{\frac{1}{q}}|u|_1 + \left(\frac{Q}{\alpha^{\frac{1}{p}}}\right)^{p'} r\,|f|_{(m,\frac{q}{p-1})}^{\frac{1}{p-1}}\,.$$

Proof We write $|u|_{r,q}^q = I_1 + I_2$ and $v = |u|$ where $I_1 = \displaystyle\int_0^{\frac{|\Omega|}{2}} \sigma^{\frac{q}{r}-1} v_*(\sigma)^q d\sigma\,.$

Then, the estimate of $I_2^{\frac{1}{q}}$ follows from Proposition 5.3.1. As for I_1 we deduce from the pointwise estimate of v'_* that:

$$v_*(s) - v_*\left(\frac{|\Omega|}{2}\right) \leqslant \left(\frac{Q}{\alpha^{\frac{1}{p}}}\right)^{p'} \int_s^{\frac{|\Omega|}{2}} t^{-\frac{p'}{N'}}\left(\int_0^t |f|_{*v}(\sigma)d\sigma\right)^{\frac{1}{p-1}}, \quad s < \frac{|\Omega|}{2}.$$

As $\displaystyle\int_0^t |f|_{*v}(\sigma)d\sigma \leqslant \int_0^t |f|_*(\sigma)d\sigma$, we deduce

$$I_3 = \int_0^{\frac{|\Omega|}{2}} \sigma^{\frac{q}{r}-1} \left[v_*(\sigma) - v_*\left(\frac{|\Omega|}{2}\right) \right]^q d\sigma$$

$$\leqslant \left(\frac{Q}{\alpha^{\frac{1}{p}}}\right)^{qp'} \int_0^{\frac{|\Omega|}{2}} \sigma^{\frac{q}{r}-1} \left[\int_0^{\frac{|\Omega|}{2}} t^{-\frac{p'}{N'}} \left(\int_0^t |f|_*(\sigma) d\sigma \right)^{\frac{1}{p-1}} dt \right]^q d\sigma .$$

If $q > 1$, we apply Hardy's inequality to the second integral to obtain the following bound

$$I_3 \leqslant \left(\frac{Q}{\alpha^{\frac{1}{p}}}\right)^{qp'} r^q \int_{\Omega*} \left[|f|_{**}(t) \right]^{\frac{q}{p-1}} t^{\frac{q}{p-1}+\frac{q}{r}+q-1-\frac{qp'}{N'}} dt.$$

But we have $\dfrac{q}{p-1} + \dfrac{q}{r} + q - \dfrac{qp'}{N'} = \dfrac{q}{p-1}\left[\dfrac{p}{N} + \dfrac{p}{r} - \dfrac{1}{r}\right] = \dfrac{q}{m(p-1)}$ with $r = \dfrac{mN(p-1)}{N-p}$. Hence

$$I_3 \leqslant \left(\frac{Q}{\alpha^{\frac{1}{p}}}\right)^{qp'} r^q \int_{\Omega_*} \left[|f|_*(t) t^{\frac{1}{m}} \right]^{\frac{q}{p-1}} \frac{dt}{t} : I_3^{\frac{1}{q}} \leqslant \left(\frac{Q}{\alpha^{\frac{1}{p}}}\right)^{p'} r\, |f|^{\frac{1}{p-1}}_{(m,\frac{q}{p-1})},$$

which implies (comparing I_1 and I_2)

$$I_1^{\frac{1}{q}} \leqslant \left(\frac{Q}{\alpha^{\frac{1}{p}}}\right)^{p'} r\, |f|^{\frac{1}{p-1}}_{(m,\frac{q}{p-1})} + v_*\left(\frac{|\Omega|}{2}\right)\left(\frac{|\Omega|}{2}\right)^{\frac{1}{r}}\left(\frac{r}{q}\right)^{\frac{1}{q}}.$$

Since $v_*\left(\dfrac{|\Omega|}{2}\right) \leqslant \dfrac{2}{|\Omega|}\,|u|_1$, we deduce the estimate of $I_1^{\frac{1}{q}}$.

Noting further that $|u|_{r,q} \leqslant I_1^{\frac{1}{q}} + I_2^{\frac{1}{q}}$, we obtain the estimate of the theorem. ◇

Note If $m > p', q < p^*$ then $L^{r,q}(\Omega) \subsetneq L^{p^*}(\Omega)$.

If we consider $m = \dfrac{q}{p-1} = 1$ ($p \geqslant 2$), then we find

$$u \in L^{\frac{N}{N-p}(p-1),1}(\Omega) \subset L^s(\Omega), \quad \forall s < \frac{N}{N-p}(p-1) \text{ for } f \in L^1(\Omega).$$

5.3.2 *Gradient Estimates of the Solution*

The same technique as used for Theorem 5.2.5 prevails here. We state the theorem without giving its proof

Theorem 5.3.3 *Consider the solution u of the variational problem $(\mathcal{P}_{vN})$ if $f \in L^1(\Omega)$. Then, $\nabla u \in L^{\frac{N}{N-1}(p-1),\infty}(\Omega)$ and there exists a constant $c > 0$ s.t*

$$|\nabla u|_{L^{\frac{N}{N-1}(p-1),\infty}(\Omega)} \leqslant c|f|_{L^1(\Omega)}^{\frac{1}{p-1}}.$$

To vary the systems to be studied, let us consider in the following a nonlinear eigenvalue problem from plasma physics.

5.4 A Nonlinear Eigenvalue Problem in Plasma Physics

Let Ω be a connected bounded open set of class C^∞ in $\mathbb{R}^2$. Consider the following closed subspace of $H^1(\Omega)$:

$$V = \left\{ v \in H^1(\Omega) : v = \text{constant on } \partial\Omega \right\}.$$

Theorem 5.4.1 *There exists a function $u \in H^2(\Omega) \cap V$ satisfying for given $\lambda > 0$ and I*

$$[T] \qquad \begin{cases} -\Delta u + \lambda u_- = 0 & \text{in } \Omega, \\ u = \gamma = (constant) & \text{on } \partial\Omega, \\ \displaystyle\int_{\partial\Omega} \frac{\partial u}{\partial n} d\ell = I > 0. \end{cases}$$

(see Chap. 11, plasma problems for a proof due to M. Sermange).

Proposition 5.4.1 *Any solution u of the problem $[T]$ is twice continuously differentiable.*

Theorem 5.4.2 *Let u be a solution of the system $[T]$. Then,*

(a) $\lambda > \lambda_1 \Longleftrightarrow u|_\Gamma > 0$,
(b) $\lambda = \lambda_1 \Longleftrightarrow u|_\Gamma = 0$,
(c) $\lambda < \lambda_1 \Longleftrightarrow u|_\Gamma < 0$

where λ_1 is the first eigenvalue on Ω of the Dirichlet problem.

Proof First, note that if $u|_\Gamma \leqslant 0$ then according to the strict maximum principle (see Harnack's inequalities, for example in Gilbarg–Trudinger's book [94]) we have $u < 0$. If $u|_\Gamma > 0$ then u is an eigenfunction of the Laplacian in $\Omega_p = \left\{u < 0\right\}$ associated with the eigenvalue λ. Since u does not vanish in Ω_p, λ is therefore the first eigenvalue in Ω_p.

Since the first eigenvalue of the Dirichlet problem is a decreasing function of the domain, this can only happen if $\lambda > \lambda_1$.

Similarly if $u|_\Gamma = 0$ then $\Omega = \Omega_p$, and consequently $\lambda = \lambda_1$.

We have thus proved that

$$\text{If } u|_\Gamma > 0 \; then \; \lambda > \lambda_1 \; and \; u|_\Gamma = 0 \Longleftrightarrow \lambda = \lambda_1.$$

Let us show that if $\lambda > \lambda_1$ then $u|_\Gamma > 0$.

Indeed, let φ_1 be the first eigenvalue function of the Laplacian associated with λ_1, then we can choose $\varphi_1 \geqslant 0$ (because $|\nabla\varphi_1| = |\nabla\,|\varphi_1||$). Using Green's formula, we have

$$- u|_\Gamma \int_{\partial\Omega} \frac{\partial\varphi_1}{\partial n} d\ell = \int_\Omega \Big(\varphi_1 \Delta u - u\Delta\varphi_1\Big) dx = \int_\Omega \Big(\lambda\varphi_1 u_- + \lambda_1\varphi_1 u\Big) dx,$$

$$> \int_\Omega \lambda_1\varphi_1(u_- + u)dx (\, car \; \lambda > \lambda_1 \; and \; |u < 0| > 0) = \lambda_1 \int_\Omega \varphi_1 u_+ dx \geqslant 0.$$

Thus

$$u|_\Gamma \int_\Omega \lambda_1\varphi_1 = -u|_\Gamma \int_{\partial\Omega} \frac{\partial\varphi_1}{\partial n} d\ell > 0 : u|_\Gamma > 0. \qquad \diamond$$

In what follows we assume that $\lambda > \lambda_1$.

Lemma 5.4.1 *Let φ_1 be the first eigenfunction of the following Dirichlet problem*

(continued)

Lemma 5.4.1 (continued)

$$\begin{cases} -\Delta\varphi_1 = \lambda_1\varphi_1, \\ \varphi_1 = 0 \; on \; \partial\Omega \\ \int_\Omega \varphi_1 dx = \dfrac{1}{\lambda_1}. \end{cases}$$

Then, for any solution u of the problem $[T]$, we have $0 < u|_\Gamma \leqslant \frac{\lambda - \lambda_1}{\lambda} I \operatorname*{Max}_{\overline{\Omega}} \varphi_1 + \lambda_1 \int_\Omega \varphi_1 u_+ dx$.

Proof We have seen that from Green's formula, we have

$$-\, u|_\Gamma \int_{\partial\Omega} \frac{\partial\varphi_1}{\partial n} d\ell = \int_\Omega \Big(\lambda\varphi_1 u_- + \lambda_1\varphi_1 u\Big) dx$$

$$= (\lambda - \lambda_1) \int_\Omega \varphi_1 u_- dx + \lambda_1 \int_\Omega \varphi_1 u_+ dx.$$

By Green's formula and the choice of φ_1

$$-\int_{\partial\Omega} \frac{\partial\varphi_1}{\partial n} d\ell = -\int_\Omega \Delta\varphi_1 dx = \lambda_1 \int_O \varphi_1 dx = 1$$

hence the next identity:

$$u|_\Gamma - \lambda_1 \int_\Omega \varphi_1 u_+ dx = (\lambda - \lambda_1) \int_\Omega \varphi_1 u_- dx. \tag{5.20}$$

Since $u \in K = \Big\{v \in V : \lambda \int_\Omega v_-(x) dx = I\Big\}$:

$$(\lambda - \lambda_1) \int_\Omega \varphi_1 u_- dx \leqslant \frac{\lambda - \lambda_1}{\lambda} I \operatorname*{Max}_{\overline{\Omega}} \varphi_1,$$

hence the result. ◇

How to estimate $\max\limits_{\overline{\Omega}} \varphi_1$?

Lemma 5.4.2 *Under the same assumptions as Lemma 5.4.1, we have:*

(i) $\underset{\overline{\Omega}}{\mathrm{Max}}\, \varphi_1 \leqslant \dfrac{\lambda_1 \, |\Omega|^{\frac{1}{2}}}{2\pi} \, |\varphi_1|_{L^2(\Omega)}$,

(ii) $|\varphi_1|_{L^2(\Omega)} \leqslant \dfrac{1}{2\sqrt{\pi}\lambda_1}$.

Proof Since $-\Delta\varphi_1 = \lambda_1\varphi_1,\ \varphi_1 \in H_0^1(\Omega)$, using Theorem 5.3.1 statement 1. and the Cauchy–Schwarz inequality:

$$\left[|\nabla\varphi_1|^2\right]_{*\varphi_1}(s) \leqslant \lambda_1 s^{\frac{1}{2}} \, |\varphi_1|_{L^2} \cdot \left(-\frac{d\varphi_{1*}}{ds}\right)(s).$$

But, according to the pointwise inequalities for the relative rearrangement:

$$-\frac{d\varphi_{1*}}{ds}(s) \leqslant \frac{s^{-\frac{1}{2}}}{2\sqrt{\pi}}\left[|\nabla\varphi_1|^2\right]_{*\varphi_1}^{\frac{1}{2}}(s),$$

thus the last two inequalities imply:

$$-\frac{d\varphi_{1*}}{ds}(s) \leqslant \frac{\lambda_1 s^{-\frac{1}{2}}}{4\pi} \, |\varphi_1|_{L^2(\Omega)} \, . \tag{5.21}$$

We integrate this last inequality to obtain:

$$\underset{\overline{\Omega}}{\mathrm{Max}}\, \varphi_1 \leqslant \frac{\lambda_1}{2\pi} \, |\Omega|^{\frac{1}{2}} \, |\varphi_1|_{L^2(\Omega)} \, . \tag{5.22}$$

To obtain the second inequality, using the same argument as in Theorem 5.3.1, we have:

$$-s\frac{d\varphi_{1*}}{ds}(s) \leqslant \frac{\lambda_1}{4\pi}\left(\int_0^s \varphi_{1*}(t)dt\right). \tag{5.23}$$

We multiply this equation by φ_{1*},

$$-s\frac{d}{ds}\varphi_{1*}^2 \leqslant \frac{\lambda_1}{4\pi}\frac{d}{ds}\left\{\int_0^s \varphi_{1*}(t)dt\right\}^2. \tag{5.24}$$

By writing

$$-s\frac{d}{ds}\varphi_{1*}^2 = -\frac{d}{ds}(s\varphi_{1*}^2) + \varphi_{1*}^2, \tag{5.25}$$

we deduce after integration that

$$\int_0^{|\Omega|} \varphi_{1*}^2(t)dt \leqslant \frac{\lambda_1}{4\pi}\left(\int_0^{|\Omega|} \varphi_{1*}(t)dt\right)^2 \tag{5.26}$$

or by equimeasurability

$$|\varphi_1|_{L^2} \leqslant \frac{1}{2\sqrt{\lambda_1}\sqrt{\pi}}. \tag{5.27}$$

◇

Corollary 5.4.1 (Of Lemma 5.4.2)

$$\operatorname*{Max}_{\overline{\Omega}} \varphi_1 \leqslant \frac{1}{4\pi}\sqrt{\frac{\lambda_1}{\pi}}\,|\Omega|^{\frac{1}{2}}.$$

The previous method can be generalized in the form:

Lemma 5.4.3 *Let $u \geqslant 0,\ u \in H_0^1(\Omega)$ be the solution of $\Delta u + f(u) = 0$ with $f : \mathbb{R}_+ \to \mathbb{R}_+$. Then*

$$8\pi \int_\Omega F(u)dx \leqslant \{\int_\Omega f(u)dx\}^2 \text{ with } F(u) = \int_0^u f(t)\,dt.$$

Proof Since $-\Delta u = f(u)$, we have: $(f \geqslant 0)$

$$\left(|\nabla u|^2\right)_{*u} \leqslant \left(\int_0^s f(u_*)d\sigma\right)\left(-\frac{du_*}{ds}\right).$$

Using the pointwise Poincaré–Sobolev inequality, we deduce:

$$-\frac{du_*}{ds} \leqslant \frac{s^{-1}}{4\pi}\left\{\int_0^s f(u_*)d\sigma\right\} \tag{5.28}$$

and since $f \geqslant 0$,

$$-sf(u_*)\frac{du_*}{ds} \leqslant \frac{1}{4\pi} f(u_*)\{\int_0^s f(u_*)d\sigma\}, \tag{5.29}$$

therefore,

$$-s\frac{d}{ds}\int_0^{u_*(s)} f(t)dt \leqslant \frac{1}{8\pi}\frac{d}{ds}\left\{\int_0^s f(u_*)(t)dt\right\}^2 \tag{5.30}$$

$$\int_0^{u_*(s)} f(t)dt - \frac{d}{ds}\left\{s\int_0^{u_*(s)} f(t)(t)dt\right\} \leqslant \frac{1}{8\pi}\frac{d}{ds}\left\{\int_0^s f(u_*(t))dt\right\}^2 \tag{5.31}$$

By integrating this relation we obtain the result. ◇

Lemma 5.4.4 *In the open Ω_p, we have:*

1. $|u|_{L^2(\Omega_p)} \leqslant \dfrac{I}{2\sqrt{\lambda}\sqrt{\pi}}$,
2. $\underset{\Omega_p}{\text{Max}}\,|u| \leqslant \dfrac{I}{4\pi}\sqrt{\dfrac{\lambda}{\pi}}\,|\Omega_p|^{\frac{1}{2}}$.

Proof The function $\varphi = |u|$ satisfies the following problem, analogous to that of φ_1 following

$$-\Delta\varphi = \lambda\varphi \text{ in } \Omega_p,\ \varphi = 0 \text{ on } \partial\Omega_p \text{ and } \lambda\int_{\Omega_p}\varphi dx = I.$$

(The equation is defined in the variational sense.)

Applying Lemma 5.4.3 or repeating the proof of Lemma 5.4.2, we have

$$\int_{\Omega_p}\varphi^2 dx \leqslant \frac{1}{4\lambda\pi}\{\lambda\int_{\Omega}\varphi\, dx\}^2 = \frac{I^2}{4\lambda\pi},$$

whence . As for 2. we know $\underset{\Omega_p}{\text{Max}}\,|u| \leqslant \dfrac{\lambda}{2\pi}\,|\Omega_p|^{\frac{1}{2}}\,|u|_{L^2(\Omega_p)}$ (see 5.22)

$$\leqslant \frac{\lambda}{2\pi}\times\frac{I}{2\sqrt{\lambda}\sqrt{\pi}}\,|\Omega_p|^{\frac{1}{2}} \leqslant \frac{I}{4\pi}\sqrt{\frac{\lambda}{\pi}}\,|\Omega_p|^{\frac{1}{2}}.$$ ◇

Lemma 5.4.5 (Lower Bound for the Plasma Size)

$$2^{4/3}\frac{\pi}{\lambda} \leqslant |\Omega_p|.$$

Proof

$$\frac{1}{|\Omega_p|}\int_{\Omega_p}|u|\,dx = \frac{1}{|\Omega_p|}\int_{\Omega} u_- = \frac{I}{\lambda\,|\Omega_p|}$$

hence

$$\frac{I}{\lambda\,|\Omega_p|} \leqslant \operatorname*{Max}_{\Omega_p}|u| \leqslant \frac{I}{4\pi}\sqrt{\frac{\lambda}{\pi}}\,|\Omega_p|^{\frac{1}{2}}$$

$$4\frac{\pi^{3/2}}{\lambda^{3/2}} \leqslant |\Omega_p|^{3/2}\,.$$

◇

5.5 Some Subsidiary Remarks

The consequences of the previous studies are numerous, here is an example:

Theorem 5.5.1 (Regularity Without Weight) *Let Ω be an open bounded, Lipschitzian set of $\mathbb{R}^2$ and $V = \left\{v \in W^{1,2}(\Omega),\ \Delta v \in L^\infty(\Omega) \text{ and } \gamma_0 v \leqslant 0\right\}$. Then,*

$$\forall\, v \in V,\ v_{+*} \in W^{1,\infty}(\Omega_*),\ -v'_{+*}(s) \leqslant \frac{1}{4\pi}\,|\Delta v|_{**}(s)\,.$$

Proof Let $w = v_+$ and write for all $s \in \Omega_*$

$$\int_\Omega \Delta v\cdot\big(w - w_*(s)\big)_+ ds = \int_\Omega \Delta v\big(w - w_*(s)\big)_+ dx,\ \ w \in W_0^{1,2}(\Omega)\,.$$

Hence $-\displaystyle\int_{\{w>w_*(s)\}} |\nabla w|^2\,ds = \int_\Omega \Delta v\big(w - w_*(s)\big)_+ dx$. By differentiating in s we deduce:

$$\Big(|\nabla w|^2\Big)_{*w}(s) = +\frac{dw_*}{ds}(s)\int_0^s (\Delta v)_{*w}(\sigma)d\sigma\,.$$

As $w \in W_0^{1,2}(\Omega)$, $w \geqslant 0$ from which, using the inequality (PSR) (Proposition 4.1.1) $-w_*'(s) \leqslant \frac{s^{-\frac{1}{2}}}{2\sqrt{\pi}}\left[\left(|\nabla w|^2\right)_{*w}(s)\right]^{\frac{1}{2}}$. By combining these last two relations, we deduce:

$$-w_*'(s) \leqslant \frac{1}{4\pi}\frac{1}{s}\int_0^s \left|(\Delta v)_{*w}\right|(\sigma)d\sigma \leqslant \frac{1}{4\pi}\frac{1}{s}\int_0^s |\Delta v|_*(\sigma)d\sigma\,.$$

◇

Note $\left|(\Delta v)_{*w}\right| \leqslant |\Delta v|_{*w}$ because

$$\left|\frac{1}{h}\int_{\{w_*(s+h)<w<w_*(s)\}} \Delta v\, dx\right| \leqslant \frac{1}{h}\int_{\{w_*(s+h)<w<w_*(s)\}} |\Delta v|\, dx \qquad ,$$

on $\{w = w_*(s)\}$, $\Delta v = 0$ a.e.

5.6 An Example of Application of the Rearrangement Method to an Equation from MechaNics and Physics with Potential $V \geqslant 0$

In this section, we will focus on estimating the solution to the equation $\begin{cases}-\Delta\phi - \overrightarrow{u}\cdot\nabla\phi + V\phi = T \; in \; \Omega,\\ \phi = 0 \; on \; \partial\Omega.\end{cases}$

Here V is a potential which is at least non negative, and the vector field $\overrightarrow{u}$ is a data satisfying

$$(H5)\begin{cases}\overrightarrow{u} \in L^N(\Omega)^N, \quad \operatorname{div}(\overrightarrow{u}) = 0 \; in \; \Omega,\\ \overrightarrow{u}\cdot\overrightarrow{n} = 0 \; on \; \partial\Omega.\end{cases}$$

Therefore we have the following property:

Lemma 5.6.1 *For any Lipschitzian function*

$$G : \mathbb{R} \to \mathbb{R},\ \forall\phi \in W_0^1 L^{N'}(\Omega),\ N' = \frac{N}{N-1},\ \textit{we have}$$

(continued)

Lemma 5.6.1 (continued)

1. $\int_\Omega (\vec{u} \cdot \nabla \phi) G(\phi) dx = 0.$
2. $\int_\Omega (\vec{u} \cdot \nabla \overline{\omega}) \phi\, dx = - \int_\Omega (\vec{u} \cdot \nabla \phi) \overline{\omega} dx \quad \forall \overline{\omega} \in H_0^1(\Omega).$

The proof is based just on an integration by parts and the density of regular functions in $L^p(\Omega)$ spaces.

Using the Lax–Milgram Theorem (see [65] for details), we have

Proposition 5.6.1 *Let $T \in H^{-1}(\Omega)$ (dual space of $H_0^1(\Omega)$), $\vec{u}$ satisfying*

$$\begin{cases} \vec{u} \in L^N(\Omega)^N, \ \text{div}\, \vec{u} = 0 & \text{in } \mathcal{D}'(\Omega) \text{ and} \\ \vec{u} \cdot \vec{n} = 0 & \text{on } \partial\Omega \end{cases} \tag{5.32}$$

and let $V \in L^0(\Omega)$ satisfy $V \geqslant -\lambda$ for some $\lambda \in [0, \lambda_1)$ where λ_1 is the first eigenvalue of $-\Delta$ with the Dirichlet boundary condition. We define $W = \left\{\varphi \in H_0^1(\Omega) : (V + \lambda)\varphi^2 \in L^1(\Omega)\right\}$, and denote by W' its space dual.

Then, there exists a unique $\phi \in H_0^1(\Omega)$, with $(V + \lambda)\phi^2 \in L^1(\Omega)$, such that

$$(\mathcal{P})_{V,T} \qquad -\Delta\phi - \vec{u} \cdot \nabla\phi + V\phi = T \text{ in } W'. \tag{5.33}$$

Additionally,

$$||\phi||_{H_0^1(\Omega)} = \left(\int_\Omega |\nabla\phi|^2 dx\right)^{frac12} \leqslant \frac{\lambda_1}{\lambda_1 - \lambda} ||T||_{H^{-1}(\Omega)},$$

$$\left(\int_\Omega (V + \lambda)\phi^2 dx\right)^{\frac{1}{2}} \leqslant \left(\frac{\lambda_1}{\lambda_1 - \lambda}\right)^{\frac{1}{2}} ||T||_{H^{-1}(\Omega)}, \quad V\phi \in L^1_{loc}(\Omega).$$

And if $V \in L^1_{loc}(\Omega)$, then (5.33) exists in the sense of distributions in $\mathcal{D}'(\Omega)$

As we have emphasized, the interest of the method of rearrangements is not only its simplicity also that it gives an explicit estimate.

Proposition 5.6.2 **(L^∞-Estimates)** *Let ϕ be the solution of (5.33) when $T \in L^{\frac{N}{2},1}(\Omega)$, $V \geqslant 0$. Then $\phi \in L^\infty(\Omega)$ and there exists a constant $K_N(\Omega)$ independent of $\overrightarrow{u}$ and V s.t*

$$||\phi||_{L^\infty(\Omega)} \leqslant K_N(\Omega)||T||_{L^{\frac{N}{2},1}(\Omega)}.$$

Proof Since the equation is linear we can assume that $T \geqslant 0$ and by density we can also take $V \in L^\infty_+(\Omega)$. Under these conditions $v = |\phi| = \phi$. We argue as before. Let $G_s(\sigma) = (\sigma - v_*(s))_+ \operatorname{sign}(\sigma)$, $\sigma \in \mathbb{R}$, $s \in \Omega_*$. The map $\sigma \mapsto G_s(\sigma)$ is Lipschitzian. Then by applying Lemma 5.6.1

$$\int_\Omega (\overrightarrow{u} \cdot \nabla\phi)G_s(\phi)\,dx = 0.$$

Consequently, we have

$$\int_\Omega \nabla\phi \cdot \nabla G_s(\phi) = \int_{v>v_*(s)} |\nabla\phi|^2 dx = \int_\Omega T(x)G_s(\phi)(x)dx - \int_\Omega V(x)G_s(\phi)dx.$$

By differentiating with respect to s, and using the Hardy–Littlewood's inequality

$$\frac{d}{ds}\int_{v>v_*(s)} |\nabla\phi|^2 dx = -v_*'(s)\int_{v>v_*(s)} \big(T(x) - V(x)\big)dx \leqslant -v_*'(s)\int_0^s T_*(t)dt$$

where T_* is the monotone rearrangement of T (we used the fact $V \geqslant 0$).

Consequently, we arrive at

$$\big[\,|\nabla\phi|^2\big]_{*v}(s) \leqslant -v_*'(s)\int_0^s T_*(\sigma)d\sigma. \tag{5.34}$$

Since

$$|\nabla\phi| = |\nabla v|, \text{ and } \; -v_*'(s) \leqslant \frac{s^{\frac{1}{N}-1}}{N\alpha_N^{\frac{1}{N}}}|\nabla v|_{*v}(s)$$

(thanks to Theorem 4.1.1 and $|\nabla v|_{*v} \leqslant \big[\,|\nabla v|^2\big]_{*v}^{\frac{1}{2}}$), we deduce, from the relation (5.34)

$$-v_*'(s) \leqslant \frac{s^{\frac{2}{N}-2}}{(N\alpha_N^{\frac{1}{N}})^2}\int_0^s T_*(\sigma)d\sigma. \tag{5.35}$$

Integrating the inequality (5.35) between 0 and $|\Omega|$, we find

$$||\phi||_{L^\infty} \leqslant c_N \int_0^{|\Omega|} s^{\frac{2}{N}} T_{**}(s)\frac{ds}{s} = c_N ||T||_{L^{\frac{N}{2},1}(\Omega)}.$$

◇

A similar result can be obtained with $T = -\mathrm{div}\,(\overrightarrow{F})$, where $\overrightarrow{F} \in L^{N,1}(\Omega)^N$.

Proposition 5.6.3 *Let $N \geqslant 2$, and ϕ a solution of (5.33) where $T = -\mathrm{div}\,(\overrightarrow{F})$,*

$\overrightarrow{F} \in L^r(\Omega)^N$, $r > N$, if $N \geqslant 3$, $\overrightarrow{F} \in L^{2+\varepsilon}(\Omega)^2$ if $N = 2$.

Then $\phi \in L^\infty(\Omega)$ and there exists a constant $K_N(\Omega) > 0$ independent of $\overrightarrow{u}$ and V s.t

$$||\phi||_{L^\infty(\Omega)} \leqslant K_N(\Omega)||\overrightarrow{F}||_{L_V} \text{ with } L_V = \begin{cases} L^r(\Omega)^N \text{ if } N \geqslant 3,\ r > N \\ L^{2+\varepsilon}(\Omega)^2 \text{ if } N = 2. \end{cases}$$

Proof We agree to write F for $\overrightarrow{F}$. Thanks to Lemma 5.6.1, we can use $G_s(\phi)$ as in Proposition 5.6.2. Then

$$\int_\Omega \nabla\phi \cdot \nabla G_s(\phi)dx + \int_\Omega V(x)G_s(\phi)dx = \int_\Omega F \cdot \nabla G_s(\phi)dx.$$

We differentiate this equation with respect to s, to obtain almost everywhere in Ω_*,

$$\left[\,|\nabla v|^2\right]_{*v}(s) - v_*'(s)\int_{v>v_*(s)} V(x)dx = \left[\,F \cdot \nabla\phi\,\right]_{*v}(s). \tag{5.36}$$

Since, $V \geqslant 0$ and $v_*'(s) \leqslant 0$, we obtain

$$\left[\,|\nabla v|^2\right]_{*v}(s) \leqslant \left[\,|\,F\,|^2\right]_{*v}^{\frac{1}{2}}\left[\,|\nabla v|_{*v}^2\right]^{\frac{1}{2}}(s), \tag{5.37}$$

$$\left[\,|\nabla v|^2\right]_{*v}^{\frac{1}{2}}(s) \leqslant \left[\,|\,F\,|^2\right]_{*v}^{\frac{1}{2}}(s). \tag{5.38}$$

As before, we have

$$-\,v_*'(s) \leqslant \frac{s^{\frac{1}{N}-1}}{N\alpha_N^{\frac{1}{N}}}\left[\,|\nabla v\,|^2\right]_{*v}^{\frac{1}{2}}(s), \tag{5.39}$$

which implies for almost all s

$$-v_*'(s) \leqslant \frac{s^{\frac{1}{N}-1}}{N\alpha_N^{\frac{1}{N}}}\left[|F|^2\right]_{*v}^{\frac{1}{2}}. \tag{5.40}$$

by integrating between 0 and $|\Omega|$ and using the Hardy–Littlewood inequality

$$||\phi||_{L^\infty} \leqslant \begin{cases} c_N \int_{\Omega_*} \sigma^{\frac{1}{N}-1}(|F|^2)_{**}^{\frac{1}{2}}(\sigma)d\sigma, & \text{if } N \geqslant 3, \\ c_{2,\varepsilon}\|F\|_{L^{2+\varepsilon}(\Omega)^2}, & \text{if } N = 2. \end{cases}$$

We conclude as in Proposition 5.2.2. ◇

To conclude, here is how we apply the gradient estimation method to obtain the regularity of ϕ and in $W_0^1 L^{N',\infty}(\Omega)$ when $T \in L^1(\Omega)$

Theorem 5.6.1 *Let ϕ be the solution to Eq. (5.33) with $T \in L^1(\Omega) \cap W'$. Then we have*

1. $\forall \varphi \in L^\infty(\Omega_*, \varphi \geqslant 0$

$$\int_{\Omega_*} \varphi^*(t)|\nabla\phi|_*^2(t)dt \leqslant \left(\frac{1}{N\alpha_N^{\frac{1}{N}}}\right)^2 \left(\int_{\Omega_*} \varphi(t)t^{-2N'}dt\right)|T|_{L^1}^2$$

2. $s^{\frac{1}{N'}}|\nabla\phi|_*(s) \leqslant c_N'|T|_{L^1}$, $\forall s \in \Omega_*$. *i.e.* $|\nabla\phi|_{L^{N',\infty}} \leqslant c_N|T|_{L^1}$, *where c_N depends only on the dimension N.*

Proof The proof follows the same argument as that of Theorem 5.2.5, since we obtained the relation:

$$\left(|\nabla\phi|^2\right)_{*v}(s) \leqslant -v_*'(s)\int_0^s T_*(\sigma)d\sigma.$$

(see (10.38), which implies using the (PSR) inequality (Theorem 4.1.1) that

$$\left(|\nabla\phi|^2\right)_{*v}(s) \leqslant -\frac{s^{-\frac{2}{N'}}}{(N\alpha_N^{\frac{1}{N}})^2}\left(\int_0^s T_*(\sigma)d\sigma\right)^2$$

which corresponds to the same inequality (5.14) of the proof of Theorem 5.2.5. This allows us to conclude with the same argument ◇

Pre-Bibliographical Notes

The majority of the proofs given in this chapter are due to the author and can be found in the various articles cited in [128, 130]. For the results themselves, those of Sects. 5.1 and 5.2 are from the author and originate from the articles published in 1986–1987 see [121, 122, 145].

As for the results of Sect. 5.3, they are due to different authors on the Grad–Mercier models that were introduced into the mathematical literature by R. Temam [166, 167], see also [22, 27, 120, 155]. For qualitative properties see the book by J. Mossino and the references cited in this book.

Theorem 5.5.1 is stated in Dìaz–Padial–Rakotoson [63].

As explained, Theorem 5.2.5 is due to Cianchi–Maz'ja (see also [15]).

The last Sect. 5.6 is taken from the article [65].

Chapter 6
Continuity of the Derivative Map of the Monotone Rearrangement

In this chapter, we will study the continuity of the map, $u \in W^{1,p}(\Omega) \to u'_* \in L^p(\Omega_*, k)$. The motivations for this question have already been discussed in Chap. 2, but in addition, the results of this chapter will be essential for the next chapters. The answer to the question was provided by Coron, Almgren and Lieb and is quite surprising.

If $N = 1$ (so $k = 1$), it is always continuous.

If $N \geqslant 2$, then the Radon measure associated with the derivative distribution of $t \to m_{o,u}(t) = measure\Big\{x : u(x) > t, \nabla u(x) = 0\Big\}$ must be totally singular (i.e. must not have a part that is absolutely continuous with respect to the Lebesgue measure) for the map to be continuous at the point u. In this case, we will say that the function is co-area regular, this is the case for functions in $W^{N,p}_{loc}(\Omega),\ p > 1$.

The link with dimension 1 is that any function in $W^{1,1}_{loc}(\Omega)$ is co-area regular for Ω an interval of $\mathbb{R}$.

We will introduce some general results for this.

6.1 Some General Results: Convergence in $W^{1,1}$ and Length of an Arc of a Monotone Function

We have the following theorem:

Theorem 6.1.1 *Let $\{v,\ v_j\}$ be a sequence of decreasing functions of $W^{1,1}(\Omega_*)$. We assume that we have the following convergences:*

1. $\lim\limits_{j} \displaystyle\int_{\Omega_*} \big|v_j(t) - v(t)\big|\,dt = 0,$

(continued)

J. M. Rakotoson, *Relative Rearrangement*, Lecture Notes in Mathematics 2376,
https://doi.org/10.1007/978-3-032-02228-8_6

Theorem 6.1.1 (continued)

2. $\lim_j \int_{\Omega_*} \sqrt{1+v_j'(t)^2}\, dt = \int_{\Omega_*} \sqrt{1+v'(t)^2}\, dt.$

Then

$$\lim_j \int_{\Omega_*} \left| v_j'(t) - v'(t) \right| dt = 0$$

and for all $0 < \alpha < \beta < |\Omega|$,

$$\lim_j \int_\alpha^\beta \sqrt{1+v_j'(t)^2}\, dt = \int_\alpha^\beta \sqrt{1+v'(t)^2}\, dt\,.$$

Proof of the Theorem

1st Step Let us show that for all $\alpha \in]0,1[$, we have:

$$\int_{\Omega_*} \left| v_j' - v' \right|^\alpha (t) dt \underset{j}{\to} 0\,.$$

Let $n \in \mathbb{N}^*$ be $\widetilde{v}_n'(x) = \begin{cases} v'(x) & if\ |v'(x)| \leqslant n \\ 0 & otherwise \end{cases}$. According to Lusin's theorem (see Rudin's book [150]), there exists $\varphi_n \in C_c(\Omega_*)$ s.t

$$0 \leqslant -\varphi_n \leqslant n, \ \left| \left\{ x \in \Omega_* : \widetilde{v}_n'(x) \neq \varphi_n(x) \right\} \right| \leqslant \frac{1}{2^n}. \tag{6.1}$$

If $a \geqslant 0 and\ b \geqslant 0$ then we have:

$$\sqrt{1+b^2} \geqslant \sqrt{1+a^2} + \frac{a}{\sqrt{1+a^2}}(b-a) + \frac{1}{2}\frac{(b-a)^2}{\left[1+\max(a,b)^2\right]^{\frac{3}{2}}}\,.$$

and since $-v_j' \geqslant 0,\ -\varphi_n \geqslant 0$, we deduce:

$$\int_{\Omega_*} \sqrt{1+v_j'(t)^2}\, dt \geqslant \int_{\Omega_*} \sqrt{1+\varphi_n^2(t)}\, dt$$

$$+\int_{\Omega_*} \frac{\varphi_n(t)\big(v_j'-\varphi_n\big)(t)}{\sqrt{1+\varphi_n^2(t)}}\, dt + \frac{1}{2}\int_{\Omega_*} \frac{\left(v_j'-\varphi_n\right)^2 dt}{\left[1+\max\left(v_j',\varphi_n\right)^2\right]^{\frac{3}{2}}}. \tag{6.2}$$

Since $\lim\limits_{j\to+\infty}\int_{\Omega_*}\frac{\varphi_n}{\sqrt{1+\varphi_n^2}}(v'_j-v')(t)dt = 0$ (according to hypotheses 1. and 2.) and we have:

$$|v'-\varphi_n|_1 \leqslant |v'-\widetilde{v}'_n|_1 + |\widetilde{v}'_n-\varphi_n|_1 \leqslant \int\limits_{|v'|\geqslant n}|v'|\,dt + \frac{n}{2^{n-1}} = o(1)_{n\to+\infty},$$

we conclude,

$$\int_{\Omega_*}\frac{\varphi_n(t)(v'_j-\varphi_n)(t)}{\sqrt{1+\varphi_n^2(t)}}dt \leqslant \int\limits_{|v'|\geqslant n}|v'|dt + \frac{n}{2^{n-1}} + o(1)_{j\to+\infty}.$$

Moreover,

$$\int_{\Omega_*}\frac{\left(v'_j-\varphi_n\right)^2(t)dt}{\left[1+\max\left(v'_j,\varphi_n\right)^2\right]^{\frac{3}{2}}}$$

$$\geqslant \int_{\Omega_*}\frac{\left(v'_j-v'\right)^2(t)dt}{\left[1+\max\left(v'_j,v'\right)^2\right]^{\frac{3}{2}}} - 4\left|\left\{|v'|\geqslant n\right\}\right| - \frac{1}{2^{n-2}} \tag{6.3}$$

(indeed, $\Omega_* = \left\{|v'|\leqslant n, v'=\varphi_n\right\}\cup\left\{|v'|\leqslant n,\ \widetilde{v}'_n\neq\varphi_n\right\}\cup\left\{|v'|\geqslant n\right\}$ and $\frac{(b-a)^2}{\left[1+\max(a,b)^2\right]^{\frac{3}{2}}}\leqslant 2$, for $a\geqslant 0,\ b\geqslant 0$, from which a simple decomposition of the integrals allows us to deduce the above inequality by taking into account the upper bound (6.1)). By making j tend towards infinity in the relation (6.2), we deduce, from these inequalities and from 2.,

$$\limsup_j \frac{1}{2}\int_{\Omega_*}\frac{\left(v'_j-v'\right)^2(t)dt}{\left[1+\max\left(v'_j,v'\right)^2\right]^{\frac{3}{2}}} \leqslant 2\left|\left\{|v'|\geqslant n\right\}\right|$$

$$+\frac{1}{2^{n-1}} + \frac{n}{2^n} + \int\limits_{\{|v'|\geqslant n\}}|v'|\,dt + \int_{\Omega_*}\left(\sqrt{1+v'^2(t)}-\sqrt{1+\varphi_n^2(t)}\right)dt. \tag{6.4}$$

But

$$\int_{\Omega_*}\left(\sqrt{1+v'^2(t)}-\sqrt{1+\varphi_n^2}\right)dt \leqslant \int_{\Omega_*}|v'-\varphi_n|\,dt \leqslant \int_{\{|v'|\geqslant n\}}|v'|\,dt+\frac{n}{2^{n-1}}\,. \tag{6.5}$$

By combining (6.5) and (6.4) and making n tend towards infinity we arrive at:

$$\lim_j \int_{\Omega_*}\frac{\left(v'_j-v'\right)^2(t)dt}{\left[1+\max(v'_j,v')^2\right]^{\frac{3}{2}}}=0. \tag{6.6}$$

We deduce from this relation (6.6) that $\forall\,\theta>0$

$$\lim_j \int_{\Omega_*}\left|v'_j(t)-v(t)\right|\chi_{|(|v'_j)\leqslant\theta\}}(t)\chi_{|v'|\leqslant\theta\}}(t)dt=0, \tag{6.7}$$

and as $\displaystyle\int_{\Omega_*}\left|v'_j\right|dt\leqslant c$, c independent of j, we deduce

$$\left|\left\{t\in\Omega_*:\left|v'_j(t)\right|>\theta\right\}\right|\leqslant\frac{c}{\theta}\ and\ \left|\left\{\left|v'\right|>\theta\right\}\right|\leqslant\frac{c}{\theta}. \tag{6.8}$$

The relations (6.7) and (6.8) imply that $\displaystyle\lim_j\int_{\Omega_*}\left|v'_j-v'\right|^\alpha=0$.

2nd Step Let us show that

$$\lim_j \int_{\Omega_*}\left|\sqrt{1+v'_j(t)^2}-\sqrt{1+v'(t)^2}\right|\,dt=0\,. \tag{6.9}$$

Let $\displaystyle A_j=\int_{\Omega_*}\sqrt{1+v_j'^2(t)}dt$, $f_j=\dfrac{\sqrt{1+v_j'^2(t)}}{A_j}$, $\displaystyle A=\int_{\Omega_*}\sqrt{1+v'^2(t)}dt$ and $f=\dfrac{\sqrt{1+v'^2(t)}}{A}$. Suppose that the convergence of (6.9) does not take place, then there exists $\eta>0$ and a subsequence denoted by (v'_j) such that:

$$\int_{\Omega_*}\left|\sqrt{1+v'_j(t)^2}-\sqrt{1+v'^2(t)}\right|\,dt\geqslant\eta>0 \qquad \forall\,j\,.$$

Using the first step, we can extract from this subsequence another subsequence (k_j) which converges simply (consequence of the 1st step, i.e $v'_{k_j}(t)\underset{j}{\to}v'(t)$, $a.e\ t\in\Omega$.).

Then $f_{k_j}(t) \underset{j}{\to} \dfrac{\sqrt{1+v'^2(t)}}{A} = f(t)$ for a.a $t \in \Omega_*$,

$$\int_{\Omega_*} f_{k_j}(t)\,dt = \int_{\Omega_*} f(t)\,dt = 1.$$

Let's show that $\left|f_{k_j} - f\right|_{L^1(\Omega_*)} \underset{j}{\to} 0$

Indeed, if $\varepsilon > 0$, $\exists\, \delta_\varepsilon > 0$ s.t if $|E| \leqslant \delta_\varepsilon then \qquad \int_E f(t)\,dt \leqslant \varepsilon.$

According to Egoroff's theorem, there exists $A_\varepsilon \subset \Omega_*$ s.t $|\Omega_* \backslash A_\varepsilon| \leqslant \delta_\varepsilon$ and $\sup\limits_{x \in A_\varepsilon} \left|f_{k_j}(x) - f(x)\right| \xrightarrow[j\to+\infty]{} 0.$

As a result, we have:

$$\limsup_j \int\limits_{\Omega_*\backslash A_\varepsilon} f_{k_j}(t)\,dt \leqslant 1 - \lim_j \int\limits_{A_\varepsilon} f_{k_j}(t)\,dt = \int\limits_{\Omega_*\backslash A_\varepsilon} f(t)\,dt \leqslant \varepsilon.$$

Consequently

$$\limsup_j \int_{\Omega_*} \left|f_{k_j}(t) - f(t)\right|\,dt \leqslant 2\varepsilon : \lim_j \int_{\Omega_*} \left|f_{k_j}(t) - f(t)\right| dt = 0\,.$$

This contradicts our initial hypothesis. We deduce among other things:

$\forall\,(\alpha, \beta) : 0 < \alpha < \beta < |\Omega|$

$$\lim_j \int_\alpha^\beta \sqrt{1 + v_j'(t)^2}\,dt = \int_\alpha^\beta \sqrt{1 + v'^2(t)}\,dt\,.$$

3rd Step $\lim\limits_j \int_{\Omega_*} \left|v_j' - v'\right| dt = 0.$

Using that for $a \geqslant 0$ and $b \geqslant 0$

$$0 \leqslant |a-b| \left(1 - \frac{2}{\sqrt{1+a^2} + \sqrt{1+b^2}}\right) \leqslant \left|\sqrt{1+a^2} - \sqrt{1+b^2}\right|$$

knowing that $-\, v' \geqslant 0,\ -v_j' \geqslant 0$, we deduce relations (6.7), (6.8) and (6.9) that $\forall \theta > 0$

$$\limsup_j \int_{\Omega_*} \left|v_j' - v'\right| dt \leqslant \frac{c}{\theta} \xrightarrow[\theta\to+\infty]{} 0\,.$$

◇

6.1.1 I-Functions and Global Regularity of the Associated Rearrangement

To obtain functions whose monotonic rearrangement is absolutely continuous in Ω_*, we need the definition:

Definition 6.1.1 Let $I = (\alpha, \beta)$, $\alpha \leqslant \beta$. To $f : \Omega \to \mathbb{R}$, we associate $f^I : \Omega \to \mathbb{R}_+$:

$$f^I(x) = \begin{cases} 0 & if\ f(x) \leqslant \alpha, \\ f(x) - \alpha & if\ \alpha \leqslant f(x) \leqslant \beta, \\ \beta - \alpha & if\ f(x) \geqslant \beta, \end{cases}$$

f^I is called the I-function of f.

We define:

$$\mathfrak{I}_f = \Big\{ I = (\alpha, \beta) \in \mathbb{R}^2 : |f > \beta| > 0, \quad |f > \alpha| < |\Omega|, \quad \alpha \leqslant \beta \Big\}.$$

Property 6.1.1 (of I-Functions)

1. If we introduce $T_{\alpha,\beta}(\sigma) = \begin{cases} \beta & if\ \sigma \geqslant \beta, \\ \sigma & if\ \alpha \leqslant \sigma \leqslant \beta, \\ \alpha & if\ \sigma \leqslant \alpha, \end{cases}$

   then $f^I = T_{\alpha,\beta}(f) - \alpha$, for any function $f : \Omega \to \mathbb{R}$.
2. $0 \leqslant f^I \leqslant \beta - \alpha$, $\left(f^I\right)_* = (f_*)^I$ (which we will denote by f_*^I).
3. If $I_j = (\alpha_j, \beta_j) \underset{j}{\to} I = (\alpha, \beta)$ in $\mathbb{R}^2$, then $f_j^{I_j} \to f^I$ in $L^1(\Omega)$ if $f_j \to f$ in $L^1(\Omega)$.

Proof Assertion 1. is immediately verified. Since the map $\sigma \to T_{\alpha,\beta}(\sigma)$ is increasing, we deduce $\left(f^I\right)_* = T_{\alpha,\beta}(f_*) - \alpha = (f_*)^I$. Since $T_{\alpha,\beta}(\sigma) \geqslant \alpha$, we have $f^I \geqslant 0$. Since $T_{\alpha,\beta}(f) \leqslant \beta$: $f^I \leqslant \beta - \alpha$. We have $\left|f_j^{I_j} - f^I\right|(x) \leqslant \left|f_j(x) - f(x)\right| + \left|\beta - \beta_j\right| + \left|\alpha - \alpha_j\right|$ hence assertion 3. ◇

Lemma 6.1.1 (Of Regularity of Rearrangements of I-Functions)

1. Let $f \geqslant 0,\ f \in W_0^{1,p}(\Omega),\ 1 \leqslant p \leqslant +\infty$.
Then, $\forall I \in \mathfrak{I}_f,\ f_*^I \in W^{1,p}(\Omega_*)$. In addition, we have:

$$\left|\left(f_*^I\right)'\right|_p \leqslant c_I(f) \| \nabla f|_{*f}|_p \text{ with } c_I(f) = \frac{\left(N\alpha_N^{\frac{1}{N}}\right)^{-1}}{|f > \beta|^{1-\frac{1}{N}}}.$$

2. If Ω is a Lipschitzian bounded connected open set then

$$\forall I \in \mathfrak{I}_f,\ f_*^I \in W^{1,p}(\Omega_*) \text{ if } f \in W^{1,p}(\Omega).$$

Moreover $\left|\left(f_*^I\right)'\right|_p \leqslant c_I^1(f) \| \nabla f|_{*f}|_p$, *with* $c_I^1 = \dfrac{Q}{\min\left(|f > \beta|, |f \leqslant \alpha|\right)^{1-\frac{1}{N}}}$, *$Q$ the relative isoperimetric constant associated with Ω.*

Let $m(t) = |f > t|$.

Proof Since $T_{\alpha,\beta}$ is locally Lipschitzian function with $\left|T'_{\alpha,\beta}\right| \leqslant 1$, we deduce that $f_*^I \in W_{loc}^{1,p}(\Omega_*)$ as soon as $f_* \in W_{loc}^{1,p}(\Omega_*)$.

If $f \geqslant 0,\ f \in W_0^{1,p}(\Omega)$, we know that $N\alpha_N^{\frac{1}{N}} \left|\sigma^{1-\frac{1}{N}} f_*'\right|_p \leqslant \| \nabla f|_{*f}|_p$, thus, for $p < +\infty$:

$$\int_{\Omega_*} \left|\left(f_*^I\right)'\right|^p d\sigma = \int\limits_{\alpha < f_* < \beta} |f_*'(\sigma)|^p\, d\sigma \leqslant \int\limits_{m(\beta) \leqslant \sigma \leqslant m(\alpha)} \left(\frac{\sigma}{m(\beta)}\right)^{p-\frac{p}{N}} |f_*'(\sigma)|^p\, d\sigma\,.$$

Hence

$$\left|\left(f_*^I\right)'\right|_p \leqslant \frac{1}{N\alpha_N^{\frac{1}{N}} m(\beta)^{1-\frac{1}{N}}} N\alpha_N^{\frac{1}{N}} \left|\sigma^{1-\frac{1}{N}} f_*'\right|_p \leqslant \frac{\| \nabla f|_{*f}|_p}{N\alpha_N^{\frac{1}{N}} m(\beta)^{1-\frac{1}{N}}}.$$

If $f \in W^{1,+\infty}(\Omega) \cap W_0^{1,1}(\Omega),\ f \geqslant 0$, then letting $p \to +\infty$ in the previous relation, we have:

$$\left|\left(f_*^I\right)'\right|_\infty \leqslant c_I(f) \| \nabla f|_{*f}|_\infty\,.$$

Relation 2 is proved in the same way using the relation

$$\frac{1}{Q}\left|\min\left(\sigma, |\Omega| - \sigma\right)^{1-\frac{1}{N}} f_*'\right|_p \leqslant ||\nabla f|_{*f}|_p \,. \qquad \diamond$$

Note Recall that $||\nabla f|_{*f}|_p \leqslant |\nabla f|_p$.

Lemma 6.1.2 (Convergence of I-Functions) *Let $\{f_j, f\}$ be a sequence in $W^{1,p}(\Omega)$ s.t $f_j \underset{j}{\rightarrow} f$ in $W^{1,p}(\Omega)$.*

Then $\forall\, I \in \mathbb{R}^2$, we have $f_j^I \xrightarrow[j\to+\infty]{} f^I$ in $W^{1,p}(\Omega)$. Here, $1 \leqslant p < +\infty$.

Proof Let $I = (\alpha, \beta) \in \mathbb{R}^2$. Then, a.e. $x \in \Omega$,

$$\left|f_j^I(x) - f^I(x)\right| = \left|T_{\alpha,\beta}(f_j)(x) - T_{\alpha,\beta}(f)(x)\right| \leqslant \left|f_j(x) - f(x)\right| \,.$$

Therefore $f_j^I \to f^I$ in $L^p(\Omega)$-strong.

Suppose that there exists $\varepsilon > 0$ and a subsequence denoted by f_j such that

$$\int_\Omega \left|\nabla\left(T_{\alpha,\beta}(f_j) - T_{\alpha,\beta}(f)\right)\right|^p dx \geqslant \varepsilon > 0, \qquad \forall\, j \,. \tag{6.10}$$

We can find a function $h \in L^p(\Omega)$ and a subsequence $f_{\sigma(j)}$ s.t

$$\nabla f_{\sigma(j)}(x) \xrightarrow[j\to+\infty]{} \nabla f(x)\ a.e., \quad f_{\sigma(j)}(x) \xrightarrow[j\to+\infty]{} f(x)\ a.e.$$

$$\left|\nabla f_{\sigma(j)}(x)\right| \leqslant h(x),\ a.e.\,.$$

We deduce

$$\nabla T_{\alpha,\beta}\left(f_{\sigma(j)}(x)\right) \xrightarrow[j\to+\infty]{} \nabla T_{\alpha,\beta}\left(f(x)\right)\ a.e.$$

$$\left|\nabla T_{\alpha,\beta}\left(f_{\sigma(j)}(x)\right)\right| \leqslant h(x)\ a.e.$$

For this subsequence we have:

$$\lim_j \int_\Omega \left|\nabla\left(T_{\alpha,\beta}\left(f_{\sigma(j)}\right) - T_{\alpha,\beta}(f)\right)\right|^p dx = 0$$

which contradicts the relation (6.10). $\diamond$

Here is the result that links the convergence of the derivatives of the rearrangements of the I-functions and their associated functions.

Theorem 6.1.2 *Let $\{f, f_j\}$ be a sequence of functions of $W^{1,1}(\Omega)$ s.t $f_j \xrightarrow[j\to+\infty]{} f$ in $W^{1,1}(\Omega)$. We suppose that $\forall I \in \mathfrak{I}_f$, $\forall I_j \in \mathfrak{I}_{f_j}$ s.t $I_j \to I$ in $\mathbb{R}^2$, $\left(f_{j*}^{I_j}\right)' \underset{j}{\to} (f_*^I)'$ in $L^1(\Omega_*)$.*

Then, $f_{j*} \underset{j}{\to} f_*$ in $W^{1,1}_{loc}(\Omega_*)$-strong.

Proof of the Theorem Since $f_{j*} \to f_*$ in $L^1(\Omega_*)$, it suffices to show that $f'_{j*} \to f'_*$ in $L^1_{loc}(\Omega_*)$. Let $0 < \varepsilon < \omega < |\Omega|$. Suppose that there exists a subsequence (f_n) of (f_j) and a number $\eta > 0$ such that

$$\Delta_n = \int_\varepsilon^\omega \left|f'_{n_*}(t) - f'_*(t)\right| dt \geqslant \eta > 0, \qquad \forall n .$$

If we are to extract a subsequence of f_n, we can find

$$(\varepsilon_1, \omega_1) : 0 < \varepsilon_1 \leqslant \varepsilon < \omega \leqslant \omega_1 < |\Omega|$$

and

$$\alpha_n = f_{n*}(\omega_1) \underset{n}{\to} f_*(\omega_1) = \alpha,\ \beta_n = f_{n_*}(\varepsilon_1) \to f_*(\varepsilon_1) = \beta.$$

Let $I_n = (\alpha_n, \beta_n)$, $I = (\alpha, \beta)$. Note that $\alpha_n \leqslant \beta_n$ and $\alpha \leqslant \beta$.

Additionally, $|f_n > \alpha_n| < |\Omega|$ (resp. $|f > \alpha| < |\Omega|$) and $|f_n > \beta_n| > 0$ (resp. $|f > \beta| > 0$) by the choice of $0 < \varepsilon_1 < \omega_1 < |\Omega|$.

Thus $I_n \in \mathfrak{I}_{f_n}$, $I \in \mathfrak{I}_f$ and $I_n \xrightarrow[n\to+\infty]{} I$. As a result

$$\left|f_{n*}^{I_n} - f_*^I\right|_{W^{1,1}(\Omega_*)} \xrightarrow[n\to+\infty]{} 0 . \tag{6.11}$$

Furthermore, we have

$$\left\{t \in \Omega_* : \varepsilon < t < \omega\right\} \subset f_{n_*}^{-1}([\alpha_n, \beta_n]) \cap f_*^{-1}(([\alpha, \beta]) = F_n ,$$

from which

$$\int_\varepsilon^\omega \left|f'_{n_*}(t) - f'_*(t)\right| dt \leqslant \int_{F_n} \left|\left(f_{n*}^{I_n}\right)(t) - \left(f_*^I\right)'(t)\right| dt \leqslant \left|f_{n*}^{I_n} - f_*^I\right|_{W^{1,1}(\Omega_*)} \to 0$$

(from the relation 6.11). This yields a the contradiction. ◇

If Ω is a connected Lipschitzian bounded open set or if the functions f and f_j have zero trace, so we can improve the previous theorem:

Theorem 6.1.3 *Under the same conditions as Theorem 6.1.2, if we define, for* $\sigma \in \Omega_*$,

$$k(\sigma) = \begin{cases} N\alpha_N^{\frac{1}{N}}\sigma^{1-\frac{1}{N}} & \text{if } f,\ f_j \text{ are in } W_{0+}^{1,1}(\Omega), \\ \dfrac{1}{Q}\min(\sigma, |\Omega|-\sigma)^{1-\frac{1}{N}} & \text{if } \Omega \text{ is a open connected} \\ & \text{Lipchitzian set} \\ & \text{or if } f,\ f_j \text{ are unsigned in} \\ & W_0^{1,1}(\Omega) \text{ with } Q = \left(N\alpha_N^{\frac{1}{N}}\right)^{-1}, \end{cases}$$

then,

$$\left|k\left(f'_{j*} - f'_*\right)\right|_{L^1(\Omega_*)} \xrightarrow[j\to+\infty]{} 0\,.$$

Proof of Theorem 6.1.3 Suppose that there exists $\eta > 0$ and a subsequence also denoted by (f_j) : $\left|k\left(f'_{j*} - f'_*\right)\right|_1 \geqslant 2\eta > 0$.

Let us set $m = \operatorname*{ess\,inf}_\Omega f$, $m_j = \operatorname*{ess\,inf}_\Omega f_j$, $M = \operatorname{ess\,sup}_\Omega f$, $M_j = \operatorname{ess\,sup}_\Omega f_j$ and let us consider $\delta > 0$ s.t $4\displaystyle\int_0^\delta |\nabla f|_*\,dt \leqslant \frac{\eta}{3}$ then, there exists (α,β), (α_j,β_j) such that

$$0 < |m < f \leqslant \alpha| \leqslant \delta, \quad 0 < \left|m_j < f_j \leqslant \alpha_j\right| \leqslant \delta,$$

$$0 < \left|M_j > f_j > \beta_j\right| \leqslant \delta, \quad 0 < |M > f > \beta| \leqslant \delta,$$

with $\alpha \leqslant \beta$, $\alpha_j \leqslant \beta_j$, $\forall\, j$ and $\lim_j \alpha_j = \alpha$, $\lim_j \beta_j = \beta$.

Let $I_j = (\alpha_j, \beta_j)$, $I = (\alpha, \beta)$. So, we have:

$$\left|k\left(f'_{j*} - f'_*\right)\right|_1$$

$$\leqslant \left|k\left(f'_{j*} - \left(f_{j*}^{I_j}\right)'\right)\right|_1 + \left|k\left(\left(f_{j*}^{I_j}\right)' - \left(f_*^{I}\right)'\right)\right|_1 + \left|k\left(\left(f_*^{I}\right)' - f'_*\right)\right|_1$$

$$= A_{1j} + A_{2j} + A_3$$

we have: $A_{1j} = \int_{\{f_{j*}\leqslant\alpha_j\}\cup\{f_{j*}>\beta_j\}} k(\sigma)\left|f'_{j_*}(\sigma)\right| d\sigma$. By introducing the truncation functions $f_j^{H_j}(x) = \min\left(f_j(x), \alpha_j\right)$, and $f_j^{Z_j}(x) = \max\left(f_j(x), \beta_j\right)$, we then have:
$f_{j_*}^{H_j} = \left(f_j\right)_*^{H_j}$, $f_{j_*}^{Z_j} = \left(f_j\right)_*^{H_j}$. So, we deduce:

$$A_{1j} \leqslant \left|k\left(f_{j_*}^{H_j}\right)'\right|_1 + \left|k\left(f_{j_*}^{Z_j}\right)'\right|_1 \leqslant \left|\nabla f_j^{H_j}\right|_1 + \left|\nabla f_j^{Z_j}\right|_1 .$$

Since $\nabla f_j = 0$ almost everywhere on $\left\{f_j = m_j\right\}$, using a simple decomposition and the Hardy–Littlewood theorem:

$$\left|\nabla f_j^{H_j}\right|_1 \leqslant \left|\nabla\left(f_j - f\right)\right|_1 + \int_0^{|m_j<f_j\leqslant\alpha_j|} |\nabla f|_*(\sigma)d\sigma$$

$$\leqslant |\nabla(f_1 - f)|_1 + \int_0^\delta |\nabla f|_*(t)dt .$$

Likewise we have:

$$\left|\nabla f_j^{Z_j}\right|_1 \leqslant \left|\nabla\left(f_j - f\right)\right|_1 + \int_0^\delta |\nabla f|_*(t)dt .$$

Therefore,

$$A_{1j} \leqslant 2\left|\nabla\left(f_j - f\right)\right|_1 + 2\int_0^\delta |\nabla f|_*(t)dt , \tag{6.12}$$

by the PSR inequality and the pointwise inequality (see Corollary 2.2.1):

$$A_3 = \left|k\left(\left(f_*^I\right)' - f_*'\right)\right|_1 \leqslant 2\int_0^\delta |\nabla f|_*(t)dt , \tag{6.13}$$

and finally the term

$$A_{2j} \leqslant \left(\operatorname*{Max}_{\Omega_*} k\right)\left|\left(f_{j_*}^{I_j}\right)' - \left(f_*^I\right)'\right|_1 \xrightarrow[j\to+\infty]{} 0 .$$

By the choice of $\delta > 0$ and for $j \geqslant j_n$ we have:

$$2\left|\nabla\left(f_j - f\right)\right|_1 + 4\int_0^\delta |\nabla f|_*(t)dt + \left(\underset{\overline{\Omega_*}}{\mathrm{Max}}\, k\right)\left|\left(f_{j*}^{I_j}\right)' - \left(f_*^{I}\right)'\right|_1$$

$$\leqslant 3\frac{\eta}{3} = \eta\,.$$

Under these conditions we have for $j \geqslant j_n$

$$0 < 2\eta \leqslant \left|k\left(f'_{j*} - f'_*\right)\right|_1 \leqslant A_{1j} + A_{2j} + A_3 \leqslant \eta\,,$$

hence the contradiction. ◇

Here are the observations that follow from the proofs of the regularity of the I-functions.

Lemma 6.1.3 *Let $f \in W^{1,1}(\Omega)$ s.t $f_* \in W^{1,1}_{loc}(\Omega_*)$ and $\left|k(\cdot)f'_*(\cdot)\right|_1 \leqslant |\nabla f|_1$ where $k : \overline{\Omega_*} \to \mathbb{R}$ is continuous with $k(0) = 0$. If $|f = f_*(0)| > 0$ then $f_* \in W^{1,1}(0,b)$ $\forall b < |\Omega|$. Moreover, if $0 < \gamma = \min\left\{k(\sigma),\ 0 < |f > f_*(0)| < \sigma < b\right\}$, then*

$$|f'_*|_{L^1(0,b)} \leqslant \frac{1}{\gamma}|\nabla f|_1\,.$$

Corollary 6.1.1 (Of the Corollary of Lemma 6.1.3) *Under the same conditions as Lemma 6.1.3, if $|f = f_*(0)| > 0$ and $|f = f_*(|\Omega|)| > 0$ then $f_* \in W^{1,1}(\Omega_*)$. If we note*
$0 < \mu = \min\left\{k(\sigma) : 0 < |f > f_*(0)| < \sigma < |f > f_*(|\Omega|)| < |\Omega|\right\}$, *then*

$$|f'_*|_{L^1(\Omega_*)} \leqslant \frac{1}{\mu}|\nabla f|_1\,.$$

We can extend Lemma 6.1.2, as follows:

Lemma 6.1.4 *Under the same conditions as Lemma 6.1.2, if* $I = (\alpha,\beta)$, $I_j = (\alpha_j,\beta_j)$ *s.t* $I_j \underset{j}{\to} I$ *in* $\mathbb{R}^2$, *then*

$$f_j^{I_j} \to f^I \text{ in } W^{1,p}(\Omega)\text{-strongly.}$$

Proof Indeed, $\left|\nabla f_j^{I_j} - \nabla f^I\right|_p^p \leqslant$

$$\leqslant c \int\limits_{\min(\alpha_j,\alpha)\leqslant f_j\leqslant\max(\alpha_j,\alpha)} |\nabla f_j|^p\,dx \quad + c\left|\nabla f_j^I - \nabla f^I\right|_p^p + c\int\limits_{\min(\beta_j,\beta)\leqslant f_j\leqslant\max(\beta_j,\beta)} |\nabla f_j|^p\,dx \quad \xrightarrow[j\to+\infty]{} 0\,.$$

◇

To complete Theorems 6.1.2 and 6.1.3, we must show

Theorem 6.1.4 *Under the same conditions as Theorem 6.1.3, if* $f_j \to f$ *in* $W^{1,p}(\Omega)$, $1 \leqslant p < +\infty$, *then*

$$\left|k\left(f'_{j*} - f'_*\right)\right|_p \xrightarrow[j\to+\infty]{} 0\,.$$

Proof Under the same conditions as Theorem 6.1.3, we apply the pointwise Poincaré–Sobolev inequality, to deduce for all measurable E

$$E \subset \Omega_*,\quad \int_E \left(k\left|f'_{j*}\right|\right)^p dt \leqslant \int_E \left(|\nabla f_j|_{*f_j}\right)^p(t)dt \leqslant \int_0^{|E|} |\nabla f_j|_*^p(t)dt.$$

Hence, for all measurable $E \subset \Omega_*$

$$\left|k\left(f'_{j*} - f'_*\right)\right|_{L^p(E)} \leqslant \left||\nabla f_j|_*\right|_{L^p(0,|E|)} + \left||\nabla f|_*\right|_{L^p(0,|E|)}\,. \tag{6.14}$$

Thus, if $f'_{j*}(\sigma) \xrightarrow[j\to+\infty]{} f'_*(\sigma)$ a.e., then by applying Vitali's theorem (thanks to the relation (6.14)), we deduce $\left|k\left(f'_{j*} - f'_*\right)\right|_p \underset{j}{\to} 0$.

Let us then suppose that there exists $\eta > 0$ and a subsequence denoted by (f'_{j*}) such that $\left|k\left(f'_{j*} - f'_*\right)\right|_p \geqslant \eta > 0$, then according to Theorem 6.1.3, we can extract a subsequence denoted (f_{k_j}) such that $f'_{k_{j*}}(\sigma) \to f'_*(\sigma)$ a.e. and the previous remark implies

$$\left| k\left(f'_{k_{j*}} - f'_*\right)\right|_p \xrightarrow[j\to+\infty]{} 0.$$

This contradicts the fact $\left| k\left(f'_{k_{j*}} - f'_*\right)\right|_p \geqslant \eta > 0.$ ◇

6.1.2 Arc Length and Inherent Properties

Our goal is to use the convergence theorem (see Theorem 6.1.1) with $v_j = u_j^{I_j}$, $\forall\, I_j \in \mathfrak{I}_{u_j}$. To verify the hypotheses of the theorem, we will need the notion of the length of an arc of a curve.

Definition 6.1.2 (Of an Arc Length) Let $f : \mathbb{R} \to [0, +\infty[$ be a decreasing function. Its derivative distribution is a Radon measure that can be decomposed as $df = g\,dt - d\nu$ where $g \in L^1_{loc}(\mathbb{R})$ and $\nu \geqslant 0$, the singular part, is a locally bounded measure. We call the length of the arc of the graph of f on the compact interval $[a, b]$ the number :

$$\ell_f(a, b) = \int_a^b \sqrt{1 + g^2(t)}\, dt + \int_{[a,b]} d\nu\,.$$

We begin with the following proposition:

Proposition 6.1.1 *Let $\{h, h_j\}$ be a sequence of integrable functions from Ω into $[0, +\infty[$ s.t*

(a) $\liminf_j \int_\Omega h_j(x)dx \leqslant \int_\Omega h\, dx,$

(b) $\liminf_j h_j(x) \geqslant h(x)$ *a.e. in* Ω.

Then, there exists a subsequence $j(1), j(2), \ldots$ s.t. $\lim_{k\to+\infty} h_{j(k)}(x) = h(x)$.

Moreover, $\left|h_{j(k)} - h\right|_{L^1(\Omega)} \xrightarrow[k\to+\infty]{} 0.$

Proof of the Proposition Let $\varphi(x) = \liminf_j h_j(x)$. We have according to Fatou's theorem and hypothesis (a) that:

$$\int_\Omega \varphi(x)\, dx \leqslant \liminf_j \int_\Omega h_j dx \leqslant \int_\Omega h\, dx. \tag{6.15}$$

Furthermore, hypothesis (b) implies that:

$$\varphi(x) \geqslant h(x) : \int_\Omega \varphi(x)\, dx \geqslant \int_\Omega h(x)\, dx. \tag{6.16}$$

Hence the equality of φ and h. Using hypothesis (b), we deduce from this equality that $\lim_j \left(h_j(x) - h(x)\right)_- = 0$ a.e., indeed, we have

$$a_n(x) = \inf_{j \geqslant n} \left(h_j(x) - h(x)\right) \nearrow_{n \to +\infty} \liminf_n (h_n(x) - h(x)) = 0$$

and

$$b_n(x) = \sup_{j \geqslant n} \left(h_j(x) - h(x)\right) \searrow_{n \to +\infty} \limsup_n (h_n(x) - h(x)) \geqslant 0\,.$$

As $a_n(x) \leqslant h_k(x) - h(x) \leqslant b_n(x)$ if $k \geqslant n$ then

$$b_{n-}(x) \leqslant (h_k - h)_-\,(x) \leqslant a_{n-}(x)$$

and $a_{n-} \xrightarrow[n \to +\infty]{} 0$, similarly $b_{n-}(x) \xrightarrow[n \to +\infty]{} 0 as\ (h_k - h)_-\,(x) \xrightarrow[k \to +\infty]{} 0.$

But $0 \leqslant \left(h_j - h\right)_-(x) \leqslant h(x)$, from which by the theorem of dominated convergence, we have $\lim_j \int_\Omega \left(h_j - h\right)_-(x) dx = 0$.

Therefore,

$$\liminf_j \int_\Omega \left(h_j - h\right)_+(x) \leqslant \liminf_j \int_\Omega \left(h_j - h\right) dx \leqslant 0\,.$$

We deduce that $\liminf_j \int_\Omega \left(h_j - h\right)_+(x) dx = 0$, consequently there exists a subsequence $j(k) : \lim_{k \to +\infty} \int_\Omega \left(h_{j(k)} - h\right)_+(x)\, dx = 0$ and

$\left(h_{j(k)}(x) - h(x)\right)_+ \xrightarrow[k \to +\infty]{} 0\ a.e.$

As $\left|h_{j(k)} - h\right|(x) = \left(h_{j(k)}(x) - h(x)\right)_+ + \left(h_{j(k)} - h(x)\right)_- \xrightarrow[k \to +\infty]{} 0$ we have $h_{j(k)} \to h(x)$ a.e.

Similarly,

$$\lim_k \int_\Omega \left|h_{j(k)}(x) - h(x)\right| dx = \lim_k \int_\Omega \left(h_{j(k)} - h\right)_+ (x)\, dx$$
$$+ \lim_k \int_\Omega \left(h_{j(k)} - h\right)_- (x)\, dx = 0\,.$$

◇

Here is one of the key theorems on the convergence of arc lengths:

Theorem 6.1.5 (Convergence of Curve Lengths) *Let $\omega_0, \omega_2, \ldots$ be a sequence of increasing functions from $\mathbb{R}$ to $[0, +\infty[$ and consider the following decomposition of $d\omega_j$, for $j \geqslant 0$*

$$d\omega_j = u_j\, dt + d\nu_j \text{ where } u_j \in L^1_{loc}(\mathbb{R}),\ d\nu_j \geqslant 0\,.$$

We assume that, for almost all $t \in \mathbb{R}$, $\liminf_j u_j(t) \geqslant u_0(t)$ and that $\left|\omega_j - \omega_0\right|_{L^1_{loc}(\mathbb{R})} \xrightarrow[j\to+\infty]{} 0$. Then, $\forall\, \varphi \in C_c(\mathbb{R})$ we have

$$\lim_{j\to+\infty} \left(\int \varphi \sqrt{1+u_j^2}\, dt + \int \varphi\, d\nu_j \right) = \int \varphi \sqrt{1+u_0^2}\, dt + \int \varphi\, d\nu_0.$$

Proof The monotonicity of the ω_j as well as the convergence in $L^1_{loc}(\mathbb{R})$ imply that $\omega_j(t) \xrightarrow[j\to+\infty]{} \omega_0(t)$ at any point t of continuity of ω_0 , so for almost all t.

For all $\psi \in C_c^\infty(\mathbb{R})$, we have:

$$\int \psi u_j\, dt + \int \psi\, d\nu_j = -\int \psi' \omega_j\, dt\,. \tag{6.17}$$

Since $\omega_j \to \omega_0$ in $L^1_{loc}(\mathbb{R})$ we have:

$$\lim_j \int \psi u_j\, dt + \int \psi\, d\nu_j = -\int \psi' \omega_0\, dt = \int \psi u_0\, dt + \int \psi\, d\nu_0\,. \tag{6.18}$$

But $d\omega_j$ is a locally uniformly bounded Radon measure (since ω_j is monotone, $\forall\, j \geqslant 0$). Since $C_c^\infty(\mathbb{R})$ is dense in $C_c(\mathbb{R})$, we deduce that (6.18) is still true for

$\psi \in C_c(\mathbb{R})$. To prove the theorem, it is then sufficient to prove that $\forall \varphi \in C_c(\mathbb{R})$

$$\int \varphi G(u_0)\, dt = \lim_j \int \varphi G(u_j)\, dt \tag{6.19}$$

where $G(u) = \sqrt{1+u^2} - u \leqslant 1$. Indeed, it suffices to subtract (6.18) and (6.19).

Let I be an interval containing the support of φ. Since ν_0 is singular, there exists a measurable set A such that $\nu_0(A) = 0$, $|A| = |I|$, and $A \subset I$. $\forall \varepsilon > 0$, there exists a compact $B_\varepsilon \subset A$ such that $|B_\varepsilon| \geqslant |I| - \varepsilon$.

In fact we have,

$$\liminf_j \int_{B_\varepsilon} u_j\, dt. \leqslant \int_{B_\varepsilon} u_0\, dt. \tag{6.20}$$

Let us assume (6.20) for a moment, then there exists a subsequence denoted u_j such that $u_j(t) \underset{j}{\to} u(t)$ a.e. on B_ε according to the previous proposition. By the theorem of dominated convergence

$$\lim_j \int_{B_\varepsilon} \varphi G(u_j) = \int_{B_\varepsilon} \varphi G(u_0)\, dt\ .$$

Let's put $a_j = \displaystyle\int_I \varphi G(u_j)\, dt - \int_I \varphi G(u_0)\, dt$.

We have $a_j = \displaystyle\int\limits_{A\backslash B_\varepsilon} \varphi\big(G(u_j) - G(u_0)\big)\, dt + \int_{B_\varepsilon} \varphi\left(G(u_j) - G(u_0)\right)\, dt.$

Hence

$$|a_j| \leqslant \varepsilon\, |\varphi|_\infty + \left| \int_{B_\varepsilon} \varphi\left(G(u_j) - G(u_0)\right)\, dt \right| : \limsup_j |a_j| \leqslant \varepsilon\, |\varphi|_\infty\ .$$

So (for a subsequence):

$$\lim_j \int \varphi G(u_j) = \int \varphi G(u_0)\ .$$

To prove (6.20), we consider $\varphi : \mathbb{R} \to [0, 1]$ continuous with support compact s.t $\varphi(t) = 1$ for all $t \in B_\varepsilon$ and $0 \leqslant \varphi(t) < 1$ for $t \notin B_\varepsilon$. For any integer m, we have the relation (6.18) for φ^m, from which:

$$\liminf_j \int_{B_\varepsilon} u_j dt \leqslant \liminf_j \int \varphi^m u_j\, dt$$

$$\leqslant \liminf_j \left(\int \varphi^m u_j\, dt + \int \varphi^m\, d\nu_j \right) = \int \varphi^m u_0\, dt + \int \varphi^m\, d\nu_0.$$

When $m \to +\infty$, we deduce from the convergence theorem that

$$\lim_m \int \varphi^m u_0 \, dt + \int \varphi^m \, d\nu_0 = \int_{B_\varepsilon} u_0 \, dt.$$

Thus, we have:

$$\liminf_j \int_{B_\varepsilon} u_j \, dt \leqslant \int_{B_\varepsilon} u_0 \, dt \ ,$$

hence (6.20) (note $\nu_0(B_\varepsilon) = 0$). Arguing by contradiction, we see that the whole sequence satisfies (6.19). ◇

We will now use the previous convergence result to give the link between the length of the distribution function curve $f(t) = m_u^I(t) = \left|u^I > t\right|$ and its generalized inverse $g(s) = u_*^I(s)$. Given the properties of I-functions we will assume that f and g are positive and bounded. The strategy consists first of looking for the polygonal lines, i.e. let (f_k) be an approximation of f (well chosen), f_k by piecewise affine and (g_k) an approximation of g obtained by the same process.

We will see that the lengths of the arcs of f_k and g_k are identical, i.e. ℓ_k. We will show that ℓ_k is bounded, and that $f_k'(x) \xrightarrow[k\to\infty]{} f'(x)$, $g_k'(x) \xrightarrow[k\to\infty]{} g'(x)$, $f_k(x)k \xrightarrow[k\to\infty]{} f(x)$, $g_k(x)k \xrightarrow[k\to\infty]{} g(x)$ a.e. The result follows by the previous theorem. ◇

To state the general theorem, here are the hypotheses on f and its generalized inverse g:

We assume $f,\ g : \mathbb{R} \to \mathbb{R}_+$ are decreasing and satisfy

(a) $f(x) = a > 0$ if $-\infty < x \leqslant 0$, $f(x) = 0$ if $0 < b \leqslant x < +\infty$,
$g(y) = b$ if $-\infty < y \leqslant 0$, $g(y) = 0$ if $a \leqslant y < +\infty$.

(b) f (resp. g) is right continuous on $[0, b]$ (resp. $[0, a]$).

(c) f and g are related by the formulas:

$$g(y) = \text{Min}\left\{x : f(x) \leqslant y\right\} = \sup\left\{x : f(x) > y\right\},\ y \in]0, a[,$$

$$f(x) = \text{Min}\left\{x : g(y) \leqslant x\right\} = \sup\left\{x : g(y) > x\right\},\ x \in]0, b[.$$

We see that f and g will have identical shapes:

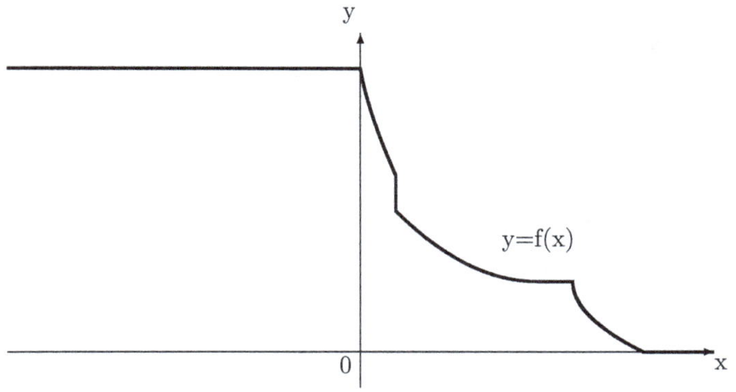

In fact, the graph of f is the image of the graph of g when we permute x and y (i.e. by symmetry with respect to the bisector). More precisely, we have:

Proposition 6.1.2 *Let $\rho : \mathbb{R}^2 \to \mathbb{R}^2$ s.t $\rho(x, y) = (y, x)$. Then, the graph of f in $[0, b] \times [0, a]$ given by:*

$$\Gamma = [0, b]\times[0, a] \cap \left\{(x, y) : \liminf_{t\to x} f(t) \leqslant y \leqslant \limsup_{t\to x} f(t)\right\}$$

coïncides with the image of the graph of g in $[0, a] \times [0, b]$ i.e

$$\sum = [0, a] \times [0, b] \cap \left\{(y, x) : \liminf_{s\to y} g(s) \leqslant x \leqslant \limsup_{s\to y} g(s)\right\}$$

under the action of ρ : $\rho(\sum) = \Gamma$.

Proof By reason of symmetry of f, g, it suffices to show that $\rho(\sum) \subset \Gamma$. Let $0 < y < a$ and x be such that $\liminf_{s\to y} g(s) \leqslant x \leqslant \limsup_{s\to y} g(s)$. This is to show that $0 \leqslant x \leqslant b$ and that $\liminf_{t\to x} f(t) \leqslant y \leqslant \limsup_{t\to x} f(t)$. It is clear that $0 \leqslant x \leqslant b$. Suppose $y < \liminf_{t\to x} f(t)$, then there exists a sequence $t_n \to x$ such that for all s (close to y) satisfying $y < s < \liminf_{t\to x} f(t)$ we have $f(t_n) > s$ for all n. Thus, $t_n \in \{\sigma : f(\sigma) > s\}$, which implies by definition of g (see above) that $g(s) > t_n$. This necessarily implies that $\liminf_{s\to y} g(s) > \lim_n t_n = x$, which contradicts the initial hypothesis for (x, y). ◇

Moreover, we have the following observation, given the monotonicity of f.

If $0 \leqslant t < x \leqslant b$ and $0 \leqslant y < z \leqslant a$ then at most one of the points (t, y) and (x, z) can belong to the graph Γ of f. Moreover, for each $x : -a \leqslant x \leqslant b$ there is exactly one point (x, y) of Γ s.t $y = x - s$ (i.e. which belongs to an axis parallel to the first bisector.) Thanks to this observation, if we consider the rotation θ of angle $\frac{\pi}{4} = 45^o$ in the counterclockwise direction, then the image Γ^* of Γ, i.e. $\Gamma^* = \theta(\Gamma)$, can be represented by the graph of a curve $h : \left[-\frac{a}{\sqrt{2}}, \frac{b}{\sqrt{2}}\right] \to \left[0, \frac{a+b}{2}\right]$ s.t $\left|h'(\sigma)\right| \leqslant 1$, for almost all $\sigma \in \left[-\frac{a}{\sqrt{2}}, \frac{b}{\sqrt{2}}\right]$. Indeed, the steps (i.e plateaus) of f become segments parallel to the first bisector, the "jumps" of f become segments parallel to the second bisector, the monotonicity implies the rest.

To obtain the piecewise affine map f_k, an approximation of f, whose arc length will be bounded, we start by discretizing the interval $\left[-\frac{a}{\sqrt{2}}, \frac{b}{\sqrt{2}}\right]$ by taking a constant value for each step, for $j = 0, \ldots, 2^k$, k being a given integer, i.e. $s_j = -\frac{a}{\sqrt{2}} + \frac{j}{2^k} \cdot \frac{a+b}{\sqrt{2}}$.

Let us then consider the points of Γ^*, $p_j^* = \left(s_j, h(s_j)\right)$, $j = 0, \ldots, 2^k$, and let us denote by Γ_k^* the union of the segments $\left[p_{i-1}^*, p_i^*\right]$ that join the points p_{i-1}^* and p_i^*. Since, $\left|h'(\sigma)\right| \leqslant 1$ a.e., the length of each segment does not exceed $\frac{a+b}{2^k}$. Consequently, the total length of Γ_k^*, ℓ_k, does not exceed $2^k \cdot \frac{a+b}{2^k} = a+b$. Hence $0 \leqslant \ell_k \leqslant a + b$.

Let $\Gamma_k = \theta^{-1}(\Gamma_k^*)$, $(x_i, y_i) = p_i = \theta^{-1}(p_i^*)$, $i = 0, \ldots, 2^k$, $K_k = \left\{x_0, x_1, \ldots, x_{2k}\right\}$ (these points can be confused). We then deduce:

Proposition 6.1.3 *For each* $i = 0, 1, \ldots, 2^k$ *we have* $\liminf_{t \to x_i} f(t) \leqslant y_i \leqslant \limsup_{t \to x_i} f(t)$ *(since* $(x_i, y_i) \in \Gamma$*). In addition, we have* $x_{i-1} \leqslant x_i$ *and* $x_i - x_{i-1} \leqslant \frac{a+b}{2^k}$, $y_{i-1} \geqslant y_i$ *and* $y_{i-1} - y_i \leqslant \frac{a+b}{2^k}$.

For each $x \in (0, b) \backslash K_k$, *there exists a unique index*

$$i \in \{1, \ldots, 2^k\} s.t x_{i-1} < x < x_i.$$

We will choose the x_i as discretization nodes.

We define $f_k : \mathbb{R} \to \mathbb{R}$:

(d) $f_k(t) = f(t)$ if $t \notin [0, b]$, $\left(t, f_k(t)\right) \in \Gamma_k$ if $t \in [0, b]$, f_k right continuous on $]0, b[$,

(continued)

(e) f_k is given for each $i = 0, \ldots, 2^k$ $x_{i-1} < x < x_i$ by

$$f_k(x) = y_{i-1} + \frac{y_i - y_{i-1}}{x_i - x_{i-1}}(x - x_{i-1}).$$

We see that f_k is decreasing.

Similarly, we define $g_k : \mathbb{R} \to \mathbb{R}$ by replacing f by g. If we then denote the different decompositions of the measures $df,\ df_k,\ dg,\ dg_k$ by

$$df = f'(t)\,dt - d\mu,$$

$$df_k = f_k'(t)\,dt - d\mu_k,$$

$$dg = g'(t)\,dt - d\nu,$$

$$dg_k = g_k'(t)\,dt - d\nu_k,$$

and by $\sum_k$ the polygonal line defined by the graph of g_k on $[0, a)$, we have

(f) $\sum_k = \rho(\Gamma_k)$, $\sum_k$ the image of Γ_k by ρ .
(g) Since arc lengths are preserved by reflections ρ we deduce that:

$$\ell_k = \int_{[0,b]} \sqrt{1 + f_k'(t)^2}\,dt + \int_{[0,b]} d\mu_k = \int_{[0,a]} \sqrt{1 + g_k'(t)^2}\,dt + \int_{[0,a]} d\nu_k\,.$$

Moreover, since $K_k \subset K_{k+1}$ are finite, $K = \bigcup_{k \geqslant 0} K_k \subset [0, b]$ is countable.

Since f and g are almost everywhere differentiable, we deduce that for almost all $x \in \mathbb{R} \backslash K$ we have:

(h) $\lim_{j \to +\infty} f_j(x) = f(x)$, $\lim_{j \to +\infty} f_j'(x) = f'(x)$,
$\lim_{j \to +\infty} g_j(x) = g(x)$, $\lim_{j \to +\infty} g_j'(x) = g'(x)$.

In fact, at any point x of differentiability, $x_{i-1} < x < x_i$,

$$|f_k(x) - f(x)| \leqslant \mathrm{Max}\left\{ |f(t) - f(x)|\,,\ |t - x| \leqslant x_i - x_{i-1} \right\} \xrightarrow{k\to+\infty} 0\,.$$

Additionally, there is a function $\varepsilon(\cdot) : \lim_{h\to 0} \varepsilon(h) = 0$ s.t

$$\left|f_k'(x) - f'(x)\right| \leqslant 2\varepsilon\left(\frac{a+b}{2^k}\right) \xrightarrow{k\to+\infty} 0\,.$$

We now come to the main theorem of this section:

Theorem 6.1.6 (Equality of Arc Lengths) *Assume that $f, g : \mathbb{R} \to \mathbb{R}$ are decreasing functions that are inverse to each other satisfying (a), (b), (c). Then, the length of f on $[0, b]$ is equal to that of g on $[0, a]$ i.e.*

$$\int_{[0,b]} \sqrt{1 + f'^2(t)}\, dt + \int_{[0,b]} d\mu = \int_{[0,a]} \sqrt{1 + g'^2(t)}\, dt + \int_{[0,a]} d\nu$$

$$df = f'(t)\, dt - d\mu,\ dg = g'(t)\, dt - d\nu.$$

Proof Since f_k (resp. g_k) is a bounded sequence which converges pointwise to f (resp. g), then $f_k \to f$ (resp. $g_k \to g$) in $L^1_{loc}(\mathbb{R})$.

As $f_k'(x) \xrightarrow{k} f'(x)$ (resp. $g_k'(x) \xrightarrow{k} g'(x)$) almost everywhere, the conditions of Theorem 6.1.5 are fufilled, and we can conclude $\forall \varphi \in C_c^\infty(\mathbb{R})$

$$\lim_k \int \varphi\sqrt{1 + f_k'(t)^2}\, dt + \int \varphi\, d\mu_k = \int \varphi\sqrt{1 + f'(t)^2}\, dt + \int \varphi\, d\mu\,. \quad (6.21)$$

(likewise for g_k, $|g|$).

Also, since $f_k(t) = f(t)\ \forall t > a,\ \forall t < 0$, we deduce: $\int_{\mathbb{R}\setminus[0,a]} \varphi\, df_k = \int_{\mathbb{R}\setminus[0,a]} \varphi\, df$

from where $\int_{\mathbb{R}\setminus[0,a]} \varphi\, d\mu_k = \int_{\mathbb{R}\setminus[0,a]} \varphi\, d\mu = 0.$

So the relation (6.21) above becomes: $\forall \varphi \in C_c^\infty(\mathbb{R})$

$$\lim_k \left(\int_{\mathbb{R}} \varphi \sqrt{1 + f_k'^2(t)}\, dt + \int_{[0,a]} \varphi\, d\mu_k \right) = \int_{\mathbb{R}} \varphi \sqrt{1+f'^2(t)}\, dt + \int_{[0,a]} \varphi\, d\mu\,. \tag{6.22}$$

Now let $\varphi \in C_c^\infty(\mathbb{R})$ s.t $\varphi(x) = 1\ \forall x \in |0, a|$, then the relation (6.22) implies that, since

$$\ell_k = \int_{[0,a]} \sqrt{1 + f_k'^2}\, dt + \int_{[0,a]} d\mu_k\,,$$

$$\lim_{k \to +\infty} \ell_k = \int_{[0,a]} \sqrt{1 + f'^2(t)}\, dt + \int_{[0,a]} d\mu\,.$$

Similarly for g_k and g, proving the result. ◇

6.2 Co-area Regular Functions and Continuity of Mapping $u \in W^{1,p}(\Omega)$(Resp. $W_0^{1,p}(\Omega)$) $\to u'_* \in L^p(\Omega_*, k)$

Notation In this section, $k(\sigma) = N\alpha_N^{\frac{1}{N}}\sigma^{1-\frac{1}{N}}$ when working with a positive function with zero trace and $k(\sigma) = \dfrac{1}{Q}\min(\sigma, \Omega - \sigma)^{1-\frac{1}{N}}$ otherwise (the same condition as Theorem 6.1.3). We define

$$L^p(\Omega_*, k) = \left\{ g \in L^1_{loc}(\Omega_*) : |kg|_{L^p(\Omega_*)} < +\infty \right\}.$$

For the function $u : \Omega \to \mathbb{R},\ I = (\alpha, \beta) \in \mathbb{R}^2$ we will set: u^I the associated I-function and the distribution function for u^I is denoted:

$$m_u^I(t) = \left| u^I > t \right|,\ t \in \mathbb{R}\,.$$

We define $V_p(\Omega) = \begin{cases} W_0^{1,p}(\Omega) & \\ W^{1,p}(\Omega) & \text{if } \Omega \text{ isan open Lipschitzian,} \\ & \text{bounded, connected set}\,. \end{cases}$

Throughout this section we will consider sequences u, u_j in $V_p(\Omega)$ s.t $u_j \xrightarrow[j\to+\infty]{} u$ in $V_p(\Omega)$. If $I_j \in \mathfrak{I}_{u_j}$, $I \in \mathfrak{I}_u$ s.t $I_j \to I$, we define:

$$f_j(t) = m_{u_j}^{I_j}(t), \qquad f(t) = m_u^I(t), \quad t \in \mathbb{R}$$

$$g_j(s) = \begin{cases} u_{j*}^{I_j}(s) & if\ s \in [0, |\Omega|], \\ 0 & if \sigma \geqslant |\Omega|, \\ \beta_j - \alpha_j & if\ s \leqslant 0\,. \end{cases}$$

Likewise

$$g(s) = \begin{cases} u_*^I(s) & if\ 0 \leqslant s \leqslant |\Omega|, \\ 0 & if \sigma \geqslant |\Omega|, \\ \beta - \alpha & if \sigma \leqslant 0\,. \end{cases}$$

We have $g(|\Omega|) = 0$. If $I = (\alpha, \beta) \in \mathfrak{I}_u$, $\operatorname*{ess\,inf}_{\Omega} u \leqslant \alpha \leqslant \beta \leqslant \operatorname*{ess\,sup}_{\Omega} u$, then, $g(s) = T_{\alpha,\beta}\big(u_*(s)\big) - \alpha = 0$, if $m_u^I(\alpha) < s \leqslant |\Omega|$. Similarly, $g(0) = \beta - \alpha$, since $u_*(0) = \operatorname*{ess\,sup}_{\Omega} u \geqslant \beta$.

With $a = |\Omega|$, $b_j = \beta_j - \alpha_j$, $b = \beta - \alpha$, we see that:

$$f_j(t) = \begin{cases} |\Omega| & if\ t \leqslant 0, \\ \text{decreasing, right continuous} & on\ [0, b_j[, \\ 0 & if\ t \geqslant b_j = \beta_j - \alpha_j. \end{cases}$$

The same is true for f. Note that $f_j(t) = f(t)$, if $t \leqslant 0$ or $t \geqslant \max(b, b_j)$.

The following proposition follows from the results related to the I-functions given in the first and second sections.

Proposition 6.2.1 *Under the previous notation, we have:*

1. g, g_j *are in* $W^{1,1}_{loc}(\mathbb{R})$ *so that in* $W^{1,1}(\Omega_*)$,
2. $|g_j - g|_{L^p(\Omega_*)} \xrightarrow[j\to+\infty]{} 0$,
3. $|f_j - f| \xrightarrow[j\to+\infty]{} 0$ *in* $L^1_{loc}(\mathbb{R})$,
4. *If* $df_j = f_j'(t)\,dt - d\mu_j$, $(df = f'(t) - d\mu)$, *then*

(continued)

Proposition 6.2.1 (continued)

$$\int_{[0,b_j]} \sqrt{1+f_j'^2(t)}\,dt + \int_{[0,b_j]} d\mu_j = \int_{\Omega_*} \sqrt{1+g_j'(s)^2}\,ds$$

$$\int_{[0,b]} \sqrt{1+f'^2(t)}\,dt + \int_{[0,b]} d\mu = \int_{\Omega_*} \sqrt{1+g'(s)^2}\,ds.$$

Proof Statement 1. follows from Lemma 6.1.1. For statement 2. we use Lemma 6.1.4 and the contraction property that

$$\left|g_j - g\right|_{L^p(\Omega_*)} = \left|u_{j*}^{I_j} - u_*^I\right|_{L^p(\Omega_*)} \leqslant \left|u_j^{I_j} - u^I\right|_{L^p(\Omega)} \xrightarrow[j\to+\infty]{} 0\,.$$

Therefore $g_j \to g$ in $L^p_{loc}(\mathbb{R})$.

3. For all t, we have:

$$f(t) \leqslant \liminf_j f_j(t) \leqslant \limsup_j f_j(t) \leqslant \left|u^I \geqslant t\right|\,,$$

thus for almost all $t \in \mathbb{R}$, we have $f(t) = \left|u^I \geqslant t\right|$, whence $\lim_j f_j(t) = f(t)$, and as $0 \leqslant f_j(t) \leqslant |\Omega|$, we deduce the result.

For statement 4./, since $g,\ g_j$ are in $W^{1,1}(\Omega_*)$, we have $dg_j = g_j'(s)\,ds$ (resp. $dg = g'\,ds$), by the invariance of arc lengths (Theorem 6.1.6) . ◇

We now make the following observation, if:

$$\left|f'(t)\right| \leqslant \liminf_j \left|f_j'(t)\right| \quad a.e.\ t \qquad \text{“principal inequality”}$$

then Theorem 6.1.5 is applicable i.e. $\forall\,\varphi \in C_c^\infty(\mathbb{R})$

$$\lim_j \int_{\mathbb{R}} \varphi\sqrt{1+f_j'^2(t)}\,dt + \int_{\mathbb{R}} \varphi\,d\mu_j = \int_{\mathbb{R}} \varphi\sqrt{1+f'^2(t)}\,dt + \int_{\mathbb{R}} \varphi\,d\mu.$$

By choosing a suitable function φ ($\varphi = 1$ on $[0, b+1]$), since $b_j \to b$ we deduce from the Proposition 6.2.1 that

$$\lim_j \int_{\Omega_*} \sqrt{1+g_j'^2(s)}\,ds = \int_{\Omega_*} \sqrt{1+g'^2(s)}\,ds$$

Using assertion (2) and Theorem 6.1.1, we deduce $g'_j \to g'$ in $L^1(\Omega_*)$. We apply Theorem 6.1.3 to obtain $\left|k\left(u'_{j*}-u'_*\right)\right|_1 \xrightarrow[j\to+\infty]{} 0$. Finally we conclude from Theorem 6.1.4:

$$\left|k\left(u'_{j_*}-u'_*\right)\right|_p \xrightarrow[j\to+\infty]{} 0.$$

Thus, we must have the main inequality: $\left|f'(t)\right| \leqslant \liminf_j \left|f'_j(t)\right|$. This hypothesis is satisfied only by a class of functions called co-area regular functions whose Definition 6.2.1 is in Sect. 6.2.2.

6.2.1 Decomposition of a Distribution Function and Various Properties

Let $u \in W^{1,p}(\Omega)$, $1 \leqslant p < +\infty$. We consider the following associated decreasing functions, for all $t \in \mathbb{R}$:

$$m_u(t) = |u > t|\,,\ m_{o,u}(t) = \left|\left\{x \in \Omega : u(x) > t,\ \nabla u(x) = 0\right\}\right|$$

and

$$m_{1,u}(t) = m_u(t) - m_{o,u}(t) = \left|\left\{x \in \Omega : u(x) > t,\ \nabla u(x) \neq 0\right\}\right|\,.$$

Here is a proposition that follows from Federer's formulas:

Proposition 6.2.2

$m_{1,u}$ is absolutely continuous on $\mathbb{R}$.

Proof Indeed, Federer's formula is written as: $\displaystyle\int_\Omega g\,|\nabla u|\,dx$

$$= \int_{-\infty}^{+\infty} dt \int_{\{u=t\}} g(x)d\mathcal{L}^{N-1}(x)\,,\ (if\ N \geqslant 2),\ g : \mathbb{R} \to \mathbb{R}_+ \text{ measurable.}$$

(Here, $\mathcal{L}^{N-1} = H^{N-1}$, the Hausdorff measure of dimension $N-1$.)

By choosing $g(x) = \frac{1}{|\nabla u(x)|}\chi_{\{|\nabla u(x)|\neq 0\ ,t<u<t+h\}}(x)$, $h > 0$, we deduce $m_{1,u}(t+h) - m_{1,u}(t) = \int_t^{t+h} d\theta \int_{\{u=0,\ \nabla u\neq 0\}} \frac{d\mathcal{L}^{N-1}(x)}{|\nabla u(x)|}$, and as the function $\theta \to \int_{\{u=\theta,\ \nabla u\neq 0\}} \frac{d\mathcal{L}^{N-1}(x)}{|\nabla u(x)|}$ is integrable on $\mathbb{R}$, we obtain the result.

The same proof applies for $N = 1$. ◇

Another consequence of this same Federer formula that can be useful is the following:

Proposition 6.2.3 *Let $B \subset \mathbb{R}$ s.t $|B| = 0$. Then, the gradient of u is almost everywhere zero on* $\Big\{x : u(x) \in B\Big\}$.

Proof We choose $g(x) = \chi_B\Big(u(x)\Big)$ in Federer's formula. Hence $\int_{\{x:u(x)\in B\}} |\nabla u|\,(x)dx = 0$, from which the result follows. ◇

Here is another proof of this same proposition without using Federer's formula. Let $U_1 \supset U_2 \supset \ldots$ be a sequence of open sets of $\mathbb{R}$ such that $B \subset \bigcap_{j=1}^{+\infty} U_j = V$, $\big|U_j\big| \xrightarrow[j\to+\infty]{} 0$.

Let $M_j(t) = \int_0^t \chi_{U_j}(s)ds$, then we have $\big|M_j(t)\big| \leqslant |t|$ and $M_j(t) \xrightarrow[j\to+\infty]{} 0$. Let us then show $\forall\, \varphi \in C_c^\infty(\Omega; \mathbb{R}^N)$, for each j,

$$\int \big[u - M_j(u)\big]\, div(\varphi)dx = -\int \Big[1 - \chi_{\{u\in U_j\}}\Big]\nabla u\cdot \varphi dx. \tag{6.23}$$

Let us admit this last formula for a moment. Since $\big|M_j(u)\big| \leqslant |u|$, and

$$M_j\big(u(x)\big) \xrightarrow[j\to+\infty]{} 0 \qquad \text{for almost all } x\ ,$$

the left-hand side of the equality (6.23) tends to $\int u\, div(\varphi)dx$.

As for the right-hand side , since $0 \leqslant 1 - \chi_{\{u \in U_j\}} \leqslant 1$ and $\chi_{\{u \in U_j\}}(x) \to \chi_{\{u \in V\}}(x)$ a.e., we deduce that it converges to $\int \left[1 - \chi_{\{u \in V\}}(x)\right] \nabla u \cdot \varphi dx$. From this we have $\forall \varphi \in C_c^{\infty}$,

$$-\int (\nabla u \cdot \varphi)\, dx = -\int \left[1 - \chi_{\{u \in V\}}\right] \nabla u \cdot \varphi$$

which then implies $\nabla u = \left[1 - \chi_{\{u \in V\}}\right] \nabla u$ a.e.

Let $\nabla u(x) = 0$ if $x \in u^{-1}(V) \supset u^{-1}(B)$.

The proof of the formula (6.23) amounts to showing that for any open subset U of $\mathbb{R}$, of finite measure, we have:

$$\int \Big[u - M(u)\Big] div(\varphi) dx = \int \Big[1 - \chi_{\{u \in U\}}\Big] \nabla u \cdot \varphi dx \tag{6.24}$$

where $M(t) = \int_0^t \chi_U(s) ds,\ t \in \mathbb{R}$.

Let $1 \geqslant g_j \geqslant 0$ be a sequence of continuous functions s.t $0 \leqslant g_1(t) \leqslant g_2(t) \leqslant \cdots \leqslant g_j(t) \xrightarrow[j \to +\infty]{} \chi_U(t)$ everywhere. So for j fixed, let us set $N_j(t) = \int_0^t g_j(s)\, ds$, since $\nabla (N_j \circ u) = (g_j \circ u)(\nabla u)$ we then deduce:

$$\int \Big[u - N_j(u)\Big] div\varphi\, dx = -\int \Big[1 - g_j(u)\Big] \nabla u \cdot \varphi\, dx. \tag{6.25}$$

By the dominated convergence theorem, since $N_j(t) \to M(t)\ \forall t$, we have

$$\lim_j \int \Big[1 - N_j(u)\Big] div\varphi\, dx = \int \Big[1 - M(u)\Big] div\varphi\, dx$$

and

$$\lim_j \int \Big[1 - g_j(u)\Big] \nabla u \cdot \varphi\, dx = \int \Big[1 - \chi_{\{u \in U\}}\Big] \nabla u \cdot \varphi\, dx.$$

We then deduce (6.24) from (6.25). ◇

Since $m_{1,u}$ is absolutely continuous on $\mathbb{R}$, we deduce:

Corollary 6.2.1 (Of Proposition 6.2.2) *For $A \subset \mathbb{R}$ measurable*

$$m'_{1,u}(A) = \int_A dm_{1,u}(t) = -\int_\Omega \chi_{\{u \in A\}} \cdot \chi_{\{\nabla u \neq 0\}} dx.$$

Proof Let a, b be two real numbers $]a, b[\subset I(u)$. We have

$$\int_{[a,b]} dm_{1,u}(t) = m_{1,u}(b) - m_{1,u}(a) = -\left|\left\{x : a < u(x) < b,\ \nabla u \neq 0\right\}\right|$$

$$= -\int_\Omega \chi_{\{u\in]a,b[\}} \cdot \chi_{\{\nabla u\neq 0\}} dx.$$

So the formula is true for an open set and therefore true for a G_δ set (for the definition, see the books by G. Choquet or the article [40]). Moreover, if E is a set of measure zero, we have: $\left|\left\{x : u(x) \in E : \nabla u(x) \neq 0\right\}\right| = 0$ according to Proposition 6.2.3. As for any measurable set A, there exists a G_δ set V such that $A \subset V$ and $|A| = |V|$, we have:

$$\left|\left\{x : u(x) \in A : \nabla u(x) \neq 0\right\}\right| = \left|\left\{x : u(x) \in V : \nabla u(x) \neq 0\right\}\right| + 0\,.$$

Hence

$$m'_{1,u}(A) = -\int_\Omega \chi_{\{u\in A\}} \cdot \chi_{\{\nabla u\neq 0\}} dx\,.$$

◇

Note Recall that according to the Fleming–Rishel formula (see Theorem 3.3.1) the map $t \to \int_\Omega \chi_{\{u>t\}}(x) \cdot |\nabla u|\, dx$ is absolutely continuous on $\mathbb{R}$ because

$$\int_{\{u>t\}} |\nabla u|\, dx = \int_t^{+\infty} P_\Omega(\{u > \theta\}) d\theta \text{ for } u \in W^{1,1}(\Omega).$$

Theorem 6.2.1 (Convergence of Truncations of m_{1,u_j}) *Let $\{u, u_1, u_2, \ldots\}$ be a sequence of elements of $W^{1,1}(\Omega)$ such that $u_j \xrightarrow[j\to+\infty]{} u$ in $W^{1,1}(\Omega)$.*

We consider, for $\varepsilon > 0$, the function $\Phi_\varepsilon(\xi) = \dfrac{|\xi|^2}{\varepsilon^2 + |\xi|^2}$, where $\xi \in \mathbb{R}^N$, $|\xi|$ being the Euclidean norm of ξ. We define for $t \in \mathbb{R}$,

$$m^\varepsilon_{1,u_j}(t) = \int_\Omega \chi_{\{u_j>t\}} \cdot \Phi_\varepsilon(\nabla u_j) dx \text{ (same for } u).$$

(continued)

Theorem 6.2.1 (continued)
1. *$t \in \mathbb{R} \to m^\varepsilon_{1,u}(t)$ (resp. $m^\varepsilon_{1,u_j}(t)$) is absolutely continuous. Moreover, the sequence $(m^\varepsilon_{1,u_j})_j$ is equi-absolutely continuous.*
2. *For a Borel subset $A \subset \mathbb{R}$ of finite measure, the functions m^ε_{1,u_j} converge in measure to $m^\varepsilon_{1,u}$ on A when $j \to +\infty$.*
   In particular, there exists a subsequence $u_{j(k)}$ such that:
   $(m^\varepsilon_{1,u_{j(k)}})'(t) \to (m^\varepsilon_{1,u})'(t)$ a.e. in $t \in A$.

Proof For assertion 1., we have for any interval $]a,b[$, according to the Fleming–Rishel formula,

$$0 \leqslant m^\varepsilon_{1,u}(a) - m^\varepsilon_{1,u}(b) \leqslant \frac{1}{\varepsilon}\int_{a<u<b} |\nabla u|\,dx = \frac{1}{\varepsilon}\int_a^b P_\Omega(u>t)dt,$$

which implies absolute continuity. Let us also denote u by u_0.

The equi-absolute continuity of the m^ε_{1,u_j} follows from this last formula because for a sequence of pairwise disjoint intervals $(]a_i,b_i[)_{i=1,\dots,m}$ such that $\sum_{i=1}^m (b_i - a_i) \leqslant \delta$, we have $\forall\, j$

$$0 \leqslant \sum_{i=1}^m \left(m^\varepsilon_{1,u_j}(a_i) - m^\varepsilon_{1,u_j}(b_i)\right) \leqslant \frac{1}{\varepsilon}\int_{\bigcup_{i=1}^m]a_i,b_i[} |\nabla u_j|\,dx \leqslant \frac{1}{\varepsilon}\int_0^\delta |\nabla u_j|_*(t)dt\ .$$ Since $u_j \to u$ in $W^{1,1}(\Omega)$, it follows that $\int_0^\delta |\nabla u_j|_*(t)dt$

$$\leqslant \left|\nabla\left(u_j - u\right)\right|_1 + \int_0^\delta |\nabla u|_*(t)dt \leqslant 2\int_0^\delta |\nabla u|_*(t)dt,\ j \geqslant j_\delta.$$

For the second assertion, consider a Borel set A of finite measure. For j fixed, and $\eta > 0$ we let

$$A_j = A \cap \left\{t : \left(m^\varepsilon_{1,u_j}\right)'(t) - \left(m^\varepsilon_{1,u}\right)'(t) > \eta\right\} \tag{6.26}$$

(The proof is identical for the case $< -\eta$). We want to show that $|A_j| \to 0$.

Suppose that this is not true. Then for a subsequence, again denoted (A_j), there exists $\alpha > 0$ such that $|A_j| > 5\alpha > 0\ \forall\, j$. Since $(m^\varepsilon_{1,u_j})_{j\geqslant 0}$ is equi-absolute continuous, there exists $\delta > 0$ s.t $B \subset \mathbb{R}$ then:

$$|B| \leqslant \delta \textit{ implies } \int_B \left|dm^{\varepsilon}_{1,u_j}(t)\right| \leqslant \alpha\eta, \forall j\,. \tag{6.27}$$

By Lusin's theorem, for each j there exists a continuous function $L_j : \mathbb{R} \to [0, 1]$ $s.t$ $B_j = \left\{t \in \mathbb{R} : L_j(t) \neq \chi_{A_j}(t)\right\}$ has measure less than δ. Then, for $j \geqslant 1,\ k \in \mathbb{N}$:

$$\int \left|L_j - \chi_{A_j}\right|(t) \left|dm^{\varepsilon}_{1,u_k}(t)\right| \leqslant \int_{B_j} \left|\left(m^{\varepsilon}_{1,u_k}\right)'(t)\right| dt \leqslant \eta\alpha. \tag{6.28}$$

Moreover, since $\left|A_j\right| > 5\alpha > 0$ and by definition of A_j, we have:

$$\int \chi_{A_j}\left[\left(m^{\varepsilon}_{1,u_j}\right)'(t) - \left(m^{\varepsilon}_{1,u}\right)'(t)\right]dt > \eta\left|A_j\right| > 5\alpha\eta. \tag{6.29}$$

Furthermore, using Federer's formula (or a simple extension of the Corollary 6.2.1 of Proposition 6.2.2), we have:

$$\int \chi_{A_j}\left(m^{\varepsilon}_{1,u_j}\right)'(t)dt = -\int_{\Omega} \chi_{A_j}(u_j)\Phi_{\varepsilon}(\nabla u_j)dx. \tag{6.30}$$

We combine relations (6.28) to (6.30) to obtain:

$$I_j = \int \left[L_j(u)\Phi_{\varepsilon}(\nabla u) - L_j(u_j)\Phi_{\varepsilon}(\nabla u_j)\right]dx > 3\eta\alpha\,. \tag{6.31}$$

Indeed, by Federer's formula, we have:

$$\int L_j(u)\Phi_{\varepsilon}(\nabla u)dx = -\int L_j(t)\left(m^{\varepsilon}_{1,u}\right)'(t)dt.$$

Similarly for u_j. Hence

$$I_j = \int \left(L_j - \chi_{A_j}\right)\left(m^{\varepsilon}_{1,u_j}\right)'(t) - \int \left(L_j - \chi_{A_j}\right)\left(m^{\varepsilon}_{1,u}\right)'(t)dt$$

$$+\int \chi_{A_j}\left[\left(m^{\varepsilon}_{1,u_j}\right)'(t) - \left(m^{\varepsilon}_{1,u}\right)'(t)\right]dt > -\eta\alpha - \eta\alpha + 5\eta\alpha = 3\eta\alpha.$$

Let us set, for fixed $j, t \in \mathbb{R}$, $M_j(t) = \displaystyle\int_0^t L_j(s)ds$, $\rho_j = M_j \circ u_j$, $\varphi_j = M_j \circ u$, then $\nabla\rho_j = L_j(u_j)\nabla u_j$ and $\nabla\varphi_j = L_j(u)\nabla u$. If we then define the vector field $U_j = \dfrac{\nabla u_j}{\varepsilon^2 + \left|\nabla u_j\right|^2}$, then each I_j can be written as: $I_j = \displaystyle\int \left[U \cdot \nabla\varphi_j - U_j \cdot \nabla\rho_j\right]dx$

$$= \int U \cdot \Big(\nabla\varphi_j - \nabla\rho_j\Big)dx + \int \Big(U - U_j\Big) \cdot \nabla\rho_j dx.$$

Since $\nabla u_j \to \nabla u$ in $L^1(\Omega)^N$, there exists a subsequence that converges simply and a function $G \in L^1(\Omega)$ s.t. $\left|\nabla u_j(x)\right| \leqslant G(x)$.

Since $0 \leqslant M_j' = L_j \leqslant 1$ we have $\left|\nabla\varphi_j\right| \leqslant G$, $\left|\nabla\rho_j\right| \leqslant G$. Since $U_j \to U$ a.e., and $\left|U_j\right| \leqslant \dfrac{1}{\varepsilon}$, it follows that $\displaystyle\int \Big(U - U_j\Big) \cdot \nabla\rho_j \xrightarrow[j\to+\infty]{} 0$ (by the dominated convergence theorem). Thus since $I_j > 3\eta\alpha$, for j large enough we have $\displaystyle\int U \cdot \left(\nabla\varphi_j - \nabla\rho_j\right) dx > 2\eta\alpha$. Since $U \in L^\infty(\Omega)^N$, there exists a function $W \in C_c^\infty(\Omega)^N$ s.t. $|U|_\infty \leqslant |W|_\infty$ and

$$2\int |U - W|\, G\, dx < \eta\alpha.$$

This then implies $\displaystyle\int (U - W) \cdot \left(\nabla\varphi_j - \nabla\rho_j\right) dx < \eta\alpha$, and by the inequality $\displaystyle\int U \cdot \left(\nabla\varphi_j - \nabla\rho_j\right) dx > 2\eta\alpha$, we deduce that

$$-\int \left(\varphi_j - \rho_j\right) div\, W\, dx = \int \Big(\nabla\varphi_j - \nabla\rho_j\Big) \cdot W\, dx > \eta\alpha. \tag{6.32}$$

Since $\left|M_j(a) - M_j(b)\right| \leqslant \left|L_j\right|_\infty |b - a| \leqslant |b - a|$ and $u_j \to u$ a.e., we deduce $\left|\rho_j - \varphi_j\right| \to 0$ a.e. and by the theorem of dominated convergence, we obtain a contradiction with (6.32), and we derive the theorem. ◇

From the previous result, we then have:

Theorem 6.2.2 (Convergence of m_{1,u_j}) *Let $\{u, u_1, u_2, \ldots\}$ be a sequence in $W^{1,1}(\Omega)$ such that $u_j \to u$ in $W^{1,1}(\Omega)$. Then there exists a subsequence $j(1), j(2), \ldots$ such that a.e. $t \in \mathbb{R}$,*

$$\limsup_k m'_{1,u_{j(k)}}(t) \leqslant m'_{1,u}(t)$$

or equivalently that

$$\liminf_k \left|m'_{1,u_{j(k)}}(t)\right| \geqslant \left|m'_{1,u}(t)\right|.$$

Proof In order to use the diagonal process of Cantor, we will choose in the framework of the previous theorem $\varepsilon = \frac{1}{2}, \frac{1}{3}, \frac{1}{4}, \ldots$. We continue to use the notation Φ_ε, m^ε_{1,u_j}, $m^\varepsilon_{1,u}$. For each ε and each Borel A (of finite measure or not), we have a subsequence

$$j(k) : \left(m^\varepsilon_{1,u_{j(k)}}\right)'(t) \to \left(m^\varepsilon_{1,u}\right)'(t) \text{a.e. on A.}$$

By the diagonal Cantor process, there exists a subsequence $j(l)$ such that

$$\left(m^\varepsilon_{1,u_{j(l)}}\right)'(t) \xrightarrow[l\to+\infty]{} \left(m^\varepsilon_{1,u}\right)'(t)\,,$$

for almost all t in A, and each $\varepsilon = \frac{1}{2}, \frac{1}{3}, \frac{1}{4}, \ldots$.

We denote by $\mathcal{B}$ the set of points t satisfying this convergence for all $\varepsilon = \frac{1}{2}, \frac{1}{3}, \frac{1}{4}, \ldots$, $\mathcal{B} \subset A$. We also know that:

1. $m_{1,u}$, m_{1,u_j} are absolutely continuous on $\mathbb{R}$ (see Proposition 6.2.2).
2. If we set $m_{j\varepsilon}(t) = m^\varepsilon_{1,u_{j(l)}}(t)$, $m_\varepsilon(t) = m^\varepsilon_{1,u}(t)$ then

$$\lim_j m'_{j\varepsilon}(t) = m'_\varepsilon(t),\ \forall t \in \mathcal{B},\ \forall \varepsilon\,.$$

Since Φ_ε is bounded, we deduce:

$$\lim_j m_{j\varepsilon}(t) = m_\varepsilon(t) \qquad \forall t \in \mathcal{B}\,.$$

Moreover, since $\varepsilon \to \dfrac{|\xi|^2}{\varepsilon^2 + |\xi|^2}$ is decreasing, we have as soon as $\varepsilon < \delta$:

$$m'_{1,u_j}(t) \leqslant m'_{j\varepsilon}(t) \leqslant m'_{j\delta}(t)$$

and

$$m'_{1,u}(t) \leqslant m'_\varepsilon(t) \leqslant m'_\delta(t)\,.$$

Let us now show that $\lim\limits_{\varepsilon\to 0} m'_{j\varepsilon}(t) = m'_{1,u_j}(t)$ (respectively, $m'_{1,u}(t) = \lim\limits_{\varepsilon\to 0} m'_\varepsilon(t)$). We have, for all t

$$-\int_t^{+\infty} m'_{1,u_j}(s)ds = m_{1,u_j}(t) = \lim_{\varepsilon\to 0} m_{j\varepsilon}(t) = -\lim_{\varepsilon\to 0}\int_t^{+\infty} m'_{j\varepsilon}(s)ds\,.$$

Due to the previous monotonicity, this last limit satisfies:

$$-\int_t^{+\infty} \lim_{\varepsilon \to +\infty} m'_{j\varepsilon}(s)ds = -\int_t^{+\infty} m'_{1,u_j}(s)ds, \text{ as } m'_{1,u_j}(s) \leqslant \lim_{\varepsilon \to 0} m'_{j\varepsilon}(s)$$

and we have

$$m'_{1,u_j}(s) = \lim_{\varepsilon \to 0} m'_{j\varepsilon}(s) \; a.e. \; in \; s \; .$$

Due to the countability of $\left\{m_{1,u_j}, m_{1,u}\right\}_{j \geqslant 1}$ we deduce that there exists $\mathcal{B}_0 \subset \mathcal{B} \subset A$ such that $|\mathcal{B}_0| = |A|$ and $\forall\, t \in \mathcal{B}_0, \forall\, j$

$$m'_{1,u_j}(t) = \lim_{\varepsilon \to 0} m'_{j\varepsilon}(t), \;\; m'_{1,u}(t) = \lim_{\varepsilon \to 0} \left(m^{\varepsilon}_{1,u}\right)'(t) \; .$$

Thus, we have: $\forall\, t \in \mathcal{B}_0$

$$m'_{1,u}(t) = \lim_{\varepsilon \to 0} m'_{\varepsilon}(t) = \lim_{\varepsilon \to 0} \lim_{j \to +\infty} m'_{j\varepsilon}(t) \; ,$$

but,

$$m'_{j\varepsilon}(t) \geqslant m'_{1,u_j}(t) : \lim_{j \to +\infty} m'_{j\varepsilon} \geqslant \limsup_{j \to +\infty} m'_{1,u_j}(t) \; .$$

Hence

$$m'_{1,u}(t) \geqslant \limsup_{j \to +\infty} m'_{1,u_j}(t) \; .$$

Hence the result. ◇

Note

In the previous proof, we actually used three important properties:

Property 6.2.1

1. $\lim_{j \to +\infty} m'_{j\varepsilon}(t) = m'_{\varepsilon}(t)$.
2. $If \; \varepsilon < \delta then m'_{1,u_j}(t) \leqslant m'_{j\varepsilon}(t) \leqslant m'_{j\delta}(t) \; \forall\, j$, and almost all t, which implies $m'_{1,u}(t) \leqslant m'_{\varepsilon}(t) \leqslant m'_{\delta}(t)$.
3. Using (2) and the fact that $\lim_{\varepsilon \to 0} m^{\varepsilon}_{1,u}(t) = m_{1,u}(t)$ we have:
   $m'_{1,u}(t) = \lim_{\varepsilon \to 0} m'_{\varepsilon}(t)$.

(continued)

Property 6.2.1 (continued)
From (3) and (1): $m'_{1,u}(t) = \lim_{\varepsilon\to 0}\lim_{j\to+\infty} m'_{j\varepsilon}(t)$
and from (2) $\lim_{j\to+\infty} m'_{j\varepsilon}(t) \geqslant \limsup_j m'_{1,u_j}(t)$. Hence

$$m'_{1,u}(t) \geqslant \limsup_j m'_{1,u_j}(t)\,.$$

We need to characterize the functions u s.t $m'_u(t) \geqslant \limsup_j m'_{u_j}(t)$. This inequality holds only for co-area regular functions u.

6.2.2 *Definition of a Co-area Regular Function and Continuity*

Definition 6.2.1 (Of a Co-area Regular Function) Let $u \in W^{1,1}_{loc}(\Omega)$. The function u is said to be co-area regular if the measure derivative (in the sense of distribution) of the following function

$$m_{o,u}(t) = \left|\left\{x \in \Omega : u(x) > t,\ \nabla u(x) = 0\right\}\right| \text{ is totally singular}$$

(i.e. the absolutely continuous part of $dm_{o,u}$ is zero): $dm_{o,u}(t) = -d\nu, \quad \nu \geqslant 0$.

Theorem 6.2.3 *Let $u \in W^{1,1}(\Omega)$ be co-area regular and let u_j be a sequence in $W^{1,1}(\Omega)$ such that $u_j \to u$ in $W^{1,1}(\Omega)$. Then,*

$$\liminf_j \left|m'_{u_j}(t)\right| \geqslant \left|m'_u(t)\right| \quad a.e.$$

Proof Since $m_u(t) = |u > t| = m_{1,u}(t) + m_{o,u}(t)$, for u co-area regular, we have $dm_u(t) = m'_{1,u}(t)dt - d\nu$. Hence, using the previous result:

$$\left|m'_u(t)\right| = \left|m'_{1,u}(t)\right| \leqslant \liminf_j \left|m'_{1,u_j}(t)\right|$$

but,

$$\liminf_j \left|m'_{1,u_j}(t)\right| \leqslant \liminf_j \left|m'_{u_j}(t)\right|$$

hence the result. ◇

Proposition 6.2.4 *If u is co-area regular then $\forall I = (\alpha, \beta) \in \mathbb{R}^2$ the I-function u^I is also co-area regular.*

Proof Indeed, let us put:

$$m^I_{o,u}(t) = \left|\left\{x \in \Omega : u^I(x) > t,\ \ \nabla u^I(x) = 0\right\}\right|.$$

Since $0 \leqslant u^I \leqslant \beta - \alpha$,
if $t \leqslant 0$ $\quad m^I_{o,u}(t) = \left|\left\{x \in \Omega : \nabla u^I(x) = 0\right\}\right| = \text{constant},$
if $t \geqslant \beta - \alpha$ $\quad m^I_{o,u}(t) = 0,$
if $0 < t < \beta - \alpha$, then

$$m^I_{o,u}(t)$$
$$= \left|\left\{u > \beta\right\}\right| + \left|\left\{x \in \Omega : \alpha \leqslant u(x) \leqslant \beta,\ u(x) > t + \alpha,\ \nabla u(x) = 0\right\}\right|.$$

As $m'_{0,u}(t) = 0$ a.e. $t \in \mathbb{R}$, we deduce that the absolutely continuous part of $m^I_{0,u}$ satisfies

$$\left(m^I_{o,u}\right)'(t) = 0.$$

(Note that $0 \leqslant m^I_{0,u}(t) - m^I_{0,u}(t+h) \leqslant m_{0,u}(t+\alpha) - m_{0,u}(t+\alpha+h)$, for $h > 0$, $0 < t < \beta - \alpha$.) ◇

The main convergence theorem is

Theorem 6.2.4 (Continuity at the Co-area Regular Point) *Let $\{u, u_j\}$ be a sequence of elements of $W^{1,p}(\Omega)$, $1 \leqslant p < +\infty$. Assume that $u_j \underset{j}{\to} u$ in $W^{1,p}(\Omega)$. If u is co-area regular then:*

(continued)

Theorem 6.2.4 (continued)

$$\left|k\left(u'_{j*}-u'_*\right)\right|_p \xrightarrow[j\to+\infty]{} 0$$

$$\textit{where } k(\sigma)=\begin{cases} N\alpha_N^{\frac{1}{N}}\sigma^{1-\frac{1}{N}} & \textit{if } u,\ u_j\geqslant 0 \textit{ with zero trace,}\\ \dfrac{1}{Q}\min\left(\sigma,|\Omega|-\sigma\right)^{1-\frac{1}{N}} & \textit{if } \Omega \textit{ is bounded,}\\ & \textit{connected, Lipchitzian}\\ & \textit{or if } u,\ u_j \textit{ are not signed in}\\ & W_0^{1,p}(\Omega), \textit{with in this case}\\ & Q=\left(N\alpha_N^{\frac{1}{N}}\right)^{-1}. \end{cases}$$

Proof Let $I=(\alpha,\beta),\ I_j=(\alpha_j,\beta_j)$ s.t $I_j\to I$, I (resp. I_j) in $\mathfrak{I}_u$ (resp. $\mathfrak{I}_{u_j}$). Then, we know that $u_j^{I_j}\to u^I$ in $W^{1,p}(\Omega)$. By adopting the same notations as previously, we have:

$$f_j(t)=\begin{cases} 0 & if\ t\geqslant\beta_j-\alpha_j,\\ \left|u_j^{I_j}>t\right| & if\ 0\leqslant t\leqslant\beta_j-\alpha_j,\\ |\Omega| & if\ t\leqslant 0, \end{cases}\quad \text{(same for } f(t)\text{)},$$

$$g_j(s)=\begin{cases} \beta_j-\alpha_j & if\,\sigma\leqslant 0,\\ u_{j*}^{I_j}(s) & if\ 0\leqslant s\leqslant|\Omega|,\\ 0 & if\ s>|\Omega|, \end{cases}\quad \text{(same for } g(s)\text{)}.$$

Then, since u^I is co-area regular, we deduce from Theorem 6.2.3 that:

$$\left|f'(t)\right|\leqslant\liminf_j\left|f_j'(t)\right|.$$

Moreover $\left|f_j-f\right|_{L^1_{loc}(\mathbb{R})}\xrightarrow[j]{}0$. Then according to Theorem 6.1.5

$$\lim_j\int_{[0,\beta_j-\alpha_j]}\sqrt{1+f_j'^2}\,dt+\int_{[0,\beta_j-\alpha_j]}d\mu_j\ =\int_{[0,\beta-\alpha]}\sqrt{1+f'^2}\,dt+\int_{[0,\beta-\alpha]}d\mu$$

if $df_j = f_j' dt - d\mu_j$, $df = f'(t)dt - d\mu$, from which, using Proposition 6.2.1, we deduce that:

$$\lim_j \int_{\Omega_*} \sqrt{1 + g_j'(s)^2}\, ds = \int_{\Omega_*} \sqrt{1 + g'(s)^2}\, ds.$$

Since

$$\left|g_j - g\right|_{L^1(\Omega_*)} \xrightarrow[j\to+\infty]{} 0\,,$$

we deduce from Theorem 6.1.1 that $\lim_j \left|g_j' - g'\right|_{L^1(\Omega_*)} = 0$. We conclude by Theorems 6.1.3 and 6.1.4 that

$$\left|k\left(u_{j_*}' - u_*'\right)\right|_p \xrightarrow[j\to+\infty]{} 0\,.$$

◇

The natural question is to determine which functions are co-area regular. The following theorem gives an answer:

Theorem 6.2.5 (Co-area Regular Functions)

1. *If $u \in W_{loc}^{N,p}(\Omega)$, $p > 1$ then u is co-area regular.*
2. *If $\left\{x : \nabla u(x) = 0\right\}$ is of measure zero, $u \in W_{loc}^{1,1}(\Omega)$, then u is co-area regular.*
3. *If $u \in W_{loc}^{1,1}(\mathbb{R})$ then u is co-area regular.*

Proof Assertion 2. follows from the definition. To prove assertion 1., start with the case where $u \in C^N(\Omega)$. By the Morse–Sard–Federer theorem we have for $B = \left\{x : \nabla u(x) = 0\right\}$, $|u(B)| = 0$. So, for all $U \subset \mathbb{R}$, we have:

$$m_{o,u}'(U) = -\int \chi_{\{u\in U\}} \cdot \chi_{\{\nabla u=0\}} dx = -\int \chi_{\{u\in U\}} \cdot \chi_{\{u\in u(B)\}} \cdot \chi_{\{\nabla u=0\}} dx$$

(where $\left\{u \in U\right\} = \left\{x : u(x) \in U\right\}$,
$\left\{u \in u(B)\right\} = \left\{x : u(x) \in u(B)\right\}$). Since $|u(B)| = 0$, the support of $dm_{o,u}$ is of measure zero, thus it is singular.

If $u \in W_{loc}^{N,p}(\Omega)$, then according to a type of Lusin's theorem , we can modify u into a function of class C^N except on a set of arbitrarily small measure. We deduce that necessarily the support of $dm_{o,u}$ is of measure zero, therefore it can only be singular.

For case 3., if $u \in W^{1,p}_{loc}(\Omega)$, $p > 1$, we deduce it from assertion 1. The case where $p = 1$ is done directly. ◇

To close this chapter, we will need the following theorem:

Theorem 6.2.6 *Let $\{u_j,\ u\}$ be a sequence in $W^{1,p}(\Omega)$, with $1 \leqslant p < +\infty$. We assume that u_j tends to u in $W^{1,p}(\Omega)$ and almost everywhere in Ω and that u is co-area regular. Then,*

$$(ku'_{j*})\left(\left|u_j > u_j(\cdot)\right|\right) \to (ku'_*)\left(|u > u(\cdot)|\right) \text{ in } L^p(\Omega)\text{-strong.}$$

Here

$$k(\sigma) = \begin{cases} \sigma^{1-\frac{1}{N}} & \text{if } u,\ u_j \in W^{1,p}_{0+}(\Omega) \\ \min(\sigma, |\Omega| - \sigma)^{1-\frac{1}{N}} & \text{otherwise,} \end{cases}$$

the open set Ω being as in Theorem 6.2.4.

Proof Let us set $K_j = (ku'_{j*})\left(\left|u_j > u_j(\cdot)\right|\right)$, $K = (ku'_*)\left(|u > u(\cdot)|\right)$.

By equimeasurability and because of Theorem 6.1.4, we know that $ku'_{j*} \to ku'_*$ in $L^p(\Omega_*)$-strong, and we deduce that:

$$\lim_j |K_j|_{L^p(\Omega)} = \lim_j \left|ku'_{j*}\right|_{L^p(\Omega_*)} = \left|ku'_*\right|_p = |K|_p \ .$$

We only need to show that $K_j \rightharpoonup K$ in $L^p(\Omega)$-weak if $1 < p < +\infty$. Let $\varphi \in L^{p'}(\Omega)$ and let $\theta = \chi_{\Omega_* \backslash P(u_*)}$. Then by the formula of the mean operators, we have:

$$\int_\Omega \varphi K_j dx = \int_{\Omega_*} \varphi_{*u_j} \cdot ku'_{j*} dt = \int_{\Omega_*} \theta \varphi_{*u_j} \left(ku'_{j*}\right) dt + \int_{\Omega_*} (1-\theta)\varphi_{*u_j} \left(ku'_{j*}\right) dt.$$

By Hölder's inequality, we have:

$$\int_{\Omega_*} \left|(1-\theta)\varphi_{*u_j} \left(ku'_{j*}\right)\right| dt \leqslant |\varphi|_{p'} \left|(1-\theta)\left(ku'_{j*}\right)\right|_p \underset{j}{\to} 0$$

(because $\left|(1-\theta)\left(ku'_*\right)\right|_p = 0$). Moreover, Theorem 2.6.1 implies $\theta \varphi_{*u_j} \rightharpoonup \theta \varphi_{*u}$ in $L^{p'}(\Omega_*)$-weak. Therefore,

$$\lim_j \int_\Omega \varphi K_j dx = \int_{\Omega_*} \theta \varphi_{*u} \left(ku'_*\right) dt = \int_{\Omega_*} \varphi_{*u} \left(ku'_*\right) dt = \int_\Omega \varphi K \, dx \ .$$

The facts that $K_j \rightharpoonup K$ in $L^p(\Omega)$-weak and $\lim \left|K_j\right|_p = |K|_p$ imply $K_j \to K$ in $L^p(\Omega)$-strong if $1 < p < +\infty$.

For the case $p = 1$, we need additional results.

First, consider truncation at the level $\alpha > 0$ i.e. $T_\alpha(t) = \min(|t|, \alpha)\,\mathrm{sign}(t)$. Let us then set,

$$K_{j,\alpha}(x) = T_\alpha(ku'_{j*})\left(\left|u_j > u_j(x)\right|\right),\ K_{,\alpha}(x) = T_\alpha(ku'_*)\left(|u > u(x)|\right)\ .$$

Then $\left|K_{j,\alpha}\right|_\infty \leqslant \alpha$, $\left|K_{,\alpha}\right|_\infty \leqslant \alpha$. Arguing in the same way as before, we have:

Proposition 6.2.5 *For all $r \in]1, +\infty[$, we have:*

(a) $K_{j,\alpha} \rightharpoonup K_{,\alpha}$ *in L^r-weak,*
(b) $\lim\limits_j \left|K_{j,\alpha}\right|_{L^r(\Omega)} = \left|K_{,\alpha}\right|_{L^r(\Omega)}$,
(c) $\lim\limits_j \left|K_{j,\alpha} - K_{,\alpha}\right|_{L^1(\Omega)} = 0.$

Moreover, we have:

$$\left|K_j - K\right|_{L^1(\Omega)} \leqslant \left|K_j - K_{j,\alpha}\right|_{L^1(\Omega)} + \left|K_{j,\alpha} - K_{,\alpha}\right|_{L^1(\Omega)} + \left|K_{,\alpha} - K\right|_{L^1(\Omega)}\ .$$

By equimeasurability we have:

$$\left|K_j - K_{j,\alpha}\right|_{L^1(\Omega)} = \left|T_\alpha(ku'_{j*}) - ku'_{j*}\right|_{L^1(\Omega_*)} \underset{j}{\to} \left|K_{,\alpha} - K\right|_{L^1(\Omega)}$$

and

$$\left|K_{,\alpha} - K\right|_{L^1(\Omega)} = \left|T_\alpha(ku'_*) - ku'_*\right|_{L^1(\Omega_*)} \xrightarrow[\alpha\to+\infty]{} 0\,.$$

From the statement of the Proposition 6.2.5, we then have

$$\limsup_j \left|K_j - K\right|_{L^1(\Omega)} \leqslant 2\left|K_{,\alpha} - K\right|_{L^1(\Omega)} \xrightarrow[\alpha\to+\infty]{} 0\,.$$

◇

Here are examples of convergence in $L^1(\Omega_*)$ for the derivative of the monotone rearrangement.

Theorem 6.2.7 *Let Ω be a connected Lipschitzian bounded open subset of $\mathbb{R}^N$ and let $W^1L^{N,1}(\Omega) = \left\{v \in L^1(\Omega) : |\nabla v| \in L^{N,1}(\Omega)\right\}$. Then, if $\{u_j\ u\}$ is a sequence in $W^1L^{N,1}(\Omega)$ such that $u_j \to u$ in $W^1L^{N,1}(\Omega)$, u co-area regular, then,*

(i) $u'_{j*} \to u'_*$ in $L^1(\Omega_*)$-strong,
(ii) $u'_{j*}\left(|u_j > u_j(\cdot)|\right) \to u'_*(|u > u(\cdot)|)$ in $L^1(\Omega)$-strong.

Proof Let us first note that $(u'_{j*})_j$ remains bounded in $L^1(E_*)$
$\forall E \subset \Omega_*$, $E_* = (0, |E|)$. Indeed, according to the PSR property

$$-u'_{j*}(t) \leqslant Q\max(t, |\Omega| - t)^{\frac{1}{N}-1}\left|\nabla u_j\right|_{*u_j}(t)\,.$$

Thus, by applying the Hardy–Littlewood inequality, we obtain

$$\begin{aligned}\int_E \left|u'_{j*}\right| dt &\leqslant Q\int_E \max(t, |\Omega| - t)^{\frac{1}{N}-1}\left|\nabla u_j\right|_{*u_j}(t)dt\\ &\leqslant Q2^{1-\frac{1}{N}}\int_{E^*} t^{\frac{1}{N}-1}\left(\left|\nabla u_j\right|_{*u_j}\right)_*(t)dt\,.\end{aligned}$$

But, $\left(\left|\nabla u_j\right|_{*u_j}\right)_* \leqslant \left|\nabla u_j\right|_{**}$, so the last relation then leads to

$$\int_E \left|u'_{j*}\right| dt \leqslant Q2^{1-\frac{1}{N}}\left|\nabla u_j\right|_{L^{(N,1)}(E_*)} \tag{6.33}$$

$$\leqslant Q2^{1-\frac{1}{N}}\left|\nabla(u - u_j)\right|_{L^{(N,1)}(\Omega_*)} + Q2^{1-\frac{1}{N}}\left|\nabla u\right|_{L^{(N,1)}(E)}$$

(where $|f|_{L^{(N,1)}(E_*)} = \int_{E_*} t^{\frac{1}{N}}\,|f|_{**}(t)\frac{dt}{t}$.)

Since $k(u'_{j*} - u'_*) \xrightarrow[j\to+\infty]{} 0$ in $L^1(\Omega_*)$ with $k(\sigma) = \min(\sigma, |\Omega| - \sigma)^{1-\frac{1}{N}}$, using the relation (6.33), $\forall\,\alpha \in]0, |\Omega|[$, we have:

$$\left|u'_{j*} - u'_*\right|_{L^1(\Omega_*)} \leqslant \int_0^\alpha \left|u'_{j*}\right| dt + \int_0^\alpha \left|u'_*\right| dt \tag{6.34}$$

$$+\operatorname{Max}(\alpha, |\Omega| - \alpha)^{\frac{1}{N}-1}\left|k(u'_{j*} - u'_*)\right|_1 + \int_{|\Omega|-\alpha}^{|\Omega|}\left|u'_{j*}\right| dt + \int_{|\Omega|-\alpha}^{|\Omega|}\left|u'_*\right| dt\,,$$

$$\limsup_j \left|u'_{j*} - u'_*\right|_{L^1(\Omega_*)} \leqslant 2\left(\int_0^\alpha \left|u'_*\right|_* dt + Q2^{1-\frac{1}{N}} \int_0^\alpha t^{\frac{1}{N}} \left|\nabla u\right|_{**}(t)\frac{dt}{t}\right). \tag{6.35}$$

When $\alpha \to 0$ the right-hand side of this inequality tends to zero, which then implies:

$$\lim_j \left|u'_{j*} - u'_*\right|_{L^1(\Omega_*)} = 0\,.$$

Part (ii) follows from (i) by following the same proof as Theorem 6.2.7 by replacing the weight k by 1.

We have an analogous result for functions with zero trace except that the open Ω is no longer necessarily connected and regular.

Theorem 6.2.8 *Let Ω be a bounded open set of $\mathbb{R}^N$ and let*

$$W_0^1 L^{N,1}(\Omega) = \left\{v \in W_0^{1,1}(\Omega),\ |\nabla v| \in L^{N,1}(\Omega)\right\}.$$

Then, if $u_j \to u$ in $W_0^1 L^{N,1}(\Omega)$, u co-area regular then:

(i) $u'_{j*} \to u'_*$ in $L^1(\Omega_*)$-strong,
(ii) $u'_{j*}\left(|u_j > u_j(\cdot)|\right) \to u'_*\left(|u > u(\cdot)|\right)$ in $L^1(\Omega)$-strong.

Pre-bibliographical Notes

All the results given in this chapter are taken from the article by Almgrem and Lieb [4] and also from Rakotoson [131].

Lusin's theorem for functions of $W_{loc}^{N,p}(\Omega)$ can be found in the book by Ziemer [171], and for measurable functions, see the book by W. Rudin [150].

Chapter 7
Strong Continuity of the Relative Rearrangement Map $u \to b_{*u}$ and Consequences

We pointed out in Chap. 2 that the equilibrium equation of a plasma confined in a "Stellarator" machine can be written in the form: $-\Delta u = G(x, u_*, u_*', b_{*u})$. The same is true in a Tokomak machine, the equilibrium equation can be written in the form of a fixed point: find u such that $u \doteq T(u - g(u)) + \varphi_0$ and the function g depends on u, the level sets of u and its rearrangements (relative and monotone). These equations are resolved by using the Galerkin method or by using topological degrees in all cases, it is necessary to study the continuity of the map $u \to G(\cdot, u_*, u_*', b_{*u})$ or of $u \to g(u)$.

In this chapter, we will give some answers regarding the continuity of the map $(u, b) \to b_{*u}$. As the map $b \to b_{*u}$ is a contraction, the study comes back to that of the continuity of $u \to b_{*u}$, and for functions b in a dense subspace of $L^1(\Omega)$, for example the bounded functions.

We will rely largely on the results of the previous chapter, since the quantity b_{*u} is closely linked to the derivative of the rearrangement of u, that is to say u_*'.

First, we will show that if $u \in W^{1,1}(\Omega)$ and $b \in L^\infty_+(\Omega) \cap W^{1,1}(\Omega)$ then we have the formula:

$$\left[b\,|\nabla u|\,\right]_{*u}(s) = -u_*'(s) P_{\Omega,b}\left(\left\{u > u_*(s)\right\}\right)$$

where $P_{\Omega,b}(E) = \sup\left\{\int_E div(\Phi \cdot b)dx,\ |\Phi|_\infty \leqslant 1,\ \Phi \in C_c^1(\Omega)^N\right\}$ for E measurable in Ω, $u_* \in W^{1,1}_{loc}(\Omega_*)$.

This formula will lead us to the following conclusion:

Let $b \in L^\infty(\Omega)$, and let u_n be a sequence converging strongly to u in $W^{1,1}(\Omega)$. If u is co-area regular then:

$$\left[b\,|\nabla u|\,\right]_{*u_n} \to \left[b\,|\nabla u|\,\right]_{*u} \ in\ L^1(\Omega_*)\text{–}strong\ .$$

This convergence will lead us to the main theorem of this chapter:

J. M. Rakotoson, *Relative Rearrangement*, Lecture Notes in Mathematics 2376,
https://doi.org/10.1007/978-3-032-02228-8_7

Theorem (Strong Continuity of the Relative Rearrangement Restricted to $W^{1,1}$) *Let $u \in W^{1,1}(\Omega)$ and u_n be a sequence of $W^{1,1}(\Omega)$, with Ω a connected Lipschitzian open bounded set if $\gamma_0 u \not\equiv 0$, and any open set Ω otherwise. Assume that $u_n \to u$ in $W^{1,1}(\Omega)$ and that u is co-area regular. Then, for $b \in L^p(\Omega)$, $1 \leqslant p < +\infty$, we have:*

$$\left(b\chi_{\{\nabla u \neq 0\}}\right)_{*u_n} \xrightarrow[n \to +\infty]{} \left(b\chi_{\{\nabla u \neq 0\}}\right)_{*u} \text{ in } L^p(\Omega_*)\text{-strong} .$$

Here are two examples of consequences of this theorem:

Corollary *Let $u \in W^{1,1}(\Omega)$, u_n be a sequence of $W^{1,1}(\Omega)$, with Ω an open bounded Lipschitzian set if $\gamma_0 u \not\equiv 0$, and any Ω otherwise.*

It is assumed that $\operatorname{meas}\left\{x : \nabla u(x) = 0\right\} = 0$.

If $u_n \to u$ in $W^{1,1}(\Omega)$ and $b \in L^p(\Omega)$, $1 \leqslant p < +\infty$, then

1. *$b_{*u_n} \to b_{*u}$ in $L^p(\Omega_*)$-strong,*
2. *$b_{*u_n}\big(|u_n > u_n(\cdot)|\big) \to b_{*u}\big(|u > u(\cdot)|\big)$ in $L^p(\Omega)$-strong if* $\operatorname{meas}\Big(P(u_n)\Big) = 0$.

We will show that for any constant function a, there exists a sequence of $C^\infty(\overline{\Omega})$, $u_n : \left|\{\nabla u_n = 0\}\right| = 0$, $u_n \to a$ and b_{*u_n} does not tend to $b_{*a} = b_*$. This suggests that our previous condition is useful.

The second corollary is:

Corollary *Let $(\lambda_k, \varphi_k)_{k \geqslant 1}$ be a sequence of eigenvalues and eigenfunctions of the Laplacian with Dirichlet conditions and let V_m denote the subspace generated by the first m eigenfunctions: $\left\{\varphi_1, \ldots, \varphi_m\right\}$. Then for $1 \leqslant p < +\infty$, the following maps are continuous:*

1. *$u \in V_m \backslash \left\{0\right\} \to b_{*u} \in L^p(\Omega_*)$-strong,*
2. *$u \in V_m \backslash \left\{0\right\} \to b_{*u}\Big(|u > u(\cdot)|\Big) \in L^p(\Omega)$-strong, provided that $b \in L^p(\Omega)$.*

7.1 Some Auxiliary Formulas

Throughout this chapter, Ω will be an open set.

Definition 7.1.1 Let $b \in L^\infty_+(\Omega) \cap W^{1,1}(\Omega)$. For $E \subset \Omega$ measurable, we define:

$$P_{\Omega,b}(E) = \sup\left\{ \int_E div\,(b\Phi)\,dx,\ |\Phi|_\infty \leqslant 1,\ \Phi \in C^1_c(\Omega)^N \right\}$$

which we will call the perimeter of E in Ω relative to b.

This perimeter has the same properties as the usual perimeter obtained with $b = 1$. In particular,

Property 7.1.1

1. If $\{E,\ E_n\}$ is a sequence of measurable sets of Ω and if $\chi_{E_n}(x) \to \chi_E(x)$ a.e. then $\liminf_n P_{\Omega,b}(E_n) \geqslant P_{\Omega,b}(E)$,
2. If $\{b,\ b_j\}$ is a sequence $L^\infty_+(\Omega) \cap W^{1,1}(\Omega)$ s.t. $b_j \to b$ in $W^{1,1}(\Omega)$, then, for all measurable E of Ω
$$\liminf_j P_{\Omega,b_j}(E) \geqslant P_{\Omega,b}(E)\,,$$
3. $P_{\Omega,b}(\Omega) = P_{\Omega,b}(E) = 0$ if $|E| = 0$.

Proof Note that $div\,(b\Phi) = \nabla b \cdot \Phi + b\,div(\Phi)$ if $\Phi \in C^1_c(\Omega)^N$, thus $div(b\,\Phi) \in L^1(\Omega)$. Therefore $\chi_{E_n} div(b\Phi) \to \chi_E div(b\Phi)$ a.e. and by the dominated convergence theorem, we have:

$$\int_E div(b\Phi))dx = \lim_n \int_{E_n} div(b\Phi)dx \leqslant \liminf P_{\Omega,b}(E_n)$$

hence 1.

If $b_j \underset{j}{\to} b$ in $W^{1,1}(\Omega)$ then $div\,(b_j\Phi) \to div\,(b\Phi)$ in $L^1(\Omega)$. Hence for all measurable subsets $E \subset \Omega$,

$$\int_E div\,(b\Phi)\,dx = \lim_j \int_E div\,(b_j\Phi)\,dx \leqslant \liminf_j P_{\Omega,b_j}(E)\,.$$

Hence 2. As for 3. we deduce it from the definition. ◇

The Fleming–Rishel–Federer formula therefore holds true:

Lemma 7.1.1 (Fleming–Rishel–Federer Formula) *Let* $b \in L^{\infty}_{+}(\Omega) \cap W^{1,1}(\Omega)$. *For* $u \in W^{1,1}(\Omega)$, *we have:*

$$\int_{\Omega} |\nabla u|\, b\, dx = \int_{\mathbb{R}} P_{\Omega,b}\left(\{u > t\}\right) dt\,.$$

Proof First of all, we always have:

$$\int_{\Omega} |\nabla u|\, b\, dx \leqslant \int_{\mathbb{R}} P_{\Omega,b}\left(\{u > t\}\right) dt\,. \tag{7.1}$$

Considering $\Phi_j \in C_c^{\infty}(\Omega)^N$ s.t.

$$\left|\Phi_j\right| \leqslant 1,\ \Phi_j(x) \to \begin{cases} \dfrac{\nabla u}{|\nabla u|} & if\ \nabla u(x) \neq 0, \\ 0 & otherwise \end{cases} \quad \text{almost everywhere in } x.$$

By integration by parts and by definition of $P_{\Omega,b}$, we have:

$$\int_{\Omega} \nabla u \cdot \Phi_j \cdot b\, dx = -\int_{\mathbb{R}} \int_{u>t} div\left(\Phi_j b\right) dx \leqslant \int_{\mathbb{R}} P_{\Omega,b}\left(\{u > t\}\right) dt\,.$$

By the dominated convergence theorem, we know that

$$\lim_{j} \int_{\Omega} \nabla u \cdot \Phi_j b\, dx = \int_{\Omega} |\nabla u|\, b\, dx\,.$$

These last two relations lead to (7.1).

As for the inverse inequality, we first consider $b \in C^{\infty}(\Omega) \cap L^{\infty}(\Omega)$, $b > 0$, $u \in C^{\infty}(\Omega) \cap W^{1,1}(\Omega)$. Then, as in the case of the De Giorgi perimeter, we have:

Lemma 7.1.2 *If* $u \in C^{\infty}(\Omega)$, *then for almost all* t

$$P_{\Omega,b}\left(\{u > t\}\right) = \int_{u^{-1}(t)} b\, dH_{N-1}\ if\ N \geqslant 2$$

(continued)

Lemma 7.1.2 (continued)
and

$$P_{\Omega,b}\left(\{u>t\}\right)=\sum\nolimits_{x\in u^{-1}(t)}b(x)\; if\; N=1.$$

More generally, if $E\subset\Omega$ s.t. $\partial E\cap\Omega$ is of class C^2 then

$$P_{\Omega,b}(E)=\int\limits_{\partial E\cap\Omega} b\,dH_{N-1}\,.$$

From this in combination with Federer's formula, we deduce:

Lemma 7.1.3 *Under the conditions of Lemma 7.1.2, if $u\in W^{1,1}(\Omega)$ then*

$$\int_{\mathbb{R}} P_{\Omega,b}\left(\{u>t\}\right)dt=\int_{\Omega} b\,|\nabla u|\,dx\,.$$

Indeed, by Federer's theorem,

$$\int_{\Omega} b\,|\nabla u|\;dx=\int_{\mathbb{R}}\int\limits_{u^{-1}(t)} b\,dH_{n-1}=\int_{\mathbb{R}} P_{\Omega,b}\left(\{u>t\}\right)dt$$

if $N\geqslant 2$, and the same is true for $N=1$.

If $b\in W^{1,1}(\Omega)\cap L_{+}^{\infty}(\Omega)$, then there exists a sequence $b_j\in C^{\infty}(\Omega)\cap L^{\infty}(\Omega)$, $b_j>0$, such that $b_j\to b$ in $W^{1,1}(\Omega)$, $\left|b_j\right|_{\infty}\leqslant|b|_{\infty}+1$, $b_j(x)\to b(x)$ a.e. Then, by statement 2. of Property 7.1.1, Fatou's lemma and the dominated convergence theorem, we have:

$$\int_{\mathbb{R}} P_{\Omega,b}\left(\{u>t\}\right)dt\leqslant\int_{\mathbb{R}}\liminf_j P_{\Omega,b_j}\left(\{u>t\}\right)dt$$

$$\leqslant\liminf_j\int_{\mathbb{R}} P_{\Omega,b_j}\left(\{u>t\}\right)dt=\lim_j\int_{\Omega} b_j\,|\nabla u|\,dx=\int_{\Omega} b\,|\nabla u|\,.$$

Now if we consider $u_n \in C^\infty(\Omega) \cap W^{1,1}(\Omega)$ such that $u_n \to u$ in $W^{1,1}(\Omega)$ then the same argument leads, from this last inequality, to:

$$\int_{\mathbb{R}} P_{\Omega,b}\left(\{u > t\}\right) dt \leqslant \liminf_n \int_{\mathbb{R}} P_{\Omega,b}\left(\{u_n > t\}\right) dt = \int_\Omega b\,|\nabla u|\,dx \qquad (7.2)$$

for $u \in W^{1,1}(\Omega),\ b \in L^\infty_+(\Omega) \cap W^{1,1}(\Omega)$.

The inequalities (7.1) and (7.2) give the lemma. ◇

As a consequence of this lemma, we have:

Theorem 7.1.1 *Let $b \in W^{1,1}(\Omega) \cap L^\infty_+(\Omega)$ and let $u \in W^{1,1}(\Omega_*)$ s.t. $u_* \in W^{1,1}_{loc}(\Omega)$. So, for almost every $s \in \Omega_*$,*

$$\left[b\,|\nabla u|\,\right]_{*u}(s) = -u_*'(s) P_{\Omega,b}\left(\left\{u > u_*(s)\right\}\right).$$

Proof As $\left(u - u_*(s)\right)_+ \in W^{1,1}(\Omega)$ then the Fleming–Rishel–Federer formula leads to, $\forall\, s \in \Omega_*$

$$\int_{u>u_*(s)} b\,|\nabla u|\,dx = \int_{u_*(s)}^{+\infty} P_{\Omega,b}\left(\{u > t\}\right) dt\,.$$

By differentiating with respect to s, knowing that $u_* \in W^{1,1}_{loc}(\Omega_*)$ and $b\,|\nabla u| \in L^1(\Omega)$, we deduce the formula using the definition of the relative rearrangement. ◇

Theorem 7.1.2 (Strong Continuity of the Restricted Relative Rearrangement to $W^{1,1}$) *Let $u \in W^{1,1}(\Omega)$ with Ω a connected, Lipschitzian and bounded open set, if $\gamma_0 u \not\equiv 0$ and any Ω otherwise. We assume that u is co-area regular. If u_n is a sequence that converges to u in $W^{1,1}(\Omega)$ (for simplicity, we will assume that $\gamma_0 u_n = 0$ if $\gamma_0 u = 0$) and b an element of $W^{1,1}(\Omega) \cap L^\infty_+(\Omega)$ then:*

1. There exists a subsequence denoted by $n(j)$ s.t.

$$\liminf_j \left[b\,\left|\nabla u_{n(j)}\right|\,\right]_{*u_{n(j)}}(s) \geqslant \left[b\,|\nabla u|\,\right]_{*u}(s)\ \textit{a.e. in}\ s \in \Omega_*\,.$$

(continued)

Theorem 7.1.2 (continued)

2. *The entire sequence* $\left[b\,|\nabla u_n|\right]_{*u_n}$ *converges to* $\left[b\,|\nabla u|\right]_{*u}$ *in* $L^1(\Omega_*)$*-strong.*
3. *The sequence* $\left[b\,|\nabla u|\right]_{*u_n}$ *converges to* $\left[b\,|\nabla u|\right]_{*u}$ *in* $L^1(\Omega_*)$.

Proof Since u is co-area regular and $u_n \to u$ in $W^{1,1}(\Omega)$, there exists a subsequence $\left(u_{n(j)}\right)_{j\geqslant 0}$ such that

(a) $u_{n(j)}(x) \to u(x)$ a.e. in Ω,
(b) $u_{n(j)_*}(s) \to u_*(s)$ a.e. in s, $u'_{n(j)_*}(s) \to u'_*(s)$ a.e. in s.

Let $s \in \Omega_*$ satisfy (b). If $s \in P(u_*)$ then $u'_*(s) = 0$. Thus, using the formula of Theorem 7.1.1, we conclude that, a.e. on $P(u_*)$

$$0 = \left[b\,|\nabla u|\right]_{*u}(s) \leqslant \liminf_j \left[b\,\left|\nabla u_{n(j)}\right|\right]_{*u_{n(j)}}(s).$$

If $s \in \Omega_*\backslash P(u_*)$, s satisfies (b) then,
if $v_j(x) = u_{n(j)}(x) - u_{n(j)_*}(s)$, $v(x) = u(x) - u_*(s)$, we have

$$\lim_j \chi_{\{v_j>0\}}(x) = \chi_{\{v>0\}}(x)\ a.e.\ in\ \Omega$$

which implies $\liminf_j P_{\Omega,b}(v_j > 0) \geqslant P_{\Omega,b}(v > 0)$.

Knowing that $u'_{n(j)_*}(s) \to u'_*(s)$, we conclude:

$$\liminf_j \left[-u'_{n(j)_*}(s) P_{\Omega,b}\left(u_{n(j)} > u_{n(j)_*}(s)\right)\right] \geqslant -u'_*(s) P_{\Omega,b}\left(u > u_*(s)\right).$$

We conclude using Theorem 7.1.1 to obtain 1.

Setting $h_j(\sigma) = \left[b\,\left|\nabla u_{n(j)}\right|\right]_{*u_{n(j)}}(\sigma)$, $h(\sigma) = \left[b\,|\nabla u|\right]_{*u}(\sigma)$ for $\sigma \in \Omega_*$, let us show:

Lemma 7.1.4

$$\lim_j \left|h_j - h\right|_1 = 0\,.$$

To prove this lemma we will need the following variant of Proposition 6.1.1

Proposition 7.1.1 *Let $\{h,\ h_j\}$ be a sequence of integrable functions with values in $[0, +\infty[$ satisfying*

(i) $\lim_j \int_\Omega h_j(x)dx = \int_\Omega h(x)dx,$
(ii) $\liminf_j h_j(x) \geqslant h(x)$ *a.e. in* $x \in \Omega$.

Then

$$\lim_j |h_j - h|_1 = 0\,.$$

Proof The proof of Proposition 6.1.1 indicates

$$\lim_j \int_\Omega (h_j - h)_-\,(x)dx = 0\,.$$

We then deduce,

$$\limsup_j \int_\Omega (h_j - h)_+\,dx \leqslant \lim_j \int_\Omega (h_j - h)\,dx = 0\,.$$

Hence $\lim_j \int_\Omega (h_j - h)_+\,dx = 0$, which entails:

$$\lim_j \int_\Omega |h_j - h|\,(x)dx = \lim_j \int_\Omega (h_j - h)_+\,dx + \lim_j \int_\Omega (h_j - h)_-\,dx = 0\,.$$

◇

We then apply this lemma to the sequence $\{h_j,\ h\}$ by replacing Ω with Ω_*. Condition (ii) of this Proposition 7.1.1 is satisfied thanks to the first assertion 1. of Theorem 7.1.2. To verify (i), we write

$$\int_{\Omega_*} h_j(\sigma)d\sigma = \int_\Omega b\Big[\,|\nabla u_{u(j)}| - |\nabla u|\,\Big]dx + \int_\Omega b\,|\nabla u|\,dx \underset{j}{\to} \int_\Omega b\,|\nabla u|\,dx = \int_{\Omega_*} h\,d\sigma$$

and deduce that $h_j \to h$ in $L^1(\Omega_*)$.

Assertion (2) of Theorem 7.1.2 follows from this, since if there exists a subsequence that does not converge in $L^1(\Omega_*)$ towards this limit h, then we can extract another subsequence that satisfies assertion 1. and therefore the use of

Proposition 7.1.1 for this new subsequence would lead to a contraction, because it would converge towards $h(\sigma) = \left[b\,|\nabla u|\right]_{*u}(\sigma),\ \sigma \in \Omega_*$.

As for the last assertion, it follows from 2. using the contraction property, which is written:

$$\left|\left[b\,|\nabla u_n|\right]_{*u_n} - \left[b\,|\nabla u|\right]_{*u_n}\right|_1 \leqslant |b|_\infty\, |\nabla(u_n - u)|_1 \xrightarrow[n\to+\infty]{} 0\,.$$

◇

We have the following corollary:

Corollary 7.1.1 *Under the same conditions as Theorem 7.1.2, if b is only in $L^\infty(\Omega)$ then we still have*

$$\left[b\,|\nabla u|\right]_{*u_n} \to \left[b\,|\nabla u|\right]_{*u} \text{ in } L^1(\Omega_*)\text{-strong}\,.$$

Proof Begin with the case where $b \geqslant 0$, $b \in L^\infty(\Omega)$. Then there exists a sequence $b_j \in C_c^\infty(\Omega)$ s.t. $b_j(x) \to b(x)$ a.e. $b_j \geqslant 0$ and $\left|b_j\right|_\infty \leqslant |b|_\infty$. Thus, $\left(b_j - b\right)|\nabla u| \xrightarrow[j\to+\infty]{} 0$ in $L^1(\Omega)$-strong.

Then, by the contraction property, we have:

$$\left|\left[b\,|\nabla u|\right]_{*u_n} - \left[b\,|\nabla u|\right]_{*u}\right|_1 \leqslant 2\left|\left(b_j - b\right)|\nabla u|\right|_1 + \left|\left[b_j\,|\nabla u|\right]_{*u_n} - \left[b_j\,|\nabla u|\right]_{*u}\right|_1.$$

Hence

$$\limsup_n \left|\left[b\,|\nabla u|\right]_{*u_n} - \left[b\,|\nabla u|\right]_{*u}\right|_1 \leqslant 2\left|\left(b_j - b\right)|\nabla u|\right|_1 \xrightarrow[j\to+\infty]{} 0.$$

If $b \in L^\infty(\Omega)$ then, writing $b = b_+ - b_-$, we have

$$\left[b\,|\nabla u|\right]_{*u_n} = \left[b_+\,|\nabla u|\right]_{*u_n} - \left[b_-\,|\nabla u|\right]_{*u_n} \xrightarrow[n\to+\infty]{} \left[b\,|\nabla u|\right]_{*u}$$

in $L^1(\Omega_*)$*−strong* . ◇

Theorem 7.1.3 *Let $u \in W^{1,1}(\Omega)$ with Ω a connected bounded Lipschitzian open set if $\gamma_0 u \not\equiv 0$ and any Ω otherwise. We assume that u is co-area regular. Then, for any sequence u_n that converges to u in $W^{1,1}(\Omega)$, (we suppose that*

(continued)

Theorem 7.1.3 (continued)
$\gamma_0 u_n = 0$ if $\gamma_0 u = 0$) for all $b \in L^p(\Omega)$, $1 \leqslant p < +\infty$, and $\varepsilon > 0$ we have:

(i) $\left[\dfrac{b\,|\nabla u|}{\varepsilon + |\nabla u|}\right]_{*u_n} \xrightarrow[n \to +\infty]{} \left[\dfrac{b\,|\nabla u|}{\varepsilon + |\nabla u|}\right]_{*u}$ *in $L^p(\Omega_*)$-strong.*

(ii) $\left(b\chi_{\{\nabla u \neq 0\}}\right)_{*u_n} \to \left(b\chi_{\{\nabla u \neq 0\}}\right)_{*u}$ *in $L^p(\Omega_*)$-strong.*

Proof Begin with the case where $b \in L^\infty(\Omega)$. If $p = 1$ then part (i) follows from Corollary 7.1.1 with b replaced by $\dfrac{b}{\varepsilon + |\nabla u|}$. If $1 < p < +\infty$, let us set $B^\varepsilon = \dfrac{b\,|\nabla u|}{\varepsilon + |\nabla u|} \in L^\infty(\Omega)$, we then have:

$$\left|B^\varepsilon_{*u_n} - B^\varepsilon_{*u}\right|_p \leqslant |2b|_\infty^{1-\frac{1}{p}} \left|B^\varepsilon_{*u_n} - B^\varepsilon_{*u}\right|_1^{\frac{1}{p}} \xrightarrow[n \to +\infty]{} 0$$

($\left|B^\varepsilon_{*u_n}\right|$ (and $\left|B^\varepsilon_{*u}\right|$) is smaller than $|B|^\varepsilon_\infty \leqslant |b|_\infty$).

If $b \in L^p(\Omega)$, $1 \leqslant p < +\infty$, then we approach b in $L^p(\Omega)$ by a sequence $b_j \in L^\infty(\Omega)$, and since $\dfrac{|\nabla u|}{\varepsilon + |\nabla u|} \leqslant 1$, we have:

$$\left|B^\varepsilon_{*u_n} - B^\varepsilon_{*u}\right|_p \leqslant 2\left|b - b_j\right|_p + \left|B^\varepsilon_{j*u_n} - B^\varepsilon_{j*u}\right|_p$$

where $B^\varepsilon_j = \dfrac{b_j\,|\nabla u|}{\varepsilon + |\nabla u|}$. We deduce from this inequality and from the first case that:

$$\limsup_n \left|B^\varepsilon_{*u_n} - B^\varepsilon_{*u}\right|_p \leqslant 2\left|b - b_j\right|_p \xrightarrow[j \to +\infty]{} 0\,.$$

As for (ii), we note that $B^\varepsilon \to b\chi_{\{\nabla u \neq 0\}}$ in $L^p(\Omega)$-strong when $\varepsilon \to 0$. Since

$$\left|\left(b\chi_{\{\nabla u \neq 0\}}\right)_{*u_n} - \left(b\chi_{\{\nabla u \neq 0\}}\right)_{*u}\right|_p \leqslant 2\left|B^\varepsilon - b\chi_{\{\nabla u \neq 0\}}\right|_p + \left|B^\varepsilon_{*u_n} - B^\varepsilon_{*u}\right|_p,$$

we conclude that

$$\limsup_n \left|\left(b\chi_{\{\nabla u \neq 0\}}\right)_{*u_n} - \left(b\chi_{\{\nabla u \neq 0\}}\right)_{*u}\right|_p \leqslant o(1)_{\varepsilon \to 0},$$

and the result follows. ◇

Corollary 7.1.2 *Under the same conditions as Theorem 7.1.3, we assume that*

$$\text{meas}\Big\{x \in \Omega : \nabla u(x) = 0\Big\} = 0.$$

Then, for $b \in L^p(\Omega)$, $1 \leqslant p < +\infty$,

(i) $b_{*u_n} \to b_{*u}$ *in* $L^p(\Omega_*)$,

(ii) $b_{*u_n}\Big(|u_n > u_n(\cdot)|\Big) \underset{n\to+\infty}{\longrightarrow} b_{*u}\left(\big|u > u(\cdot)\big|\right)$ *in* $L^p(\Omega)$*-strong, if* $\text{meas}\big(P(u_n)\big)=0$.

Proof Part (i) follows from Theorem 7.1.3 part (ii) since $b\chi_{\{\nabla u \neq 0\}} = b$. As for (ii), starting with the case $1 < p < +\infty$, we have

$$b_{*u_n}\Big(|u_n > u_n(\cdot)|\Big) = M_{u_n}(b_{*u_n}) \rightharpoonup b_{*u}\left(\big|u > u(\cdot)\big|\right) \ in\ L^p(\Omega)-weak.$$

Indeed if $\varphi \in L^{p'}(\Omega)$, then according to the properties of the mean operators:

$$\int_\Omega M_{u_n}(b_{*u_n})\varphi\, dx = \int_{\Omega_*} b_{*u_n}\varphi_{*u_n} ds\ .$$

According to (i) $\varphi_{*u_n} \to \varphi_{*u}$ in $L^{p'}(\Omega_*)$ and $b_{*u_n} \to b_{*u}$ in $L^p(\Omega_*)$. As a result we have:

$$\lim_n \int_\Omega M_{u_n}\left(b_{*u_n}\right)\varphi\, dx = \int_{\Omega_*} b_{*u}\cdot\varphi_{*u} = \int_{\Omega_*} \varphi(x) b_{*u}\left(\big|u > u(x)\big|\right) dx.$$

Furthermore, $\left|M_{u_n}\left(b_{*u_n}\right)\right|^p \leqslant M_{u_n}\left(\left|b_{*u_n}\right|^p\right)$ hence

$$\left|M_{u_n}\left(b_{*u_n}\right)\right|_p \leqslant \left|b_{*u_n}\right|_p \Longrightarrow \limsup_n \left|M_{u_n}\left(b_{*u_n}\right)\right|_p \leqslant |b_{*u}|_p\, .$$

But $|b_{*u}|_{L^p(\Omega_*)} = |b_{*u}\left(|u > u(\cdot)|\right)|_{L^p(\Omega)}$ (by equimeasurability), and since $L^p(\Omega)$ is uniformly convex, we deduce that

$$M_{u_n}\left(b_{*u_n}\right) \to b_{*u}\left(|u > u(\cdot)|\right) \ in\ L^p(\Omega)-strong.$$

If $p = 1$, we consider $b_j \in L^\infty(\Omega)$ s.t. $b_j \to b$ in $L^1(\Omega)$. Since the "mean operator" map is linear continuous of norm 1, and since the map $b \to b_{*u}$ is contractive, we have:

$$\left|M_{u_n}\left(b_{*u_n}\right) - M_u\left(b_{*u}\right)\right|_1 \leqslant 2\left|b - b_j\right| + \left|M_{u_n}\left(b_{j*u_n}\right) - M_u\left(b_{j*u}\right)\right|_1$$

$(M_u(b_{*u})(x) = b_{*u}(|u > u(\cdot)|))$. Since

$$\lim_n \left|M_{u_n}\left(b_{j*u_n}\right) - M_u\left(b_{j*u}\right)\right|_1 = 0,$$

we conclude

$$\limsup_n \left|M_{u_n}\left(b_{*u_n}\right) - M_u\left(b_{*u}\right)\right|_1 = 0.$$

Hence the result. ◇

Other variants of these theorems can be given. For example, we have:

Theorem 7.1.4 *Let $\Phi : \mathbb{R}^N \to \mathbb{R}$ with $\Phi(0) = 0$ and let $\{u_n,\ u\}$ be a sequence of $W^{1,1}(\Omega)$ such that $\Phi(\nabla u_n) \xrightarrow[n\to+\infty]{} \Phi(\nabla u)$ in $L^p(\Omega)$, $u_n \to u$ in $W^{1,1}(\Omega)$, the bounded open set Ω being as in Theorem 7.1.3. If $b \in L^{p'}(\Omega)$. $\dfrac{1}{p} + \dfrac{1}{p'} = 1$ and u is co-area regular, then*

$$\left[b\Phi\left(\nabla u_n\right)\right]_{*u_n} \to \left[b\Phi(\nabla u)\right]_{*u} \ in\ L^1(\Omega_*)\text{−strong}.$$

Here is a corollary of Theorem 7.1.4:

Corollary 7.1.3 *Let $1 < p < +\infty$, $b \in L^p(\Omega)$, and $\Phi : \mathbb{R}^N \to \mathbb{R}$ continuous with*

$$|\Phi(\xi)| \leqslant c_p\,|\xi|^p,\ 1 = \frac{1}{p} + \frac{1}{p'},\ \forall \xi \in \mathbb{R}^N.$$

Let $\{u,\ u_n\}$ be a sequence of $W^{1,1}(\Omega)$ such that $u_n \to u$ in $W^{1,p'}(\Omega)$ (then $b\Phi(\nabla u_n) \to b\Phi(\nabla u)$ in $L^1(\Omega)$-strong). Assume that u is co-area regular. Then,

$$\left[b\Phi(\nabla u_n)\right]_{*u_n}\left(|u_n > u_n(\cdot)|\right) \text{ tends to } \left[b\Phi(\nabla u)\right]_{*u}\left(|u > u(\cdot)|\right)$$

in L^1-strong.

Proof of the Corollary Since $\Phi(0) = 0$ then $b\Phi(\nabla u_n) = 0$ on $P(u_n)$. Consequently, if $\sigma \in P(u_{n*})$ we have $\big[b\Phi(\nabla u_n)\big]_{*u_n}(\sigma) = 0$ (in particular, this is true for $\sigma = |u_n > u_n(x)|$, $x \in P(u_n)$). Consequently, we have for all measurable φ

$$M_{u_n,\varphi}\Big(\big[b\Phi(\nabla u_n)\big]_{*u_n}\Big) = \big[b\Phi(\nabla u_n)\big]_{*u_n}\big(|u_n > u_n(\cdot)|\big). \tag{7.3}$$

Let us set, for convenience of writing,

$$K_n^b(x) = \big[b\Phi(\nabla u_n)\big]_{*u_n}\big(|u_n > u_n(x)|\big)$$

and

$$K^b(x) = \big[b\Phi(\nabla u)\big]_{*u}\big(|u > u(x)|\big).$$

Let us show that $K_n^b \to K^b$ in $L^1(\Omega)$-strong. We start with the case where $b \in L^\infty(\Omega)$. Then, in this case $b\Phi(\nabla u_n) \to b\Phi(\nabla u)$ in $L^{p'}(\Omega)$ and by Theorem 7.1.4 we deduce

$$\big[b\Phi(\nabla u_n)\big]_{*u_n} \to \big[b\Phi(\nabla u)\big]_{*u} \text{ in} L^{p'}(\Omega_*)\text{- strong.}$$

According to the previous relation, we have by equimeasurability:

$$\left|K_n^b\right|_{L^{p'}(\Omega)} = \left|\big[b\Phi(\nabla u_n)\big]_{*u}\right|_{L^{p'}(\Omega_*)}. \tag{7.4}$$

Hence,

$$\lim_n \left|K_n^b\right|_{L^{p'}(\Omega)} = \left|\big[b\Phi(\nabla u)\big]_{*u}\right|_{L^{p'}(\Omega_*)} = \left|K^b\right|_{L^{p'}(\Omega_*)}.$$

If $\varphi \in L^\infty(\Omega)$, then according to (7.3), we deduce

$$\int_\Omega \varphi K_n^b dx = \int_{\Omega_*} \varphi_{*u_n} \cdot \big[b\Phi(\nabla u_n)\big]_{*u_n} d\sigma. \tag{7.5}$$

If $\theta = \chi_{\Omega_*\backslash P(u_*)}$ then $\theta\varphi_{*u_n} \rightharpoonup \theta\varphi_{*u}$ in $L^\infty(\Omega_*)$-weak-star (see Theorem 2.6.1). Consequently,

$$\int_{\Omega_*} \theta\varphi_{*u_n}\big[b\Phi(\nabla u_n)\big]_{*u_n} d\sigma \underset{n}{\to} \int_{\Omega_*} \theta\varphi_{*u}\big[b\Phi(\nabla u)\big]_{*u} d\sigma$$

$$= \int_{\Omega_*} \varphi_{*u}\big[b\Phi(\nabla u)\big]_{*u} d\sigma$$

and

$$\int_{\Omega_*} (1-\theta)\varphi_{*u_n}\big[b\Phi(\nabla u_n)\big]_{*u_n} d\sigma = 0.$$

These last two relations together with the relation (7.5) imply

$$\lim_n \int_{\Omega} \varphi K_n^b dx = \int_{\Omega_*} \varphi_{*u}\big[b\Phi(\nabla u)\big]_{*u} d\sigma = \int_{\Omega} \varphi K^b dx. \tag{7.6}$$

By density, we conclude that $K_n^b \rightharpoonup K^b$ in $L^{p'}$-weak and $\lim_n \left|K_n^b\right|_{L^{p'}(\Omega)} = \left|K^b\right|_{L^{p'}(\Omega)}$, we deduce that $K_n^b \to K^b$ in $L^{p'}(\Omega)$-strong.

If $b \in L^p(\Omega)$ then for $b_j \in L^\infty(\Omega)$ s.t. $b_j \to b$ in $L^p(\Omega)$, the relation (7.3) leads to

$$\left|K_n^{b_j} - K_n^b\right|_1 \leqslant \left|b_j - b\right|_p |\Phi(\nabla u_n)|_{p'} \leqslant c\left|b_j - b\right|_p \underset{j}{\to} 0$$

and similarly,

$$\left|K^{b_j} - K^b\right|_1 \leqslant c\left|b_j - b\right|_p \underset{j}{\to} 0.$$

Thus,

$$\limsup_n \left|K_n^b - K^b\right|_1 \leqslant c\left|b_j - b\right|_p + \limsup_n \left|K_n^{b_j} - K_n^{b_j}\right|_1$$

which implies

$$\lim_n \left|K_n^b - K^b\right|_1 = 0.$$

◇

The condition that $\Big\{\nabla u = 0\Big\}$ is of zero measure seems to be necessary, as shown by the following counterexample given in dimension 1.

Let $\Omega =]-2, 2[$, b be an integrable and odd function. Let $\{u, u_n\}$ be a sequence of Lipschitz and even functions defined on $[-2, 0]$ by:

$$u(x) = \begin{cases} 2 & if\ -1 \leqslant x \leqslant 0, \\ 2(x+2) & if\ -2 \leqslant x \leqslant -1, \end{cases}$$

$$u_n(x) = \begin{cases} 2 + \dfrac{1}{n}(1 - x^2) & if\ -1 \leqslant x \leqslant 0, \\ 2(x+2) & if\ -2 \leqslant x \leqslant -1. \end{cases}$$

Proposition 7.1.2 *For almost all $s \in [0, 4]$, we have*

$$b_{*u}(s) = \begin{cases} 0 & if\ 2 \leqslant s \leqslant 4, \\ \left(b|_{\{u=2\}}\right)_*(s) & if\ 0 \leqslant s < 2, \end{cases}$$

$$b_{*u_n}(s) = 0.$$

Furthermore, we have:

$$u_n \to u,\ strongly\ in\ W^{1,1}(-2,2),\ and\ \left|u' = 0\right| = 2,\ \left|u'_n = 0\right| = 0\,.$$

The proof follows from the definition of relative rearrangement.

It suffices to choose b s.t. $b|_{\{u=2\}} = b|_{[-1,1]} \neq 0$ to deduce that b_{*u_n} does not tend to b_{*u} even weakly.

7.2 Approximation of b_{*u} for $u \in L^1(\Omega)$.

The previous section provides us with approximations of b_{*u} when meas$\left\{x : \nabla u(x) = 0\right\} = 0$. We will now give special approximations for any function u of $L^1(\Omega)$ using simple functions u_n that converge to u in $L^1(\Omega)$. We have:

Theorem 7.2.1 *Let $b \in L^p(\Omega)$, $p \in [1, +\infty]$ and let $u \in L^1(\Omega)$. Then, there exists a sequence u_n, of simple functions, converging to u a.e. and in $L^1(\Omega)$ such that:*

(i) $b_{*u_n} \rightharpoonup b_{*u}$ *in $L^p(\Omega)$-weak if $1 \leqslant p < +\infty$, $L^\infty(\Omega_*)$-weak-*.*
(ii) $\left(b_{*u_n}\right)_* = b_*$.

The first lemma concerns the relative rearrangement of simple functions.

Lemma 7.2.1 *Let $u : \Omega \to \mathbb{R}$ be a simple function. Then for all $b \in L^1(\Omega)$, we have:*

$$(b_{*u})_* = b_*.$$

Proof Since the map $b \to (b_*, (b_{*u})_*)$ is Lipschitzian, it suffices to consider $b \in L^\infty(\Omega)$. Let $Im(u) = \left\{a_1, \cdots, a_m\right\}$ s.t. $a_1 \geqslant \ldots \geqslant a_m$. Let $c_k = \sum_{j=1}^{k} \left|u = a_j\right|$, $c_0 = 0$.

Then $u_*(s) = a_k$ if $s \in [c_{k-1}, c_k[$, $k = 1, \ldots, m$.

Hence $|u > u_*(s)| = c_{k-1}$. By definition of the relative rearrangement:

$$b_{*u}(s) = \left(b|_{\{u=u_*(s)\}}\right)_* (s - c_{k-1}), \ \ s \in [c_{k-1}, c_k[\,.$$

Thus, for any polynomial P, and by equimeasurability:

$$\int_{\Omega_*} P\,(b_{*u})\,ds = \sum_{k=1}^{m} \int_{c_{k-1}}^{c_k} P\left(\left(b|_{\{u=u_*(s)\}}\right)_* (s - c_{k-1})\right)\,ds$$

$$= \sum_{k=1}^{m} \int_{\{u=a_k\}} P(b)\,dx = \int_{\Omega} P(b)dx : b_* = (b_{*u})_*\,.$$

◇

From this lemma, we have part of the theorem, as shown by the following corollary:

Corollary 7.2.1 *Let $u \in L^1(\Omega)$ be without a plateau. For any sequence u_n of simple functions, we have for $b \in L^1(\Omega)$:*

(i) $\left(b_{*u_n}\right)_* = b_*$.

(ii) $b_{*u_n} \rightharpoonup b_*$ *in $L^1(\Omega_*)$-weak if u_n tends to u a.e.*

Proof Assertion (i) follows from the lemma while assertion (ii) follows from Proposition 2.6.1.

For the general case, in order to obtain weak convergence when u has plateaus, we will modify u_n.

Proof Let $u \in L^1(\Omega)$. Let $P_t(u) = \Big\{u = t\Big\}$ be a plateau of u,

$P(u) = \bigcup_{t \in D_u} P_t(u)$, D_u is at most countable. If D_u is countable then we will set $D_u = \Big\{t_0, t_1, \ldots, t_k, \ldots\Big\}$, $\quad t_i \neq t_j$ and $D_u^m = \Big\{t_k,\ k \leqslant m\Big\}$. Since the case where D_u is empty was discussed in the previous corollary, we will now only consider the case where $D_u \neq \emptyset$.

We will start by proving the following lemma:

Lemma 7.2.2 *Let $u \in L^1(\Omega)$. There exists a sequence (u_n) of simple functions such that:*

(i) $u_n(x) \xrightarrow[n \to +\infty]{} u(x)$ *a.e.*

(ii) There exists a sequence (m_n) with $m_n \nearrow +\infty$ if D_u is countable, $D_u^{m_n} = D_u$ if D_u is finite and n large, such that

$$\forall t \in D_u^{m_n} : \big\{u_n > t\big\} = \big\{u > t\big\},\ \big\{u_n = t\big\} = \big\{u = t\big\}.$$

Proof of the Lemma It suffices to consider the case where $u \geqslant 0$. For the general case we will decompose into $u = u_+ - u_-$ and we will apply the same method separately to u_+ and u_-.

Let n then be an integer $\geqslant \max(1, t_0)$. Consider an integer m_n such that $\max\Big\{t_k,\ k \leqslant m_n\Big\} \leqslant n$. Let $t_{\sigma(0)} < t_{\sigma(1)} < \cdots < t_{\sigma(m_n)}$ denote a rearrangement of $D_u^{m_n}$. Let $\delta_n = \min\Big\{t_{\sigma(j+1)} - t_{\sigma(j)},\ j = 0, \ldots, m_n\Big\}$ and let's replace $D_u^{m_n}$ with:

$$D(u, n) = \Big\{t'_{\sigma(i)} \in \mathbb{R} - (D_u \cup \mathbb{Q}),\ i = 0, \ldots, m_n,$$

$$0 < t'_{\sigma(i)} - t_{\sigma(i)} < \min\left(\frac{1}{2^{n+1}}, \delta_n\right) \textit{ and } i \to t'_{\sigma(i)} \text{ is strictly increasing}\Big\}$$

$\mathbb{Q}$ denoting the set of rational numbers.

Consider also the following set:

$$\widehat{D}(u, n) = D(u, n) \cup \left(\left\{\frac{i}{2^n},\ i = 0, \ldots, n2^n\right\} - D_u^{m_n}\right)$$

Let us rearrange the elements of $\widehat{D}(u,n)$:

$$0 \leqslant \alpha_{n,0} < \alpha_{n,1} < \cdots < \alpha_{n,q},\ q \leqslant m_n + n2^n + 1,\ \alpha_{n,j} \in \widehat{D}(u,n)\,.$$

Note that the inequality is strict by the choice of $D(u,n)$ and $\alpha_{n,j+1} - \alpha_{n,j} \leqslant \dfrac{1}{2^n}$ and $\alpha_{n,q} \geqslant n$.

We associate with $(\alpha_{n,j})_j$ the following sets:

$$E_{n,j} = \Big\{x : \alpha_{n,j-1} \leqslant u(x) < \alpha_{n,j}\Big\},\ j = 1,\dots,q$$

$$F_n = \Big\{x : u(x) \geqslant \alpha_{n,q}\Big\}.$$

For j fixed in $\Big\{0,\dots,m_n\Big\}$, there exists only one index $q(j) \in \Big\{1,\dots,q\Big\}$ s.t.$\alpha_{n,q(j)-1} \leqslant t_j < \alpha_{n,q(j)}$. Let $Z_u = \Big\{q(j),\ j = 0,\dots,m_n\Big\}$. For $k = q(j) \in Z_u$ we write:

$$E_{n,q(j)} = E^1_{n,q(j)} \cup E^2_{n,q(j)}$$

where $E^2_{n,q(j)} = \Big\{t_j < u(x) < \alpha_{n,q(j)}\Big\}$. Let's define the following simple function:

$$\overline{u}_n(x)$$

$$= \sum_{i=0,\ i\notin Z_u}^{q} \alpha_{n,i-1}\chi_{E_{n,i}}(x) + \sum_{k\in Z_u}\alpha_{n,k-1}\chi_{E^1_{n,k}}(x) + \sum_{k\in Z_u}\alpha_{n,k}\chi_{E^2_{n,k}} + \alpha_{n,q}\chi_{F_n}(x).$$

$$\textit{For } x \in \Omega,\ \textit{we have } |\overline{u}_n(x) - u(x)| \leqslant \frac{1}{2^n},\ \textit{for n large.} \tag{7.7}$$

Furthermore, for $t_j \in D^{m_n}_u$, we have:

$$\Big\{\overline{u}_n > t_j\Big\}$$

$$= \Big\{\overline{u}_n > \alpha_{n,q(j)}\Big\} \cup \Big\{t_j < \overline{u}_n \leqslant \alpha_{n,q(j)}\Big\} = \Big\{u \geqslant \alpha_{n,q(j)}\Big\} \cup E^2_{n,q(j)}$$

$$= \Big\{u > t_j\Big\}.$$

Finally let's define u_n as follows:

$$u_n(x) = \begin{cases} \overline{u}_n|_{\Omega \backslash P(u)}(x) & if\ x \in \Omega \backslash P(u), \\ u(x) & if\ x \in P(u). \end{cases}$$

From (7.7) we deduce (i). By construction $\left\{\overline{u}_n = t\right\} = \emptyset$ if $t \in D_u^{m_n}$, hence:

$$\left\{u_n = t\right\} = \left\{u = t\right\}\ for\ t \in D_u^{m_n}\ .$$

According to the above, we have, for $t \in D_u^{m_n}$,

$$\left\{u_n > t\right\} = \left\{\overline{u}_n|_{\Omega \backslash P(u)} > t\right\} \cup \left\{x \in P(u) : u_n(x) > t\right\}$$

$$= \left\{u|_{\Omega \backslash P(u)}(\cdot) > t\right\} \cup \left\{x \in P(u) : u(x) > t\right\} = \left\{u > t\right\}.$$

In the case where u is unsigned, we decompose $u = u_+ - u_-$ and apply the previous method to u_+ and to u_-. If we denote by, u_n^+ the sequence associated with u_+ and by u_n^- that of u_-, then the sequence $u_n = u_n^+ - u_n^-$ is suitable. Note that $support(u_n^+) \subset \left\{u \geqslant 0\right\}$ and $support(u_n^-) \subset \left\{u \leqslant 0\right\}$.

Let us return to the proof of the theorem and consider the sequence $(u_n)_{n \geqslant 0}$ constructed in Lemma 7.2.2. By Lemma 7.2.1, we know that $(b_{*u_n})_* = b_*$. Therefore b_{*u_n} belongs to a bound of $L^p(\Omega_*)$ as soon as $b \in L^p(\Omega)$. Moreover, by the Hardy–Littlewood lemma, we have for all measurable $E \subset \Omega_*$

$$\int_E |b_{*u_n}| \leqslant \int_0^{|E|} |b_{*u_n}|_*(t)dt \leqslant \int_0^{|E|} |b|_*(t)dt\,,$$

(see Chap. 1 for this property of the relative rearrangement). We deduce that the sequence $(b_{*u_n})_n$ is weakly compact in $L^p(\Omega_*)$,

$1 \leqslant p < +\infty$ and in $L^\infty(\Omega_*)$-weak-star otherwise. It is then enough to show $\forall \varphi \in C(\overline{\Omega}_*)$,

$$\lim_n \int_{\Omega_*} (\varphi b_{*u_n})(t)dt = \int_{\Omega_*} (\varphi b_{*u})(t)dt.$$

Since for all $t \in D_u^{m_n}$, we have:

$$\left\{u_n = t\right\} = \left\{u = t\right\}, \quad |u_n > t| = |u > t|\ ,$$

then for almost all $x \in P^{m_n}(u) = \bigcup_{t \in D_u^{m_n}} P_t(u)$,

$$M_{u_n,b}(\varphi)(x) = M_{u,b}(\varphi)(x). \tag{7.8}$$

Furthermore, since $u_n(x) \xrightarrow[n \to +\infty]{} u(x)$, a.e. then for almost all $x \in \Omega \backslash P(u)$, we have

$$\lim_n M_{u_n,b}(\varphi)(x) = M_{u,b}(\varphi)(x). \tag{7.9}$$

If D_u is finite then for n large, we have $P^{m_n}(u) = P(u)$, so according to (7.8), (7.9) and the formula for the mean operator:

$$\lim_n \int_{\Omega_*} \varphi b_{*u_n} dt = \lim_n \int_{\Omega \backslash P(u)} M_{u_n,b}(\varphi) b dx + \int_{P(u)} M_{u,b}(\varphi) b(x) dx = \int_{\Omega_*} \varphi b_{*u}. \tag{7.10}$$

If D_u is countable then as m_n is increasing in n, we have $P^{m_n}(u) \subset P^{m_{n+1}}(u)$, and we deduce that:

$$\int_{P(u)} M_{u,b}(\varphi) b \, dx = \lim_n \int_{P^{m_n}(u)} M_{u_n,b}(\varphi) b \, dx \,. \tag{7.11}$$

Thus with the help of (7.8) and (7.9), the relation (7.11) leads to the same conclusion as (7.10). ◇

Theorem 7.2.2 *Let u, v be in $L^1(\Omega)$ and $\Phi : \mathbb{R} \to \mathbb{R}$ be convex increasing s.t. $\Phi(v) \in L^p(\Omega)$, $1 \leqslant p \leqslant +\infty$, and the map $v \to \Phi(v)$ is continuous on $L^1(\Omega_*)$.*

Then for all $\varphi \in L^{p'}(\Omega_*)$, $\dfrac{1}{p} + \dfrac{1}{p'} = 1$, $\varphi \geqslant 0$, we have

$$\int_{\Omega_*} \Phi(v_{*u}) \varphi \, d\sigma \leqslant \int_{\Omega_*} \Phi(v_*) \cdot \varphi_* d\sigma.$$

Proof According to Theorem 7.2.1, there exists a sequence of simple functions $(u_n)_{n\geqslant 0}$ such that:

1. $u_n \to u$ in $L^1(\Omega)$,
2. $v_{*u_n} \rightharpoonup v_{*u}$ in $L^p(\Omega_*)$-weak (or weak-* if $p = +\infty$),
3. $\left(v_{*u_n}\right)_* = v_*$.

By Mazur's lemma, there exists a sequence of integers $(m_n)_n$ and a sequence $(\lambda_j)_{j=1,\dots,m_n}$ s.t. $\sum_{j=1}^{m_n} \lambda_j v_{*u_j} \xrightarrow[n\to+\infty]{} v_{*u}$ almost everywhere and in $L^1(\Omega_*)$-strong with $\sum_{j=1}^{m_n} \lambda_j = 1,\ \lambda_j \geqslant 0$. By convexity of Φ and the Hardy–Littlewood inequality, we deduce for bounded φ:

$$\int_{\Omega_*} \Phi(v_{*u})\varphi\, d\sigma = \lim_n \int_{\Omega_*} \Phi\left(\sum_{j=1}^{m_n} \lambda_j v_{*u_j}\right)\varphi \leqslant \lim_n \sum_{j=1}^{m_n} \lambda_j \int_{\Omega_*} \Phi(v_{*u_j})_* \varphi_* d\sigma \tag{7.12}$$

but $\Phi(v_{*u_j})_* = \Phi(v_*)$ (because Φ is increasing), so the inequality (7.12) implies the result (by completing via a density argument).

Note If $v \geqslant 0$ it is sufficient that $\Phi : \mathbb{R}_+ \to \mathbb{R}$ is convex and increasing. If Φ is concave then we replace φ by $-\varphi$ so that we obtain

$$\int_{\Omega} \varphi^*(s)\Phi(v_*)(s)ds \leqslant \int_{\Omega_*} \Phi(v_{*u})\varphi ds.$$

with $\varphi^*(s) = -(-\varphi)_*(s)$ (non-decreasing rearrangement of φ).

7.3 Strong Convergence of the Directional Derivative $u \to u_*$

An interesting question relating to the approximation of the relative rearrangementis the following: When does the differential quotient $\dfrac{(u+tv)_* - u_*}{t}$ converge strongly to v_{*u} in $L^p(\Omega_*)$, $1 \leqslant p < +\infty$, as $t \searrow 0$?

The study of the strong continuity of the map $u \to v_{*u}$ will lead us to answers to this question.

We will prove the following theorem (and its variants).

Theorem 7.3.1 *Let* $u \in W^{1,1}(\Omega)$ *s.t.* measure$\left\{x : \nabla u(x) = 0\right\} = 0$. *We assume that* Ω *is connected and Lipschitz if* $\gamma_0 u \neq 0$ *and* Ω *is arbitrary otherwise.*

Let $v \in L^p(\Omega)$, $1 \leqslant p < +\infty$. *Then*

$$\left|\frac{(u+\lambda v)_* - u_*}{\lambda} - v_{*u}\right|_p \xrightarrow[\lambda \to 0]{} 0.$$

The key lemma to show this strong convergence is the following integral formula:

Lemma 7.3.1 *Let* $u \in L^1(\Omega)$ *and let* $v \in L^\infty(\Omega)$. *Then,* $\forall \lambda \in \mathbb{R}$ *and for almost all* $s \in \Omega_*$, *we have:*

$$(u+\lambda v)_*(s) - u_*(s) = \int_0^\lambda v_{*(u+tv)}(s)dt.$$

Proof of the Lemma Let us set for $(t, s) \in \mathbb{R} \times \Omega_*$, $G(t, s) = (u + tv)_*(s)$. $\forall s \in \overline{\Omega}_*$ (fixed), $t \in \mathbb{R} \to G(t, s)$ is Lipschitzian because according to the contraction property, we have ; $\forall t,\ t'$ of $\mathbb{R}$,

$$\left|G(t,s) - G(t',s)\right| \leqslant |v|_\infty \left|t - t'\right|, \quad \forall s \in \overline{\Omega}_*.$$

In particular, we deduce $\forall \lambda \in \mathbb{R}$, $\forall s \in \overline{\Omega}_*$,

$$(u+\lambda v)_*(s) - u_*(s) = \int_0^\lambda \frac{\partial G}{\partial t}(t,s)dt. \tag{7.13}$$

To calculate $\dfrac{\partial G}{\partial t}$, consider $H(t,s) = \displaystyle\int_0^s (u+tv)_*(\sigma)d\sigma$. Then we have $H \in W^{1,1}_{loc}(\mathbb{R} \times \Omega_*)$. Indeed, $\forall s \in \overline{\Omega_*}$, $t \to \displaystyle\int_0^s (u+tv)_*(\sigma)d\sigma$ is Lipschitzian, $\forall t \in \mathbb{R}$, $s \to \displaystyle\int_0^s (u+tv)_*(\sigma)d\sigma$ is absolutely continuous and

$$\frac{\partial H}{\partial s} = (u+tv)_* \in L^1_{loc}(\mathbb{R} \times \Omega_*), \qquad \left|\frac{\partial H}{\partial t}\right|_\infty \leqslant |\Omega|\,|v|_\infty .$$

In particular, for almost all (t, s):

$$\begin{aligned}\frac{\partial H}{\partial t}(t,s) &= \lim_{\delta t \to 0\ \delta t>0} \frac{H(t+\delta t, s) - H(t,s)}{\delta t} \\ &= \lim_{\delta t \to 0\ \delta t>0} \int_0^s \frac{(u+tv+(\delta t)v)_*(\sigma) - (u+tv)_*(\sigma)}{\delta t} d\sigma \\ &= \int_0^s v_{*(u+tv)}(\sigma) d\sigma,\end{aligned}$$

(by Theorem 2.1.1 of the directional derivative). We deduce that in the sense of distributions $\dfrac{\partial^2 H}{\partial s \partial t} = v_{*(u+tv)}$, but in the sense of distributions we also have $\dfrac{\partial^2 H}{\partial s \partial t} = \dfrac{\partial^2 H}{\partial t \partial s} = \dfrac{\partial G}{\partial t}$ from which

$$\frac{\partial G}{\partial t} = v_{*(u+tv)}. \tag{7.14}$$

The relations (7.13) and (7.14) give the result. ◇

Proof of Theorem 7.3.1 We begin with the case where $v \in W^{1,\infty}(\Omega)$. Since u is co-area regular, we deduce by Corollary 7.1.2 that

$$v_{*(u+tv)} \xrightarrow[t \to 0]{} v_{*u} \text{ in } L^p(\Omega_*).$$

Thus,

$$\lim_{\lambda \to 0} \frac{1}{\lambda} \int_0^\lambda \left| v_{*(u+tv)} - v_{*u} \right|_p dt = 0.$$

Using Lemma 7.3.1, we deduce that:

$$\lim_{\lambda \to 0} \left| \frac{(u+\lambda v)_* - u_*}{\lambda} - v_{*u} \right|_p = 0.$$

Case Where $v \in L^p(\Omega)$ Here we consider a sequence $v_n \in C_c^\infty(\Omega)$ such that $v_n \to v$ in $L^p(\Omega)$-strong. Since the map $v \to v_{*u}$ is a contraction on L^p, we deduce:

$$\left| \frac{(u+\lambda v)_* - u_*}{\lambda} - v_{*u} \right|_p \leqslant 2\,|v - v_n|_p + \left| \frac{(u+\lambda v_n)_* - u_*}{\lambda} - (v_n)_{*u} \right|_p.$$

Hence the result. ◇

If we remove the condition $\text{mes}\{\nabla u = 0\} = 0$ then we can assume that $u \in W^{2,1}(\Omega)$ is co-area regular, to obtain the same type of result.

Theorem 7.3.2 *Let $u \in W^{2,1}(\Omega)$ be a co-area regular function, $v \in L^p(\Omega)$, $1 \leqslant p < +\infty$. Assume that Ω is Lipschitz connected if $\gamma_0 u \neq 0$ and Ω is arbitrary otherwise. Then,*

$$\left| \frac{(u + \lambda v \chi_{\{\nabla u \neq 0\}})_* - u_*}{\lambda} - (v\chi_{\{\nabla u \neq 0\}})_{*u} \right|_p \xrightarrow[\lambda \to 0]{} 0.$$

This is a consequence of the following Theorem 7.3.3.

Theorem 7.3.3 *Let $\Phi : \mathbb{R}^N \to \mathbb{R}$ be Lipschitzian with $\Phi(0) = 0$. Let $u \in W^{2,1}(\Omega)$ be a co-area regular function, $v \in L^p(\Omega)$, $1 \leqslant p < +\infty$. Assume that Ω is Lipschitz connected if $\gamma_0 u \neq 0$ and any Ω otherwise. Then*

$$\left| \frac{\big(u + \lambda v \Phi(\nabla u)\big)_* - u_*}{\lambda} - \big[v\Phi(u)\big]_{*u} \right|_p \xrightarrow[\lambda \to 0]{} 0.$$

This last theorem is proved in the same way as Theorem 7.3.1. Here are the main points. For Φ bounded:

If $v \in C_c^\infty(\Omega)$ then $u + \lambda v \Phi(\nabla u) \xrightarrow[\lambda \to 0]{} u$ in $W^{1,1}(\Omega)$-strong.

Thus, $\big[v\Phi(\nabla u)\big]_{*(u+\lambda v \Phi(\nabla u))} \xrightarrow[\lambda \to 0]{} \big[v\Phi(\nabla u)\big]_{*u}$ in $L^p(\Omega_*)$-strong. Since

$$\frac{\big(u + \lambda v \Phi(\nabla u)\big)_* - u_*(s)}{\lambda} = \frac{1}{\lambda} \int_0^\lambda \big[v\Phi(\nabla u)\big]_{*(u+tv\Phi(\nabla u))}(s)dt,$$

we deduce the result. We conclude by density for the case $v \in L^p(\Omega)$ and Φ unbounded.

As for the proof of Theorem 7.3.2, we consider for $0 < \varepsilon < 1$,

$$\psi_\varepsilon(t) = \begin{cases} 1 & if\ t \geqslant \varepsilon, \\ 0 & if\ t \leqslant 0, \\ \text{continuous affine} & \text{on } [0, \varepsilon]. \end{cases}$$

Then $\psi_\varepsilon \in W^{1,\infty}(\mathbb{R})$ and $\psi_\varepsilon(t) \xrightarrow[\varepsilon\to 0]{} \chi_{]0,+\infty[}(t)\ \forall t$. Thus the function $\Phi_\varepsilon(\xi) = \psi_\varepsilon(|\xi|)$, $\xi \in \mathbb{R}^N$ is Lipschitzian and $\Phi_\varepsilon(\nabla u) \xrightarrow[\varepsilon\to 0]{} \chi_{\{\nabla u\neq 0\}}$ a.e., $v\Phi_\varepsilon(\nabla u) \to v\chi_{\{\nabla u\neq 0\}}$ in $L^p(\Omega)$-strong. We conclude by using the contraction properties and Theorem 7.3.3. ◇

We have the following theorem:

Theorem 7.3.4 *Let u be a simple function on Ω.*

(i) If v is a simple function, then

$$\frac{(u+\lambda v)_* - u_*}{\lambda}(s) = v_{*u}(s),\ for 0 < \lambda < \overline{\lambda}(u,v),\ \forall s \in \Omega_*.$$

(ii) If $v \in L^p(\Omega)$, $1 \leqslant p < +\infty$, then $\dfrac{(u+\lambda v)_* - u_*}{\lambda} \xrightarrow[\lambda\to 0]{} v_{*u}$ in $L^p(\Omega_*)$-strong.

Proof Let $u = \sum_{j=1}^{j=m} a_j\chi_{E_j}$ with $a_1 > \cdots > a_m$ where the E_j are pairwise disjoint. Let v be a simple function on Ω. For each $j \in \{1,\ldots,m\}$, we write the restriction to E_j of the function v by: $v|_{E_j} = \sum_{k=1}^{k=m} v_{kj}\chi_{F_{kj}}$ with the F_{kj} pairwise disjoint for $k \in \{1,\cdots m\}$, $E_j = \bigcup_{k=1}^{k=m} F_{kj}$, and, $v_{1j} \geqslant \cdots \geqslant v_{mj}$ Thus, we have for $\lambda \geqslant 0$:

$$(u+\lambda v) = \sum_{j=1}^{j=m}\sum_{k=1}^{k=m}(a_j + \lambda v_{kj})\chi_{F_{kj}}.$$

We reindex this double sum by introducing a bijection, denoted n from

$$\{1,\cdots,m^2\} \text{ to } \{1,\cdots,m\}\times\{1,\cdots,m\}$$

as follows

$$\begin{cases} n(1) = (1,1),\cdots, & \cdots, n(m) = (m,1),\\ n(m+1) = (1,2),\cdots, & \cdots, n(2m) = (m,2),\\ \cdots = \cdots & \\ n((m-1)m+1) = (1,m),\cdots, & \cdots, n(m^2) = (m,m).\end{cases}$$

We set for $1 \leqslant j \leqslant m$, and $1 + (j-1)m \leqslant i \leqslant jm$, $c^{\lambda}_{n(i)} = a_j + \lambda v_{kj}$ with $k = i - (j-1)m$. Choosing $0 \leqslant \lambda < \overline{\lambda}(u, v) = \dfrac{\inf_{j \neq k} |a_j - a_k|}{2|v|_\infty}$, we then have $c^{\lambda}_{n(i)} \geqslant c^{\lambda}_{n(i+1)}$ for $1 \leqslant i \leqslant m^2 - 1$. As we have $(u + \lambda v)_* = \sum_{i=1}^{i=m^2} c^{\lambda}_{n(i)} \chi_{[b_{i-1}, b_i)}$ with $b_i = \sum_{k=1}^{k=i} |F_{n(k)}|$, $b_0 = 0$ for $\lambda \geqslant 0$, we then deduce for $0 < \lambda < \overline{\lambda}(u, v)$

$$\frac{(u + \lambda v)_\star - u_\star}{\lambda} = \sum_{i=1}^{i=m^2} \frac{c^{\lambda}_{n(i)} - c^{0}_{n(i)}}{\lambda} \chi_{[b_{i-1}, b_i)}$$

$$= \sum_{j=1}^{j=m} \sum_{i=1+(j-1)m}^{i=jm} v_{(i-(j-1)m)j} \chi_{[b_{i-1}, b_i)}.$$

We deduce in particular that for all s: $\dfrac{(u + \lambda v)_\star - u_\star}{\lambda}(s) = v_{\star u}(s)$.

By density of the simple functions, we obtain the theorem, that is, the strong convergence in $L^p(\Omega_\star)$ of $\dfrac{(u + \lambda v)_\star - u_\star}{\lambda}$ to $v_{\star u}$. ◇

Remark A new proof of Theorem 7.3.4, statement (ii), is given in Exercise 11.3.2. It is based on Lemma 7.2.1 and the fact that $(v_{*u})_* = v_* : |v_{*u}|_{L^p} = |v|_{L^p}$.

Pre-bibliographical Notes

The majority of the proofs and results in this chapter are due to the author and/or collaborators, see Rakotoson [131], Rakotoson and Seoane [139], B. Simon [156], Rakotoson and Simon [141–143].

Theorem 7.2.1 is inspired by a result of Alvino and Trombetti [7].

As for the plasma physics equations associated with Tokomak machines, one can consult the articles of R. Temam [166, 167] or the book of B. Saramito [151]. A resolution by the topological degree method and the Galerkin method of a Grad–Shaframov model is given in Ferone, Jalal et al. [77, 79].

Chapter 8
Some Problems Related to Relative Rearrangement

In this chapter, we will solve some problems directly involving monotone and/or relative rearrangement. These problems will serve to illustrate the use of the results of the previous chapters.

For this, we have chosen two abstract theorems that seem significant and general to us.

The first concerns Euler's equations for multi-constraint optimization problems.

The second concerns the resolution of semi-linear problems of the form: $Au + G(u) = 0$, u belonging to a linear vector space V and G is continuous only on a subset of V. In the applications of this theorem, we will see functions $G(u)$ depending on b_{*u}, u'_* and the operator $Au = -\Delta u$ or any other second-order operator with regular coefficients. The motivation for such a consideration is essentially related to models in plasma physics.

8.1 Optimization

We will focus on the search for Euler equation problems of the type

$$\text{Min}\left\{J(v),\ S_0 v \in K\right\}$$

where J is a differentiable coercive function on a vector space X, S_0 is a map of X into a space Y and K is a cone of Y.

J. M. Rakotoson, *Relative Rearrangement*, Lecture Notes in Mathematics 2376,
https://doi.org/10.1007/978-3-032-02228-8_8

The example that we will treat entirely will be the case where

$$J(v) = \frac{1}{2}\int_\Omega |\nabla v|^2\,dx - \int_\Omega f v\,dx,\ f \in L^\infty(\Omega)\,, \quad X = H_0^1(\Omega),$$

$$S_0 v(t) = \Big(\int_0^t h_*(s)ds - \int_0^t v_*(s)ds;\ \int_\Omega v\,dx - \int_\Omega h(x)dx\Big)$$

$$K = K_1 \times \mathbb{R}_+, \quad K_1 = \Big\{\varphi \in L^\infty(\Omega_*) : \varphi(t) \geqslant 0\ a.e.\Big\}$$

and Y is the product space $L^\infty(\Omega) \times \mathbb{R}$. To do this, we will need some preliminary notions.

8.1.1 An Abstract Theorem for the Existence of Lagrange Multipliers

We will consider X and Y two normed spaces.

Definition 8.1.1 (Convex Cone) A subset K of Y is called a convex cone if the following two properties are satisfied:

(a) if $x,\ y \in K,\ \alpha \geqslant 0, \beta \geqslant 0$ then $\alpha x + \beta y \in K$,
(b) if $x \in K$ and $-x \in K$ then $x = 0$.

Definition 8.1.2 (Tangent Cone) Let K be a convex cone. The tangent cone at a point $y \in K$ is the set defined by

$$T_K(y) = \text{Closure}\left[\bigcup_{h>0} \frac{1}{h}(K - y)\right].$$

The closure of a subset A of a topological space Y is denoted $\overline{A} = \text{Closure}A = Cl(A)$.

Proposition 8.1.1 *Let K be a convex cone. Then:*

(i) $K - \mathbb{R}_+ y \subset T_K(y), \quad \mathbb{R}_+ = [0, +\infty[,$

(ii) $T_K(y) = \left\{x \in Y : x = \lim_{t_n \to 0\, t_n > 0} \frac{x_n - y}{t_n}, x_n \in K\right\},$

(iii) $T_K(y)$ *is also a convex cone.*

Proof

(i) If $v \in K$ and $h \in \mathbb{R}_+$ then for $\alpha > 0$ s.t. $\alpha h < 1$ we have:

$\alpha v + (1 - \alpha h)y \in K : v - hy \in \frac{1}{\alpha}(K - y)$. Hence $K - \mathbb{R}_+ y \subset T_K(y)$.

(ii) Let $A = \left\{x \in Y : x = \lim_{t_n > 0\, t_n \to 0} \frac{x_n - y}{t_n},\ x_n \in K\right\}$.

Since $\frac{x_n - y}{t_n} \in \frac{1}{t_n}(K - y)$ we have $A \subset T_K(y)$.

Conversely if $x \in T_K(y)$ then there exists a sequence $h_n > 0$, $x_n \in Y$ s.t. $x_n h_n + y \in K$ and $x_n \xrightarrow[n \to +\infty]{} x$. Let t_n be a sequence s.t. $0 < t_n < h_n$ and $t_n \to 0$ then: $\frac{t_n}{h_n}(x_n h_n + y) + \left(1 - \frac{t_n}{h_n}\right) y \in K$, we have $t_n x_n + y \in K$ and $\frac{t_n x_n + y - y}{t_n} = x_n \to x$. This proves that $x \in A$.

(iii) This assertion follows from (ii).

◇

Definition 8.1.3 Let K be a subset of Y. We call the dual cone or polar cone of K the set

$$K^* = \left\{\ell^* \in Y^*,\ \forall x \in K,\ < \ell^*, x > \geqslant 0\right\}$$

where Y^* is the dual of Y and $< \cdot, \cdot >$ denotes the duality between Y^* and Y.

We will now denote by $< \cdot, \cdot > by (\cdot, \cdot)$.

Let X and Y be two normed vector spaces with respective (topological) duals X^* and Y^*. In Y, we are given a closed convex cone with non-empty interior denoted by K. We are then interested in the optimal solution u_0 of the problem:

$$J(u_0) = \inf\left\{J(u) : u \in X \quad Su \in -K\right\} \qquad (\mathcal{P}_a)$$

where J is a map from X to $\mathbb{R}$ and S a map from X to Y.

We assume that J and S satisfy:

(H1) $\forall v \in X$, there exists a directional derivative $J'(u_0; v)$ in $\mathbb{R}$ i.e.

$$\lim_{t\to 0,\ t>0} \frac{J(u_0 + tv) - J(u_0)}{t} = J'(u_0; v),$$

and the map $X \ni v \mapsto J'(u_0; v)$ is convex.

(H2) $\forall v \in X$, there exists an element $S'(u_0; v) \in Y$ such that:

$$\lim_{t\to 0\ t>0} \frac{S(u_0 + tv) - Su_0}{t} = S'(u_0; v)\ .$$

Moreover, the map $X \ni v \mapsto S'(u_0; v) \in Y$ is convex, meaning that $\forall t \in [0, 1]$, $\forall v_1 \in X$, $\forall v_2 \in X$,

$$S'\big(u_0; tv_1 + (1-t)v_2\big) - tS'(u_0; v_1) - (1-t)S'(u_0; v_2) \in -K\ .$$

We then have the main theorem of this section:

Theorem 8.1.1 (Existence of Lagrange Multipliers for Multiconstraint Problems) *Assume (H1) and (H2). Then there exists a positive or zero real number c_0 and an element λ^* belonging to the polar cone K^* such that: $\forall v \in X$.*

$$c_0 J'(u_0; v) + \big(\lambda^*, S'(u_0; v)\big) \geqslant 0 \tag{8.1}$$

with $(c_0, \lambda^*) \neq (0, 0)$. Moreover, we have the following orthogonality relation:

$$(\lambda^*, Su_0) = 0. \tag{8.2}$$

Proof Consider the tangent cone to K at the point $-Su_0$, i.e.

$$T_K(-Su_0) = \mathrm{Cl}\left[\bigcup_{h>0} \frac{1}{h}(K + Su_0)\right]$$

(Cl denotes the closure of the set) and define in $\mathbb{R} \times Y$:

$$B(u_0)$$

$$= \left\{\Big(J'(u_0; v) + \alpha,\ S'(u_0; v) + k\Big) : \alpha \geqslant 0,\ v \in X \text{ and } k \in T_K(-Su_0)\right\}$$

and the connected open set

$A =]-\infty, 0[\times(-\overset{\circ}{K})$. We have $A \neq \emptyset$ because $\overset{\circ}{K} \neq \emptyset$.

◇

Lemma 8.1.1

$B(u_0)$ is a non-empty convex set.

Proof Let $b_i = \big(J'(u_0; v_i) + \alpha_i,\ S'(u_0; v_i) + k_i\big)$, $i = 0, 1$ be two elements of $B(u_0)$. The convexity of $J'(u_0; \cdot)$ and $S'(u_0; \cdot)$ ensure that for all $t \in [0, 1]$:

$$\alpha_2 = tJ'(u_0; v_0) + (1-t)J'(u_0; v_1) - J'(u_0; tv_0 + (1-t)v_1\big) \geqslant 0\,,$$

$$k_2 = tS'(u_0; v_0) + (1-t)S'(u_0; v_1) - S'\big(u_0; tv_0 + (1-t)v_1\big) \in K\,.$$

Defining $k_3 = k_2 + tk_0 + (1-t)k_1$ we have that k_3 belongs to $T_K(-Su_0)$ and $\alpha_3 = \alpha_2 + t\alpha_0 + (1-t)\alpha_1 \geqslant 0$. Hence, we have that

$$tb_0 + (1-t)b_1 = \Big(J'(u_0; tv_0 + (1-t)v_1\big) + \alpha_3,\ S'\big(u_0; tv_0 + (1-t)v_1\big) + k_3\Big)$$

belongs to $B(u_0)$.

◇

Lemma 8.1.2

The two non-empty convex sets A and $B(u_0)$ are disjoint.

Proof Suppose that $A \cap B(u_0) \neq \emptyset$. So, by definition, there exists $\alpha \geqslant 0$, $v \in X$, $k_0 \in T_K(-Su_0)$, $\beta > 0$ and $k_1 \in \overset{\circ}{K}$ such that:

$$\Big(J'(u_0; v) + \alpha,\ S'(u_0; v) + k_0\Big) = (-\beta,\ -k_1)\,.$$

Therefore

$$I(u_0; v) = -\beta - \alpha < 0, \tag{8.3}$$

$$S'(u_0; v) = -k_1 - \frac{k_0^n + Su_0}{t_n} + o(1) \tag{8.4}$$

where $k_0^n \in K$ and $\lim_{t_n \to 0} \dfrac{k_0^n + Su_0}{t_n} = k_0$. By definition of $S'(u_0; v)$, we also have:

$$S'(u_0; v) = \frac{S(u_0 + t_n v) - Su_0}{t_n} + o(1). \tag{8.5}$$

The relations (8.4) and (8.5) then imply that:

$$S(u_0 + t_n v) = -k_0^n - t_n\big(k_1 + o(1)\big). \tag{8.6}$$

Since $k_1 \in \overset{\circ}{K}$, there exists $\tau > 0$ s.t. for $0 < t_n < \tau$ we have $k_1 + o(1) \in \overset{\circ}{K}$. Since K is a cone, we deduce from (8.6) that

$$S(u_0 + t_n v) \in -K. \tag{8.7}$$

Moreover, the relation (8.3) implies that there exists $\tau_1 < \tau$ such that if $0 < t_n < \tau_1$ we have:

$$J(u_0 + t_n v) - J(u_0) = t_n\big(J'(u_0; v) + o(1)\big) < 0. \tag{8.8}$$

Thus, from the previous relations (8.7) and (8.8) we obtain for $0 < t_n < \tau_1$:

$$\begin{cases} S(u_0 + t_n v) \in -K, \\ J(u_0 + t_n v) < J(u_0). \end{cases}$$

This contradicts the optimality of u_0.

Lemma 8.1.2 then allows us to apply the Hahn–Banach theorem, i.e. there exists a real c_0 and an element λ^* of Y^* such that

$\forall \alpha \geqslant 0,\ \forall k \in T_K(-Su_0),\ \forall v \in X,\ \forall k' \in \overline{\overset{\circ}{K}} = K$
(since K is convex closed) and $\forall \beta \geqslant 0$ we have:

$$-c_0\beta - (\lambda^*, k') \leqslant c_0\big(J'(u_0; v) + \alpha\big) + \big(\lambda^*, S'(u_0; v) + k\big)\,. \tag{8.9}$$

This relation necessarily implies that $c_0 \geqslant 0$ and then the relation (8.9) reduces to: $\forall k \in T_K(-Su_0),\ \forall v \in X$

$$-(\lambda^*, k) \leqslant c_0 J'(u_0; v) + \big(\lambda^*, S'(u_0; v)\big)\,. \tag{8.10}$$

Since $T_K(-Su_0)$ is a cone, the relation (8.10) implies that:

$$(\lambda^*, k) \geqslant 0\ \forall k \in T_K(-Su_0)\,, \tag{8.11}$$

which implies in particular that $\lambda^* \in K^*$. By, replacing k by $k + tSu_0$, $t > 0$, the relation (8.11) implies that $(\lambda^*, Su_0) = 0$. Thus, we obtain:

$$\begin{cases} c_0 J'(u_0; v) + \big(\lambda^*, S'(u_0; v)\big) \geqslant 0 \, \forall\, v \in X, \\ (\lambda^*, Su_0) = 0, \\ \lambda^* \in K^*, \quad c_0 \geqslant 0 \end{cases} \tag{8.12}$$

◇

8.1.2 A Concrete Application

Let $f \in L^\infty(\Omega)$ and $h \in L^1(\Omega)$. We consider the unique optimal solution u_0 of:

$$(\mathcal{P}) \qquad\qquad J(u_0) = \inf\Big\{J(u) : u \in \mathcal{K}(h)\Big\}$$

where $J(u) = \dfrac{1}{2}\displaystyle\int_\Omega |\nabla u|^2\,dx - \int_\Omega fu\,dx$ and

$$\mathcal{K}(h) = \Big\{v \in H_0^1(\Omega) : \int_\Omega \varphi(v)dx \leqslant \int_\Omega \varphi(h)dx$$

$$\forall\,\varphi(\text{Lipschitz convex}) : \mathbb{R} \to \mathbb{R}\Big\}.$$

We want to obtain information on the regularity of u_0. We then have:

Theorem 8.1.2 *Under the previous hypotheses, the optimal solution u_0 belongs to $W^{2,p}(\Omega)$ for all $p \in [1, +\infty[$. More precisely, there exists a constant $c_h \geqslant 0$ and a positive Radon measure λ^* belonging to the dual of $L^\infty(0, |\Omega|)$ such that if we define the following elements of $L^\infty(\Omega)$:*

$$\psi_1(x) = \int_{|u_0 \geqslant u_0(x)|}^{|\Omega|} d\lambda^*(\sigma) - c_h\,,\ \psi_2(x) = \psi_1(x) + \chi_{P(u_0)}(x)\int_{P(u_{0*})} d\lambda^*(\sigma)\,,$$

we have in the sense of distributions:

$$-\Delta u_0 - f \in [-\psi_2; -\psi_1] \tag{8.13}$$

(continued)

Theorem 8.1.2 (continued)
and moreover λ^* *satisfies:*

$$\int_0^{|\Omega|} \left[\int_0^{\sigma} h_*(\tau)d\tau - \int_0^{\sigma} u_{0*}(\tau)d\tau\right] d\lambda^*(\sigma) = 0\,. \tag{8.14}$$

In particular: support $\lambda^* \subset \left\{\sigma \in [0, |\Omega|\,] : \int_0^{\sigma} h_*(\tau)d\tau = \int_0^{\sigma} u_{0*}(\tau)d\tau\right\}$.

For convenience of writing, we will assume throughout this section that $|\Omega| = 1$.

The passage from the problem $(\mathcal{P})$ to abstract problem $(\mathcal{P}_a)$ in order to apply Theorem 8.1.1 requires Corollary 1.2.3 of Proposition 1.2.1, which we recall .

Lemma 8.1.3 *Let* $v \in L^1(\Omega)$*; we have the following equivalence:*

$$\begin{cases} \forall\, \varphi \text{ convex and Lipschitzian } \mathbb{R} \to \mathbb{R} \\ \int_\Omega \varphi(v)(x)dx \leqslant \int_\Omega \varphi(h)(x)dx \end{cases}$$

$$\iff \begin{cases} \forall\, t \in [0, 1] \\ \int_0^t v_*(\sigma)d\sigma \leqslant \int_0^t h_*(\sigma)d\sigma \\ \text{and } \int_\Omega v(x)dx = \int_\Omega h(x)dx. \end{cases}$$

We consider the following closed convex cones:

$$K_1 = \left\{\ell \in L^\infty(0, 1) : \ell(t) \geqslant 0\, a.e.\, t \in [0, 1]\right\}$$

$K = K_1 \times [0, +\infty[$ thus $\overset{\circ}{K} \neq \emptyset$. For $v \in H_0^1(\Omega)$, we define:

$$(S_1 v)(t) = \int_0^t v_*(\sigma)d\sigma - \int_0^t h_*(\sigma)d\sigma,\ t \in [0, 1],\ S_2 v = \int_\Omega h(x)dx - \int_\Omega v(x)dx.$$

We denote by S the map: $\begin{array}{ccc} H_0^1(\Omega) & \longrightarrow & L^\infty(\Omega_*) \times \mathbb{R} \\ v & \longmapsto & Sv = (S_1 v, S_2 v) \end{array}$. Since the rearrangement preserves the integral, Lemma 8.1.3 and the definition of S imply:

Lemma 8.1.4

$$Sv \in -K \Longleftrightarrow \begin{cases} \forall \varphi \text{ convex, Lipschitzian} \\ \int_\Omega \varphi(v)dx \leqslant \int_\Omega \varphi(h)dx. \end{cases}$$

In particular, the problem $(\mathcal{P})$ *is equivalent to:*

$$J(u_0) = \inf\Big\{J(u) : u \in H_0^1(\Omega) \ s.t. \ Su \in -K\Big\}.$$

Lemma 8.1.5 *For* $u, \ v$ *in* $L^1(\Omega)$ *and* $\sigma \in [0, 1]$ *we for convenience, we denote by* $w(u; v)(\sigma)$ *the function:*

$$w(u; v)(\sigma) = \begin{cases} \displaystyle\int_{u>u_*(\sigma)} v(x)dx & \text{if } |u = u_*(\sigma)| = 0 \\ \displaystyle\int_{u>u_*(\sigma)} v(x)dx + \int_0^{\sigma-|u>u_*(\sigma)|} \big(v|_{P(\sigma)}\big)_*(\tau)d\tau & \text{otherwise} \end{cases}$$

with $P(\sigma) = \{x \in \Omega : u(x) = u_*(\sigma)\}$, $v|_{P(\sigma)}$ *is the restriction of* v *to* $P(\sigma)$. *Then:*

(i) $w(u; v) \in C[0, 1]$,

(ii) $\displaystyle\lim_{t\to 0, t>0} \left\| \frac{S_1(u+tv) - S_1 u}{t} - w(u, v) \right\|_\infty = 0;$

(iii) for $v_1, \ v_2$ *in* $L^1(\Omega)$, $\alpha \geqslant 0, \ \beta \geqslant 0$ *we have:*

$$w(u; \alpha v_1 + \beta v_2) - \alpha w(u, v_1) - \beta w(u, v_2) \in -K_1.$$

Proof Parts (i) and (ii) follow from Theorem 2.1.1 and its Corollary 2.1.1 since

$$\frac{S_1(u+tv)(\sigma)-S_1u(\sigma)}{t}=\frac{1}{t}\left[\int_0^\sigma\left[(u+tv)_*(\tau)-u_*(\tau)\right]d\tau\right].$$

The proof of (iii) follows directly from the following Hardy–Littlewood property: for $g\in L^1(\Omega)$ and for all $\sigma\in[0,1]$

$$\int_0^\sigma g_*(\tau)d\tau=\text{Max}\left\{\int_E g(x)dx : |E|=\sigma\right\}. \qquad \diamond$$

Lemma 8.1.5 ensures that the operator $S'(u_0;\cdot)$ is given by

$$S'(u_0;v)=\left(w(u_0;v);-\int_\Omega v(x)dx\right);$$

moreover, it is a convex operator in the sense of Theorem 8.1.2.

Theorem 8.1.3 (Optimality Condition for u_0) *Let u_0 be the optimal solution $(\mathcal{P})$ which is not identically zero. Then there exists a nonnegative Radon measure $\lambda^*\in K_1^*$ (dual cone of K_1) and a real number $c_h\geqslant 0$ such that for all $v\in H_0^1(\Omega)$:*

$$\int_\Omega \nabla u_0\nabla v dx-\int_\Omega f v\,dx+\int_0^1 w(u_0;v)(\sigma)d\lambda^*(\sigma)-c_h\int_\Omega v(x)dx\geqslant 0.$$

Note If u_0 is identically zero ($u_0=0$) then Theorem 8.1.2 is trivial. In this Theorem 8.1.3, this assumption has been made for reasons of convenience of the following proof.

Proof We apply Theorem 8.1.1 with $X=H_0^1(\Omega)$, $Y=L^\infty(0,1)\times\mathbb{R}$, $K=K_1\times[0,+\infty[$ and J and S defined as above. There then exist two real numbers $c_0\geqslant 0$, $c_1\geqslant 0$ and a positive measure $\lambda_1^*\in K_1^*$ such that $(c_0,c_1,\lambda_1^*)\neq(0,0,0)$ satisfying, for all $v\in H_0^1(\Omega)$:

$$c_0J'(u_0)\cdot v+\left(\lambda_1^*,w(u_0;v)\right)-c_1\int_\Omega v(x)dx\geqslant 0. \tag{8.15}$$

Lemma 8.1.6 (Fundamental)

$$c_0 \neq 0.$$

Proof If $c_0 = 0$ the relation (8.15) implies that for all $v \in L^1(\Omega)$:

$$\int_0^1 w(u_0, v)(\sigma) d\lambda_1^*(\sigma) - c_1 \int_\Omega v(x)dx \geqslant 0. \tag{8.16}$$

(Indeed, the map $L^1(\Omega) \ni v \mapsto w(u_0; v)$ is 1-Lipschitzian.) In particular, if F is Borel with $F(u_0) \in L^1(\Omega)$, the relation (8.16) gives:

$$\int_0^1 d\lambda_1^*(\sigma) \left(\int_0^\sigma F(u_{0*})(\tau)d\tau \right) = c_1 \int_0^1 F(u_{0*})d\tau \tag{8.17}$$

hence

$$\int_0^1 \sigma d\lambda_1^*(\sigma) = c_1. \tag{8.18}$$

1st Case If $s_m = |u_0 = \text{ess sup } u_0| > 0$ then ess sup $u_0 \in \mathbb{R}$.

Note that, since $u_0 \neq 0$ then $s_m < 1$. Replacing F by $\chi_{\{\text{ess sup } u_0\}}(\sigma)$, Eqs. (8.17) and (8.18) give:

$$\int_0^1 \inf(\sigma, s_m) d\lambda_1^*(\sigma) = c_1 \int_0^{s_m} d\tau = \int_0^1 s_m \sigma \, d\lambda_1^*(\sigma) \tag{8.19}$$

(because $F(u_{0*})(\tau) = \chi_{\{\text{ess sup } u_0\}}(u_{0*})(\tau) = \begin{cases} 1 & si \ 0 \leqslant \tau \leqslant s_m \\ 0 & else. \end{cases}$).

Thus, if we set $\psi(\sigma) = \inf(\sigma, s_m) - \sigma s_m \geqslant 0$, the relation (8.19) results in:

$$\int_0^1 \psi(\sigma) d\lambda_1^*(\sigma) = 0\,. \tag{8.20}$$

Since λ_1^* is a positive measure, the relation infers that:

$$support \, \lambda_1^* \subset \Big\{ \sigma \in [0, 1] : s_m \sigma = \inf(\sigma, s_m) \Big\} = A_0\,.$$

Let us study this last set. Let $\sigma \in A_0$. If $0 \leqslant \sigma \leqslant s_m$ then necessarily $\sigma = 0$ (because $s_m < 1$). If $1 \geqslant \sigma > s_m$ then $\sigma = 1$.

Thus *support* $\lambda_1^* \subset \{0, 1\}$, and consequently λ_1^* cannot be an element of the dual of $L^\infty(0, 1)$ unless $\lambda_1^* = 0$; but in this case $c_1 = 0$ and this would contradict the fact that $(c_0, c_1, \lambda_1^*) \neq (0, 0, 0)$.

2nd Case $|u_0 = \text{ess sup}\, u_0| = 0$.

Since $u_{0*}(0) = \text{ess sup}\, u_0$, we have $0 \notin P(u_{0*})$. Moreover, for $v \in L^1(\Omega)$, if we define $\overline{v}(x)$ by $\overline{v}(x) = \chi_{\Omega \setminus P(u_0)}(x) \cdot v(x)$ then

$$w(u_0; \overline{v})(\sigma) = \int\limits_{u_0 > u_{0*}(\sigma)} \overline{v}(x)dx \qquad \forall \sigma \in [0, 1] .$$

In this case we see that the map $v \mapsto w(u_0; \overline{v})$ is linear; hence the relation (8.16) becomes: $\forall v \in L^1(\Omega)$

$$c_1 \int_\Omega \overline{v}(x)dx = \int_0^1 w(u_0; \overline{v})d\lambda_1^*(\sigma) = \int_\Omega v(x)\chi_{\Omega \setminus P(u_0)}(x) \int\limits_{|u_0 \geqslant u_0(x)|}^{1} d\lambda_1^*(\sigma) \tag{8.21}$$

(here, we have used Fubini's theorem) thus for almost all $x \in \Omega$

$$c_1 \chi_{\Omega \setminus P(u_0)}(x) = \chi_{\Omega \setminus P(u_0)}(x) \int\limits_{|u_0 \geqslant u_0(x)|}^{1} d\lambda_1^*(\sigma) . \tag{8.22}$$

By equimeasurability the relation (8.22) becomes: for all $s \in \overline{\Omega}_* \setminus P(u_{0*})$

$$\begin{cases} c_1 = \displaystyle\int_s^1 d\lambda_1^*(\sigma) \\ 0 \in \overline{\Omega}_* \setminus P(u_{0*}) \end{cases} \Longrightarrow c_1 = \int_0^1 d\lambda_1^*(\sigma) . \tag{8.23}$$

The relations (8.18) and (8.23) then give $\displaystyle\int_0^1 (1 - \sigma)d\lambda_1^*(\sigma) = 0$, which implies that either $\lambda_1^* = 0$ and therefore $c_1 = 0$ (contradiction), or λ_1^* is a Dirac measure at the point $\sigma = 1$, contradicting the fact that $\lambda_1^* \in L^\infty(0, 1)^*$.

In all cases we have a contradiction, so $c_0 \neq 0$. ◇

We set $\lambda^* = \dfrac{\lambda_1^*}{c_0}$ and $c_h = \dfrac{c_1}{c_0}$, we have Theorem 8.1.3. ◇

Proof of Theorem 8.1.2 It remains to interpret Theorem 8.1.3 in the sense of distributions: let $v \in \mathcal{D}(\Omega),\ v \leqslant 0$, then as in the relation (8.21) we have

$$\int_0^1 w(u_0; v)(\sigma) d\lambda^*(\sigma) \leqslant \int_0^1 d\lambda^*(\sigma) \int_\Omega v(x) \chi_{u_0 > u_{0*}(\sigma)}(x) dx \tag{8.24}$$

$$= \int_\Omega v(x) \left(\int_{|u_0 \geqslant u_0(x)|}^{1} d\lambda^*(\sigma) \right) dx.$$

The inequality of Theorem 8.1.3 and the relation (8.24) yield the inequality

$$\int_\Omega \nabla u_0 \cdot \nabla v \, dx - \int_\Omega f v \, dx + \int_\Omega v(x) \psi_1(x) dx \geqslant 0 \tag{8.25}$$

for all $v \in \mathcal{D}(\Omega),\ v \leqslant 0$, where

$$\psi_1(x) = \int_{|u_0 \geqslant u_0(x)|}^{1} d\lambda^*(\sigma) \ - c_h$$

hence, in $\mathcal{D}'(\Omega)$:

$$-\Delta u_0 - f + \psi_1 \leqslant 0. \tag{8.26}$$

Similarly if $v \in \mathcal{D}(\Omega),\ v \geqslant 0$, then

$$\int_0^1 w(u_0; v)(\sigma) d\lambda_*(\sigma) \leqslant \int_\Omega v(x) \psi_1(x) dx$$

$$+ \int_\Omega v(x) \chi_{P(u_0)}(x) \left(\int_{P(u_{0*})} d\lambda^*(x) \right) dx$$

$$+ c_h \int_\Omega v(x) dx$$

which, combined with Theorem 8.1.3, gives in $\mathcal{D}'(\Omega)$:

$$-\Delta u_0 - f + \psi_1 + \chi_{P(u_0)} \int_{P(u_{0*})} d\lambda^*(\sigma) \geqslant 0. \tag{8.27}$$

As for the orthogonality relation, it translates to:

$$\int_0^1 \left[\int_0^\sigma h_*(\tau)d\tau - \int_0^\sigma u_{0*}(\tau)d\tau \right] d\lambda^*(\sigma) \tag{8.28}$$

$$= c_h \left[\int_\Omega h(x)dx - \int_\Omega u_0(x)dx \right] = 0 \,.$$

Since $\displaystyle\int_0^\sigma h_*(\tau)d\tau \geqslant \int_0^\sigma u_{0*}(\tau)d\tau, \quad \sigma \in [0, 1]$, the relation (8.28) implies that the support of λ^* is contained in the set

$$\left\{ \sigma \in [0, 1] : \int_0^\sigma h_*(\tau)d\tau = \int_0^\sigma u_{0*}(\tau)d\tau \right\} .$$

Since ψ_1 and ψ_2 are in $L^\infty(\Omega)$, we deduce that $-\Delta u_0 - f \in L^\infty(\Omega)$ from which we conclude that $u_0 \in W^{2,p}(\Omega)\ \forall\, p \in [1, +\infty[$. ◊

8.2 On an Abstract Semilinear Problem and its Applications to Nonlocal Problems

Various equilibrium equations in problems from physics can be written in the form $-\Delta u = F(u)$. This is the case, for example, for problems of confinement of a plasma in a Tokamak machine or a Stellarator machine. In either of these models, the function $F(u)$ can be a non-local function of u depending, for example, on the relative rearrangement of u with respect to a given value b : b_{*u} or on the derivatives of the monotone rearrangement $u'_*,\ u''_* \ldots$. As we saw in Chaps. 6 and 7, in this case the function $u \to F(u)$ will only be continuous on a small subset that is not necessarily linear. We will begin by giving a resolution method for this type of problem.

8.2.1 Abstract Theorems for Nonlocal Problems

Consider a Hilbert space $(V, \|\cdot\|)$ and a separable Banach space $(H, |\cdot|)$ and let G be an operator of V in the dual $(H', |\cdot|_*)$ of H. Assume

(H1) $V \subsetneq H$, compact injection (thus $\inf_{|v|=1} \|v\| > 0$).
(H2) There exists a family of finite elements $V_h \subset V$ and a family of linear operators Π_h s.t.

$$\lim_{h \to 0} \|v - \Pi_h v\| = 0, \quad \forall v \in V .$$

(H3) G is continuous from V-strong to H'-weak-* (i.e. equipped with the topology weak $* \sigma(H', H)$).

Theorem 8.2.1 *Assume (H1) to (H3) and let* $B : V \times V \to \mathbb{R}$ *be a coercive bilinear form (i.e.* $\alpha = \inf_{\|v\|=1} B(v, v) > 0$*) which is continuous (i.e.* $\sup_{\|u\|=\|v\|=1} B(u, v) = M < +\infty$*). We further assume that G satisfies the following growth:*

(H4) There exists $(\lambda_0, \lambda_1) \in]0, +\infty[\times]0, +\infty[$ *s.t.*

$$0 < \lambda_0 < \alpha \inf_{|v|=1} \|v\|, \ |G(v)|_* \leqslant \lambda_0 \|v\| + \lambda_1 \, \forall v \in V .$$

(i) there exists $u_h \in V_h$ *such that:*

$$B(u_h, v_h) = \langle G(u_h), v_h \rangle, \quad \forall v_h \in V_h,$$

(ii) there exists $u \in V$*, (a subsequence)* $u_h \in V$ *s.t.* $u_h \to u$ *in* V*-strong and*

$$B(u, v) = \langle G(u), v \rangle, \quad \forall v \in V .$$

Proof Let $m = \dim V_h$ and $\{\varphi_1, \ldots, \varphi_m\}$ be a basis of V_h. Let us define the following scalar product on V_h:

$$if \ v = \sum_{j=1}^{m} v_j \varphi_j, \quad w = \sum_{j=1}^{m} w_j \varphi_j, \ then \ [v, w] = \sum_{j=1}^{m} v_j w_j .$$

We then introduce the operator $T_m : V_h \to V_h$ defined by

$$T_m v = \sum_{j=1}^{m} \left[B(v, \varphi_j) - \langle G(v), \varphi_j \rangle \right] \varphi_j \, .$$

We have for all $v \in V_h$:

$$[T_m v, v] = B(v, v) - \langle G(v), v \rangle \geqslant \alpha \, \|v\|^2 - \lambda_0 \, \|v\| \, |v| - \lambda_1 \, |v| \tag{8.29}$$

$$\geqslant \left(\alpha \inf_{|z|=1} \|z\| - \lambda_0 \right) \|v\| \, |v| - \lambda_1 \, |v| \, .$$

We then deduce that $[T_m v, v] \to +\infty$ if $[v, v] \to +\infty$. Moreover, T_m is continuous due to the continuity of B and G. By Brouwer's fixed point theorem, we then conclude the existence of $u_h \in V_h$ s.t. $T_m u_h = 0$, hence assertion i). Moreover, using the estimate (8.29), we deduce

$$\|u\|_h \leqslant constant = \frac{\lambda_1}{\alpha \inf_{|z|=1} \|z\| - \lambda_0} = c_1 . \tag{8.30}$$

Let us then consider $u \in V$, $\ell_u \in H'$ and a subsequence also denoted by $u_h \in V_h$ s.t.

(a) $u_h \rightharpoonup u$ in V-weak,
(b) $u_h \to u$ in H-strong,
(c) $G(u_h) \rightharpoonup \ell_u$ in H'-weak-star.

Let $v \in V$. Then, we have:

$$B\,(u_h, \Pi_h v) = \langle G(u_h), \; \Pi_h v \rangle \, , \tag{8.31}$$

$$|B\,(u_h, \Pi_h v) - B\,(u_h, v)| \leqslant c_1 M \, \|v - \; \Pi_h v\| \tag{8.32}$$

and

$$|\langle G(u_h), \; \Pi_h v \rangle - \langle G(u_h), \; v \rangle| \leqslant c_0 \, \|v - \Pi_h v\| \; . \tag{8.33}$$

From the relations (8.31) to (8.33), we deduce:

$$B(u, v) = \lim_{h \to 0} B(u_h, \; \Pi_h v) = \lim_{h \to 0} \; < G(u_h), \; v > = < \ell_u, \; v > \; .$$

Let us now show that $\lim_{h \to 0} \|u_h - u\| = 0$.

Since $B(u_h - u, u_h - u) \geqslant \alpha \|u_h - u\|^2$, it suffices to show that $\lim_{h \to 0} B(u_h, u_h) = B(u, u)$. Now, since $|u_h - u| \underset{h\to 0}{\longrightarrow} 0$, we have:

$$\lim_{h\to 0} B(u_h, u_h) = \lim_{h\to 0} \langle G(u_h), u_h\rangle = \langle \ell_u, u\rangle = B(u, u) \,.$$

By continuity of G, we deduce that $\ell_u = G(u)$. ◇

Note that statement (i) only uses hypotheses (H1), (H2) and (H4).

To take into account that the function G can be continuous only on a subset $\mathcal{V}$ of V, we modify the hypothesis (H3) and to do so:

Let us consider the continuous linear operato rA of V in V' defined by

$$\langle Av, w\rangle = B(v, w), \text{ for all} v, \ w \text{of} V$$

and let us denote by $D(A) = \left\{ v \in V, \ Av \in H \right\}$ the domain of A.

We assume instead of (H3) the following hypothesis:

(H5) We assume that $H = H'$ and that G restricted to $\mathcal{V} = D(A) \cup \left(\bigcup_{h>0} V_h \right)$ is continuous from $(\mathcal{V}, \|\cdot\|)$-strong to H-weak.

Theorem 8.2.2 *We assume (H1), (H2), (H4) and (H5). If B is the same bilinear form as in Theorem 8.2.1, then we have the same conclusions as Theorem 8.2.1. Moreover, $u \in D(A)$.*

Proof The proof of statement (i) is the same as above. As for the proof of the statement (ii), since the function $u = \lim_h u_h$ in V-strong, and $B(u, v) = \langle \ell_u, v\rangle \ \forall v \in V$ is equivalent to $Au = \ell_u \in H$, hence $u \in D(A)$, we can then apply hypothesis (H5) to conclude that $G(u_h) \rightharpoonup G(u)$ in H-weak, which implies that $\ell_u = G(u)$. ◇

The set of co-area regular functions do not form a vector space. Thus, to account for other cases of maps where $D(A) \not\subset \mathcal{V}$, we give here a variant of the previous theorem by replacing (H5) with the following hypothesis:

(H6) We assume that $H = H' \subset V'$. Let $\mathcal{V}$ be such that $W = vect\left(\bigcup_{h>0} V_h\right)$ = set of finite linear combinations of $\bigcup_{h>0} V_h$ is contained in $\mathcal{V}$. Let $G : \mathcal{V} \to H$ be continuous from $(\mathcal{V}, \|\cdot\|)$ to H-weak. We assume that the closure of $G(W)$ in H-weak, denoted by $\overline{G(W)}^{\sigma(H)}$, satisfies

$$A^{-1}\left(\overline{G(W)}^{\sigma(H)}\right) \subset \mathcal{V}.$$

Theorem 8.2.3 *We assume (H1), (H2), (H4) and (H6). We then have the same conclusions as in Theorem 8.2.1. Moreover, $u \in \mathcal{V}$.*

Proof Since $u_h \in W$ and $G(u_h) \rightharpoonup \ell_u$ in H-weak, we deduce $\ell_u \in \overline{G(W)}^{\sigma(H)}$. Since A is an isomorphism of V in V', we conclude $u = A^{-1}\ell_u \in \mathcal{V}$ (by H6) ◇

8.2.2 *Applications to Some Nonlocal Problems*

(H1) Consider a bounded open subset Ω of $\mathbb{R}^N$ with C^1 boundary , and for $j = 1, 2$, continuous $\Phi_j : \mathbb{R}^N \to \mathbb{R}$, s.t. there exists $c_1 > 0$ s.t.

$$\left|\Phi_j(\xi)\right| \leqslant c_1 |\xi|^2, \quad \forall \xi \in \mathbb{R}^N,\ j = 1, 2.$$

(H2) Let $F : \Omega \times L^1(\Omega)^2 \times L^1(\Omega_*)^2 \to]\varepsilon, +\infty[$, $\varepsilon > 0$, be bounded and Carathéodory in the sense that:

(i) For x fixed, if $X_n = (v_{1n}, v_{2n}, w_{1n}, w_{2n})$ is a sequence in $L^1(\Omega)^2 \times L^1(\Omega_*)^2$ which converges to X in $L^1(\Omega)^2 \times L^1(\Omega_*)^2$ then $F(x, X_n) \to F(x, X)$ in $\mathbb{R}$.

(ii) $\forall X \in L^1(\Omega)^2 \times L^1(\Omega_*)$, the map

$$x \in \Omega \to F(x, X) \text{ is measurable.}$$

Theorem 8.2.4 *Let b_1 and b_2 be two elements of $L^\infty(\Omega)$. Then, there exists $u \in H_0^1(\Omega) \cap W^{2,p}(\Omega)$, $\forall\, p \in [1, +\infty[$ s.t.*

$$-\Delta u = F\Big(x; [b_1\Phi_1(\nabla u)]_{*v}\,(|u > u(\cdot)|),$$

$$u'_*(|u > u(\cdot)|),\ u'_*, [b_2\Phi_2(\nabla u)]_{*u}\Big).$$

Proof Let $X(v) = \Big([b_1\Phi_1(\nabla v)]_{*v}\,(|v > v(\cdot)|),\ v'_*(|v > v(\cdot)|), v'_*, [b_2\Phi_2(\nabla v)]_{*v}\Big)$ for $v \in H_0^1(\Omega) \cap W^1L(\Omega, |\cdot|_{N,1})$. We have $X(v) \in L^1(\Omega)^2 \times L^1(\Omega_*)^2$. To apply the previous theorem , let us then consider $V = H_0^1(\Omega)$, $H = L^2(\Omega)$ and $\mathcal{V} = \Big\{v \in V \cap W^1L^{N,1}(\Omega) : v \text{ is co-area regular}\Big\}$. Consider the sequence of eigenfunctions $(\varphi_j)_{j\geqslant 1}$ associated with the following Dirichlet problem:

$$-\Delta\varphi_j = \lambda_j\varphi_j,\ \varphi_j \in H_0^1(\Omega) \cap H^2(\Omega) \cap C^\infty(\Omega)\,.$$

For $h = \dfrac{1}{m}$, $m \in \mathbb{N}^*$, we define

$$V_h \equiv V_m = vect\Big\{\varphi_1, \ldots, \varphi_m\Big\} \subset \mathcal{V},\ V_m \subset V_{m+1}\,.$$

We then define a function $G : \mathcal{V} \to H$ by

$$\big(G(v), \varphi\big) = \int_\Omega \varphi(x)F\big(x; X(v)\big)dx,\ \ for\ v \in \mathcal{V},\ \forall\varphi \in H\,.$$

If we introduce the closed convex $C_\varepsilon = \Big\{f \in L^2(\Omega) : f \geqslant \varepsilon\Big\}$, then by the hypothesis on F, we have $G(v) \in C_\varepsilon$.

Proposition 8.2.1 *If $\{v, v_n\}$ is a sequence in $\mathcal{V}$ s.t. $v_n \to v$ in $H_0^1(\Omega)$-strong, then:*

(i) $X(v_n) \to X(v)$ *in* $L^1(\Omega)^2 \times L^1(\Omega_*)^2$*-strong,*
(ii) $G(v_n) \to G(v)$ *in* $L^2(\Omega)$*-strong.*

Proof According to Theorems 6.2.8, 7.1.4 and Corollary 7.1.3, we deduce

$$X(v_n) \to X(v) \text{in} L^1(\Omega)^2 \times L^1(\Omega_*)^2\text{-strong},$$

by hypothesis $F\big(x, X(v_n)\big) \to F\big(x, X(v)\big)$ in $\mathbb{R}$ for almost all x and by the dominated convergence theorem we have:

$F\big(\cdot; X(v_n)\big) \to F\big(\cdot; X(v)\big)$ in $L^2(\Omega)$-strong, in particular

$$G(v_n) \to G(v) inL^2(\Omega) - \text{strong}.$$

Consider $(-\Delta)^{-1} : L^2(\Omega) \to H^2(\Omega) \cap H_0^1(\Omega)$ defined by

$$\begin{cases} (-\Delta)^{-1} f = u \\ f \in L^2(\Omega) \end{cases} \iff \begin{cases} u \in H^2(\Omega) \cap H_0^1(\Omega) \\ \displaystyle\int_\Omega \nabla u \cdot \nabla \varphi = \int_\Omega f\varphi, \ \forall \varphi \in H_0^1(\Omega) \ . \end{cases}$$

If $f \in C_\varepsilon$ then $-\Delta\Big((-\Delta)^{-1} f\Big) = f \geqslant \varepsilon$.

Subsequently the set $\Omega_f = \Big\{x \in \Omega : \nabla\big((-\Delta)^{-1} f\big)(x) = 0\Big\}$ has measure zero (otherwise $-\Delta\big((-\Delta)^{-1} f\big) = 0 \geqslant \varepsilon$ a.e. on Ω_f).

This shows $(-\Delta)^{-1}(C_\varepsilon) \subset \mathcal{V}$.

But as, $G(\mathcal{V}) \subset C_\varepsilon$ and $W = vect\Big(\bigcup\limits_{m \geqslant 1} V_m\Big) \subset \mathcal{V}$, we deduce that

$$\overline{G(W)}^{\sigma(H)} \text{ (closure in } L^2(\Omega)\text{-weak)} \subset C_\varepsilon$$

(C_ε is weakly closed in $L^2(\Omega)$). Hence

$$(-\Delta)^{-1}\Big(\overline{G(W)}^{\sigma(H)}\Big) \subset (-\Delta)^{-1}(C_\varepsilon) \subset \mathcal{V}$$

and the hypotheses of Theorem 8.2.3 are satisfied.

We conclude that there exists $u \in \mathcal{V}$ such that $-\Delta u - G(u) = 0$, i.e. Theorem 8.2.3. The regularity follows from the standard Agmon–Douglas–Nirenberg theorems. ◇

The next applications will require the following lemmas:

Lemma 8.2.1 *Let $v \in L^1_{loc}(\Omega)$ and $\theta = \chi_{\Omega_* \backslash P(v_*)}$ be the characteristic function of $\Omega_* \backslash P(v_*)$. Let v_n be a sequence in $L^1_{loc}(\Omega)$ that converges to v almost everywhere. For $x \in \Omega$, we denote by $I(v_n)(x)$ (resp $I(v)(x)$ for v)*

(continued)

Lemma 8.2.1 (continued)
the interval $\left[m_{v_n}(v_{n+}(x)), m_{v_n}(0)\right]$ *where* m_{v_n} *is the distribution function of* v_n.

Then, for all $\sigma \in \Omega_*$ $\sigma \neq |v > 0|$, $\sigma \neq |v > v_+(x)|$, *we have:*

$$\lim_{n \to +\infty} \theta(\sigma)\chi_{I(v_n)(x)}(\sigma) = \theta(\sigma)\chi_{I(v)(x)}(\sigma)\,. \tag{8.34}$$

Proof Let us note for any real number t, and a sequence $t_n \to t$, we have:

$$|v > t| \leqslant \liminf_n |v_n > t_n| \leqslant \limsup_n |v_n > t_n| \leqslant |v \geqslant t|\ .$$

Let $\sigma \in \Omega_*, \sigma \neq |v > 0|, \sigma \neq |v > v_+(x)|$. If $\sigma < |v > 0|$ or $\sigma > |v \geqslant v_+(x)|$ then for n large enough, we have
$\chi_{I(v_n)(x)}(\sigma) = \chi_{I(v)(\sigma)}(x) = 0$ so we have (8.34).

If $\sigma \in]\,|v \geqslant 0|\,,\ |v > v_+(x)|\,[$ then for n enough large $\chi_{I(v_n(x)}(\sigma) = \chi_{I(v)(x)}(\sigma) = 1$ and therefore we have (8.34).

If $\sigma \in \big]\,|v > 0|\,,\ |v \geqslant 0|\,\big[$ and $|v{=}0|{>}0$ or $\sigma \in \big]\,|v{>}v_+(x)|\,,\,|v{\geqslant}v_+(x)|\,\big[$ and $|v = v_+(x)| > 0$ then $\theta(\sigma) = 0$ and we have (8.34). ◇

Lemma 8.2.2 *Under the same conditions as in Lemma 8.2.1, if* v *and* v_n *are in a bounded set of* $W^{1,q}(\Omega)$, $q > N$, *and such that* $v_n \to v$ *in* $W^1L^{N,1}(\Omega)$-*strong then for all* $b \in L^\infty(\Omega)$, $p \in C^1(\mathbb{R})$ *with* $|p'(t)| \leqslant c_3\,|t| + c_4$, $\forall t \in \mathbb{R}$, *for almost all* $x \in \Omega$:

$$\lim_n \int_{\Omega_*} \left(\theta\, \chi_{I(v_n)(x)}[p(v_{n*})]' b_{*v_n}\right)(\sigma)d\sigma = \int_{\Omega_*} \left(\theta \chi_{I(v)(x)}[p(v_*)]' b_{*v}\right)(\sigma)d\sigma.$$

Proof Since $p' \in C(\mathbb{R})$ and $|p'(t)| \leqslant c_3\,|t| + c_4$ (c_i constant > 0), we deduce from Theorem 6.2.7 that

$$[p(v_{n*})]' \to [p(v_*)]' \ in\ L^1(\Omega_*)\text{-strong.}$$

(Indeed $v'_{n*} \to v'_*$ in $L^1(\Omega_*)$ and consequently $v_{n*} \to v_*$ in $C(\overline{\Omega}_*)$-strong.)

But since v, v_n are in a bounded set of $W^{1,q}(\Omega)$, $q > N$, there exists $r' \in]1, +\infty[$ s.t. $[p(v_{n*})]' \to [p(v_*)]'$ in $L^{r'}(\Omega_*)$-strong. Moreover, by Lemma 8.2.1, we deduce that for almost all x, we have:

$$\theta\chi_{I(v_n)(x)} \to \theta\chi_{I(v)(x)}\ in\ L^r(\Omega_*), \quad (r \text{ the conjugate of } r').$$

From these last two convergences, we conclude

$$\theta \chi_{I(v_n)(x)}[p(v_{n*})]' \to \theta \chi_{I(v)(x)}[p(v_*)]' \ in \ L^1(\Omega_*)\text{-strong}. \tag{8.35}$$

Since $b \in L^\infty(\Omega)$, we know from Theorem 2.6.1 that

$$\theta b_{*u_n} \rightharpoonup \theta b_{*u} \text{in} L^\infty(\Omega_*)\text{-weak-}^*.$$

This weak convergence and the strong convergence of (8.35) entail the result. ◇

As a consequence of the last two lemmas, we have

Corollary 8.2.1 *Let $a \in L^\infty(\Omega)$ and $q > N$. Under the same conditions as in Lemma 8.2.2 if we define for $x \in \Omega$, $v \in W^{1,q}(\Omega)$*

$$F(v)(x) = a(x) \left[F_0^2 - \int_{m_v(0)}^{m_v(v_+(x))} \Big([p(v_*)]' b_{*v}\Big)(\sigma) d\sigma \right]_+^{\frac{1}{2}} .$$

Then the mapping $v \in W^{1,q}(\Omega) \to F(v) \in L^q(\Omega)$ is continuous in the strong topologies.

Proof Let $\theta = \chi_{\Omega_* \setminus P(v_*)}$, $v_n \in W^{1,q}(\Omega)$. We have, for x fixed,

$$J_n = \int_{m_{v_n}(0)}^{m_{v_n}(v_{n+}(x))} [p(v_{n*}]' b_{*v_n} d\sigma \ = J_{n,1}(x) + J_{n,2}(x)$$

with

$$J_{n,1}(x) = \int_{\Omega_*} (1-\theta) \chi_{I(v_n)(x)} [p(v_{n*}]' b_{*v_n} d\sigma \ .$$

If $v_n \to v$ in $W^{1,q}(\Omega)$ then we know $v'_{n_*} \to v'_*$ in $L^1(\Omega_*)$ and $v_{n_*} \to v_*$ in $C(\overline{\Omega}_*)$,

hence $\lim_n \int_{\Omega_*} (1-\theta) \left|v'_{n_*}\right| (\sigma) d\sigma = \int_{P(v_*)} \left|v'_*\right| (\sigma) d\sigma \ = 0.$

As

$$\left|J_{n,1}(x)\right| \leqslant |b|_\infty \left(c_3 \left|v_{n*}\right|_\infty + c_4\right) \int_{\Omega_*} (1-\theta) \left|v'_{n_*}\right| d\sigma \xrightarrow[n\to+\infty]{} 0 \tag{8.36}$$

according to Lemma 8.2.2, the term $J_{n,2}(x)$ satisfies

$$\lim_n J_{n,2}(x) = \int_{\Omega_*} \left(\theta \chi_{I(v)(x)}[p(v_*)]' b_{*v}\right)(\sigma) d\sigma. \tag{8.37}$$

Using the last two relations (8.36) and (8.37), we have

$$\lim_n F(v_n)(x) = F(v)(x) \; a.e. \tag{8.38}$$

Since we have:

$$|F(v_n)(x)| \leqslant |a|_\infty \left[F_0^2 + |b|_\infty \left(\int_{\Omega_*} \left|v'_{n*}\right| dt\right)\left(c_3 \left|v_{n*}\right|_\infty + c_4\right)\right]^{\frac{1}{2}}, \tag{8.39}$$

we deduce that there exists $c_5 > 0$ s.t. $|F(v_n)(x)| \leqslant c_5$ a.e. The result then follows from the dominated convergence theorem. ◇

Theorem 8.2.5 *Let* $N = 2$ *be assumed. Let* $\Phi_\varepsilon(\xi) = \dfrac{|\xi|}{\varepsilon + |\xi|}$, $\varepsilon > 0$, $\xi \in \mathbb{R}^2$. *Under the same conditions as the previous Corollary 8.2.1, if* $b \geqslant 0$, $p' \geqslant 0$ *and* $0 < c_3 < \inf\left\{\int_\Omega |\nabla\varphi|^2 dx,\ |\varphi|_2 = 1,\ \varphi \in H_0^1(\Omega)\right\}$ *then there exists a function* $u \in W^{2,p}(\Omega) \cap H_0^1(\Omega), \forall\, p \in]1, +\infty[$, *satisfying*

$$-\Delta u(x) = F(u)(x) + p'(u)(x)\Big[b(x) - \left[b\Phi_\varepsilon(\nabla u)\right]_{*u}\big(|u > u(x)|\big)\Big].$$

Proof Let $(\varphi_j)_{j\geqslant 0}$ be the sequence of eigenfunctions of the Laplacian and $V_m = vect\{\varphi_1, \ldots, \varphi_m\}$.
We have: $D(-\Delta) = H^2(\Omega) \cap H_0^1(\Omega)$, $V = H_0^1(\Omega)$, $H = L^2(\Omega)$,

$$vect\left(\bigcup_{m\geqslant 0} V_m\right) \subset D(-\Delta) = \mathcal{V}.$$

We define $G : \mathcal{V} \to L^2(\Omega) = H$ by

$$\big(G(v), \varphi\big) = \int_\Omega \varphi(x)\Big[F(v)(x) + K_\varepsilon(v)(x)\Big]dx, \quad \forall \varphi \in L^2(\Omega)$$

where $K_\varepsilon(v)(x) = p'(v)(x)\Big[b(x) - \big(b\Phi_\varepsilon(\nabla v)\big)_{*v}\big(|v > v(x)|\big)\Big]$.

According to Theorem 6.2.5 the elements of $\mathcal{V}$ are co-area regular.

According to the Corollary 8.2.1 of Lemmas 8.2.1 and 8.2.2, if v_n, v remain within a bound set of $D(-\Delta)$ and $v_n \to v$ in $H_0^1(\Omega)$ then, $v_n \to v$ in $W^{1,q}(\Omega)$, for all $q > 2$. Consequently $F(v_n) \to F(v)$ strongly in $L^2(\Omega)$. By Corollary 7.1.3, we deduce $K_\varepsilon(v_n) \to K_\varepsilon(v)$ converges strongly in $L^2(\Omega)$. Therefore, the sequence $G(v_n)$ converges strongly in $L^2(\Omega)$ to $G(v)$. From a simple modification of Theorem 8.2.1 we deduce that there exists $u \in D(-\Delta)$ s.t. $-\Delta u(x) = F(u)(x) + K_\varepsilon(u)(x)$. The regularity comes from the Agmon–Douglas–Nirenberg theorem. $\diamond$

This last application concerns the equations of plasma physics when we have a plasma confined in a machine called a Stellarator. The stationary model obtained from the magnetohydrodynamic equations is as follows: For $\Omega \subset \mathbb{R}^2$, open bounded connected of class C^1, $\gamma \in]-\infty, 0[$, we seek $u \in H_0^1(\Omega) \cap W^{2,p}(\Omega)$, $p < +\infty$

$$\begin{cases} -\Delta u = F(u) + K(u), \\ u = \gamma \ on \ \partial\Omega, \end{cases}$$

$K(u)(x) = p'(u)\Big[b(x) - b_{*u}\big(|u > u(x)|\big)\Big]$, $\quad p'(t) = \lambda t_+$.

We can solve this last problem by using the previous approximate problem, i.e. replace K by K_ε. We will not ptove it here (it can be done as an exercise with the same arguments as above).

Pre-bibliographical Notes

The results of Sect. 8.1 are due to Rakotoson and Serre [140], and those of Sect. 8.2 are essentially the work of Rakotoson and Seoane [138, 139]. The methods used in Theorem 8.2.4 find their origin in Rakotoson [126, 127], Dìaz et al. [63].

Chapter 9
Time-Dependent Functions and Evolution Equations

Both monotone rearrangement and relative rearrangement find their applications in evolution problems. Indeed, comparison theorems similar to those of G. Talenti have been proven by Catherine Bandle for linear equations governed by a Laplacian. She compared the regular solution of $\dfrac{\partial u}{\partial t} - \Delta u = f \in L^2_+(Q)$ in an arbitrary cylinder $Q =]0, T[\times\Omega$ (with Dirichlet boundary conditions) with the regular solution U of $\dfrac{\partial U}{\partial t} - \Delta U = \underset{\sim}{f}$ in the "regular" cylinder of the same measure $\widetilde{Q} =]0, T[\times\widetilde{\Omega}$ with comparable initial data, showing that $\int_0^s u_*(t,\sigma)d\sigma \leqslant \int_0^s U_*(t,\sigma)d\sigma$, $\forall t \in [0, T[$, $\forall s \in \Omega_*$. Later, J. Mossino and J.M. Rakotoson [116] were led to study the case where the operator Δ is replaced by a linear operator with discontinuous coefficients. This led to the study of the regularity of $\partial_t u_*$. In particular, we will see that the relative rearrangement intervenes in the study of the derivative of the monotone rearrangement, by showing the following formulas: *if* $u : Q =]0, T[\times\Omega \to \mathbb{R}$, $u \in W^{1,r}\Big(0, T; L^p(\Omega)\Big)$ ($u \geqslant 0$ if Ω is unbounded) then

1. $u_* \in W^{1,r}\Big(0, T; L^p(\Omega_*)\Big)$ with $u_*(t,s) = u(t)_*(s), \theta \in]0, T[, \sigma \in \Omega_*$,
2. $\dfrac{\partial u_*}{\partial t}(t,s) = \left(\dfrac{\partial u}{\partial t}(t)\right)_{*u(t)}(s) \doteq \left(\dfrac{\partial u}{\partial t}\right)_{*u}(t,s)$.

We will deduce from these formulas, for example, the study of the continuity of $u \to \dfrac{\partial u_*}{\partial t}$.

As in Chap. 5, we will use these regularities and properties of monotone rearrangement to obtain a priori estimates for the quasilinear parabolic equations. We will also present theorems for comparing solutions that will lead, in the case of the Chemotaxis system , to the study of the time behavior of the system.

J. M. Rakotoson, *Relative Rearrangement*, Lecture Notes in Mathematics 2376,
https://doi.org/10.1007/978-3-032-02228-8_9

9.1 Relative Rearrangement of a Family of Functions

For the sake of simplicity, we will assume that Ω is a bounded open set of $\mathbb{R}^N$. For a measurable function $u :]0, T[\times\Omega = Q \to \mathbb{R}$, we define, for t fixed in $]0, T[$, $u(t) : \Omega \to \mathbb{R}$ by $u(t)(x) = u(t, x)$ for $x \in \Omega$. To this function $u(t)$ we can apply all the theory of the previous chapters, namely the monotone rearrangement $u(t)_*$. Hence the following definition:

Definition 9.1.1 (Symmetrization of a Family of Functions) For measurable $u : Q =]0, T[\times\Omega \to \mathbb{R}$, we define $u_* : Q_* =]0, T[\times\Omega_* \to \mathbb{R}$ by

$$u_*(t, s) = u(t)_*(s), \qquad (t, s) \in Q_*.$$

Definition 9.1.2 (Relative Rearrangement of a Family of Functions) Let u, b be two integrable functions of $L^1(Q)$. We define $b_{*u} : Q_* \to \mathbb{R}$ by:

$$b_{*u}(t, s) = b(t)_{*u(t)}(s), \qquad (t, s) \in \Omega_*.$$

Property 9.1.1 Let u and b be in $L^r\big(0, T; L^p(\Omega)\big)$ with $1 \leqslant r \leqslant +\infty$, $1 \leqslant p \leqslant +\infty$. Then the measurable functions u_*, b_{*u} satisfy

1. u_* and b_{*u} are in $L^r\big(0, T; L^p(\Omega_*)\big)$.
2. $|u_*|_{L^r\left(0,T;L^p(\Omega_*)\right)} = |u|_{L^r\left(0,T;L^p(\Omega)\right)}$ and $|b_{*u}|_{L^r\left(0,T;L^p(\Omega_*)\right)} \leqslant |b|_{L^r\left(0,T;L^p(\Omega)\right)}$.

Note $L^r\big(0, T; L^p(\Omega)\big)$

$$= \left\{u : Q \to \mathbb{R}\, measurable\ \|u\|^r = \int_0^T |u(t)|^r_{L^p(\Omega)}\, dt < +\infty\right\} \text{ if } r \text{ is finite.}$$

$L^\infty\big(0, T; L^p(\Omega)\big) =$

$$\left\{u : Q \to \mathbb{R}\, measurable\ \operatorname*{ess\,sup}_{[0,T]} |u(t)|_{L^p(\Omega)} = \|u\| < +\infty\right\}.$$

They are equipped with the usual norms $|u|_{L^r\left(0,T;L^p(\Omega)\right)} = \|u\|$.

As we have mentioned the relative rearrangement, let us recall that it is obtained by computing the following limit:

$$\lim_{h\to 0\ h>0} \int_0^s \frac{\left(u(t)+hb(t)\right)_* - u(t)_*}{h} d\sigma = \int_0^s b_{*u}(t,\sigma)d\sigma.$$

Thus, all the properties obtained in the previous chapters are valid by just fixing the parameter t. In particular, the pointwise inequalities of Polyà–Szegö or Poincaré–Sobolev.

Theorem 9.1.1 *Let $u \in L^1\left(0, T; W_0^{1,1}(\Omega)\right)$, $u \geqslant 0$. Then, $\forall s \in \overline{\Omega}_*$, a.e. t*

1. $\int_0^s \left|\nabla \underset{\sim}{u}\right|_*(t,\sigma)d\sigma \leqslant \int_0^s |\nabla u|_{*u}(t,\sigma)d\sigma \leqslant \int_0^s |\nabla u|_*(t,\ s)d\sigma,$
2. *If ρ is a Fatou norm, invariant under rearrangement on $L^0_+(\Omega_*)$, then:*

$$\rho\left(\left|\nabla \underset{\sim}{u}\right|_*(t)\right) \leqslant \rho\left(|\nabla u|_{*u}(t)\right) \leqslant \rho\left(|\nabla u|_*(t)\right).$$

Here $\underset{\sim}{u}$ is the spherical rearrangement defined by

$\underset{\sim}{u}(t,x) = u_*(t, \alpha_N |x|^N)$, α_N =measure of the unit ball and x is in the ball of the same measure as Ω centered at the origin.

We are not going to report all the properties instead, we go directly to the application, to the regularity of $\dfrac{\partial u_*}{\partial t}$.

9.2 Regularity in Time of the Rearrangement $u_*(t, s)$

Theorem 9.2.1 *Let $u \in W^{1,q}(0, T; L^p(\Omega))$, with $1 \leqslant q \leqslant +\infty$ $1 \leqslant p \leqslant +\infty$. Then*

(i) $u_* \in W^{1,q}\left(0, T; L^p(\Omega_*)\right)$,

(ii) $\dfrac{\partial u_*}{\partial t} = \dfrac{\partial w}{\partial s}$ *in $\mathcal{D}'(Q_*)$ where $w : Q_* \to \mathbb{R}$ is defined by*

(continued)

Theorem 9.2.1 (continued)

$$w(t,s)=\int_{\{u(t)>u(t)_*(s)\}}\frac{\partial u}{\partial t}(t,x)dx+\int_0^{s-|u(t)>u(t)_*(s)|}\left(\frac{\partial u}{\partial t}(t)|_{\{u(t)=u(t)_*(s)\}}\right)_*(\sigma)d\sigma.$$

The proof of this theorem requires the following lemma:

Lemma 9.2.1 *Let $u\in W^{1,q}\big(0,T;L^p(\Omega)\big)$, $1\leqslant q\leqslant+\infty$, $1\leqslant p\leqslant+\infty$.*

For $h>0$, let $g_h(t,x)=\dfrac{u(t+h)-u(t)}{h}(x)$, $x\in\Omega$.

Then, g_h converges to $\dfrac{\partial u}{\partial t}$ in $L^\alpha_{loc}\Big([0,T),L^\alpha(\Omega)\Big)$ with $\alpha=$
$$\begin{cases}\min(p,q) & if\ p\ or\ q\ is\ finite\\ \alpha<+\infty & if\ p=q=+\infty.\end{cases}$$

Proof Let α be defined as in the above lemma, then $\dfrac{\partial u}{\partial t}\in L^\alpha(Q)$ as soon as $u\in W^{1,q}\big(0,T;L^p(\Omega)\big)$. From this, we deduce for $0<h<\delta<T$ that:

$$\left(\int_0^{T-\delta}\int_\Omega|g_h(t,x)|^\alpha\,dxdt\right)^{\frac{1}{\alpha}}\leqslant\left(\int_0^{T-\delta+h}\int_\Omega\left|\frac{\partial u}{\partial t}(t,x)\right|^\alpha dxdt\right)^{\frac{1}{\alpha}}\leqslant\left|\frac{\partial u}{\partial t}\right|_{L^\alpha(Q)}.$$

Consequently, if $Q_\delta=]0,T-\delta[\times\Omega$, we have

$$\limsup_{h\to0}|g_h|_{L^\alpha(Q_\delta)}\leqslant\left|\frac{\partial u}{\partial t}\right|_{L^\alpha(Q_\delta)}. \tag{9.1}$$

Thus for $\alpha>1$, since g_h converges weakly to $\dfrac{\partial u}{\partial t}$ in $L^\alpha(Q_d)$, we deduce from the fact that $L^\alpha(Q_\delta)$ is uniformly convex that $g_h\to\dfrac{\partial u}{\partial t}$ in $L^\alpha(Q_\delta)$.

If $\alpha=1$, we approach u by a sequence of elements u_n for example of $W^{1,2}\big(0,T;L^2(\Omega)\big)$, and which converges in $W^{1,1}\big(0,T;L^1(\Omega)\big)$.

if

$$\varepsilon^n(h)(t) = \frac{u_n(t+h) - u_n(t)}{h} - \frac{\partial u_n}{\partial t}(t)$$

then $\varepsilon^n(h) \xrightarrow[h \to 0]{} 0$ in $L^1(Q_\delta)$.

If we put

$$\varepsilon(h)(t) = \frac{u(t+h) - u(t)}{h} - \frac{\partial u}{\partial t}(t)$$

then we have:

$$|\varepsilon(h)|_{L^1(Q_\delta)} \leqslant \left|\varepsilon^n(h)\right|_{L^1(Q_\delta)} + \left|\varepsilon^n(h) - \varepsilon(h)\right|_{L^1(Q_\delta)}$$

and

$$\left|\varepsilon^n(h) - \varepsilon(h)\right|_{L^1(Q_\delta)} \leqslant 2\left|\frac{\partial}{\partial t}(u_n - u)\right|_{L^1 Q_\delta)} \xrightarrow[n \to +\infty]{} 0.$$

Thus, these relations combined together lead to

$$\limsup_{h \to 0} |\varepsilon(h)|_{L^1(Q_\delta)} \leqslant 2\left|\frac{\partial}{\partial t}(u_n - u)\right|_{L^1(Q_\delta)} : \lim_{h \to 0} |\varepsilon(h)|_{L^1(Q_\delta)} = 0.$$

◇

Proof Let us first show ii), that is to say that $\forall \varphi \in \mathcal{D}(Q_*)$, we have:

$$-\int_0^T \int_{\Omega_*} u_*(t, \sigma) \frac{\partial \varphi}{\partial t}(t, \sigma) dt d\sigma = \int_0^T \int_{\Omega_*} \frac{\partial w}{\partial s}(t, \sigma) \varphi(t, \sigma) dt d\sigma. \tag{9.2}$$

For this, for $h > 0$ let us consider the integral

$$I(h) = \int_0^T \int_{\Omega_*} \frac{u(t+h)_*(\sigma) - u(t)_*(\sigma)}{h} \varphi(t, \sigma) dt d\sigma$$

and introduce the following quantities:

$$\varepsilon(h) : Q \to \mathbb{R} \text{ s.t. } u(t+h) = u(t) + h\frac{\partial u}{\partial t}(t) + h\varepsilon(h)(t),$$

$$I_1\big(\varepsilon(h)\big) = \int_{Q_*} \frac{u(t+h)_* - \big(u(t) + h\frac{\partial u}{\partial t}(t)\big)_*}{h}(\sigma)\varphi(t, \sigma) dt d\sigma,$$

$$I_2(h,t) = \int_{\Omega_*} \frac{\left(u(t) + h\frac{\partial u}{\partial t}(t)\right)_*(\sigma) - u(t)_*(\sigma)}{h} \varphi(t,\sigma)d\sigma,$$

$$I_2(h) = \int_0^T I_2(h,t)dt.$$

We have of course $I(h) = I_1(\varepsilon(h)) + I_2(h)$.

Study $I_1(\varepsilon(h))$ Let $\delta > 0$ s.t. *support* $\varphi \subset]0, T-\delta[\times\Omega_* = Q_{*\delta}$. The contraction property of the rearrangement leads to:

$$\left|I_1(\varepsilon(h))\right| \leqslant |\varphi|_{L^\infty(0,T-\delta)\times\Omega_*)} \cdot |\varepsilon(h)|_{L^1(Q_\delta)} \cdot$$

Applying Lemma 9.2.1 to this last inequality, we deduce:

$$\lim_{h\to 0} I_1(\varepsilon(h)) = 0.$$

As for the term $\underline{I_2(h,t)}$, the theorem of the directional derivative of Chap. 2 leads to:

$$\lim_{h\to 0} I_2(h,t) = \int_{\Omega_*} \left(\frac{\partial u}{\partial t}\right)_{*u} (t,\sigma)\varphi(t,\sigma)d\sigma \tag{9.3}$$

for almost all t.

By the contraction property of the monotone rearrangement, we deduce:

$$|I_2(h,t)| \leqslant |\varphi|_{L^\infty(Q)} \cdot \left|\frac{\partial u}{\partial t}(t)\right|_{L^1(\Omega)}. \tag{9.4}$$

The relations (9.3) and (9.4) allow us to apply the dominated convergence theorem. Hence

$$\lim_{h\to 0} \int_0^T I_2(h,t)dt = \int_0^T \int_{\Omega_*} \left(\frac{\partial u}{\partial t}\right)_{*u} (t,\sigma)\varphi(t,\sigma)dtd\sigma. \tag{9.5}$$

As a result, we have (by changing variables)

$$\lim_{h\to 0} \int_0^T I_2(h,t)dt = \lim_{h\to 0} I(h) = -\int_0^T \int_{\Omega_*} u_*(t,\sigma)\frac{\partial\varphi}{\partial t}(t,\sigma)dtd\sigma. \tag{9.6}$$

Which proves that $\dfrac{\partial u_*}{\partial t} = \left(\dfrac{\partial u}{\partial t}\right)_{*u}$. Since $\left(\dfrac{\partial u}{\partial t}\right)_{*u} \in L^q\big(0,T;L^p(\Omega_*)\big)$, we then deduce that $\dfrac{\partial u_*}{\partial t} \in L^q\big(0,T;L^p(\Omega_*)\big)$. ◇

A first consequence of the theorem is:

Theorem 9.2.2 *Let $u \in W^{1,1}(0, T; L^1(\Omega))$. So for almost all $t \in (0, T)$, $\frac{\partial u}{\partial t}(t, \cdot)$ is constant (almost everywhere) on any level of $u(t)$, i.e. where $u(t, \cdot)$ is constant (almost everywhere).*

Proof In the proof of the previous theorem consider $h < 0$ in $I_2(h, t)$. Then

$$I_2(h, t) = -\int_{\Omega_*} \frac{\left(u(t) - h\left(-\frac{\partial u}{\partial t}\right)\right)_*(\sigma) - u(t)_*(\sigma)}{-h} \varphi(t, \sigma) d\sigma$$

which converges to $-\int_{\Omega_*} \left(-\frac{\partial u}{\partial t}\right)_* (t, \sigma)\varphi(t, \sigma) d\sigma$. Consequently, we have in $\mathcal{D}'(Q_*)$

$$\frac{\partial u_*}{\partial t} = \left(\frac{\partial u}{\partial t}\right)_{*u} = -\left(-\frac{\partial u}{\partial t}\right)_{*u}.$$

Taking the function w, we have:
$\left(\frac{\partial u}{\partial t}\right)_{*u} = \frac{\partial w}{\partial s}$ and $-\left(-\frac{\partial u}{\partial t}\right)_{*u} = \frac{\partial w'}{\partial s}$ where w' is given by

$$w'(t, s) = \begin{cases} \displaystyle\int_{\{u(t)>u_*(t,s)\}} \frac{\partial u}{\partial t} dx & if \ |u(t) = u_*(t, s)| = 0, \\ \\ \displaystyle\int_{\{u(t)>u_*(t,s)\}} \frac{\partial u}{\partial t} dx + \int_0^{s-|\{u(t)>u_*(t,s)\}|} -\left(-\frac{\partial u}{\partial t}|_{\{u(t)=u_*(t,s)\}}\right)_* (\sigma) d\sigma & else. \end{cases}$$

Consequently we have:

$$\int_0^{s-|\{u(t)>u_*(t,s)\}|} \left(\frac{\partial u}{\partial t}|_{\{u(t)=u_*(t,s)\}}\right)_* (\sigma) d\sigma = -\int_0^{s-|\{u(t)>u_*(t,s)\}|} \left(-\frac{\partial u}{\partial t}|_{\{u(t)=u_*(t,s)\}}\right)_* (\sigma) d\sigma . \tag{9.7}$$

Let us then fix $t \in (0, T)$ and a plateau of $u(t)$, i.e.
$P_\theta(t) = \{u(t) = \theta\}$ of non-zero measure. Let's ask $s_\theta = |u(t) > \theta|$,
$s'_\theta = |u(t) \geqslant \theta|$. As $w(t, 0) = 0 = w'(t, 0)$, we have:

$$w(t, s) = w'(t, s) \qquad \forall s \in \overline{\Omega}_*.$$

Consequently, for all $s \in [s_\theta, s'_\theta]$, we have according to (9.7)

$$\frac{\partial w}{\partial s}(t, s) = \left(\frac{\partial u}{\partial t}|_{P_\theta(t)}\right)_* (s - s_\theta) = -\left(-\frac{\partial u}{\partial t}|_{P_\theta(t)}\right)_* (s - s_\theta).$$

In particular

$$\left(\frac{\partial u}{\partial t}|_{P_\theta(t)}\right)_*(0) = \operatorname*{ess\,sup}_{P_\theta(t)} \frac{\partial u}{\partial t} = -\left(-\frac{\partial u}{\partial t}|_{P_\theta(t)}\right)_*(0) = \operatorname*{ess\,inf}_{P_\theta(t)} \frac{\partial u}{\partial t}.$$

◇

Corollary 9.2.1 *Let B be a Borel set of $\mathbb{R}$ and let $u \in W^{1,1}(0, T; L^1(\Omega))$. Then, for almost every $t \in (0, T)$*

$$\int_{\Omega_*} \frac{\partial u_*}{\partial t}(t, \sigma)\chi_B\big(u_*(t, \sigma)\big)d\sigma = \int_\Omega \frac{\partial u}{\partial t}(t, x)\chi_B\big(u(t, x)\big)dx.$$

This is a consequence of equimeasurability, but here is another proof:

Proof Since $\dfrac{\partial u_*}{\partial t} = \left(\dfrac{\partial u}{\partial t}\right)_{*u}$, the formula for the mean operator leads to:

$$\int_{\Omega_*} \frac{\partial u_*}{\partial t}(t, \sigma) \cdot \chi_B\big(u_*(t, \sigma)\big)d\sigma = \int_\Omega \frac{\partial u}{\partial t}(t, x) M_{u(t), \frac{\partial u}{\partial t}(t)}\Big(\chi_B\big(u_*(t, \cdot)\big)\Big)dx$$

$$= \int_{\Omega \backslash P(u(t))} \frac{\partial u}{\partial t}(t, x) \cdot \chi_B\big(u(t, x)\big)dx + \sum_{i \in D} \int_{P_i(u(t))} M_{\frac{\partial u}{\partial t}(t)|_{P_i(u(t))}}\Big(\chi_B\big(u(t)_*\big)\Big) \cdot \frac{\partial u}{\partial t}(t, x)dx. \tag{9.8}$$

According to Theorem 9.2.2, we have

$$\frac{\partial u}{\partial t}(t, \cdot) = \text{constant} = c_i, \; if \; \{u(t) = \theta_i\} = P_i\big(u(t)\big).$$

Hence

$$\int\limits_{\{u(t)=\theta_i\}} M_{\frac{\partial u}{\partial t}(t)|\{u(t)=\theta_i\}}\Big(\chi_B\big(u(t)_*\big)\Big)\cdot\frac{\partial u}{\partial t}(t,x)dx = c_i\,\chi_B(\theta_i)\,|u(t)=\theta_i|$$

$$= \int\limits_{\{u(t)=\theta_i\}}\frac{\partial u}{\partial t}(t,x)\chi_B\big(u(t,x)\big)dx\,.$$

Consequently,

$$\int\limits_{P_i(u(t))} M_{\frac{\partial u}{\partial t}(t)|_{P_i(u(t))}}\Big(\chi_B\big(u(t)_*\big)\Big)\cdot\frac{\partial u}{\partial t}(t,x)dx = \int\limits_{\{u(t)=\theta_i\}}\frac{\partial u}{\partial t}(t,x)\chi_B\big(u(t,x)\big)dx\,. \tag{9.9}$$

Hence we obtain the formula by combining the relations (9.8) and (9.9). ◇

By combining the previous theorems with Theorem 7.2.2 we have:

Theorem 9.2.3 *Let $u \in W^{1,1}(0,T;L^1(\Omega))$ and $\Phi : R \to \mathbb{R}$ be convex increasing, $B \subset \mathbb{R}$ a Borel set of $\mathbb{R}$ such that the map $v \in L^1(\Omega_*) \to \Phi(v)$ is defined and continuous. Then, for almost all t*

$$\int_{\Omega_*}\Phi\left(\frac{\partial u_*}{\partial t}(t,\sigma)\right)\chi_B\big(u_*(t,\sigma)\big)d\sigma$$

$$\leqslant\int_{\Omega_*}\Phi\left(\left(\frac{\partial u}{\partial t}\right)_*\right)\Big(\chi_B(u)\Big)_*(t,\sigma)d\sigma.$$

Proof According to Theorem 7.2.2, we have

$$\int_{\Omega}\Phi\left(\left(\frac{\partial u}{\partial t}\right)_{*u}\right)\Big(\chi_B(u_*)\Big)(t,\sigma)d\sigma\leqslant\int_{\Omega_*}\Phi\left(\left(\frac{\partial u}{\partial t}\right)_*\right)\Big(\chi_B(u_*)\Big)_*(t,\sigma)d\sigma$$

but $\Big(\chi_B(u_*)\Big)_* = \Big(\chi_B(u)\Big)_*$ hence the result. ◇

We will now focus on the convergence problems of $\dfrac{\partial u_{n*}}{\partial t}$ to $\dfrac{\partial u_*}{\partial t}$ for a sequence u_n converging to u.

9.3 Convergence and Continuity for $u \to \frac{\partial u_*}{\partial t}$

Lemma 9.3.1 *Let $\{u,\ u_n\}$ be a sequence in $W^{1,1}(0,T;L^1(\Omega))$ s.t. $u_n \to u$ in $W^{1,1}(0,T;L^1(\Omega))$. For $t \in (0,T)$, $s \in \Omega_*$ we note:*
$\theta(t,s) = \chi_{\Omega_*\backslash P(u_*(t))}(s)$, $\theta(t)(s) = \theta(t,s)$.

Then there exists a subsequence also denoted $(u_n)_n$, s.t. we have for almost every t

$$\theta(t)\cdot\frac{\partial u_{n*}}{\partial t}(t) \rightharpoonup \theta(t)\cdot\frac{\partial u_*}{\partial t}(t) \text{ in } L^1(\Omega_*)\text{-weak.}$$

Proof Since $u_n \to u$ in $W^{1,1}(0,T;L^1(\Omega))$, there exists a subsequence denoted by u_n s.t. $\dfrac{\partial u_n}{\partial t}(t) \to \dfrac{\partial u}{\partial t}(t)$ in $L^1(\Omega)$- strong, $u_n(t) \to u(t)$ in $L^1(\Omega)$, for almost all t. Moreover, according to Theorem 2.6.1, we know that

$$\theta(t)\cdot\left(\frac{\partial u}{\partial t}(t)\right)_{*u_n(t)} \rightharpoonup \theta(t)\cdot\left(\frac{\partial u}{\partial t}(t)\right)_{*u(t)} \tag{9.10}$$

in $L^1(\Omega_*)$-weak.

We have from the contraction property of the relative rearrangement,

$$\left|\frac{\partial u_{n*}}{\partial t}(t) - \left(\frac{\partial u}{\partial t}(t)\right)_{*u_n(t)}\right|_{L^1(\Omega_*)} \leqslant \left|\frac{\partial}{\partial t}(u_n-u)(t)\right|_{L^1(\Omega)} \to 0. \tag{9.11}$$

Consequently, since we have

$$\theta(t)\frac{\partial u_{n*}}{\partial t}(t) = \theta(t)\left[\frac{\partial u_{n*}}{\partial t}(t) - \left(\frac{\partial u}{\partial t}(t)\right)_{*u_n(t)}\right] + \theta(t)\left(\frac{\partial u}{\partial t}(t)\right)_{*u_n(t)}$$

the relations (9.10) and (9.11) lead to the result. ◇

Theorem 9.3.1 *Let $\{u,\ u_n\}$ be a sequence s.t.*

$$\left(\frac{\partial u_n}{\partial t}(t),\, \nabla_x u_n(t)\right) \to \left(\frac{\partial u}{\partial t}(t),\, \nabla_x u(t)\right) \text{ in } L^1(\Omega)^{N+1}\ \forall t \in \mathcal{T}_0 \subset [0, T].$$

Then, if $u(t)$ is co-area regular, t in $\mathcal{T}_0$, we have

$$\left(\frac{\partial u_n}{\partial t}\chi_{\{\nabla u(t)\neq 0\}}\right)_{*u_n(t)} \to \left(\frac{\partial u}{\partial t}(t)\chi_{\{\nabla u(t)\neq 0\}}\right)_{*u(t)}$$

in $L^1(\Omega_*)$-strong.

Proof Let

$$K_n(t,\sigma) = \left(\frac{\partial u_n}{\partial t}\chi_{\{\nabla u(t)\neq 0\}}\right)_{*u_n(t)}(\sigma)$$

$$K(t,\sigma) = \left(\frac{\partial u}{\partial t}\chi_{\{\nabla u(t)\neq 0\}}\right)_{*u(t)}(\sigma)$$

$$K_{n,1}(t,\sigma) = \left(\frac{\partial u}{\partial t}\chi_{\{\nabla u(t)\neq 0\}}\right)_{*u_n(t)}(\sigma)$$

for $t \in (0, T),\ \sigma \in \Omega_*$.

We have $\left|K_n(t) - K_{n,1}(t)\right|_{L^1(\Omega_*)} \leqslant \left|\frac{\partial}{\partial t}(u_n - u)(t)\right|_{L^1(\Omega)} \xrightarrow[n\to+\infty]{} 0$. Since $u_n(t) \to u(t)$ in $W^{1,1}(\Omega)$ and $u(t)$ is co-area regular, we deduce that

$$K_{n,1}(t) \to K(t) \text{ in } L^1(\Omega_*)\text{-strong.}$$

9.4 Applications to a Priori *Estimates* and to Regularity

This section is based on principles similar to those of Chap. 5 to obtain estimates of the gradient or a solution of a PDE. The first step is to estimate $|\nabla v|_{*v}(t,s)$ and then apply the Poincaré–Sobolev pointwise inequalities to obtain information about $v = |u|$.

We illustrate this by taking a model operator which is the p-Laplacian. Nevertheless, we can consider the same types of operators that we used in Chap. 5, i.e. $Au = -div(a(x, u, \nabla u))$. We can also study evolution equations involving $b_{*u}(t, \cdot)$ or $u'_*(t, \cdot)$. This corresponds to certain models of transient regimes in plasma physics. These equations can be written as:

$$\frac{\partial}{\partial t} H(u) - \Delta u = G(t, x, u'_*(t, \cdot), b_{*u}(t, \cdot))$$

whose stationary solutions correspond to those cited in the introduction to Chap. 2. Here H is a monotone function (see [64]).

First, here is an abstract theorem that will help us ensure that the time derivative $\frac{\partial u}{\partial t}$ is in $L^1(Q)$, which will allow us to apply the theorem on $\frac{\partial u_*}{\partial t}$.

9.4.1 An Abstract Theorem of Existence and Regularity

Let $(V, \|\cdot\|)$ be a reflexive Banach space with dual V' and $(H, |\cdot|)$ be a Hilbert space, $V \subset H$, V dense in H with a continuous injection. Therefore, we have $V \subset H \subset V'$.

Definition 9.4.1 Let A be an operator from V to V'. The operator A is said to be hemi-continuous if the map $\lambda \to \langle A(u + \lambda v), w\rangle$ is continuous from $\mathbb{R}$ to $\mathbb{R}$ for all u, v, w of V.

Here, the bracket $< \cdot, \cdot >$ denotes the duality bracket between V and V'.

We will say that A is monotone if $\forall (u, v) \in V^2$ $\langle A(u) - A(v), u - v\rangle \geqslant 0$.

We will say that A derives from a potential if there exists a function J : $V \to \mathbb{R}$ Gâteaux differentiable s.t. $J'(u) = A(u) \quad \forall u \in V$.

We will call the domain of A the set $D(A) = \left\{ v \in V,\ A(v) \in H \right\}$.

Theorem 9.4.1 *We consider $(V, \|\cdot\|)$ and $(H, |\cdot|)$ satisfying the previous conditions and let A be a hemi-continuous, monotone operator deriving from a potential J. We further assume that A and J satisfy the following growth*

(continued)

Theorem 9.4.1 (continued)
conditions: $\forall v \in V$

$$\left\langle A(v), v\right\rangle \geqslant \alpha \|v\|^p, \ \alpha > 0, \ 2 \leqslant p < +\infty,$$

$$|A(v)|_{V'} \, (\text{norm in } V') \leqslant c_1 \|v\|^{p-1},$$

$$c_2 \|v\|^p - c_3 \leqslant J(v) \leqslant c_4 \|v\|^p + c_5.$$

Then $\forall u_0 \in D(A)$, $\forall f \in H$, *there exists a unique solution*

$$u \in L^\infty(0, T; V), \ T > 0 \ of \ u' + Au = f, \ u(0) = u_0.$$

Furthermore, u' *and* $Au \in L^\infty(0, T; H)$.

The model example we will choose will be

$$Au = -div\Big(|\nabla u|^{p-2}\nabla u\Big) = -\Delta_p u, \ V = W_0^{1,p}(\Omega)$$

$H = L^2(\Omega)$. In this case, we have:

$$J(v) = \frac{1}{p}\int_\Omega |\nabla v|^p \, dx, \ v \in V, \quad p \geqslant 2.$$

More generally, by giving suitable hypotheses on a_i we can consider the following so-called Leray–Lions operators:

$$Au = -\sum_{i=1}^n \frac{\partial}{\partial x_i} a_i(x, \nabla u).$$

This theorem ensures that if $f \in L^2(\Omega)$, $u_0 \in W_0^{1,p}(\Omega)$, $2 \leqslant p < +\infty$ such that $-\Delta_p u_0 \in L^2(\Omega)$ then the unique solution of

$$\begin{cases} \dfrac{\partial u}{\partial t} - \Delta_p u = f & \text{in } (0, T) \times \Omega, \\ u = 0 & \text{on } (0, T) \times \partial\Omega, \\ u(0) = u_0, \end{cases}$$

satisfies $u \in L^\infty\big(0, T; W_0^{1,p}(\Omega)\big)$ and $\dfrac{\partial u}{\partial t} \in L^\infty\big(0, T; L^2(\Omega)\big)$.

9.4.2 Case of Quasilinear Parabolic Equations

In what follows, we will always assume that $u \in W^{1,1}\big(0, T; L^1(\Omega)\big)$. The equation will be taken in the following variational sense:

$$\int_\Omega \frac{\partial u}{\partial t}(t,x)\psi(x)dx + \int_\Omega |\nabla u|^{p-2}\nabla u \cdot \nabla\psi\, dx = \int_\Omega f(t,x)\psi(x)dx$$

$\forall \psi \in W_0^{1,p}(\Omega)$, in $\mathcal{D}'(0,T)$ and a.e. in t.

Theorem 9.4.2 *Let $f \in L^1(Q)$, u be a solution of $\dfrac{\partial u}{\partial t} - \Delta_p u = f$ s.t. $u(t) \in W_0^{1,p}(\Omega)$ for almost all t. Then, we have for $s \in \Omega_*$, $t \in (0,T)$ and $v = |u|$,*

$$\Big(|\nabla v|^p\Big)_{*v}(t,s) \leqslant \frac{s^{-\frac{p(N-1)}{N(p-1)}}}{\Big(N\alpha_N^{\frac{1}{N}}\Big)^{\frac{p}{(p-1)}}}\left[\int_0^s |f|_*(t,\sigma)d\sigma - \int_0^s \frac{\partial v_*}{\partial t}(t,\sigma)d\sigma\right]^{\frac{p}{p-1}}$$

where $\Omega_s(t) = \Big\{x \in \Omega : v(t,s) > v_*(t,s)\Big\}$.

In particular, we have

(i) $\displaystyle\int_0^s |f|_*(t,\sigma)s\sigma - \int_0^s \frac{\partial v_*}{\partial t}(t,\sigma)d\sigma \geqslant 0, \qquad \forall s \in \overline{\Omega}_*,$

(ii) $\forall \varphi \geqslant 0, \quad \varphi \in L^\infty(\Omega_*)$,

$$\int_{\Omega_*} \varphi^*(s)\, |\nabla v|_*^{\frac{p}{*}-1}(t,s)ds$$

$$\leqslant \frac{1}{N\alpha_N^{\frac{1}{N}}}\int_{\Omega_*}\frac{\varphi(s)}{s^{1-\frac{1}{N}}}\left[\int_0^s |f|_*(t,\sigma)d\sigma - \int_0^s \frac{\partial v_*}{\partial t}(t,\sigma)d\sigma\right]ds$$

where $\varphi^*(s) = -(-\varphi)_*(s)$, $\sigma \in \Omega_*$.

Proof of the Theorem For $s \in \Omega_*$, $t \in (0,T)$, consider,

$$\psi(t) = \Big(v(t) - v_*(t,s)\Big)_+ \operatorname{sign}\big(u(t)\big) \in W_0^{1,p}(\Omega).$$

Using ψ as a test function and differentiating with respect to s, we have:

$$\left[|\nabla v|^p\right]_{*v}(t,s) = \left[\int\limits_{\Omega_s(t)} f \operatorname{sign} u\, dx - \int_{\Omega_s(t)} \frac{\partial u}{\partial t} dx\right]\left(-\frac{\partial v_*}{\partial s}(t,s)\right).$$

Since $\dfrac{\partial v_*}{\partial t} = \left(\dfrac{\partial v}{\partial t}\right)_{*v}$, we deduce from this last relation that:

$$\left[|\nabla v|^p\right]_{*v}(t,s) = \left[\int_{\Omega_s(t)} f \operatorname{sign}(u) dx - \int_0^s \frac{\partial v_*}{\partial t}(t,\sigma) d\sigma\right]\left(-\frac{\partial v_*}{\partial s}(t,s)\right). \tag{9.12}$$

In fact, we generally have the following lemma:

Lemma 9.4.1 *Let* $v \in W^{1,1}(0,T;L^1(\Omega))$.
For $t \in (0,T)$*, we define* $\mu(t,\theta) = |v(t) > \theta|$, $\theta \in \mathbb{R}$*. Then,* $\forall \theta \in \mathbb{R}$

$$\int\limits_0^{\mu(t,\theta)} \frac{\partial v_*}{\partial t}(t,\sigma) d\sigma = \int\limits_{\{v(t)>\theta\}} \frac{\partial v}{\partial t}(t,x) dx .$$

Proof of the Lemma If $|v(t) = \theta| = 0$, then this relation follows from the identity

$$\int\limits_0^{\mu(t,\theta)} \frac{\partial v_*}{\partial t}(t,\sigma) d\sigma = \int\limits_0^{\mu(t,\theta)} \left(\frac{\partial v}{\partial t}\right)_{*v}(t,\sigma) d\sigma = \int\limits_{\{v(t)>\theta\}} \frac{\partial v}{\partial t}(t,x) dx .$$

Since the set of θ s.t. $|v(t) = \theta| \neq 0$ is at most countable and that the maps $\theta \to \int\limits_0^{\mu(t,\theta)} \frac{\partial v_*}{\partial t}(t,\sigma) d\sigma$, $\theta \to \int\limits_{\{v(t)>\theta\}} \frac{\partial v}{\partial t}(t,x) dx$ are right continuous, we conclude that we have equality for all θ. ◇

See Corollary 9.2.1 As a consequence of this lemma, we have:

Lemma 9.4.2 *For almost all t, all $\theta \in \mathbb{R}_+ = [0, +\infty[$, we have*

$$\int_{\{v(t)>\theta\}} f \operatorname{sign} u \, dx - \int_0^{\mu(t,\theta)} \frac{\partial v_*}{\partial t}(t,\sigma)d\sigma \geqslant 0.$$

Proof of Lemma 9.4.2 Considering as a test function, $\psi_0 = (v - \theta)_+ \operatorname{sign} u$ for $\theta \geqslant 0$, we deduce by differentiating with respect to θ

$$0 \leqslant -\frac{d}{d\theta} \int_{v(t)>\theta} |\nabla v|^p \,(t,x)dx = \left[\int_{\{v(t)>\theta\}} f \operatorname{sign}(u)dx - \int_0^{\mu(t,\theta)} \frac{\partial v_*}{\partial t}(t,\sigma)d\sigma \right].$$

This inequality holds for almost all θ and by right continuity (mentioned above), the inequality of Lemma 9.4.2 holds for all $\theta \geqslant 0$. ◇

By the Hardy–Littlewood inequality, we deduce from Lemma 9.4.2 that:

$$\int_0^{\mu(t,\theta)} |f|_*(t,\sigma)d\sigma - \int_0^{\mu(t,\theta)} \frac{\partial v_*}{\partial t}(t,\sigma)d\sigma \geqslant 0, \qquad \forall \theta \geqslant 0. \tag{9.13}$$

Defining $\overline{\mu}(t,\theta) = \left|\left\{v(t) \geqslant \theta\right\}\right|$, we deduce from this last inequality that $\forall \theta > 0$,

$$\int_0^{\overline{\mu}(t,\theta)} |f|_*(t,\sigma)d\sigma - \int_0^{\overline{\mu}(t,\theta)} \frac{\partial v_*}{\partial t}(t,\sigma)d\sigma \geqslant 0. \tag{9.14}$$

Let us then show that $\forall s \in \overline{\Omega}_*$

$$\chi(s) = \int_0^s |f|_*(t,\sigma)d\sigma - \int_0^s \frac{\partial v_*}{\partial t}(t,\sigma)d\sigma \geqslant 0.$$

If $|v(t) = v_*(t,s)| = 0$ then $\mu\big(t, v_*(t,s)\big) = s$, and the relation follows from (9.13). If $v_*(t,s) \neq 0$ and $|v(t) = v_*(t,s)| \neq 0$, then let $s' = |v(t) > v_*(t,s)|$ and $s'' = |v(t) \geqslant v_*(t,s)|$.

According to the relations (9.13) and (9.14), we have:

$$\chi(s') \geqslant 0, \qquad \chi(s'') \geqslant 0.$$

Furthermore, according to Theorem 9.2.2, we know:

$$\frac{\partial v_*}{\partial t}(t,\sigma) = c(t) = \text{constant if } \sigma \in (s', s'').$$

Since $\frac{d}{ds}\chi(s) = |f|_*(t,s) - c(t)$ is decreasing, we deduce that $\chi(s)$ is concave, from which:

$$\chi(s) \geqslant \min\Big(\chi(s'), \chi(s'')\Big) \geqslant 0.$$

If $v_*(t,s) = 0$ and $|v(t) = 0| \neq 0$, we always have $\chi(s') \geqslant 0$, $s' = |v(t) > 0|$, $s'' = |\Omega|$. Again, we have two cases to consider:

If $\text{meas}\Big\{(\tau, x) \in Q : v(\tau, x) = 0\Big\} = 0$, then for almost all τ, $|v(\tau) = 0| = 0$, (by Fubini's theorem) we can neglect the t s.t. $|v(t) = 0| \neq 0$.

If $\text{meas}\Big\{(t, x) \in Q : v(t, x) = 0\Big\} > 0$, then (using Fubini's theorem) for almost all t, $\frac{\partial v}{\partial t}(t) = 0$ a.e. on $\Big\{v(t) = 0\Big\}$. So, we have: $\frac{\partial v_*}{\partial t}(t,\sigma) = \left(\frac{\partial v}{\partial t}\right)_{*v}(t,\sigma) = 0$, for a.a. $\sigma \in \Big(s', |\Omega|\Big)$ *and all* t. In this case, we have for $s \in \Big(s', |\Omega|\Big)$

$$\chi(s) = \chi(s') + \int_{s'}^{s} \frac{\partial v_*}{\partial t}(t,\sigma)d\sigma = \chi(s') \geqslant 0.$$

It has thus been shown that

Lemma 9.4.3 *Let $v \in W^{1,1}(0, T; L^1(\Omega))$, $v \geqslant 0$. If for all $\theta \geqslant 0$ we have:*

$$0 \leqslant \chi\big(\mu(t,\theta)\big) = \int_0^{\mu(t,\theta)} |f|_*(t,\sigma)d\sigma - \int_0^{\mu(t,\theta)} \frac{\partial v_*}{\partial t}(t,\sigma)d\sigma\,, \text{ a.e. in } t,$$

then,

$$\chi(s) \geqslant 0 \qquad \forall s \in \overline{\Omega}_*.$$

End of the Proof of Theorem 9.4.2

From this lemma and the relation (9.12) we deduce:

$$\left[|\nabla v|^p\right]_{*v}(t,s) \leqslant \left[\int_0^s |f|_*(t,\sigma)d\sigma - \int_0^s \frac{\partial v_*}{\partial t}(t,\sigma)d\sigma\right]\left(-\frac{\partial v_*}{\partial s}(t,s)\right).$$

Using the Poincaré–Sobolev inequality, we know that

$$-\frac{\partial v_*}{\partial s}(t,s) \leqslant \frac{s^{\frac{1}{N}-1}}{N\alpha_N^{\frac{1}{N}}}|\nabla v|_{*v}(t,s) \qquad \text{and} \qquad (|\nabla v|_{*v})^p \leqslant \left(|\nabla v|^p\right)_{*v},$$

and we deduce, after simplification, the estimate on $\left(|\nabla v|^p\right)_{*v}$.

From the proof of Theorem 7.2.2 (see remark after the proof), we also deduce that $\forall \varphi \geqslant 0\ \varphi \in L^\infty(\Omega_*)$

$$\int_{\Omega_*} \varphi^*(\sigma)\Phi\left(|\nabla v|_*\right)d\sigma \leqslant \int_{\Omega_*} \varphi\, \Phi\left(|\nabla v|_{*v}\right)d\sigma$$

if $\Phi : \mathbb{R}_+ \to \mathbb{R}$ is increasing and concave. Since the map $t \to t^{1-\frac{1}{p}}$ is increasing by $\mathbb{R}_+$ in itself and concave, we have

$$\int_{\Omega_*} \varphi^*(\sigma)\left(|\nabla v|_*^p\right)^{1-\frac{1}{p}}d\sigma$$

$$\leqslant \frac{1}{N\alpha_N^{\frac{1}{N}}}\int_{\Omega_*}\frac{\varphi(s)}{s^{1-\frac{1}{N}}}\left[\int_0^s |f|_*(t,\sigma)d\sigma - \int_0^s \frac{\partial v_*}{\partial t}(t,\sigma)d\sigma\right]ds$$

and

$$\left(|\nabla v|^p\right)_*^{1-\frac{1}{p}} = |\nabla v|_*^{p-1},$$

hence the result. ◇

To have an estimate depending only on the initial data, we integrate the last relation (ii) of Theorem 9.4.2 with respect to t. Knowing that the relation (i) of the same Theorem implies after integration in $[0,\sigma]$ that $\forall\sigma \in \Omega_*$

$$\int_0^\sigma |u|_*(t,s)ds \leqslant \int_0^\sigma \left(|u_0|_*(\sigma) + \int_0^t |f|_*(\tau,s)d\tau\right)ds \quad fora.a.\,t. \tag{9.15}$$

We then have the following theorem and proposition:

Theorem 9.4.3 *If* $u(0) = u_0 \in L^1(\Omega)$ *then under the same conditions as Theorems 9.4.2, we have* $\forall \varphi \geqslant 0\ \varphi \in L^\infty(\Omega_*)$,

$$\int_{\Omega_*} \varphi^*(\sigma)d\sigma \int_0^t |\nabla u|_*^{p-1}(\tau,\sigma)d\tau + \frac{1}{N\alpha_N^{\frac{1}{N}}} \int_{\Omega_*} \frac{\varphi(\sigma)}{\sigma^{1-\frac{1}{N}}} d\sigma \int_0^\sigma |u|_*(t,s)ds$$

$$\leqslant \frac{1}{N\alpha_N^{\frac{1}{N}}} \int_{\Omega_*} \frac{\varphi(\sigma)}{\sigma^{1-\frac{1}{N}}} \left[\int_0^\sigma \left(|u_0|_*(s) + \int_0^t |f|_*(\tau,s)d\tau \right) ds \right] d\sigma.$$

Here is an example of a consequence of the previous estimates

Proposition 9.4.1 *If* $\int_0^t |f|_*(\tau,\cdot)d\tau \in L^{p,q}(\Omega)$, $(\forall t)$ *and* $u_0 \in L^{p,q}(\Omega)$ *then*

$$u(t) \in L^{p,q}(\Omega)\ for\ almost\ all\ t.$$

We can imitate the proof of the gradient estimate given in Chap. 5, when we know that $\dfrac{\partial u}{\partial t} \in L^1(Q)$. We then have

Theorem 9.4.4 *Under the assumptions of Theorem 9.4.2 if* $\dfrac{\partial u}{\partial t} \in L^1(Q)$, *then* $\forall \varphi \geqslant 0$, $\varphi \in L^\infty(\Omega_*)$, *and for almost all* $t \in (0,T)$, *we have*

$$\int_{\Omega_*} \varphi^*(s)|\nabla v|^p(t,s)_* ds \leqslant c_N^{p'} \left(\int_{\Omega_*} \varphi(s)s^{-\frac{p'}{N'}} ds \right) F_u(t)^{p'}$$

where $F_u(t) = \int_\Omega |f|(t,x)dx + \int_\Omega \left| \dfrac{\partial u}{\partial t} \right| (t,x)dx.$

In particular, $|\nabla u| \in L^1\left(0,T; L^{\frac{N}{N-1}(p-1),\infty}(\Omega)\right)$, *and for almost every t*

(continued)

Theorem 9.4.4 (continued)

$$s^{\frac{N-1}{N(p-1)}}|\nabla u(t)|_*(s) \leqslant c'_{N1}\left[|f(t)|_{L^1(\Omega)} + \left|\frac{\partial u}{\partial t}\right|_{L^1(\Omega)}(t)\right]$$

where c'_{N1} is the same constant as in Theorem 5.2.5 of Chap. 5.

Proof We know that $\dfrac{\partial v_*}{\partial t}(t\sigma) = \left(\dfrac{\partial v}{\partial t}\right)_{*v}(t, \sigma)$. Thus

$$\int_0^s \left|\frac{\partial v_*}{\partial t}(t,\sigma)\right| d\sigma \leqslant \int_0^s \left|\left(\frac{\partial v}{\partial t}\right)_{*v}(t)\right|_*(\sigma)d\sigma \leqslant \int_0^s \left|\frac{\partial v}{\partial t}\right|_*(t,\sigma)d\sigma$$

and as $\left|\dfrac{\partial u}{\partial t}\right| = \left|\dfrac{\partial v}{\partial t}\right|$ we then deduce

$$0 \leqslant \int_0^s |f|_*(t,s)d\sigma - \int_0^s \frac{\partial v_*}{\partial t}(t,v)d\sigma$$

$$\leqslant \int_{\Omega_*} |f|_*(t,\sigma)d\sigma + \int_{\Omega_*} \left|\frac{\partial u}{\partial t}\right|_*(t,\sigma)d\sigma = F_u(t).$$

According to Theorem 9.4.2, and using the same argument as in the proof, we deduce that for almost all $t \in (0, T)$

$$\int_{\Omega_*} \varphi^*(s)|\nabla v|_*^p(t,s)ds \leqslant c_N^{p'}\left(\int_{\Omega_*} \varphi(s)s^{-\frac{p'}{N'}}\right) F_u(t)^{p'}$$

(because $\displaystyle\int_{\Omega_*} \varphi^*(s)|\nabla v|_*^p(t,s)ds \leqslant \int_{\Omega_*} \varphi(s)(|\nabla v|^p)_{*v}(t,s)ds$).

We reproduce the proof of Theorem 5.2.5 of Chap. 5, for t fixed to obtain the estimate of the theorem. ◇

In the linear case, we can use a duality argument to find the estimate of the gradient without the hypothesis $\dfrac{\partial u}{\partial t} \in L^1(Q)$. This proof is given in [134] where $f \in L^{N'}\left(0, T; L^1(Q)\right)$, $u_0 \in L^{N',\infty}(\Omega)$.

9.4.3 Special Case of Linear Equations: Pointwise Estimates of $\int_0^s \left|\frac{\partial u}{\partial t}\right|_* (t,\sigma)d\sigma$ and $\int_0^s |u|_* (t,\sigma)$

Theorem 9.4.5 *Let* $f \in W^{1,2}\big(0,T; L^2(\Omega)\big)$, $u_0 \in H_0^1(\Omega) \cap H^2(\Omega)$. *Let* Ω *be a bounded open set of class* C^∞. *Then the unique solution* u *of*

$$\begin{cases} \dfrac{\partial u}{\partial t} - \Delta u = f, & \text{in } \mathcal{D}'(Q), \\ u \in L^2\big(0,T; H_0^1(\Omega) \cap H^2(\Omega)\big), \\ u(0) = u_0, \end{cases}$$

satisfies the following pointwise estimate: $\forall s \in \Omega_*$, $\forall t$, *we have:*

$$\int_0^s \left|\frac{\partial u}{\partial t}\right|_* (t,\sigma)d\sigma \leqslant \int_0^s |f(0)+\Delta u_0|_* d\sigma + \int_0^s d\sigma \int_0^t \left|\frac{\partial f}{\partial t}\right|_* (\tau,\sigma)d\tau.$$

Proof Let φ_k be a sequence of eigenfunctions associated to the Laplacian with Dirichlet conditions

$$-\Delta\varphi_k = \lambda_k \varphi_k, \qquad \varphi_k \in H_0^1(\Omega) \cap C^\infty(\overline{\Omega}).$$

If $V_m = vect\left\{\varphi_1, \dots, \varphi_m\right\}$, then the usual Galerkin method leads to

$$\begin{cases} \dfrac{\partial u_m}{\partial t} - \Delta u_m = P_m f(t) = \displaystyle\sum_{j=1}^m \Big(f(t), \varphi_j\Big)\varphi_j, \\ u_m \in W^{1,2}\big(0,T; V_m\big), \\ u_m(0) = u_{0m} \to u_0 \text{ in } H^2(\Omega)\text{-strong}, \ u_{0m} \in V_m. \end{cases}$$

Here $(\cdot,\cdot)$ denotes the scalar product in $L^2(\Omega)$. Let $\overline{u_m} = \dfrac{\partial u_m}{\partial t}$, then we deduce that $\overline{u_m}$ satisfies:

$$\begin{cases} \dfrac{\partial}{\partial t}\overline{u_m} - \Delta\overline{u_m} = P_m f'(t), \\ \overline{u_m}(0) = \Delta u_{0m} + P_m f(0), \\ \overline{u_m}(t) \in V_m. \end{cases}$$

Applying Theorem 9.4.2 relation (i) to $\overline{u_m}$ we deduce:

$$\int_0^s \frac{\partial}{\partial t} |\overline{u_m}| (t,\sigma)d\sigma \leqslant \int_0^s \left|P_m f'(t)\right|_*(\sigma)d\sigma.$$

We integrate this last relation in time to conclude that:

$$\int_0^s \left|\frac{\partial u_m}{\partial t}\right|_*(t,\sigma)d\sigma \leqslant$$

$$\int_0^s |P_m f(0) + \Delta u_{0m}|_*(\sigma)d\sigma + \int_0^s d\sigma \int_0^t \left|P_m f'(\tau)\right|_*(\sigma)d\sigma. \tag{9.16}$$

Since $P_m(f) \to f$ in $W^{1,2}\big(0,T;L^2(\Omega)\big)$-strong and

$\Delta u_{0m} \to \Delta u_0$ in $L^2(\Omega)$-strong it follows that $\dfrac{\partial u_m}{\partial t} \to \dfrac{\partial u}{\partial t}$ in $L^2(Q)$-strong. By passing to the limit in (9.16), we deduce:

$$\int_0^s \left|\frac{\partial u}{\partial t}\right|_*(t,\sigma)d\sigma \leqslant \int_0^s |f(0)+\Delta u_0|_*(\sigma)d\sigma + \int_0^s d\sigma \int_0^t \left|\frac{\partial f}{\partial t}\right|_*(\tau,\sigma)d\tau.$$

◇

Other consequences of the previous results are the comparison theorems, which we will simply illustrate with linear equations, i.e. $p = 2$.

Theorem 9.4.6 *Let $f \in L^2_+(Q)$, $u_0 \in L^2_+(\Omega)$. Then the unique solution u of*

$$\begin{cases} \dfrac{\partial u}{\partial t} - \Delta u = f, \\ u \in L^2(0,T;H_0^1(\Omega)) \cap C\Big(\big[0,T\big], L^2(\Omega)\Big), \\ u(0) = u_0, \end{cases}$$

satisfies, for almost all t, $\forall s \in \Omega_*$

$$\begin{cases} \dfrac{\partial k}{\partial t} - N^2\alpha_N^{\frac{2}{N}} s^{2-\frac{2}{N}} \dfrac{\partial^2 k}{\partial s^2} \leqslant F \text{ in } Q_* = \big(0,T\big) \times \Omega_*, \\ k(t,0) = 0, \quad \dfrac{\partial k}{\partial s}(t,|\Omega|) = 0 \quad \forall t \in [0,T], \\ k(0,s) = \displaystyle\int_0^s u_{0*}(\sigma)d\sigma = k_0(s), \end{cases}$$

where $k(t,s) = \displaystyle\int_0^s u_*(t,\sigma)d\sigma$, $F(t,s) = \displaystyle\int_0^s f_*(t,\sigma)d\sigma$.

Proof Since $f \geqslant 0$, $u_0 \geqslant 0$, by the maximum principle, we have $u \geqslant 0$. According to the previous Theorem 9.4.2 we have

$$\left[\left(|\nabla u|^2_{*u}\right)\right]^{\frac{1}{2}} \leqslant \frac{s^{\frac{1}{N}-1}}{N\alpha_N^{\frac{1}{N}}}\left[\int_0^s f_*(t,\sigma)d\sigma - \int_0^s \frac{\partial u_*}{\partial t}(t,\sigma)d\sigma\right]. \tag{9.17}$$

By the pointwise Poincaré–Sobolev inequality (PSR), we have:

$$-\frac{\partial u_*}{\partial s}(t,s) \leqslant \frac{s^{\frac{1}{N}-1}}{N\alpha_N^{\frac{1}{N}}}\left[\left(|\nabla u|^2\right)_{*u}\right]^{\frac{1}{2}}(t,s). \tag{9.18}$$

These last two inequalities lead to:

$$-\frac{\partial u_*}{\partial s}(t,s) \leqslant \frac{s^{\frac{2}{N}-2}}{\left(N\alpha_N^{\frac{1}{N}}\right)^2}\left[\int_0^s f_*(t,\sigma)d\sigma - \int_0^s \frac{\partial u_*}{\partial t}(t,\sigma)d\sigma\right]. \tag{9.19}$$

By introducing, $k(t,\sigma) = \int_0^s u_*(t,\sigma)d\sigma$, the relation (9.19) is then written:

$$\frac{\partial k}{\partial t}(t,s) - \left(N\alpha_N^{\frac{1}{N}}\right)^2 s^{2-\frac{2}{N}}\frac{\partial^2 k}{\partial s^2} \leqslant F(t,s) = \int_0^s f_*(t,\sigma)d\sigma.$$

Hence the theorem. ◇

If we introduce the radial solution U of

$$\begin{cases} \frac{\partial U}{\partial t} - \Delta U = \underset{\sim}{f}, \\ U \in L^2(0,T; H_0^1(\underset{\sim}{\Omega})) \cap C\left([0,T], L^2(\underset{\sim}{\Omega})\right), \\ U(0) = \underset{\sim}{u_0}, \end{cases}$$

where $\underset{\sim}{f}(t,x) = f_*(t,\alpha_N|x|^N)$, $x \in \underset{\sim}{\Omega}$ (ball of same measure as Ω), a direct calculation leads to:

Proposition 9.4.2 *The function* $K(t,s) = \int_0^s U_*(t,\sigma)d\sigma$ *satisfies*

$$\frac{\partial K}{\partial t}(t,s) - \left(N\alpha_N^{\frac{1}{N}}\right)^2 s^{2-\frac{2}{N}} \frac{\partial^2 K}{\partial s^2}(t,s) = \int_0^s f_*(t,\sigma)d\sigma.$$

Furthermore, we have $k(t,s) \leqslant K(t,s), \quad \forall t \in [0,T]$ *and* $\forall s \in \overline{\Omega}_*$.

(For this proposition see Exercise 11.1.18). Using Lemma 9.5.3 below, we have $k \leqslant K$.

9.5 Behavior, Over a Long Time, of a System of Chemotaxis Equations

Let Ω be a bounded open set of $\mathbb{R}^N$ of class $C^{0,1}$. We will be interested in the asymptotic behavior of the following elliptic-parabolic system:

$$(Ch)\begin{cases} \dfrac{\partial u}{\partial t} = div\,(\nabla u - \overline{\chi} u \nabla v) & \text{in } Q_T = (0,T)\times\Omega, \\ 0 = \Delta v - \gamma v + \alpha u & \text{in } Q_T, \\ u(0) = u_0 & \text{in } \Omega, \end{cases}$$

where $\overline{\chi},\ \gamma,\ \alpha$ are strictly positive numbers, $u_0 \geqslant 0,\ u_0 \in W_0^{1,p}(\Omega),\ p > N$. Such a system appears in the so-called chemotaxis models.

Theorem 9.5.1 *For any data* $u_0 \geqslant 0$ *of* $W_0^{1,p}(\Omega)$*, there exists a time* $T_{max} > 0$ *and a unique solution* (u,v) *of* (Ch) *satisfying*

1. $u > 0,\ v > 0$ *in* $(0, T_{max}) \times \Omega$,
2. $u \in C\big([0,T_{max}); W_0^{1,p}(\Omega)\big) \cap C^1\big([0,T_{max}); L^p(\Omega)\big)$,
   $u(t) \in W^{2,p}(\Omega)$ *for* $0 < t < T_{max}$ *and* $v \in C\big((0,T_{max}); W^{2,p}(\Omega) \cap W_0^{1,p}(\Omega)\big)$.

We will not prove this theorem but we will be interested in knowing if $T_{max} = +\infty$ or not, using the rearrangement.

Lemma 9.5.1 *Let (u, v) be the solution of (Ch).*
Let $k(t, s) = \int_0^s u_*(t, \sigma)d\sigma$, for $t \in (0, T_{max})$, $s \in \Omega_*$. Then,

$$\begin{cases} \dfrac{\partial k}{\partial t} - (N\alpha_N^{\frac{1}{N}})^2 s^{2-\frac{2}{N}} \dfrac{\partial^2 k}{\partial s^2} - \alpha\overline{\chi}k\dfrac{\partial k}{\partial s} \leqslant 0 \, ae \text{ in } Q_{T_{max}*} = (0, T_{max}) \times \Omega_*, \\ k(t, 0) = 0, \qquad \dfrac{\partial k}{\partial s}(t, |\Omega|) = 0, \qquad \forall t \in [0, T_{max}], \\ k(0, s) = \displaystyle\int_0^s u_{0*}(\sigma)d\sigma, \ \forall s \in \Omega_*. \end{cases}$$

Proof of the Lemma Note that $\dfrac{\partial u}{\partial t}(t) \in L^p(\Omega)$. Multiplying the first equation by $\Big(u(t) - u_*(t, s)\Big)_+$ for $s \in \Omega_*$ fixed, $t \in (0, T_{max})$, and integrating

$$\int_{u(t)>u_*(t,s)} |\nabla u|^2 (t, x)dx = \overline{\chi} \int_{u(t)>u_*(t,s)} u\nabla v \cdot \nabla u - \int_\Omega \frac{\partial u}{\partial t}(t)\big(u(t) - u_*(t, s)\big)_+ dx. \tag{9.20}$$

Since $\Big\{u(t) > u_*(t, s)\Big\} = \Big\{u(t)^2 > u_*^2(t, s)\Big\}$ (because $u(t) \geqslant 0$) we write:

$$\int_{u(t)>u_*(t,s)} u\nabla v \cdot \nabla u = \frac{1}{2}\int_{u(t)^2>u_*^2(t,s)} \nabla v \cdot \nabla(u^2)dx = \frac{1}{2}\int_\Omega \nabla v \cdot \nabla \Big(u(t)^2 - u_*^2(t, s)\Big)_+ dx. \tag{9.21}$$

We multiply the second equation of (Ch) by $\big(u(t)^2 - u_*^2(t, s)\big)_+$, from which we get

$$\int_\Omega \nabla v \cdot \nabla !DEu(t)^2 - u_*^2(t, s)_+ dx = \int_\Omega (\alpha u - \gamma v)\Big(u(t)^2 - u_*^2(t, s)\Big)_+ dx\,. \tag{9.22}$$

Combining these last relations, by differentiating with respect to s, we have since $\Big\{u(t) > u_*(t,s)\Big\} = \Big\{u(t)^2 > u_*^2(t,s)\Big\} = \Omega_s(t)$,

$$\left(|\nabla u|^2\right)_{*u}(t,s) =$$

$$\left[\overline{\chi} u_*(t,s)\int_{\Omega_s(t)}(\alpha u - \gamma v)dx - \int_{\Omega_s(t)} \frac{\partial u}{\partial t}(t,s)dx\right]\left(-\frac{\partial u_*}{\partial s}\right)(t,s). \tag{9.23}$$

By definition of the relative rearrangement and the theorem of regularity on $\dfrac{\partial u_*}{\partial t}$, we deduce from this relation (9.23) that:

$$|\nabla u|^2_{*u}(t,s) =$$

$$\left[\overline{\chi} u_*(t,s)\int_0^s(\alpha u_* - \gamma v_{*u})(t,\sigma)d\sigma - \int_0^s \frac{\partial u_*}{\partial t}(t,\sigma)d\sigma\right]\left(-\frac{\partial u_*}{\partial s}\right)(t,s). \tag{9.24}$$

As in Lemma 9.4.3, we have:

Lemma 9.5.2 *Let* $\mu(t,\theta) = |u(t) > \theta|$, $\theta > 0$. *Then,* $\forall\theta > 0$

$$\overline{\chi}\alpha\theta \int_0^{\mu(t,\theta)} u_*(t,\sigma)d\sigma - \int_0^{\mu(t,\theta)} \frac{\partial u_*}{\partial t}(t,\sigma)d\sigma \geqslant 0.$$

Consequently $\forall s \in \Omega_*$, *we have:*

$$\overline{\chi}\alpha u_*(t,s)\int_0^s u_*(t,\sigma)d\sigma - \int_0^s \frac{\partial u_*}{\partial t}(t,\sigma)d\sigma \geqslant 0.$$

Proof Let $\theta > 0$, then taking as test function $\big(u(t) - \theta\big)_+$ we deduce as above that

$$\int_{u(t)>\theta} |\nabla u|^2(t,x)dx = \frac{\overline{\chi}}{2}\int_\Omega \nabla v \cdot \nabla\big(u(t)^2 - \theta^2\big)_+ - \int_\Omega \frac{\partial u}{\partial t}\big(u(t) - \theta\big)_+ dx$$

(using the second equation of (Ch))

$$= \frac{\overline{\chi}}{2}\int_\Omega (\alpha u - \gamma v)(t,x)\big(u(t)^2 - \theta^2\big)_+ - \int_\Omega \frac{\partial u}{\partial t}\big(u(t) - \theta\big)_+ dx.$$

Thus, by the derivation rule seen in Chap. 5,

$$0 \leqslant -\frac{d}{d\theta}\int_{u(t)>\theta} |\nabla u|^2\,dx$$

$$= \overline{\chi}\alpha\theta \int_{u(t)>\theta} u(t,s)dx - \int_{u(t)>\theta} \frac{\partial u}{\partial t}(t,x)dx - \overline{\chi}\gamma \int_{u(t)>\theta} v(t,x)dx\,.$$

By the Hardy–Littlewood inequality and Lemma 9.4.1 since $v \geqslant 0$, we have:

$$0 \leqslant \overline{\chi}\alpha\theta \int_0^{\mu(t,\theta)} u_*(t,\sigma)d\sigma \;-\; \int_0^{\mu(t,\theta)} \frac{\partial u_*}{\partial t}(t,\sigma)d\sigma\,,$$

we conclude as in Lemma 9.4.3. ◇

From relation (9.23), since $v \geqslant 0$ (and therefore $v_{*u} \geqslant 0$), we have:

$$\left(|\nabla u|^2\right)_{*u}(t,s)$$

$$\leqslant \left[\overline{\chi}\alpha u_*(t,!s)\int_0^s u_*(t,s)d\sigma - \int_0^s \frac{\partial u_*}{\partial t}(t,\sigma)d\sigma\right]\left(-\frac{\partial u_*}{\partial s}\right)(t,s). \tag{9.25}$$

We apply the pointwise Poincaré–Sobolev inequality to obtain:

$$\left(N\alpha_N^{\frac{1}{N}}\right)^2 s^{2-\frac{2}{N}}\left(-\frac{\partial u_*}{\partial s}(t,s)\right)^2 \leqslant \left(|\nabla u|^2\right)_{*u}(t,s).$$

Therefore, the relation (9.25) leads to:

$$\left(N\alpha_N^{\frac{1}{N}}\right)^2 s^{2-\frac{2}{N}}\left(-\frac{\partial u_*}{\partial s}(t,s)\right)$$

$$\leqslant \left[\overline{\chi}\alpha u_*(t,s)\int_0^s u_*(t,\sigma)d\sigma - \int_0^s \frac{\partial u_*}{\partial t}(t,\sigma)d\sigma\right] \tag{9.26}$$

(taking into account Lemma 9.5.2).

By introducing $k(t,s) = \displaystyle\int_0^s u_*(t,\sigma)d\sigma$, we therefore have

$$\frac{\partial k}{\partial t}(t,s) = \int_0^s \frac{\partial u_*}{\partial t}(t,\sigma)d\sigma, \qquad \frac{\partial^2 k}{\partial s^2}(t,s) = \frac{\partial u_*}{\partial s}(t,s).$$

These inequalities combined with the relation (9.26) give

$$\frac{\partial k}{\partial t}(t,s) - \left(N\alpha_N^{\frac{1}{N}}\right)^2 s^{2-\frac{2}{N}} \frac{\partial^2 k}{\partial s^2} - \alpha\overline{\chi}\left(k\frac{\partial k}{\partial s}\right)(t,s) \leqslant 0. \qquad \diamond$$

We will show from this lemma the following theorem:

Theorem 9.5.2 *We have $T_{max} = +\infty$ in the following conditions:*

(C1) for $N = 1$, no condition,
(C2) for $N = 2$ if u_0 satisfies $\alpha\overline{\chi}\,|u_0|_1 < 8\pi$,
(C3) for $N \geqslant 3$ if u_0 satisfies $\alpha\overline{\chi}\,|u_0|_{L^N} < N\alpha_N^{\frac{2}{N}}\,|\Omega|^{-\frac{1}{N}}$.

The proof of the theorem is based on the following comparison lemma:

Lemma 9.5.3 (Of Comparison) *Let $f,\ g$ be two functions defined on $\overline{Q}_{T*} = [0,T]\times\overline{\Omega}_*$ and satisfying*

(i) $f,\ g$ are in
$$L^\infty(Q_{T_*}) \cap H^1(0,T;L^2(\Omega_*)) \cap \bigcap_{\delta>0} L^2(0,T;W^{2,2}(\delta,|\Omega|),$$

(ii) $\left|\frac{\partial f}{\partial s}(t,s)\right| \leqslant c(t)$ *and* $\left|\frac{\partial g}{\partial s}\right|(t,s) \leqslant c(t)\max(s^{-\ell},1)$ *where ℓ is a constant satisfying $0 \leqslant \ell < 1,\ c \in L^2(0,T)$.*

If f and g satisfy the system:

$$\begin{cases} \dfrac{\partial f}{\partial t} - \left(N\alpha_N^{\frac{1}{N}}\right)^2 s^{2-\frac{2}{N}} \dfrac{\partial^2 f}{\partial s^2} - \alpha\overline{\chi} f\dfrac{\partial f}{\partial s} \leqslant \\ \qquad \leqslant \dfrac{\partial g}{\partial t} - \left(N\alpha_N^{\frac{1}{N}}\right)^2 s^{2-\frac{2}{N}} \dfrac{\partial^2 g}{\partial s^2} - \alpha\overline{\chi} g\dfrac{\partial g}{\partial s} \text{ in } Q_{T_*}, \\ 0 = f(t,0) \leqslant g(t,0), \quad \dfrac{\partial f(t,|\Omega|)}{\partial s} \leqslant \dfrac{\partial g(t,|\Omega|)}{\partial s}, \quad \forall t \in (0,T) \\ f(0,s) \leqslant g(0,s), \quad s \in \Omega_*, \text{ and } g(t,s) \geqslant 0, \end{cases}$$

then $f \leqslant g$ in Q_{T_*}.

Proof Let $w = f - g$. Then, w satisfies

$$\begin{cases} \dfrac{\partial w}{\partial t} - (N\alpha_N^{\frac{1}{N}})^2 s^{2-\frac{2}{N}} \dfrac{\partial^2 w}{\partial s^2} - \alpha\overline{\chi}\left(f\dfrac{\partial f}{\partial s} - g\dfrac{\partial g}{\partial s}\right) \leqslant 0 \text{ in } Q_{T_*} \\ w(t,0) \leqslant 0, \qquad \dfrac{\partial w}{\partial s}(t,|\Omega|) \leqslant 0, \quad \forall t \in [0,T] \\ w(0,s) \leqslant 0 \text{ for } s \in \overline{\Omega}_*. \end{cases}$$

We multiply the first inequality by $s^{\frac{2}{N}-2}w_+$, ($w_+ = \max(w,0)$), we then obtain for $0 < \delta < |\Omega|$

$$\int_\delta^{|\Omega|} s^{\frac{2}{N}-2} \frac{\partial w}{\partial t} w_+ ds$$

$$\leqslant \left(N\alpha_N^{\frac{1}{N}}\right)^2 \int_\delta^{|\Omega|} \frac{\partial^2 w}{\partial s^2} w_+ ds + \alpha\overline{\chi} \int_\delta^{|\Omega|} \left(f\frac{\partial f}{\partial s} - g\frac{\partial g}{\partial s}\right) w_+ s^{\frac{2}{N}-2} ds. \tag{9.27}$$

By integration by parts, we have: $\displaystyle\int_\delta^{|\Omega|} \frac{\partial^2 w}{\partial s^2} w_+ ds$

$$= -\int_\delta^{|\Omega|} \left(\frac{\partial w_+}{\partial s}\right)^2 ds + \frac{\partial w}{\partial s}\Big(t,|\Omega|\Big) w_+\Big(t,|\Omega|\Big) - \frac{\partial w}{\partial s}(t,\delta) w_+(t,\delta). \tag{9.28}$$

Since, $\dfrac{\partial w}{\partial s}(t,|\Omega|) \leqslant 0$, this last inequality gives:

$$J_0 = \int_\delta^{|\Omega|} \frac{\partial^2 w}{\partial s^2} w_+ ds \leqslant -\frac{\partial w}{\partial s}(t,\delta) w_+(t,\delta) - \int_\delta^{|\Omega|} \left(\frac{\partial w_+}{\partial s}\right)^2 ds. \tag{9.29}$$

Now let's write $f = w + g$. Then:

$$J_1 = \int_\delta^{|\Omega|} \left(f\frac{\partial f}{\partial s} - g\frac{\partial g}{\partial s}\right) w_+ s^{\frac{2}{N}-2} ds$$

$$= \int_\delta^{|\Omega|} s^{\frac{2}{N}-2} w_+^2 \frac{\partial f}{\partial s} ds + \int_\delta^{|\Omega|} s^{\frac{2}{N}-2} w_+ \frac{\partial w}{\partial s} g\, ds. \tag{9.30}$$

Since $\left|\dfrac{\partial f}{\partial s}\right|(t,s) \leqslant c(t)$, we deduce:

$$J_1 \leqslant c(t) \int_\delta^{|\Omega|} s^{\frac{2}{N}-2} w_+^2 ds + \int_\delta^{|\Omega|} s^{\frac{2}{N}-2} w_+ \frac{\partial w_+}{\partial s} g\, ds. \tag{9.31}$$

Since $g \geqslant 0$ and $f(t,0) = 0$, we then have:

$$w_+(t,s) \leqslant f_+(t,s) = \int_0^s \frac{\partial f_+}{\partial \sigma}(t,\sigma)\,d\sigma \leqslant \int_0^s \left|\frac{\partial f}{\partial \sigma}\right|(t,\sigma)\,d\sigma \leqslant c(t)s \tag{9.32}$$

hence

$$s^{\frac{2}{N}-2} w_+^2(t,s) \leqslant s^{\frac{2}{N}} c(t)^2 \in L^1(\Omega_*).$$

Moreover,

$$\int_\delta^{|\Omega|} s^{\frac{2}{N}-2} w_+ \left|\frac{\partial w_+}{\partial s}\right| g\,ds \leqslant \int_\delta^{|\Omega|} s^{\frac{2}{N}-2} w_+ \left|\frac{\partial w_+}{\partial s}\right| f\,ds$$

because $w_+ = 0$ if $f \leqslant g$. From these facts, we deduce that:

$$J_1 \leqslant c(t) \int_\delta^{|\Omega|} s^{\frac{2}{N}-2} w_+^2(t,s) ds + \int_\delta^{|\Omega|} s^{\frac{2}{N}-2} w_+ \left|\frac{\partial w_+}{\partial s}\right| f\,ds. \tag{9.33}$$

By combining the relations (9.27), (9.29), (9.32) and (9.33), we obtain:

$$\frac{1}{2}\frac{d}{dt} \int_\delta^{|\Omega|} s^{\frac{2}{N}-2} w_+^2(t,s) ds + \left(N\alpha_N^{\frac{1}{N}}\right)^2 \int_\delta^{|\Omega|} \left(\frac{\partial w_+}{\partial s}\right)^2 (t,s)\,ds$$

$$\leqslant \alpha\overline{\chi}c(t) \int_\delta^{|\Omega|} s^{\frac{2}{N}-2} w_+^2(t,s) ds + \alpha\overline{\chi}c(t) \int_\delta^{|\Omega|} s^{\frac{2}{N}-2} w_+ \left|\frac{\partial w_+}{\partial s}\right| s\,ds + \tag{9.34}$$

$$+ \left(N\alpha_N^{\frac{1}{N}}\right)^2 \left|\frac{\partial w}{\partial s}(t,\delta)\right| w_+(t,\delta).$$

By applying Young's inequality, we know:

$$\alpha\overline{\chi}c(t) \int_\delta^{|\Omega|} s^{\frac{2}{N}-2} w_+ \left|\frac{\partial w_+}{\partial s}\right| s\,ds$$

$$\leqslant \frac{1}{2}\left(N\alpha_N^{\frac{1}{N}}\right)^2 \int_\delta^{|\Omega|} \left(\frac{\partial w_+}{\partial s}\right)^2 (t,s)\,ds + c_1(t) \int_\delta^{|\Omega|} s^{\frac{2}{N}-2} w_+^2(t,s) ds \tag{9.35}$$

where $c_1 \in L^1(0,T)$.

We arrive at the following Gronwall inequality:

$$\frac{1}{2} y_\delta'(t) \leqslant c_2(t) y_\delta(t) + \left(N\alpha_N^{\frac{1}{N}}\right)^2 \left|\frac{\partial w}{\partial s}(t,\delta)\right| w_+(t,\delta) \tag{9.36}$$

where $y_\delta(t) = \int_\delta^{|\Omega|} s^{\frac{2}{N}-2} w_+^2(t,s)ds$ and $c_2 \in L^1(0,T)$.

But because $\left|\frac{\partial g}{\partial t}\right| \leqslant c(t)\max(1, s^\ell)$ and $\left|\frac{\partial f}{\partial t}\right| \leqslant c(t)$, we deduce that, for $\delta < 1$

$$\left(N\alpha_N^{\frac{1}{N}}\right)^2 \left|\frac{\partial w}{\partial s}\right|(t,\delta)w_+(t,\delta) \leqslant c_3(t)\delta^{1-\ell} \underset{\delta \to 0}{\longrightarrow} 0, \qquad c_3 \in L^2(0,T).$$

Hence when $\delta \to 0$, we obtain from Gronwall's inequality that:

$$y_0(t) = \int_{\Omega_*} s^{\frac{2}{N}-2} w_+^2(t,s)ds \leqslant c y_0(0) = 0.$$

Hence $w_+ = 0 : f \leqslant g$. ◇

Proof of Theorem 9.5.2.
Case Where N $= 1$.

We consider $f = k$ and g the function defined by

$$g(t,s) = pe^{-\lambda t} th\left(\alpha\overline{\chi}p\frac{s}{4}\right) \doteq e^{-\lambda t} h_p(s)$$

(th= hyperbolic tangent function) where λ and p are chosen so that

$$\int_0^s u_{0*}(\sigma)d\sigma \leqslant h_p(s) \text{ (for} p \text{large)}$$

and $0 < \lambda \leqslant \alpha\overline{\chi}h_p'(|\Omega|)$.

Under these conditions, we have:

$$\frac{\partial g}{\partial t} - \left(N\alpha_N^{\frac{1}{N}}\right)^2 s^{2-\frac{2}{N}} \frac{\partial^2 g}{\partial s^2} - \alpha\overline{\chi}g\frac{\partial g}{\partial s} \geqslant e^{-\lambda t}h_p\left(\alpha\overline{\chi}h_p'(|\Omega|) - \lambda\right) \geqslant 0.$$

Since $f = k$ satisfies Lemma 9.5.1, we deduce from the comparison lemma that:

$$\int_0^s u_*(t,\sigma)d\sigma \leqslant e^{-\lambda t}h_p(s),$$

hence

$$|u(t)|_{L^\infty(\Omega)} \leqslant (\text{constant})e^{-\lambda t}.$$

Case Where N = 2.

We let

$$g(t,s)=e^{-\lambda t}\frac{aqs}{1+qs}$$

with $\lambda,\ a,\ q$ to be choosen.

First, we choose a : $\int_{\Omega} u_0(x)dx < a$ and $\alpha\overline{\chi}a < 8\pi$ and q large enough so that $\int_0^s u_{0*}(s)d\sigma \leqslant \frac{aqs}{1+qs}, \quad \forall s\in\overline{\Omega}_*$.

Finally, we take $0<\lambda\leqslant \frac{q}{(1+q\,|\Omega|)^2}(8\pi-\alpha\overline{\chi}a)$.

Under these conditions, we have:

$$\frac{\partial g}{\partial t}-\left(N\alpha_N^{\frac{1}{N}}\right)^2 s^{2-\frac{2}{N}}\frac{\partial^2 g}{\partial s^2}-\alpha\overline{\chi}g\cdot\frac{\partial g}{\partial s}\geqslant 0.$$

Hence

$$\int_0^s u_*(t,\sigma)\leqslant e^{-\lambda t}\frac{aqs}{1+qs},\qquad \forall s\in\overline{\Omega}_*.$$

In particular

$$|u(t)|_\infty\leqslant ce^{-\lambda t}.$$

Case Where N ⩾ 3.

We assume that u_0 satisfies $\alpha\overline{\chi}\,|u_0|_N < N\alpha_N^{\frac{2}{N}}\,|\Omega|^{-\frac{1}{N}}$. We choose

$$g(t,s)=e^{-\lambda t}\ell_q(s)$$

where $\ell_q(s)=qs^{1-\frac{1}{N}},\quad q=\frac{N}{\beta}\alpha_N^{\frac{2}{N}}\,|\Omega|^{-\frac{1}{N}}$ and $0<\lambda\leqslant(\beta-\alpha\overline{\chi})\,\ell_q'(|\Omega|)$.

Then g is suitable, i.e.

$$\frac{\partial g}{\partial t}-\left(N\alpha_N^{\frac{1}{N}}\right)^2 s^{2-\frac{2}{N}}\frac{\partial^2 g}{\partial s^2}-\alpha\overline{\chi}g\cdot\frac{\partial g}{\partial s}\geqslant 0.$$

Hence, by applying the comparison theorem, we have:

$$f(t,s)=\int_0^s u_*(t,\sigma)d\sigma\leqslant e^{-\lambda t}\ell_q(s),$$

which implies:

$$|u(t)|_\infty \leqslant ce^{-\lambda t} \qquad \forall t.$$

Pre-bibliographical Notes

The majority of the theorems cited in this chapter are due to the author or his collaborators, see for example Mossino and Rakotoson [116], Dìaz and Nagai [59], Dìaz and Nagai–Rakotoson [62], Rakotoson [128].

Theorem 9.4.1 is from Temam's book [165].

Here are some recent papers on the chemotaxis problem and using comparison techniques: Takasi and Takashi [159], Tian and Li [170], Chiacchio and Blasio [37].

Chapter 10
Relative Rearrangement and Lebesgue Spaces with Variable Exponents

As we have pointed out throughout this book, rearrangements are very suitable tools for rearrangement-invariant spaces because they allow us to go from a space of dimension $N \geq 2$ a space of dimension one over $\mathbb{R}$.

This then allows us to make an almost manual calculation to obtain explicit or even optimal estimates.

The natural question is then *what does this tool bring when working with spaces which are not invariant under rearrangement?*

In recent decades, particular attention has been paid to spaces with variable exponents $L^{p(\cdot)}(\Omega)$, which are in fact Lebesgue spaces but where the power p in the integral depends on the variables of the domain over which we integrate. In fact, we have two functions, one: the exponent, here usually denoted by $p(\cdot)$, and the other: the one that is in the Lebesgue space. We then see that if we rearrange the function, as the space is no longer the same for each point, we cannot speak of invariance under rearrangement. Nevertheless, we wish to use rearrangement techniques. So the idea is to use rearrangement both on the concerned function and the exponent $p(\cdot)$.

Definition 10.0.1 (Of the Space $L^{p(\cdot)}(\Omega)$) Let Ω be a bounded measurable set and $p : \Omega \to [1, +\infty[$ a measurable function. For measurable $u : \Omega \to \mathbb{R}$, let us set $\Phi_p(u) = \displaystyle\int_\Omega |u(x)|^{p(x)}dx$.

We define the norm $|u|_{p(\cdot)} = \inf\left\{\lambda > 0, : \Phi_p\left(\frac{u}{\lambda}\right) \leqslant 1\right\}$, and the space $L^{p(\cdot)}(\Omega) = \left\{u : \Omega \to \mathbb{R} : \text{ measurable s.t. } |u|_{p(\cdot)} < +\infty\right\}$.

(continued)

J. M. Rakotoson, *Relative Rearrangement*, Lecture Notes in Mathematics 2376,
https://doi.org/10.1007/978-3-032-02228-8_10

Definition 10.0.1 (continued)
The space $\left(L^{p(\cdot)}(\Omega), |\cdot|_{p(\cdot)}\right)$ is a Banach function space.
We have an equivalent norm, called Amemiya's norm, which is defined by

$$||u||_{p(\cdot)} = \inf_{\lambda>0} \lambda \left(1 + \Phi_p\left(\frac{u}{\lambda}\right)\right), \quad \textit{and} \quad |u|_{p(\cdot)} \leqslant ||u||_{p(\cdot)} \leqslant 2|u|_{p(\cdot)}.$$

For a complete study, we refer the reader to the book by L. Diening et al. [66].

10.1 Comparison of the Norms $|u|_{p(\cdot)}$ and $|u_*|_{m(\cdot)}$ with $m = p^*(\cdot)$ or $m = p_*(\cdot)$

For replacing the equimeasurabilities, we have

Theorem 10.1.1 *Let $u : \Omega \to \mathbb{R}_+$, $p : \Omega \to [1, +\infty[$ be two measurable functions. We denote by p^* the increasing rearrangement of p, and p_* its decreasing rearrangement. Thus*

$$p^*(s) = -(-p)_*(s), \qquad s \in \Omega_* =]0, |\Omega|[.$$

Then $$\frac{1}{1+|\Omega|}||u_*||_{p^*(\cdot)} \leqslant ||u||_{p(\cdot)} \leqslant (1+|\Omega|)|u_*|_{p_*(\cdot)}$$
From this relation we deduce

$$\frac{1}{2(1+|\Omega|)}|u_*|_{p^*(\cdot)} \leqslant |u|_{p(\cdot)} \leqslant 2(1+|\Omega|)|u_*|_{p_*(\cdot)}$$

Proof We will need the following lemma

Lemma 10.1.1 *Under the same hypotheses as Theorem 10.1.1, we have for all $\lambda > 0$:*

$$\int_{\{u_*>\lambda\}} \left(\frac{u_*}{\lambda}\right)^{p^*(s)}(s)ds \leqslant \int_{\{u>\lambda\}} \left(\frac{u}{\lambda}\right)^{p(x)}(x)dx \leqslant \int_{\{u_*>\lambda\}} \left(\frac{u_*}{\lambda}\right)^{p_*(s)}(s)ds.$$

Proof We use Hardy–Littlewood inequalities:

$$\text{for } u \geqslant 0, v \geqslant 0, \int_{\Omega_*} u_* v^* ds \leqslant \int_{\Omega} uv dx \leqslant \int_{\Omega_*} u_* v_* ds,$$

and the fact that if $\psi : \mathbb{R} \to \mathbb{R}$ is a non-decreasing function then $(\psi(u))_* = \psi(u_*)$ almost everywhere in Ω_*. We define $H(\sigma) = \begin{cases} 1 & \text{if } \sigma > 0, \\ 0 & \text{otherwise} \end{cases}$ for all $m \geqslant 0$, for all $\lambda > 0$

$$\sum_{k=0}^{m} \frac{1}{k!} \int_{\{u>\lambda\}} p(x)^k \left(\text{Log}\, \frac{u}{\lambda}\right)^k dx = \sum_{k=0}^{m} \frac{1}{k!} \int_{\Omega} p(x))^k \left[H(u-\lambda)\text{Log}\left(\frac{u}{\lambda}\right)\right]^k dx$$

$$\leqslant \sum_{k=0}^{m} \frac{1}{k!} \int_{\Omega_*} p_*(s)^k \left[H(u_*-\lambda)\text{Log}\left(\frac{u_*}{\lambda}\right)\right]^k ds \leqslant \int_{\{u_*>\lambda\}} \left(\frac{u_*}{\lambda}\right)^{p_*(s)}(s) ds.$$

By Fatou's Lemma, we have $\int_{\{u>\lambda\}} \left(\frac{u}{\lambda}\right)^{p(x)}(x)dx \leqslant$

$$\lim_{m\to+\infty} \sum_{k=0}^{m} \frac{1}{k!} \int_{\{u>\lambda\}} p(x)^k \left(\text{Log}\, \frac{u}{\lambda}\right)^k dx \leqslant \int_{\{u_*>\lambda\}} \left(\frac{u_*}{\lambda}\right)^{p_*(s)} (s) ds.$$

The same argument shows that: $\sum_{k=0}^{m} \frac{1}{k!} \int_{\Omega_*} p^*(s)^k \left[H(u_*-\lambda)\text{Log}\left(\frac{u_*}{\lambda}\right)\right]^k ds$

$$\leqslant \sum_{k=0}^{m} \frac{1}{k!} \int_{\Omega} p(x))^k \left[H(u-\lambda)\text{Log}\left(\frac{u}{\lambda}\right)\right]^k dx \leqslant \int_{\{u>\lambda\}} \left(\frac{u}{\lambda}\right)^{p(x)} (x) dx.$$

Proof of Theorem 10.1.1 We have, using Lemma 10.1.1 $\int_{\Omega} \left(\frac{u}{\lambda}\right)^{p(x)} (x)dx$

$$\leqslant |\{u \leqslant \lambda\}| + \int_{\{u_*>\lambda\}} \left(\frac{u_*}{\lambda}\right)^{p_*(s)} (s)ds \leqslant |\Omega| + \Phi_{p_*}\left(\frac{u_*}{\lambda}\right).$$

Hence for all $\lambda > 0$, $\lambda\left(1+\Phi_p\left(\frac{u}{\lambda}\right)\right)$

$$\leqslant \lambda|\Omega| + \lambda\left(1+\Phi_{p_*}\left(\frac{u_*}{\lambda}\right)\right) \leqslant \lambda\left(1+\Phi_{p_*}\left(\frac{u_*}{\lambda}\right)\right)(1+|\Omega|),$$

which implies:

$$||u||_{p(\cdot)} \leqslant (1+|\Omega|)||u_*||_{p_*(\cdot)}.$$

The same argument can be used for the second inequality:

$$\lambda\left(1+\Phi_{p^*}\left(\frac{u_*}{\lambda}\right)\right) \leqslant (1+|\Omega|)\lambda\left(1+\Phi_p\left(\frac{u}{\lambda}\right)\right) \qquad \forall\lambda > 0.$$

◇

Important Note
Here is a function $u \in L^{p(\cdot)}(\Omega)$ such that

$$||u_*||_{p_*(\cdot)} = +\infty.$$

Let $\Omega =]0, 1[$, $u(x) = x^{-\frac{1}{3}}$ and $p(x) = \begin{cases} 2 & if\ 0 \leqslant x \leqslant \frac{1}{2}, \\ 3 & if\ \frac{1}{2} < x \leqslant 1. \end{cases}$

Since $u \in L^{p(\cdot)}(\Omega)$,

we have $u_*(s) = s^{-\frac{1}{3}}$, $s \in]0, 1]$ and $p_*(s) = \begin{cases} 3 & if\ 0 \leqslant s < \frac{1}{2}, \\ 2 & if\ \frac{1}{2} \leqslant s \leqslant 1. \end{cases}$. Therefore

$$\int_0^1 u_*^{p_*(s)}(s)ds = +\infty.$$

See also Cruz and Uribe [49] for similar examples. ◇

The following inequalities are the natural substitutions properties of invariant rearrangement for usual Lebesgue spaces.

Proposition 10.1.1 *Let $u : \Omega \to \mathbb{R}_+$ and $p : \Omega \to [1, +\infty[$ be two measurable functions and $F : \mathbb{R}_+ \to \mathbb{R}_+$ non-decreasing with $F(0) > 0$. Then:*

$$\int_{\Omega_*} \left(\frac{F(u_*)}{F(0)}\right)^{p^*(s)} (s)ds \leqslant \int_{\Omega} \left(\frac{F(u)}{F(0)}\right)^{p(x)} (x)dx$$

$$\leqslant \int_{\Omega_*} \left(\frac{F(u_*)}{F(0)}\right)^{p_*(s)} (s)ds = \Phi_{p_*}\left(\frac{F(u_*)}{F(0)}\right).$$

In particular, if $\Phi_{p_*}\left(\frac{F(u_*)}{F(0)}\right) < +\infty$ then

$$F(u) \in L^{p(\cdot)}(\Omega) \text{ and } ||F(u)||_{p(\cdot)} \leqslant F(0)\left(1 + \Phi_{p_*}\left(\frac{F(u_*)}{F(0)}\right)\right).$$

Proof The proof of these inequalities follows the same arguments as before, so we leave it to the reader.

The second proposition follows the definition of the Amemiya norm. ◇

In the following, we denote by: $c > 0$ different constants

and we write c_p when we want to show the dependence of c as a function of p.

A consequence of Theorem 10.1.1 is:

Theorem 10.1.2 *Let $p :\in \Omega \to [1, +\infty[$ be a bounded measurable function.*

We assume that the increasing rearrangement of p^* of p satisfies: $1 < p^*(0)$ and that in the neighborhood of the origin 0 we have $|p^*(s) - p^*(t)|| \leqslant \dfrac{A}{|\mathrm{Log}\,|s-t||}$, for $A > 0$, and for $t,\ s$ in this neighborhood. We have: For all $v \geqslant 0, u \geqslant 0$ in $L^1(\Omega)$,

$$\text{if } v \in L^{p(\cdot)}(\Omega) \text{then} (v_{*u})_* \in L^{p^*(\cdot)}(\Omega_*).$$

Furthermore, there exists a constant $c > 0$ such that:

$$||(v_{*u})_*||_{p^*(\cdot)} \leqslant ||(v_{*u})_{**}||_{p^*(\cdot)} \leqslant c||v_*||_{p^*(\cdot)} \leqslant c||v||_{p(\cdot)}$$

Proof We recall the following result due to Kokilashvili and Samko [104]:

Lemma 10.1.2 *Under the same hypotheses on p^*, we have: $\forall f \geqslant 0$ measurable function on Ω_* such that if $f_* \in L^{p^*(\cdot)}(\Omega_*)$ then $f_{**} \in L^{p^*(\cdot)}(\Omega_*)$.*
In addition, there exists a constant $c > 0$ such that:

$$||f_*||_{p^*(\cdot)} \leqslant ||f_{**}||_{p^*(\cdot)} \leqslant c||f_*||_{p^*(\cdot)}.$$

Proof of Theorem 10.1.2 Since $v \geqslant 0$ we have $v_{*u} \geqslant 0$, so

$$||(v_{*u})_*||_{p^*(\cdot)} \leqslant ||(v_{*u})_{**}||_{p^*(\cdot)},$$

since $(v_{*u})_{**} \leqslant v_{**}$ (see Chap. 2) by Lemma 10.1.2

$$||(v_{*u})_{**}||_{p^*(\cdot)} \leqslant ||v_{**}||_{p^*(\cdot)} \leqslant c||v_*||_{p^*(\cdot)}.$$

We conclude by applying Theorem 10.1.1. ◇

10.2 A Study of the Continuity of the Maps $u \in L^{p(\cdot)}(\Omega) \to u_* \in L^{m(\cdot)}(\Omega_*)$ and $v \in L^{p(\cdot)} \to (|v|_{*u})_* \in L^{m(\cdot)}(\Omega_*)$

We only consider $m(\cdot) = p^*(\cdot)$ and $m = p_*(\cdot)$. As we saw in the previous example, if we rearrange the exponent p in a decreasing way p_* and the function u in a decreasing way then the function $u_*^{p_*}$ can be non-integrable.

Since rearranging a positive function in a decreasing way returns the singularities to the origin $t = 0$, it is for this reason that we will put a weight function, in this case $w(t) = t^\alpha$ on the neighborhood of $t = 0$. When p is bounded and $p_m = \operatorname{ess\,inf}_{x\in\Omega} p(x)$, $p_M = \operatorname{ess\,sup}_{x\in\Omega}(x)$ we will see that $\alpha = \dfrac{p_M}{p_m} - 1$ is appropriate.

The first lemma that will lead us to the contraction property in a weighted space is the following:

Lemma 10.2.1 *Let u, v be in $L^{p(\cdot)}(\Omega)$ and $p : \Omega \to]1, +\infty[$, a bounded measurable function with $p_M = \operatorname{ess\,sup}_{\Omega} p$, $p_m = \operatorname{ess\,inf}_{\Omega} p$. Suppose for simplicity that $|\Omega| = 1$, and for $0 < a \leqslant b < 1$*

(continued)

Lemma 10.2.1 (continued)

$$K(a,b) = \underset{a \leqslant \sigma \leqslant b}{\text{Max}} \{|u_*(\sigma) - v_*(\sigma)|; 1\}.$$

Then for all $a \leqslant b$, *we have:*

$$\int_a^b |u_*(\sigma) - v_*(\sigma)|^{p_*(\sigma)} d\sigma \leqslant K(a,b)^{p_M - p_m} \int_a^b |u_* - v_*|^{p_m}(\sigma) d\sigma). \tag{10.1}$$

The same result holds for p^*.

Proof We have : $\displaystyle \frac{1}{K(a,b)^{p_M}} \int_a^b |u_* - v_*|^{p_*(\sigma)} d\sigma \leqslant \int_a^b \left(\frac{|u_* - v_*|}{K(a,b)}\right)^{p_*(\sigma)} d\sigma$

$$\leqslant \int_a^b \frac{|u_* - v_*|^{p_m}(\sigma)}{K(a,b)^{p_m}} d\sigma,$$

and the lemma follows.

The same proof holds by replacing p_* by p^*. ◇

Lemma 10.2.2 *Under the same hypotheses as for Lemma 10.2.1, we have:*

1. $\displaystyle \int_a^b |\sigma^{\frac{p_M}{p_m} - 1}\big(u_*(\sigma) - v_*(\sigma)\big)|^{p_*(\sigma)} d\sigma \leqslant$

$$\big[bK(a,b)\big]^{p_M - p_m} \int_a^b |u_* - v_*|^{p_m} d\sigma.$$

2. $\displaystyle \int_a^b \left|(1-\sigma)^{\frac{p_M}{p_m} - 1}\big(u_*(\sigma) - v_*(\sigma)\big)\right|^{p_*(\sigma)} d\sigma$

$$\leqslant \Big[(1-a)K(a,b)\Big]^{p_M - p_m} \int_a^b |u_* - v_*|^{p_m} d\sigma.$$

Likewise for p^*.

Proof It suffices to show item 1, since the arguments are the same for item 2.

$\displaystyle \int_a^b |\sigma^{\frac{p_M}{p_m} - 1}(u_*(\sigma) - v_*(\sigma))|^{p_*(\sigma)} d\sigma$

$$\leqslant \int_a^b \sigma^{p_M - p_m} |u_* - v_*|^{p_*(\sigma)}(\sigma) d\sigma$$

$$\leqslant b^{p_M - p_m} \int_a^b |u_* - v_*|^{p_*(\sigma)}(\sigma) d\sigma$$

$$\leqslant \left[bK(a,b)\right]^{p_M - p_m} \int_a^b |u_* - v_*|^{p_m} d\sigma$$

(by Lemma 10.2.1). ◇

Lemma 10.2.3 *Let* $u \in L^1(\Omega)$.

1. *If* $u \geqslant 0$ *then:* $\quad tu_*(t) \leqslant |u|_{L^1(\Omega)}$.
2. *If u is a function that changes sign then for all* $t \in]0,1[$*:*

$$\text{Min}(t; 1-t)|u_*(t)| \leqslant |u|_{L^1(\Omega)}.$$

Proof Let $t \in]0,1[$. If $u_*(t) \geqslant 0$ then

$$u_*(t) \leqslant u_*(\sigma),\ \forall \sigma \leqslant t : tu_*(t) \leqslant \int_0^t u_*(\sigma) d\sigma \leqslant |u|_{L^1(\Omega)}.$$

If $u_*(t) \leqslant 0$ then $u_*(\sigma) \leqslant u_*(t)$ for all $t \leqslant \sigma$, which implies:

$$0 \leqslant -(1-t)u_*(t) \leqslant \int_t^1 |u_*(\sigma)| d\sigma \leqslant |u|_{L^1(\Omega)}.$$

This proves the lemma. ◇

Lemma 10.2.4 *Let* $u \in L^1(\Omega),\ 0 < a \leqslant b < 1$. *Then:*

$$b \operatorname*{Max}_{a \leqslant \sigma \leqslant b} |u_*(\sigma)| \leqslant \frac{b}{a} \text{Max}\left\{a|u_*(a)|; b|u_*(b)|\right\},$$

$$(1-a) \operatorname*{Max}_{a \leqslant \sigma \leqslant b} |u_*(\sigma)| \leqslant \frac{1-a}{1-b} \text{Max}\left\{(1-a)|u_*(a)|; (1-b)|u_*(b)|\right\}.$$

Proof We have: $\underset{a\leqslant\sigma\leqslant b}{\text{Max}} |u_*(\sigma)| = \text{Max}\left\{|u_*(a)|;\ |u_*(b)|\right\}$, so:

$$b \underset{a\leqslant\sigma\leqslant b}{\text{Max}} |u_*(\sigma)| = \text{Max}\left\{b|u_*(a)|;\ b|u_*(b)|\right\}.$$

We have $b|u_*(a)| = \dfrac{b}{a}a|u_*(a)|$ and $\dfrac{b}{a} \geqslant 1$, so we deduce

$$b \underset{a\leqslant\sigma\leqslant b}{\text{Max}} |u_*(\sigma)| \leqslant \frac{b}{a}\,\text{Max}\left\{a|u_*(a)|;\ b|u_*(b)|\right\}.$$

The arguments are the same for the second item. ◇

Lemma 10.2.5 *Let $p : \Omega \to [1, +\infty[$ be a measurable function. Then $\forall u \in L^{p(\cdot)}(\Omega)$*

1. $\displaystyle\int_\Omega |u(x)|^{p(x)}dx \leqslant ||u||_{p(\cdot)}^{p_m}$ *if* $||u||_{p(\cdot)} \leqslant 1$.
2. *If p is bounded, then* $2^{-p_M} \min\left(||u||_{p(\cdot)}^{p_m};\ ||u||_{p(\cdot)}^{p_M}\right)$

$$\leqslant \int_\Omega |u(x)|^{p(x)}dx \leqslant \text{Max}\left(||u||_{p(\cdot)}^{p_m};\ ||u||_{p(\cdot)}^{p_M}\right),$$

with $p_m = \underset{\Omega}{ess\ inf}\, p$, $p_M = \underset{\Omega}{ess\ sup}\, p$.

The proof of this lemma is a direct consequence of the definition (see [66]). We now introduce:

Definition 10.2.1 Let $\phi : \mathbb{R}_+ \to \mathbb{R}_+$ be a continuous map, with $\phi(0) = 0$.
Let $(X, ||\cdot||_X)$, $(Y, ||\cdot||_Y)$ be two normed spaces and let T be an operator from X to Y. We will say that T is locally ϕ-Hölderian if there is a continuous map $g : X \times X \to \mathbb{R}_+$ such that for all u in X there exists a neighborhood $\mathcal{V}_u$, open in X, such that $\forall v \in \mathcal{V}_u$, we have

$$||Tu - Tv||_Y \leqslant g(u; v)\phi\big(||u - v||_X\big).$$

Theorem 10.2.1 *Let p be as in Lemma 10.2.1. For $u,\ v \in L^{p(\cdot)}_+(\Omega)$, we have:*

$$\left[\int_0^1 |t^{\frac{p_M}{p_m}-1}(u_*(t)-v_*(t))|^{p_*(t)}dt\right]^{\frac{1}{p_m}}$$

$$\leqslant c_p\Big(|u|_{L^1(\Omega)}+|v|_{L^1(\Omega)}+1\Big)^{\frac{p_M}{p_m}-1}||u-v||_{p(\cdot)}.$$

The same inequality remains for p^*. In particular, the map

$$u \in L^{p(\cdot)}_+(\Omega) \to t^{\frac{p_M}{p_m}-1}u_* \in L^{p_*(\cdot)}_+(\Omega)(resp L^{p^*(\cdot)}_+(\Omega))$$

is locally

ϕ-Hölderian, that is to say

$$||t^{\frac{p_M}{p_m}-1}(u_*-v_*)||_{p_*(\cdot)} \leqslant g_0(|u|_{L^1(\Omega)}+|v|_{L^1(\Omega)})\phi((||u-v)||_{p(\cdot)})$$

where $\phi(\sigma)=\mathrm{Max}(\sigma;\sigma^{\frac{p_m}{p_M}})$,

$$g_0(\sigma)=c_p\,\mathrm{Max}((\sigma+1)^{\frac{p_M}{p_m}-1},(\sigma+1)^{1-\frac{p_m}{p_M}}).$$

Additionally,

$$t^{\frac{p_M}{p_m}-1}u_* \in L^{p_*(\cdot)}_+(\Omega_*)\ if\ u \in L^{p(\cdot)}_+(\Omega).$$

We have the same conclusion for p^*.

Proof Let $u \geqslant 0$, $v \geqslant 0$ be in $L^{p(\cdot)}(\Omega)$. Let n be a (large) integer ($n \geqslant 2$)) and consider a subdivision of the interval $]0,1[$, with nodes $a_j = \dfrac{j}{n}$, $j \geqslant 1$ such that $\dfrac{a_{j+1}}{a_j} = \dfrac{j+1}{j} \leqslant 2$.

Applying Lemma 10.2.2, we have for $j = 1,\dots,n-1$

$$\int_{a_j}^{a_{j+1}} |\sigma^{\frac{p_M}{p_m}-1}\big(u_*(\sigma)-v_*(\sigma)\big)|^{p_*(\sigma)}d\sigma$$

$$\leqslant [a_{j+1}K(a_j,a_{j+1})]^{p_M-p_m}\int_{a_j}^{a_{j+1}}|u_*-v_*|^{p_m}d\sigma. \tag{10.2}$$

But we have

$$\begin{aligned} a_{j+1}K(a_j, a_{j+1}) &\leqslant a_{j+1} + a_{j+1}u_*(a_j) + a_{j+1}v_*(a_j) \\ &\leqslant 1 + \frac{a_{j+1}}{a_j}[a_j u_*(a_j) + a_j v_*(a_j)] \\ &\leqslant 2(1 + |u|_{L^1(\Omega)} + |v|_{L^1(\Omega)}) \\ &\quad \text{(following Lemma 10.2.3)} \end{aligned}$$

The last inequality, relation (10.2) and $||u - v||_{p_m} \leqslant c||u - v||_{p(\cdot)}$ imply for all $n \geqslant 2$:

$$\left[\int_{\frac{1}{n}}^{1} |\sigma^{\frac{p_M}{p_m}-1}(u_*(\sigma) - v_*(\sigma)|^{p_*(\sigma)} d\sigma\right]^{\frac{1}{p_m}}$$

$$\leqslant c_p(|u|_{L^1(\Omega)} + |v|_{L^1(\Omega)} + 1)^{\frac{p_M}{p_m}-1} \cdot ||u - v||_{p(\cdot)} \tag{10.3}$$

This shows the first item of Theorem 10.2.1 by letting $n \to \infty$.

Using Lemma 10.2.5 and relation (10.3), we have:

$$\min\left(||\sigma^{\frac{p_M}{p_m}-1}(u_* - v_*)||_{p_*(\cdot)}; ||\sigma^{\frac{p_M}{p_m}-1}(u_* - v_*)||_{p_*(\cdot)}^{\frac{p_M}{p_m}}\right)$$

$$\leqslant c_p(|u|_{L^1(\Omega)} + |v|_{L^1(\Omega)} + 1)^{\frac{p_M}{p_m}-1} ||u - v||_{p(\cdot)} \tag{10.4}$$

which implies

$$||\sigma^{-1+\frac{p_M}{p_m}}(u_* - v_*)||_{p_*(\cdot)} \leqslant g_0(|u|_{L^1(\Omega)} + |v_{L^1(\Omega)})\phi(||u - v||_{p(\cdot)}), \tag{10.5}$$

with g_0 and ϕ identical to the above. Since $L^{p(\cdot)}(\Omega)$ is continuously embedded in $L^1(\Omega)$, we can conclude that: the map $u \in L_+^{p(\cdot)}(\Omega) \to t^{-1+\frac{p_M}{p_m}} u_*$ is locally ϕ-Hölderian on $X = L_+^{p(\cdot)}(\Omega)$ in $Y = L_+^{p_*(\cdot)}(\Omega_*)$ (resp $L_+^{p^*(\cdot)}(\Omega_*)$).

◇

Corollary 10.2.1 (Of Theorem 10.2.1) *If* $\left|\{x \in \Omega : p(x) = \operatorname{ess\,inf}_{\Omega} p\}\right| > 0$ *then*

the map $u \in L_+^{p(\cdot)}(\Omega) \to u_* \in L_+^{p^*(\cdot)}(\Omega_*)$

is locally ϕ*-Hölderian,*

ϕ *defined as in Theorem 10.2.1.*

Proof If $|\{p = p_m\}| > 0$ then on $[0, |p = p_m|) = [0, a)$, $p^*(s) = p_m$

$$||(u_* - v_*)||_{p^*(\cdot)} \leqslant ||(u_* - v_*)\chi_{[0,a)}||_{p^*(\cdot)} + ||(u_* - v_*)\chi_{[a,|\Omega|]}||_{p^*(\cdot)}.$$

We have:

$$||(u_* - v_*)\chi_{[0,a)}||_{p^*(\cdot)} = ||(u_* - v_*)\chi_{[0,a)}||_{p_m}$$

$$\leqslant ||u_* - v_*||_{p_m} \leqslant ||u - v||_{p_m} \leqslant c||u - v||_{p(\cdot)}.$$

For the second term:

$$||(u_* - v_*)\chi_{[a,|\Omega|}||_{p^*} \leqslant a^{-\frac{p_M}{p_m}+1} \cdot ||t^{-1+\frac{p_M}{p_m}}(u_* - v_*)||_{p^*(\cdot)}$$

$$\leqslant c_a g_0(|u|_{L^1(\Omega)} + |v|_{L^1(\Omega)}) \cdot \phi(||u - v||_{p^*(\cdot)}),$$

so

$$||u_* - v_*||_{p^*(\cdot)} \leqslant c_a g_0(|u|_{L^1(\Omega)} + |v|_{L^1(\Omega)})\phi(||u - v||_{p(\cdot)}).$$

$\diamond$

Next, we prove a theorem analogous to Theorem 10.2.1 but where the functions u and v might change sign. The idea of the proof is the same but the weight $t^{\frac{p_M}{p_m}-1}$ is replaced by $k(t) = \text{Min}(t, 1 - t)^{\frac{p_M}{p_m}-1}$.

Theorem 10.2.2 *Let* $k(t) = \text{Min}(t, 1 - t)^{\frac{p_M}{p_m}-1}$, $t \in]0.1[$. *Then for all* u *and* v *in* $L^{p(\cdot)}(\Omega)$

$$||k \cdot (u_* - v_*)||_{p_*(\cdot)} \leqslant g_0(|u|_{L^1(\Omega)} + |v|_{L^1(\Omega)})\phi(||u - v||_{p(\cdot)})$$

where g_0 *and* ϕ *are the same as in Theorem 10.2.1.*

In particular,

$$k \cdot u_* \in L^{p_*(\cdot)}(\Omega_*) \text{ if } u \in L^{p(\cdot)}(\Omega)$$

and the map

$$u \in L^{p(\cdot)}(\Omega) \to k \cdot u_* \in L^{p_*(\cdot)}(\Omega_*) \text{is locally} \phi\text{-Höldérian.}$$

We have the same conclusion for p^*.

Proof We sketch the proof, because it resembles that of Theorem 10.2.1.

On $[0, \frac{1}{2}]$, $k(t) = t^{\frac{p_M}{p_m}-1}$ and the arguments used in Theorem 10.2.1 imply:

$$\int_0^{\frac{1}{2}} \Big|k(\sigma)\big(u_*(\sigma) - v_*(\sigma)\big)\Big|^{p_*(\sigma)} d\sigma$$

$$\leqslant c_p\big(|u|_{L^1(\Omega)} + |v|_{L^1(\Omega)} + 1\big)^{p_M - p_m} ||u - v||^{p_m}_{p(\cdot)}. \tag{10.6}$$

As for $\left[\frac{1}{2}, 1\right]$, we define $b_j = \frac{1}{2} + \frac{j}{2n}$, $j = 0, \ldots, n$, and since $k(t) = (1-t)^{\frac{p_M}{p_m}-1}$ we apply Lemma 10.2.2 above to get:

$$\int_{b_j}^{b_{j+1}} \Big|(1-\sigma)^{\frac{p_M}{p_m}-1}\big(u_*(\sigma) - v_*(\sigma)\big)\Big|^{p_*(\sigma)} d\sigma$$

$$\leqslant \Big[(1 - b_j)K(b_j, b_{j+1})\Big]^{p_M - p_m} \int_{b_j}^{b_{j+1}} |u_* - v_*|^{p_m} d\sigma \tag{10.7}$$

Assuming that $j \leqslant n - 2$, we have $1 \leqslant \dfrac{1 - b_j}{1 - b_{j+1}} \leqslant 2$, $(1 - b_j)K(b_j, b_{j+1})$

$$\leqslant 1 + \frac{1 - b_j}{1 - b_{j+1}}\Big[\operatorname{Max}\big\{(1 - b_j)|u_*(b_j)|; (1 - b_{j+1})|u_*(b_{j+1})|\big\}$$

$$+ \operatorname{Max}\big\{(1 - b_j)|v_*(b_j)|; (1 - b_{j+1})|v_*(b_{j+1})|\big\}\Big]. \tag{10.8}$$

But, we have $1 - b_k = \operatorname{Min}(b_k, 1 - b_k)$, $\forall k$, so we deduce from Lemma 10.2.3 and the relation (10.8) that

$$(1 - b_j)K(b_j, b_{j+1}) \leqslant 2\big(|u|_{L^1(\Omega)} + |v|_{L^1(\Omega)} + 1\big). \tag{10.9}$$

Also, relations (10.7) and (10.9) imply: $\forall n \geqslant 2$

$$\int_{\frac{1}{2}}^{1-\frac{1}{n}} |k \cdot (u_* - v_*)|^{p_*(\sigma)} d\sigma \leqslant c_p\big(|u|_{L^1(\Omega)} + |v|_{L^1(\Omega)} + 1\big)^{p_M - p_m} ||u - v||^{p_m}_{p(\cdot)}. \tag{10.10}$$

Both relations (10.6) and (10.10) show that

$$\left[\int_0^1 |k \cdot (u_* - v_*)|^{p_*(\sigma)} d\sigma\right]^{\frac{1}{p_m}} \leqslant c_p\big(|u|_{L^1(\Omega)} + |v|_{L^1(\Omega)} + 1\big)^{\frac{p_M}{p_m}-1} ||u - v||_{p(\cdot)}. \tag{10.11}$$

The theorem follows as in Theorem10.2.1. ◇

The weight function is especially needed when using the decreasing rearrangement p_*. When we take p^* as the increasing rearrangement, we then have a contraction property as indicated by the following theorem:

Theorem 10.2.3 *Let $p : \Omega \to [1, +\infty[$ be a bounded measurable function such that the increasing rearrangement p^* of p satisfies in the neighborhood of zero*

$$p^*(0) > 1 \text{ and } |p^*(t) - p^*(\sigma)|\,|\mathrm{Log}\,|t - \sigma|| \leqslant A$$

for some constant $A > 0$. Then there exists a constant $c_L > 0$ such that (if we define $g_{**}(s) = \frac{1}{s}\int_0^s g_*(\sigma)\mathrm{d}s$ for $g \in L^1(\Omega)$)

$$\left\| \Big[(u+v)_* - u_*\Big]_{**} \right\|_{p^*(\cdot)} \leqslant c_L \|v\|_{p(\cdot)}, \quad \text{for all } u \in L^1(\Omega),\ \forall\, v \in L_+^{p(\cdot)}(\Omega).$$

Corollary 10.2.2 (Of Theorem 10.2.3) *Under the same hypotheses as for Theorem 4.4.5, and if u and v are in $L^{p(\cdot)}(\Omega)$, we have (where c_L is the same constant as in Theorem 4.4.5)*

$$\Big\|\, |u_* - v_*|_* \Big\|_{p^*(\cdot)} \leqslant \Big\|\, |u_* - v_*|_{**} \Big\|_{p^*(\cdot)} \leqslant c_L \|u - v\|_{p(\cdot)}.$$

Proof of Theorem 10.2.3 We shall proceed by approximation, noticing first that if the main inequality of the theorem is true for u and v non-negative and bounded on Ω , then the result follows. Indeed, if

$$u \in L^1(\Omega),\ v \in L_+^{p(\cdot)}(\Omega),$$

then for any $k \geqslant 1$, $u_k = T_k(u) + k \geqslant 0$ (resp. $v_k = T_k(v)$) is non-negative and bounded, therefore the inequality holds true:

$$\left\| \Big[(u_k + v_k)_* - u_{k*}\Big]_{**} \right\|_{p^*(\cdot)} \leqslant c_L \|v_k\|_{p(\cdot)} \leqslant c_L \|v\|_{p(\cdot)}.$$

Using Fatou's lemma, since

$$\Big[(u_k + v_k)_* - u_{k*}\Big]_{**}(s) \to \Big[(u+v)_* - u_*\Big]_{**}(s)$$

everywhere in Ω_*, we derive the inequality in that case. We can also assume that the open bounded domain Ω is a very smooth domain, say C^∞, since we know when using a C^∞ function equivalent of the distance function, we can produce a sequence of domains $(\Omega_n)_n$ that are as smooth as we want, such that:

$$\overline{\Omega}_n \subset \Omega_{n+1}, \Omega = \cup_{n\geqslant 0}\Omega_n$$

(see [136]). First, we may replace Ω by Ω_n. Introducing sequences $(u_n)_n$; $(v_n)_n$ of non-negative bounded functions satisfying

$u_n = u$, $v_n = v$, $p_n = p$ on Ω_n and $u_n = v_n = p_n = 0$ elsewhere, their decreasing rearrangements satisfy $u_{n*} = u_*$, $v_{n*} = v_*$, $p_{n*} = p_*$ on $]0, \Omega_{n*}[$ and $u_{n*} = v_{n*} = p_{n*} = 0$ elsewhere. Therefore we have the following equality:

$$\int_{\Omega_*} \left[\frac{(u_n + v_n)_* - u_{n*}}{\lambda}\right]^{p_{n*}} ds = \int_{\Omega_{n*}} \left[\frac{(u + v)_* - u_*}{\lambda}\right]^{p_*} ds \doteq J$$

and obviously,

$$J \leqslant \int_{\Omega_*} \left[\frac{(u + v)_* - u_*}{\lambda}\right]^{p_*} ds.$$

Since the constant c_L is independent of n, if we show that the inequality is true for the smooth domain Ω_n, the above inequalities imply the result, after letting n go to ∞. Again, using standard approximation (by mollifier for instance), we can replace the boundedness of u and v by assuming $u \in C_c^\infty(\Omega)$ and $v \in C_c^\infty(\Omega)$, $v \geqslant 0$, $u \geqslant 0$. If the inequality is true under those assumptions, then using Fatou's lemma we shall derive the inequality.

Next, using a spectral basis (see the last exercise of the next chapter), we can prove the following approximation:

Lemma 10.2.6 *Assume that the domain Ω is of class C^∞, let*

$$u \in C_c^\infty(\Omega),\ v \in C_c^\infty(\Omega),\ v \geqslant 0,\ u \geqslant 0,\ m > N.$$

Then, there is a sequence $(u_n; v_n)_n$ satisfying:

1. $u_n, v_n \in W^{m,2}(\Omega)$, $v_n \geqslant 0$, $u_n \geqslant 0$
2. measure $\left\{x \in \Omega : \nabla(u_n + tv_n)(x) = 0\right\} = 0$ *for* $t \in [0, 1]$.
3. $||u_n - u||_\infty \to 0$, $||v_n - v||_\infty \to 0$

Proof We know that there is a sequence of functions $(\phi_j)_j$ and real numbers $(\lambda_j)_j$ such that

$$(-\Delta)^m \phi_j = \lambda_j \phi_j, , \phi_j \in C^\infty(\overline{\Omega}) \cap W_0^{m,2}(\Omega), \forall j \geqslant 1.$$

Let us consider $P_n(u) = \sum_{j=1}^{j=n} ((u, \phi_j))\phi_j$, $P_n(v) = \sum_{j=1}^{j=n} ((v, \phi_j))\phi_j$ where $((\cdot, \cdot))$ denotes the usual scalar product in the Hilbert space $W^{m,2}(\Omega)$. Then, the sequences $(P_n(u))_n$ $(P_n(v))_n$ remain in a bounded set of $W_0^{m,2}(\Omega)$ By usual compactness, we deduce that

$$\varepsilon_n = ||P_n(u) - u||_\infty \to 0, \eta_n = ||P_n(v) - v||_\infty \to 0.$$

Setting $u_n = P_n(u) + \varepsilon_n$, $v_n = P_n(v) + \eta_n$, for large $n \geqslant n_0$, the right sequence satisfies the conditions of the lemmata, (see the last exercise of the next chapter for an idea the proof of the second item). Note that the ϕ_j are analytic in Ω . ◇

Thanks to this approximation lemma, it is sufficient to prove the main inequality of the theorem for $u, v \in W^{m,2}(\Omega)$, $u \geqslant 0$, $v \geqslant 0$ with measure $\left\{x \in \Omega : \nabla(u + tv)(x) = 0\right\} = 0$ for $t \in [0, 1]$.

Due to the last condition, $u + tv$ satisfies the conditions of Corollary 7.1.2 for all t and then the map

$$\begin{aligned} [0, 1] &\to L^1(\Omega_*) \\ t &\mapsto v_{*(u+tv)} \end{aligned} \text{ is uniformly continuous.}$$

For $\sigma \in \overline{\Omega}_*$, we define $g(\sigma) = \int_0^1 v_{*(u+tv)}(\sigma)dt$. We know by Lemma 7.3.1 that

$$(u + v)_*(\sigma) - u_*(\sigma) = g(\sigma) \quad \forall \sigma \in \overline{\Omega}_*. \tag{10.12}$$

Consequently, we have

$$\frac{1}{s}\int_0^s \Big[(u + v)_*(\sigma) - u_*(\sigma)\Big]_* ds = g_{**}(s). \tag{10.13}$$

We consider $m \in \mathbb{N} - \{0\}$ and $t_i = \frac{i}{m}$, $i = 0, \ldots, m$. We define

$$g_m(\sigma) = \frac{1}{m}\sum_{i=0}^{m-1} v_{*(u+t_i v)}(\sigma). \tag{10.14}$$

Lemma 10.2.7 *There exists a constant* $c_L > 0$ *such that*

$$\left\| g_{m**} \right\|_{p^*(\cdot)} \leqslant c_L \|v\|_{p(\cdot)} \qquad \text{for all } m.$$

Proof By the convexity of the map $h \to h_{**}(s)$, we have

$$g_{m**}(s) \leqslant \frac{1}{m} \sum_{i=0}^{m-1} \left(v_{*(u+t_i v)} \right)_{**}(s). \tag{10.15}$$

Consequently, we have

$$\left\| g_{m**} \right\|_{p^*(\cdot)} \leqslant \frac{1}{m} \sum_{i=0}^{m-1} \left\| \left(v_{*(u+t_i v)} \right)_{**} \right\|_{p^*(\cdot)}. \tag{10.16}$$

Using Theorem 10.1.2 we deduce

$$\left\| (v_{*(u+t_i v)})_{**} \right\|_{p^*(\cdot)} \leqslant c_L \|v\|_{p(\cdot)}. \tag{10.17}$$

By relation (10.16) and (10.17), we obtain the result. ◇

Lemma 10.2.8 *The sequence* g_m *converges strongly to* g *in* $L^1(\Omega_*)$*. In particular, for all* $s > 0$*, we have*

$$g_{m**}(s) \to g_{**}(s) \text{ as } m \to +\infty.$$

Proof The map $t \in [0,1] \to v_{*(u+tv)} \in L^1(\Omega_*)$ is uniformly continuous. Therefore, for $\varepsilon > 0$, then there exists $\delta > 0$ such that if $m \geqslant \frac{1}{\delta}$, then

$$\int_{\Omega_*} \left| v_{*(u+tv)} - v_{*(u+t_i v)} \right| (\sigma) d\sigma \leqslant \varepsilon. \qquad \text{for} |t - t_i| \leqslant \frac{1}{m}. \tag{10.18}$$

Consequently, we have

$$\int_{\Omega_*} \left| g_m(\sigma) - g(\sigma) \right| d\sigma \leqslant \sum_{i=0}^{m-1} \int_{t_i}^{t_{i+1}} \int_{\Omega_*} \left| v_{*(u+tv)} - v_{*(u+t_i v)} \right| (\sigma) d\sigma \leqslant \varepsilon. \tag{10.19}$$

This shows the first part. The second part follows from the first part and the fact that $\forall s > 0$

$$\left|g_{m**}(s) - g_{**}(s)\right| \leqslant \frac{1}{s}\int_{\Omega_*}\left|g_m(s) - g(\sigma)\right|\mathrm{d}\sigma. \tag{10.20}$$

◇

End of the Proof of Theorem 10.2.3 We have, by the Fatou property applied to the norm on the Banach function space $L^{p^*(\cdot)}(\Omega_*)$,

$$\left\|g_{**}\right\|_{p^*(\cdot)} \leqslant \liminf_{m\to+\infty}\left\|g_{m**}\right\|_{p^*(\cdot)}. \tag{10.21}$$

We conclude from relation (10.13) and Lemma 10.2.7 that

$$\left\|\left[(u+v)_* - u_*\right]_{**}\right\|_{p^*(\cdot)} \leqslant c_L\|v\|_{p(\cdot)}. \tag{10.22}$$

◇

Proof of the Corollary We replace v by $v - u$ and we note that for all $s \in \Omega_*$

$$|u_* - v_*|_*(s) \leqslant |u_* - v_*|_{**}(s),$$

so that the result follows if u and v are in $L_+^{p(\cdot)}(\Omega)$. ◇

Otherwise, we can consider $T_k(\sigma) = \min\left(|\sigma|, k\right)\operatorname{sign}(\sigma)$, $\sigma \in \mathbb{R}$ and u_k (resp. v_k) defined by $u_k = T_k(u) + k \geqslant 0$ (resp $v_k = T_k(v) + k$). Consequently

$$\left\|\,|T_k(u_*) - T_k(v_*)|_{**}\right\|_{p^*(\cdot)} \leqslant c_L\left\|T_k(u) - T_k(v)\right\|_{p(\cdot)} \text{ letting } k \to +\infty,$$

and we have the result. ◇

Corollary 10.2.3 (Of Theorem 10.2.3) *With the same hypotheses as for Theorem 4.4.5, we have for all* v_1, v_2 *and* u *in* $L^{p(\cdot)}(\Omega)$

$$\left\|\,|v_{1*u} - v_{2*u}|_*\right\|_{p^*(\cdot)} \leqslant \left\|\,|v_{1*u} - v_{2*u}|_{**}\right\|_{p^*(\cdot)} \leqslant c_L\|v_1 - v_2\|_{p(\cdot)}.$$

In particular, $v_{1*u} \in L^{p^*(\cdot)}$ *if* v_1 *and* u *are in* $L^{p(\cdot)}(\Omega)$.

Proof From Corollary 10.2.2 of Theorem 10.2.3, we know that for all $\lambda > 0$

$$\left\| \left| \frac{(u + \lambda v_1)_* - (u + \lambda v_2)_*}{\lambda} \right|_{**} \right\|_{p^*(\cdot)} \leqslant c_L \left\| v_1 - v_2 \right\|_{p(\cdot)}. \tag{10.23}$$

Using $g_\lambda(s) = \dfrac{(u + \lambda v_1)_* - (u + \lambda v_2)_*}{\lambda}(s)$, we know (see Theorem 2.1.1 and its Corollary 2.1.1) that g_λ converges weakly to $v_{1*u} - v_{2*u}$ in $L^1(\Omega_*)$. We choose $\lambda \doteq \dfrac{1}{m}$, $m \geqslant 1$. Then, by Mazur's lemma, there exists $(\alpha_{jn})_{j \geqslant n}$, and $\sum_{j=n}^{m_n} \alpha_{jn} = 1$, $\alpha_{jn} \geqslant 0$, $h_n \doteq \sum_{j=n}^{m_n} \alpha_{jn} g_{\frac{1}{j}}$ converges strongly to $v_{1*u} - v_{2*u}$ in $L^1(\Omega_*)$.

Thus $|h_n|_* \to |v_{1*u} - v_{2*u}|_*$ strongly in $L^1(\Omega_*)$ when $n \to +\infty$ and

$$|h|_{**}(s) \xrightarrow[n \to +\infty]{} |v_{1*u} - v_{2*u}|_{**}(s) \text{ for all } s > 0. \tag{10.24}$$

But, we also have

$$|h_n|_{**}(s) \leqslant \sum_{j=n}^{m_n} \alpha_{jn} |g_{\frac{1}{j}}|_{**}(s), \ \forall s \in \Omega_*. \tag{10.25}$$

By (10.23) and (10.25), we deduce

$$\left\| \ |h_n|_{**} \right\|_{p^*(\cdot)} \leqslant \sum_{j=n}^{m_n} \alpha_{jn} \left\| \ |g_{\frac{1}{j}}|_{**} \right\|_{p^*(\cdot)} \leqslant c_L \left\| v_1 - v_2 \right\|_{p(\cdot)}. \tag{10.26}$$

From (10.24) and Fatou's lemma, those two last relations gives:

$$\left\| \ |v_{1*u} - v_{2*u}|_{**} \right\|_{p^*(\cdot)} \leqslant c_L \|v_1 - v_2\|_{p(\cdot)}. \tag{10.27}$$

We always have

$$\left\| \ |v_{1*u} - v_{2*u}|_{*} \right\|_{p^*(\cdot)} \leqslant \left\| \ |v_{1*u} - v_{2*u}|_{**} \right\|_{p^*(\cdot)}, \tag{10.28}$$

thus, from (10.27) and (10.28), we obtain the result. ◇

10.3 Sobolev Injection for Spaces with Variable Exponent

Theorem 10.3.1 *Let $p : \Omega \to]1, +\infty[$ be a bounded measurable function, with*

$1 < p_m = \operatorname{ess\,inf}_{\Omega} p$*, and assume p^* satisfies the condition*

$$|p^*(t) - p^*(s)||\operatorname{Log}|s - t|| \leqslant A \text{near zero.}$$

If $||t^{-\frac{1}{N'}}||_{q*(\cdot)} < +\infty$ then $W^{1,p(\cdot)}(\Omega) \subset L^{\infty}(\Omega) \cap C(\Omega)$ provided that Ω is a connected Lipschitzian bounded open set. In addition, we have for all $x \in \Omega$, and $r > 0$ such that $B(x, r) \subset \Omega$:

$$\operatorname{osc}_{B(x,r)} u \leqslant c_N ||t^{-\frac{1}{N'}}||_{q_{r*}(\cdot)} ||\nabla u \chi_{B(x,r)}||_{p(\cdot)} \ \forall u \in W^{1,p(\cdot)}(\Omega)$$

where $\operatorname{osc}_{B(x,r)} u$ is the oscillation of u in the ball, p_r (resp. q_r) is the restriction of p to $B(x, r)$ and p_r^*(resp. $q_{r*} = \dfrac{p_r^*}{p_r^* - 1}$) is the increasing rearrangement of p_r.

Proof It suffices to show the inequality on the oscillation.

Knowing that $b_1(s) \leqslant c||t^{-\frac{1}{N'}}||_{q_*(\cdot)} < +\infty \ \forall s \in \overline{\Omega}_*$, the estimate in $L^{\infty}(\Omega)$ follows the same reasoning as in Chap. 4 (Theorem 4.2.1) or the reasoning that follows. We let f_r be the restriction to the ball $B(x, r)$ of the function f on Ω.

The proof is based on the following formulas proven in Chap. 4 (Corollary 4.2.2)

$$\operatorname{osc}_{B(x,r)} u \leqslant \frac{\alpha_N^{1-\frac{1}{N}}}{\alpha_{N-1}} \int_0^{\alpha_N r^N} t^{-\frac{1}{N'}} \left(|\nabla u_r|_{*u_r}(t)\right)_* dt$$

$$\leqslant \frac{\alpha_N^{1-\frac{1}{N}}}{\alpha_{N-1}} 2||t^{-\frac{1}{N'}}||_{q_{r*}(\cdot)} ||\left(|\nabla u_r|_{*u_r}(t)\right)_*||_{p_r^*(\cdot)} \tag{10.29}$$

where we note

$$||f_r||_{q_r(\cdot)} = \operatorname*{Inf}_{\lambda>0} \lambda \left(1 + \int_0^{\alpha_N r^N} \left(\frac{f_r}{\lambda}\right)^{q_r(\cdot)} dt\right), \ f_r : [0, \alpha_N r^N] \to \mathbb{R}.$$

But according to Theorem 10.1.2, we have

$$||(|\nabla u_r|_{*u_r})_*||_{p_r^*} \leqslant c|| \, |\nabla u_r| \, ||_{p_r(\cdot)}.$$

Finally, since we have (see [3], for a necessary-sufficient condition)

$$||\nabla u_r||_{p_r(\cdot)} = ||\nabla u \chi_{B(x,r)}||_{p(\cdot)} \xrightarrow[r\to 0]{} 0,$$

we then deduce that then $u \in C(\Omega)$. ◇

A consequence of this inequality (10.29) is:

Corollary 10.3.1 *With the same hypotheses as in Theorem 10.3.1, if we have constants* c_{1N} *and* $0 < \alpha < 1$ *such that* $||t^{-\frac{1}{N'}}||_{q_{r*}(\cdot)} \leqslant c_{1N} r^\alpha$ *then*

$$W^{1,p(\cdot)}(\Omega) \subset C^{0,\alpha}(\Omega).$$

In addition, we have:

$$\text{osc}_{B(x,r)} u \leqslant c_{1N} r^\alpha ||\nabla u \chi_{B(x,r)}||_{p(\cdot)}$$

for all $(x, r) : B(x, r) \subset \Omega$.

10.3.1 Acerbi–Mingione Type Operators

Let $p \in L^\infty(\Omega), \; p : \Omega \to]1, +\infty[$. We consider a map

$$\widehat{a} : \Omega \times \mathbb{R}^N \to \mathbb{R}^N, \text{ where } \Omega \text{ is a bounded open set of} \mathbb{R}^N,$$

satisfying at least the following condition:

C1. *There exist two constants* $\alpha_0 > 0, \; a_0 > 0$ *such that for almost all* $x \in \Omega$, *for all* $\xi \in \mathbb{R}^N, \; \xi' \in \mathbb{R}^N$

$$\Big(\widehat{a}(x, \xi) - \widehat{a}(x, \xi'), \xi - \xi'\Big) \geqslant \alpha_0 |\xi - \xi'|^2 \Big(a_0 + |\xi| + |\xi'|\Big)^{p(x)-2}.$$

Since, for all $\delta > 0$, we have,

$$|\xi - \xi'|^\delta \leqslant \Big(|\xi| + |\xi'|\Big)^\delta \leqslant \Big(a_0 + |\xi| + |\xi'|\Big)^\delta,$$

we then deduce, as in [108] that

Proposition 10.3.1 *If $\widehat{a}$ satisfies the condition **C1**. then, for all $\delta \geqslant 0$*

$$\Big(\widehat{a}(x,\xi)-\widehat{a}(x,\xi'),\xi-\xi'\Big) \geqslant \alpha_0|\xi-\xi'|^{2+\delta}\Big(a_0+|\xi|+|\xi'|\Big)^{p(x)-2-\delta}.$$

Following the proof of [108], we have

Proposition 10.3.2 *Let $\widehat{a}(x,\xi)=(1+|\xi|^2)^{\frac{p(x)-2}{2}}\xi,\ \xi\in\mathbb{R}^N, x\in\Omega$. then,*

$\widehat{a}$satisfies the condition **C1**.

Furthermore, we can choose $a_0 = 2$ if $\operatorname*{ess\,inf}_{\Omega} p(x) \geqslant 2$, otherwise $a_0 = 1$.

Next, we are interested in contraction properties involving variable exponents but with the solutions of quasilinear equations.

Theorem 10.3.2 *Let $p \in L^\infty(\Omega)$, $p:\Omega\to[2,+\infty[$, F_1, F_2 two functions such that*

$$F_i \in L^{q(\cdot)}(\Omega)^N, \quad \frac{1}{q(x)}+\frac{1}{p(x)}=1\ a.e$$

u_1, u_2 two elements in $W_0^{1,p(\cdot)}(\Omega)=W_0^{1,1}(\Omega)\cap W^{1,p(\cdot)}(\Omega)$ satisfying, $\forall\varphi\in W_0^{1,p(\cdot)}(\Omega)$

$$\int_\Omega \widehat{a}(x,\nabla u_i)\cdot\nabla\varphi \mathrm{d}x=\int_\Omega F_i\cdot\nabla\varphi \mathrm{d}x,$$

and $w=|u_1-u_2|$, $\delta F=F_1-F_2$.
So, there exists a constant $c_N(\alpha_0,a_0,p)>0$ such that

$$-\frac{dw_*}{ds}(s)\leqslant c_N(\alpha_0,a_0,p)s^{\frac{1}{N}-1}\Big[\big|\delta F\big|^2\Big]_{*w}^{\frac{1}{2}}(s),$$

(continued)

Theorem 10.3.2 (continued)
provided that $|\delta F| \in L^2(\Omega)$.

Here $c_N(\alpha_0, a_0, p) = \dfrac{1}{\alpha_0 N \alpha_N^{\frac{1}{N}}} \left| \dfrac{1}{a_0^{(x)-2}} \right|_\infty .$

Proof We set $\delta\widehat{a}(x) = \widehat{a}(x, \nabla u_1) - \widehat{a}(x, \nabla u_2)$ and $u_{12} = u_1 - u_2$. For fixed $s \in \Omega_*$, we consider the test function:

$$\varphi_s(x) = \big(w(x) - w_*(s)\big)_+ \operatorname{sign}(u_{12}(x)),\ (x) \in \Omega.$$

Then we have as in Chap. 5, Lemma 5.2.1

$$\Big[\delta\widehat{a} \cdot \nabla u_{12}\Big]_{*w}(s) = \Big[\delta F \cdot \nabla u_{12}\Big]_{*w}(s). \tag{10.30}$$

Let

$$h(x) = a_0 + |\nabla u_1(x)| + |\nabla u_2(x)|. \tag{10.31}$$

Then, by Proposition 10.3.1, almost everywhere in Ω

$$(\delta\widehat{a} \cdot \nabla u_{12})(x) \geqslant \alpha_0 |\nabla w(x)|^2 h^{(x)-2}. \tag{10.32}$$

By the relations (10.30) and (10.32), we have, still using the properties of the relative rearrangement

$$\alpha_0 \Big[|\nabla w|^2 h^{(x)-2}\Big]_{*w}(s) \leqslant \Big[|\delta F|\,|\nabla w|\Big]_{*w}(s), \tag{10.33}$$

and

$$\Big[|\delta F|\,|\nabla w|\Big]_{*w}(s) \leqslant \Big[|\delta F|^2\Big]_{*w}^{\frac{1}{2}}(s)\Big[|\nabla w|^2\Big]_{*w}^{\frac{1}{2}}. \tag{10.34}$$

Since $h \geqslant a_0$, we deduce from (10.33) and (10.34) that

$$\alpha_0 \Big[|\nabla w|^2 h^{(x)-2}\Big]_{*w}(s) \leqslant \left| \frac{1}{a_0^{p(\cdot)-2}} \right|_\infty^{\frac{1}{2}} \Big[|\nabla w|^2 h^{(x)-2}\Big]_{*w}^{\frac{1}{2}}(s)\Big[|\delta F|_{*w}^2\Big]^{\frac{1}{2}}. \tag{10.35}$$

So, we have

$$\left[|\nabla w|^2 h^{(x)-2}\right]_{*w}(s) \leqslant c(\alpha_0, a_0, p)^2 \left(|\delta F|^2\right)_{*w}(s) \tag{10.36}$$

with $c(\alpha_0, a_0, p) = \dfrac{1}{\alpha_0} \left| \dfrac{1}{a_0^{p(\cdot-2)}} \right|_\infty^{\frac{1}{2}}$.

By the PSR inequality (Poincaré–Sobolev inequality for relative rearrangement) (see Theorem 4.1.1 and Definition 4.1.1) we have

$$-\frac{dw_*}{ds}(s) \leqslant \frac{s^{\frac{1}{N}-1}}{N\alpha_N^{\frac{1}{N}}} \cdot |\nabla w|_{*w} \leqslant \frac{s^{\frac{1}{N}-1}}{N\alpha_N^{\frac{1}{N}}} \cdot \left[|\nabla w|^2\right]_{*w}^{\frac{1}{2}}(s)$$

$$\leqslant \frac{s^{\frac{1}{N}-1}}{N\alpha_N^{\frac{1}{N}}} \left| \frac{1}{a_0^{p(\cdot)-2}} \right|_\infty^{\frac{1}{2}} \left[h^{p(\cdot-2)}|\nabla w|^2\right]_{*w}^{\frac{1}{2}}. \tag{10.37}$$

With (10.36) we deduce

$$-\frac{dw_*}{ds}(s) \leqslant \frac{s^{\frac{1}{N}-1}}{N\alpha_N^{\frac{1}{N}}} \left| \frac{1}{a_0^{p(\cdot-2)}} \right|_\infty^{\frac{1}{2}} \cdot c(\alpha_0, a_0, p) \left(|\delta F|^2\right)_{*w}^{\frac{1}{2}}(s). \tag{10.38}$$

Setting $c_N(\alpha_0,\ a_0,\ p) = \dfrac{c(\alpha_0, a_0, p)}{N\alpha_N^{\frac{1}{N}}} \left| \dfrac{1}{a_0^{p(\cdot-2)}} \right|_\infty^{\frac{1}{2}}$, we deduce from (10.38) the point estimate of Theorem 10.3.2. ◇

Corollary 10.3.2 ***of Theorem 10.3.2*** *With the same hypotheses as in the Theorem 10.3.2, if $r : \Omega \to [2, +\infty[$ is a measurable function, then we have, for all $s \in \Omega_*$*

$$w_*(s) \leqslant c_N(\alpha_0, a_0, p) b_1(s) \|f_*\|_{r^*(\cdot)},$$

with

$$f(t) = \left(|\delta F|^2_{*w}\right)^{\frac{1}{2}}(t)$$

and

$$b_1(s) = \left\| \left(\chi_{[s,|\Omega|]}(t) \cdot t^{\frac{1}{N}-1}\right)_* \right\|_{\overline{r}^*(\cdot)},\ \overline{r}_*(s) = \frac{r^*(s)}{r^*(s)-1}.$$

Proof We integrate the relation (10.38) from s to $|\Omega|$,

$$w_*(s) \leqslant c_N(\alpha_0, a_0, p) \int_0^{|\Omega|} \chi_{[s,|\Omega|]}(t) t^{\frac{1}{N}-1} f(t) \mathrm{d}t. \tag{10.39}$$

By the Hardy–Littlewood inequality, we have

$$w_*(s) \leqslant c_N(\alpha_0, a_0, p) \int_{\Omega_*} \left(\chi_{[s,|\Omega|]}(t) t^{\frac{1}{N}-1} \right)_*(t) f_*(t) \mathrm{d}t.$$

By Holder's inequality, we deduce

$$w_*(s) \leqslant c_N(\alpha_0, a_0) b_1(s) \| f_* \|_{r^*(\cdot)}.$$

◇

Note $\forall \sigma \in \Omega_*,\ \forall s \in \Omega_*,\ \left(\chi_{[s,\Omega|]}(t) t^{\frac{1}{N}-1} \right)_*(\sigma) = (\sigma + s)^{\frac{1}{N}-1} \chi_{[0,|\Omega|-s]}(\sigma).$

Corollary 10.3.3 ***Of Theorem 10.3.2*** *With the same hypotheses as in Theorem 10.3.2, if $\delta F \in L^{r^*(\cdot)}(\Omega)^N$, $r^*(0) > 2$, is bounded then we have*

$$w_*(s) \leqslant \widetilde{c}_N(\alpha_0, a_0) b_1(s) \|\delta F\|_{r^*(\cdot)},$$

provided that r^* satisfies $\left| r^*(t) - r^*(\sigma) \right| |\operatorname{Log}|t - \sigma|| \leqslant A$ near zero.

Proof We have (see Corollary 2.2.1 of Proposition 2.2.1)

$$\left((|\delta F|^2)_{*w} \right)_*(t) \leqslant \left(|\delta F|^2 \right)_{**}(t) \text{ for all } t \in \Omega_*.$$

Therefore,

$$\| f_* \|_{r^*(\cdot)} \leqslant c \left\| \ |\delta F|^2_{**} \right\|_{\frac{r^*(\cdot)}{2}}.$$

By Lemma 10.1.2 (Hardy's inequality) we have

$$\begin{aligned} \left\| \ |\delta F|^2_{**} \right\|_{\frac{r^*(\cdot)}{2}} &\leqslant c \| \ |\delta F|^2_* \|_{\frac{r^*(\cdot)}{2}} \\ &\leqslant c \| \ |\delta F|_* \|_{r^*(\cdot)} \leqslant c \|\delta F\|_{r^*(\cdot)} \text{ by Theorem 10.1.1.} \end{aligned}$$

◇

A result similar to Corollary 10.3.3, where r is a constant function., can be found in [24].

if

Corollary 10.3.4 (Of Theorem 10.3.2) *With the same hypotheses as in Corollary 10.3.3 of Theorem 10.3.2, let ρ be a rearrangement-invariant norm on a Banach function space and $||w||_{L(\Omega,\rho)} = \rho(|w|_*)$. If $\rho(b_1) < +\infty$ then*

$$\left\|u_1 - u_2\right\|_{L(\Omega,\rho)} = \rho(|u_1 - u_2|_*) \leqslant \widetilde{c}_N(\alpha_0, a_0)\rho(b_1)\|F_1 - F_2\|_{r^*(\cdot)}.$$

10.3.2 Pointwise Inequalities for Anisotropic Equations with Variable Exponents

We want to deduce a result similar to Theorem 10.3.2 but for operators of the kind

$$\widehat{a}(x,\xi) = \big(a_1(x,\xi_1),\ldots,a_N(x,\xi_N)\big),$$

for $\xi = (\xi_1, \ldots, \xi_N) \in \mathbb{R}^N$, $x \in \Omega$, where each a_i satisfies the condition **C1.**, i.e.
C2. *There exist $\alpha_{0i} > 1,\ a_{0i} > 0$ such that*

$$\big(a_i(x,t) - a_i(x,\sigma)\big)(t-\sigma) \geqslant \alpha_{0i}\big(a_{0i} + |t| + |\sigma|\big)^{p_i(x)-2}|t-\sigma|^2,$$

for a.e. $x \in \Omega,\ \forall (t,\sigma) \in \mathbb{R} \times \mathbb{R}$.

Here, $p_i : \Omega \to]1, +\infty[$ is a bounded measurable function .

We start with the case where $\operatorname*{ess\,inf}_{\Omega} p_i(x) \geqslant 2$ for all $i \in \{1, \ldots, N\}$. For this, we can consider the Sobolev space, for $\vec{p} = (p_1, \ldots, p_N)$:
$\mathbf{V}_{\vec{p}} = W_0^{1,p_1(\cdot),\ldots,p_N(\cdot)}(\Omega)$

$$= \left\{v \in W_0^{1,1}(\Omega) : \int_\Omega \left|\frac{\partial v}{\partial x_i}(x)\right|^{p_i(x)} \mathrm{d}x < +\infty \text{ for } i = 1, \ldots, N\right\}.$$

If necessary, we can equip this Banach functions space with a norm

$$\|v\|_{\mathbf{V}_{\vec{p}}} = |v|_{L^1(\Omega)} + \sum_{i=1}^{N} \left\|\frac{\partial v}{\partial x_i}\right\|_{p_i(\cdot)},$$

or by the equivalent norm

$$\sum_{i=1}^{N} \left\| \frac{\partial v}{\partial x_i} \right\|_{p_i(\cdot)}.$$

We denote by $q_i(x) = \dfrac{p_i(x)}{p_i(x)-1}$ the conjugate of $p(x)$ (see [41] for a systematic study of such spaces and problems).

Theorem 10.3.3 (A Contraction Property) *Let F_1 and F_2 be in $\prod_{i=1}^{N} L^{q_i(\cdot)}(\Omega)$ such that $\delta F = F_1 - F_2$ satisfy $|\delta F| \in L^2(\Omega)$. Let u_1 and u_2 be two elements of $\mathbf{V}_{\vec{p}}$ satisfying*

$$\int_\Omega \widehat{a}(x, \nabla u_j) \cdot \nabla \varphi \mathrm{d}x = \int_\Omega F_j \cdot \nabla \varphi \mathrm{d}x, \ j = 1, 2,$$

for all $\varphi \in \mathbf{V}_{\vec{p}}$. Assume that $p_i(x) \geqslant 2$ a.e. $i = 1, \dots, N$.

Then there exists a constant $c_N(\widehat{a}) > 0$ dependent only on N and α_{0i}, a_{0i}, p_i such that for almost all $s \in \Omega_*$,

$$-\frac{dw_*}{ds}(s) \leqslant c_N(\widehat{a}) s^{\frac{1}{N}-1}\Big[\big(|\delta F|^2\big)_{*w}(s)\Big]^{\frac{1}{2}}, \text{ with } w = |u_1 - u_2|.$$

We choose

$$c_N(\widehat{a}) = \frac{1}{N \alpha_N^{\frac{1}{N}} \alpha_0 a_0}, \ \alpha_0 = \operatorname*{Min}_{1 \leqslant i \leqslant N} \alpha_{01}, \ a_0 = \operatorname*{Min}_{1 \leqslant i \leqslant N} \Big[\operatorname*{ess\,inf}_{\Omega} a_{0i}^{(x)-2}\Big].$$

Proof The idea of the proof is similar to that of Theorem 10.3.2. We introduce

$$\delta \widehat{a} = \widehat{a}(x, \nabla u_1) - \widehat{a}(x, \nabla u_2),$$
$$\delta F = F_1 - F_2, \ w = |u_1 - u_2|, \ u_{12} = u_1 - u_2.$$

For s (fixed) $\in \Omega_*$, we choose as a test function

$$\varphi_s(x) = \Big(w(x) - w_*(s)\Big)_+ \operatorname{sign}(u_{12}(x)).$$

So,

$$\left[\delta\widehat{a}\cdot\nabla u_{12}\right]_{*w}(s)=\left[\delta F\cdot\nabla u_{12}\right]_{*w}(s). \tag{10.40}$$

We consider $\alpha_0 = \underset{1\leqslant i\leqslant N}{\operatorname{Min}} \alpha_{0i}$ and $a_0 = \underset{1\leqslant i\leqslant N}{\operatorname{Min}} \left[\underset{\Omega}{\operatorname{ess\,inf}}\, a_{0i}^{p_i(x)-2}\right]$ and we define themeasurable function, $k:\Omega\to\mathbb{R}$

$$k(x)=\begin{cases}\dfrac{\displaystyle\sum_{i=1}^{N}\alpha_{0i}\big(a_{0i}+|\partial_i u_1|+|\partial_i u_2|\big)^{p_i(x)-2}|\partial_i w|^2}{|\nabla w|^2}(x) & if\ \nabla w(x)\neq 0,\\ \alpha_0 a_0 & if\ \nabla w(x)=0.\end{cases}$$

Here $\partial_i u_j=\dfrac{\partial u_j}{\partial x_i}$, $\partial_i w=\dfrac{\partial w}{\partial x_i}$.

Then, $k(x)\geqslant\alpha_0 a_0>0$ for a.e. x, and for a.e. $x\in\Omega$

$$\left[\widehat{a}(x,\nabla u_1)-\widehat{a}(x,\nabla u_2)\right]\cdot\nabla u_{12}\geqslant k(x)|\nabla w(x)|^2. \tag{10.41}$$

From the relations (10.40) and (10.41), we have:

$$\begin{aligned}\left[k|\nabla w|^2\right]_{*w}(s)&\leqslant\left|\delta F\cdot\nabla u_{12}\right]_{*w}(s)\leqslant\left[\left(|\delta F|^2\right)_{*w}(s)\right]^{\frac12}\left[\left(|\nabla w|^2\right)_{*w}(s)\right]^{\frac12}\\ &\leqslant\frac{1}{(\alpha_0 a_0)^{\frac12}}\left[\left(|\delta F|^2\right)_{*w}(s)\right]^{\frac12}\left[\left(k|\nabla w|^2\right)_{*w}(s)\right]^{\frac12}.\end{aligned} \tag{10.42}$$

Consequently, we have, for a.e. $s\in\Omega_*$

$$\left[k|\nabla w|^2\right]_{*w}(s)\leqslant\frac{1}{(\alpha_0 a_0)}\left(|\delta F|^2\right)_{*w}(s). \tag{10.43}$$

Then we use the PSR inequality

$$\begin{aligned}-\frac{dw_*}{ds}(s)&\leqslant\frac{s^{\frac1N-1}}{N\alpha_N^{\frac1N}}\cdot|\nabla w|_{*w}(s)\leqslant\frac{s^{\frac1N-1}}{N\alpha_N^{\frac1N}}\left[\left(|\nabla w|^2\right)_{*w}(s)\right]^{\frac12}\\ &\leqslant\frac{s^{\frac1N-1}}{N\alpha_N^{\frac1N}}\frac{1}{\sqrt{\alpha_0 a_0}}\left[\left(k|\nabla u|^2\right)_{*w}(s)\right]^{\frac12}.\end{aligned} \tag{10.44}$$

Combining the relations (10.43) and (10.44), we obtain, for a.e. $s \in \Omega_*$

$$-\frac{dw_*}{ds}(s) \leqslant \frac{s^{\frac{1}{N}-1}}{N\alpha_N^{\frac{1}{N}}} \frac{1}{(\alpha_0 a_0)} \left[\left(|\delta F|^2 \right)_{*w}(s) \right]^{\frac{1}{2}}. \tag{10.45}$$

◇

Then we obtain the same corollaries of Theorem 10.3.2.

In particular, we have

Corollary 10.3.5 (Of Theorem 10.3.3) *Suppose that $|\delta F| \in L^{r(\cdot)}(\Omega)$ for a bounded measurable function r with $r^* > 2$ such that r^* satisfies in the neighborhood of zero,*

$$|r^*(t) - r^*(\sigma)| \, |\mathrm{Log}\,|t - \sigma|| \leqslant A \text{ constant.}$$

Then, there exists a constant $c_\Omega > 0$ depending only Ω, $\widehat{a}$, $\overrightarrow{p}$, r, such that for a.e. $s \in \Omega_*$

$$w_*(s) \leqslant c_\Omega \cdot b_1(s) \Big\| \, \|F_1 - F_2| \, \Big\|_{r(\cdot)},$$

where b_1 is as in Corollary 10.3.2 of Theorem 10.3.2.

10.4 New Results for Spaces with Variable Exponents Related to Rearrangement

Spaces with variable exponents are *not* invariant under rearrangement. By using the relative rearrangement with a class of exponents we can obtain a theorem conservation of integrals which generalizes the notion of equimesurability from the constant exponents case.

Theorem 10.4.1 *Let q be a simple function, $b \geqslant 0$ measurable and $F : \mathbb{R} \to \mathbb{R}_+$ a Borel function, defined everywhere. We assume that $F(b) \in L^{q(\cdot)}(\Omega)$. Then*

$$\int_\Omega F(b(x))^{q(x)} dx = \int_{\Omega_*} F\left(b_{*q}(t)\right)^{q_*(t)} dt.$$

Proof Consider $Imq = \left\{a_1, \ldots, a_m\right\}$ with $0 \leqslant a_m \leqslant \ldots \leqslant a_1,\ c_0 = 0$ and $c_k = \sum_{j=1}^{k} |q = a_j|$, for $k = 1, \ldots, m$. So, we have

$$q_*(s) = a_k \ if \ s \in [c_{k-1}, c_k],\ k = 1, \ldots, m.$$

We know that $b_{*q}(s) = \left(b\Big|_{q=q_*(s)}\right)_*(s - c_{k-1})$ for $s \in [c_{k-1}; c_k[$. We then have

$$\begin{aligned}\int_\Omega \left[F\left(b_{*q}(s)\right)\right]^{q_*(s)} ds &= \sum_{k=1}^{m} \int_{q_*=a_k} \left[F\left(b_{*q}(s)\right)\right]^{q_*(s))} ds \\ &= \sum_{k=1}^{m} \int_{q_*=a_k} \left[F\left(b|_{q=a_k}\right)_*(s - c_{k-1})\right]^{a_k} ds \\ &= \sum_{k=1}^{m} \int_0^{c_k - c_{k-1}} \left[F\left(b|_{q=a_k}\right)_*(\sigma)\right]^{a_k} d\sigma.\end{aligned}$$

Using equimeasurability, we obtain

$$= \sum_{k=1}^{m} \int_{q=a_k} F\left(b(x)\right)^{a_k} dx = \int_\Omega F\left(b(x)\right)^{q(x)} dx.$$

Remark This theorem generalizes the equimeasurability for the monotone rearrangement, since if $q(x) = 1$ then $b_{*q} = b_*$.

Here is a theorem which completes those we saw previously on the estimation of the norm of the relative rearrangement in a space with variable exponents.

Theorem 10.4.2 *Let $q \in L^\infty(\Omega) : 1 < q^- = ess\,inf_\Omega\, q_- \leqslant q \leqslant q^+,\ q^+ =$ $ess\,sup_\Omega\ q,\ b \in L^{q(\cdot)}(\Omega)$. Then, we have*

1. $\displaystyle\int_{\Omega_*} (b_{*q}(s))^{q_*(s)} ds \leqslant \int_\Omega b(x)^{q(x))}\ if\ b \geqslant 0.$
2. $\left\| |b|_{*q} \right\|_{L^{q_*(\cdot)}} \leqslant ||b||_{L^{q(\cdot)}}.$

We will first assume that $b \in L^\infty(\Omega)$ and $b_0 = \operatorname{ess\,inf}_\Omega b > 0$ and finally, we will prove the following

Lemma 10.4.1 *Let $b \geqslant 0$, $b \in L^{q(\cdot)}(\Omega)$. Define, for $k \geqslant 1$, $b_k = \dfrac{1}{k} + T_k(b)$. Then*

1. $b_{*q} \in L^{q_*(\cdot)}(\Omega_*)$.
2. $\lim\limits_{k\to+\infty} |b_{k*q}|_{L^{q_*(\cdot)}} = |b_{*q}|_{L^{q_*(\cdot)}}$.
3. $\lim\limits_{k\to+\infty} |b_{k*q} - b_{*q}|_{L^{q_*(\cdot)}(\Omega_*)} = 0$.

Proof of Theorem 10.4.2 Like $b \in L^\infty(\Omega)$ then $b_{*q} \in L^\infty(\Omega_*)$ and

$$||b_{*q}(s)|| \leqslant |b|_\infty \in L^{q_*(\cdot)}(\Omega_*).$$

According to Theorem 7.2.1 (see also its proof) there exists a sequence $q_n \in L^\infty(\Omega)$ s.t.

1. q_n is a simple function, $|q_n - q|_\infty \xrightarrow[n\to+\infty]{} 0$.
2. $1 < q^- - \delta \leqslant q_n \leqslant q^+ + 1$, for $\delta > 0$, $\forall n$.
3. $b_{*q_n} \rightharpoonup b_{*q}$ in $L^\infty(\Omega_*)$-weak-*.

Thus, $||b_{*q_n}||_{L^{q_*(\cdot)}} \leqslant \Big(||b||_\infty + 1\Big)^{q+} |\Omega| = c_0 < +\infty$.

Since the space $L^{q_*(\cdot)}(\Omega_*)$ is reflexive, we deduce that b_{*q_n} converges weakly to b_{*q} in $L^{q_*(\cdot)}(\Omega_*)$.

By the contraction property $|q_{n*} - q_*|_\infty \xrightarrow[n\to+\infty]{} 0$.

Better yet, the proof of the construction of q_n shows that we have:

$$|q_{n*} - q_*|_\infty \leqslant |q_n - q|_\infty \leqslant \frac{1}{2^n}.$$

$$\lim_n \underline{\gamma_n} = \lim_n \overline{\gamma_n} = 1$$

and

$$\underline{\gamma_n} \leqslant [b(x)]^{q(x)-q_n(x)} \leqslant \overline{\gamma_n},\ x \in \Omega$$

$$\underline{\gamma_n} \leqslant [b_{*q_n}(s)]^{q_*-q_{n*}(s)} \leqslant \overline{\gamma_n},$$

$\Big($Likewise for $[b_{*q_n}(s)]^{(q_{n*}-q_*)(s)}$, $[b(x)]^{q_n(x)-q(x)}\Big)$.

Thus, by the dominated convergence theorem

$$\lim_{k\to+\infty}\int_\Omega b(x)^{q_k(x)}dx = \int_\Omega b(x)^{q(x)}dx \tag{10.46}$$

$\forall\varepsilon>0,\ \exists n_\varepsilon : \forall k \geqslant n_\varepsilon \quad \overline{\gamma_k}\leqslant 1+\varepsilon$ and

$$\int_\Omega b(x)^{q_k(x)}dx \leqslant \int_\Omega b(x)^{q(x)}dx + \varepsilon. \tag{10.47}$$

Moreover, according to Mazur's Theorem, there exists a sequence $(m_n)_{n\geqslant n_\varepsilon}$ and $(\lambda_{n,k})_{n\leqslant k\leqslant m_n}$ and $\lambda_{n,k}\in[0,1]$ such that $\sum\limits_{k=n_k}^{m_n}\lambda_{n,k}b_{*q_k}$ converges strongly in norm to b_{*q} in $L^{q_*(\cdot)}(\Omega_*)$ when $n\to+\infty$ and $\sum\limits_{k=n}^{m_n}\lambda_{n,k}=1,\ n\geqslant n_\varepsilon$.

As a result

$$\int_{\Omega_*}\left[\sum_{k=n}^{m_n}\lambda_{n,k}b_{*q_k}(s)\right]^{q_*(s)} ds \xrightarrow[n\to+\infty]{} \int_{\Omega_*} b_{*q}(s)^{q_*(s)}ds.$$

But for fixed $s\in\Omega_*$, as $q_*(s)\geqslant q^- > 1$ the map $t\in\mathbb{R}_+\to t^{q_*(s)}$ is convex, so we have

$$\left(\sum_{k=n}^{m_n}\lambda_{n,k}b_{*q_k}(s)\right)^{q_*(s)} \leqslant \sum_{k=n}^{m_n}\lambda_{n,k}\Big(b_{*q_k}(s)\Big)^{q_*(s)}. \tag{10.48}$$

As $b_{*q_k}(s)^{q_*(s)} \leqslant \overline{\gamma}_k\big(b_{*q_k}(s)\big)^{q_{k*}(s)}$, we then deduce, from the relation (10.48) and the fact that $\overline{\gamma}_k\leqslant 1+\varepsilon$, $k\geqslant n\geqslant n_\varepsilon$, that

$$\int_{\Omega_*}\left(\sum_{k=n}^{m_n}\lambda_{n,k}b_{*q_k}(s)\right)^{q_*(s)} ds \leqslant \sum_{k=n}^{m_n}\lambda_{n,k}(1+\varepsilon)\int_{\Omega_*}\Big(b_{*q_k}(s)\Big)^{q_{k*}(s)}ds \tag{10.49}$$

But according to the theorem on the conservation of integrals Theorem 10.4.1, we have

$$\int_{\Omega_*}\Big(b_{*q_k}(s)\Big)^{q_{k*}(s)}(s)ds = \int_\Omega b(x)^{q_k(x)}dx. \tag{10.50}$$

Thus, using the relation (10.47) and letting $n\to+\infty$ in (10.49), we deduce

$$\int_{\Omega_*}\Big(b_{*q}(s)\Big)^{q_*(s)}ds \leqslant (1+\varepsilon)\left(\int_\Omega b(x)^{q(x)}dx + \varepsilon\right) \quad \forall\,\varepsilon.$$

We let $\varepsilon \to 0$ and obtain the inequality

$$\int_{\Omega_*} \left(b_{*q}(s)\right)^{q_*(s)} ds \leqslant \int_{\Omega} b(x)^{q(x)} dx \quad \text{for } b \in L^{\infty}(\Omega). \tag{10.51}$$

For this general case, let's start by proving Lemma 10.4.1.

Proof of the Lemma Since $T_k(b) = \min(b; k),\ k \geqslant 0$, we have

$$\frac{1}{k} \leqslant b_k \leqslant \frac{1}{k} + b,\ b_k(x) \xrightarrow[k \to +\infty]{} b.$$

According to statement 1., that is to say (10.51), we have

$$\int_{\Omega_*} \left(b_{k*q}(s)\right)^{q_*(s)} ds \leqslant \int_{\Omega} b_k(x)^{q(x)} dx \leqslant \text{constant}$$

Consequently b_{k*q} remains within a bound of $L^{q_*(\cdot)}$. By the reflexivity of the space $L^{q_*(\cdot)}$, we deduce that b_{k*q} converges weakly to b_{*q}. Thus we have

$$|b_{*q}|_{L^{q_*(\cdot)}} \leqslant \liminf_{k} |b_{*k}|_{L^{q_*(\cdot)}}$$

and since $b_{k*q} \leqslant \frac{1}{k} + b_{*q}$ we have $\limsup_{k} |b_{k*q}| \leqslant |b_{*q}|_{L^{q_*(\cdot)}}$. Hence $\lim_{k \to +\infty} |b_{k*q}|_{L^{q_*(\cdot)}} = |b_{*q}|_{L^{q_*(\cdot)}}$. But according to Theorem 3.4.9 of Dening et al [66] the space $L^{q_*(\cdot)}$ provided with the norm $|\cdot|_{q_*(\cdot)}$ is uniformly convex, hence, we deduce that

$$\lim_{k \to +\infty} |b_{k*q} - b_{*q}|_{q_*(\cdot)} = 0.$$

◇

End of the Proof of Theorem 10.4.2 As

$$\int_{\Omega_*} b_{k*q}(s)^{q_*(s)} ds \leqslant \int_{\Omega} b_k^q dx, \tag{10.52}$$

We can pass to the limit in the inequality thanks to the previous convergences.

◇

Pre-bibliographic Notes

All results of this chapter are given in the author's papers (see [85] and also [88]) . The last section has already been announced in *Jean, Michel Rakotoson*, **Réarrangement relatif revisité**, 2021, HAL-3277063.

Chapter 11
Exercises and Problems

Unless otherwise stated, we will always assume $|\Omega| < +\infty$.

11.1 Exercises

Exercise 11.1.1 Let $u \in L^p(\Omega)$ with $1 \leqslant p < +\infty$. Calculate $\ell = \lim_{t \to +\infty} t^p m(t)$. Show that if $u \in L^{p,\infty}(\Omega)$ then $\sup_{t>0} t^p \big| |u| > t \big| < +\infty$. Deduce that $L^p(\Omega) \hookrightarrow L^{p,\infty}(\Omega)$ by specifying the inclusion constant.

Solution 11.1.1

Answer: $\ell = 0$ because $t^p m(t) \leqslant \int_{u_+ > t} u_+^p(x)dx$.

Since $A = \sup_{0<s<|\Omega|} s^{\frac{1}{p}} |u|_*(s) < +\infty$ this is equivalent to

$$A^p = \sup_{t>0} t^p \big| |u| > t \big| < +\infty.$$

If $u \in L^p(\Omega)$ then

$$t^p \big| |u| > t \big| \leqslant |u|_p^p : |u|_{L^{p,\infty}} \leqslant |u|_{L^p}.$$

Exercise 11.1.2 Let $u : \Omega \to \mathbb{R}$ be measurable, with $|\Omega| < +\infty$.

(i) Show that $|u \geqslant u_*(s)| \geqslant s \quad \forall s \in \overline{\Omega_*}$.
(ii) Show that if $meas\Big(P(u)\Big) = 0$ (u without a flat zone or plateau), then

$$\forall s \in \overline{\Omega_*} \quad |u \geqslant u_*(s)| = s = m\big(u_*(s)\big).$$

J. M. Rakotoson, *Relative Rearrangement*, Lecture Notes in Mathematics 2376,
https://doi.org/10.1007/978-3-032-02228-8_11

(iii) Show that if $u \in C(\Omega)$ (continuous on Ω) then

$$u_*\big(m(t)\big) = t \quad \forall t \in]\inf u, \sup u[.$$

Solution 11.1.2
Hint for the Proof of Exercise 2

(i) As $|u \geqslant u_*(s)| = |u_* \geqslant u_*(s)|$, the interval $\{u_* \geqslant u_*(s)\}$ contains s, therefore $s \leqslant |u \geqslant u_*(s)|$.

(ii) If measure$\Big(P(u)\Big) = 0$ then $|u_* \geqslant u_*(s)| = |u_* > u_*(s)|$.

(iii) Let $t \in \mathbb{R}$. If $\theta > t$ then $m(\theta) < m(t)$ and if $\theta < t$ then $m(\theta) > m(t)$ (because u is continuous $|\theta < u < t| > 0$ if $\{\theta, t\} \subset]\inf u, \sup u[$). On the other hand, we have seen(Property 1.1.1) that $\forall\, t,\ u_*(m(t) \leqslant t$. If $u_*(m(t)) < t$, by definition of $u_*(m(t)) = \inf\Big\{\theta : m(\theta) \leqslant m(t)\Big\}$, there would exist θ s.t.$u_*(m(t)) < \theta < t$ and $m(\theta) \leqslant m(t)$ which is impossible for continuous u according to the above result.

Exercise 11.1.3 Let us suppose that Ω is bounded in $\mathbb{R}^N$. Show that:

1. $L^p_\#(\Omega) = \big\{u \in C(\Omega) \cap L^p(\Omega) : \text{measure}\big(P(u)\big) = 0\big\}$ is dense in $L^p(\Omega)$, $1 \leqslant p < +\infty$.
2. If $p = +\infty$, *then* $\mathbb{L}^\infty_\#(\Omega) = \{u \in L^\infty(\Omega) : \text{measure}\big(P(u)\big) = 0\}$ is dense in $L^\infty(\Omega)$.

Solution 11.1.3
Hint

1. $\mathbb{R}[X] \cap L^p(\Omega)$ i.e $A = \Big\{P : \overline{\Omega} \to \mathbb{R},\ P$ polynomial$\}$ is contained in $\mathbb{L}^p_\#(\Omega)$.
2. Consider $P(u) = \displaystyle\bigcup_{t \in D}\{u = t\},\ u \in L^\infty(\Omega)$ then for $j \in \mathbb{N}$, we define

$$u_j(x) = \begin{cases} u(x),\ (x) \in \Omega \backslash P(u) & \\ te^{-\frac{|x|+\gamma}{j+1}} & if\ u(x) = t \neq 0 \\ \dfrac{|x|}{j+1} & if\ u(x) = 0 \\ \gamma \in \mathbb{R},\ \inf\limits_{x\in\Omega} |x| > -\gamma & \end{cases}$$

then for $\theta \in \mathbb{R}$, $\Big|u_j = \theta\Big| = 0 \quad \forall j \geqslant 0$,

$$\Big|u_j - u\Big|_\infty \leqslant \left(1 - e^{-\frac{\sup|x|+\gamma}{j+1}}\right)|u|_\infty + \frac{\sup|x|}{j+1} \xrightarrow[j\to+\infty]{} 0$$

Exercise 11.1.4 Let $a \in L^1(\Omega)$, $a > 0$ a.e. We define, for $u : \Omega \to \mathbb{R}$ measurable, the distribution function relative to the weight a, for $t \in \mathbb{R}$,

$$m_{u,a}(t) = \int_{\{u>t\}} a(x)dx \,.$$

We set $|\Omega|_a = \displaystyle\int_\Omega a(x)dx$ and we define the decreasing rearrangement of u (with respect to the measure $a(x)dx \doteq a$),

$$s \in]0, |\Omega|_a[, \quad u^a_{*\cdot}(s) = \text{Inf}\,\{t \in \mathbb{R},\ m_{u,a}(t) \leqslant s\}.$$

1. Give the properties of $u^a_*(s)$ ($u^a_*(0) = \text{ess}\sup_\Omega u$, $u^a_*(|\Omega|_a) = \text{ess}\inf_\Omega u$).
2. Show that u and u^a_* are equimeasurable relative to these measures, that is ($|u > t|_a = m_{u,a}(t) = m_{u^a_*}(t) = \left|u^a_* > t\right| \quad \forall t$).
3. Show that if $a_1 \leqslant a_2$ then $u^{a_1}_* \leqslant u^{a_2}_*$.
   ($a \to u^a_*$ is increasing). Similarly, $u \to u^a_*$ is increasing.
4. Show that if $u \geqslant 0$ s.t. $\displaystyle\int_\Omega u(x)^p a(x)dx < +\infty$ then

$$\int_0^{+\infty} t^p dm_{u,a}(t) = -p \int_0^{+\infty} m_{u,a}(t) t^{p-1} dt.$$

   Show directly that

$$\int_\Omega u^p(x)a(x)\,dx = \int_0^{|\Omega|_a} \left(u^a_*(s)\right)^p ds\,.$$

Solution 11.1.4 Repeat the arguments of the course. For part 4, note as in Exercise 1

$$\lim_{t\to+\infty} t^p m_{u_a}(t) = 0.$$

Use Fubini's theorem.

Exercise 11.1.5 Let $u_j(x) \to u(x)$ a.e. Show that $\forall t \in \mathbb{R}\backslash D_u$, where D_u is given in Definition 1.1.1, we have $\lim_j m_{u_j}(t) = m_u(t)$.

Solution 11.1.5

Proof We have: $m_{u_j}(t) = \int_\Omega \chi_{]t,+\infty[}(u_j(x))dx, \quad t \notin D_u$. Let $x \in \Omega$ • if $u(x) > t$, then $\chi_{]t,+\infty[}(u_j(x)) = 1$ for $j \geqslant j_x$

• if $u(x) < t$ then $\begin{cases} \chi_{]t,+\infty[}(u_j(x)) = 0 \quad \text{for} j \geqslant j_x \text{and} \lim \chi_{]t,+\infty[}(u_j(x)) = \\ \chi_{]t,+\infty[}(u(x)) \end{cases}$

By the dominated convergence theorem, $\lim_j m_{u_j}(t) = m_u(t)$.

Exercise 11.1.6 Let $u_j(x) \leqslant u_{j+1}(x) \leqslant \cdots \leqslant u(x), \quad u_j(x) \to u(x)$.
Show that $u_{j*}(\sigma) \to u_*(\sigma) \quad \forall\sigma$.

Solution 11.1.6

Proof $u_{j*}(\sigma) \leqslant u_{(j+1)_*}(\sigma) \leqslant \cdots \leqslant u_*(\sigma) \quad \forall\sigma$. So, $\lim_j u_{j*}(\sigma) \doteq v_*(\sigma)$ satisfies $v_*(\sigma) \leqslant u_*(\sigma) \quad \forall\sigma$.

$|u > t| = \lim_j \left|u_j > t\right| = |v_* > t|$ a.e. t. Therefore, since for all t because $|u > t - h| = |v_* > t - h| \Longrightarrow |u > t| = |v_* > t|$. we have $u_* = v_*$.

Exercise 11.1.7 For $1 \leqslant p \leqslant +\infty, \quad 1 \leqslant q \leqslant +\infty, \quad f \in L^{p,q}(\Omega)$, we define

$$|f|_{p,q} = \left[\int_{\Omega_*} \left[t^{\frac{1}{p}} |f|_* (t)\right]^q \frac{dt}{t}\right]^{\frac{1}{q}} \quad \textit{if } q < +\infty$$

and

$$|f|_{p,+\infty} = \sup_{0<t<|\Omega|} \left[t^{\frac{1}{p}} |f|_* (t)\right] \quad \textit{if } q = +\infty.$$

Show that there exists a constant $c > 0$ s.t.

$$|f|_{p,q} \leqslant |f|_{(p,q)} \leqslant c\,|f|_{p,q}\,,$$

$$\forall f \in L^{p,q}(\Omega),\ 1 < p < +\infty, \quad 1 \leqslant q \leqslant +\infty.$$

Solution 11.1.7

Hint

We have $|f|_* (t) \leqslant \frac{1}{t}\int_0^t |f|_* (\sigma)d\sigma$ because $|f|_*(t) \leqslant |f|_*(\sigma)$ if $0 \leqslant s \leqslant t$, then we integrate. For the other inequality, use the Hardy inequality of Lemma 1.4.2.

Exercise 11.1.8

1. Let $\{E_i\}_i$ be a partition of Ω, $a_i \geqslant 0, \ i = 0, \cdots, n$ and $u(x) = \sum_{i=0}^n a_i \chi_{E_i}(x)$.
   Calculate u_*.
2. Show that the simple functions are dense in $L^{p,q}(\Omega)$, $1 \leqslant p \leqslant +\infty$, $1 \leqslant q < +\infty$.

Solution 11.1.8
Hint

1. Let $a_{j+1}^* \leqslant a_j^* \leqslant \cdots \leqslant a_*^0$ be a rearrangement of $\{a_0, \cdots, a_n\}$, then

$$u_*(s) = \begin{cases} a_0^* & if\ s \in \left[0, |E_0^*|\right[\\ a_i^* & if\ s \in \left[\sum_{k=0}^{i-1} |E_k^*|, \sum_{k=0}^{i} |E_k^*|\right[\\ i = 1, \cdots, n & \\ 0 & if\ s \geqslant |\Omega| = \sum_{k=0}^{n} |E_k| \end{cases}$$

(with a_i^* associated with E_i^*) i.e. $E_i^* = \{x : u(x) = a_i^*\}$).
2. Just consider $f \geqslant 0$, $f \in L^{p,q}(\Omega)$. Now we know that there exists a sequence of simple functions $(f_n)_n$ defined on Ω s.t.$0 \leqslant f_1 \leqslant f_2 \leqslant \cdots \leqslant f$ and $f_n(x) \xrightarrow[n\to\infty]{} f(x)\ \forall x \in \Omega$. Using the same argument as for Exercise 11.1.6, we deduce

$$(f - f_n)_*(s) \xrightarrow[n\to+\infty]{} 0, \quad 0 \leqslant (f - f_n)_*(s) \leqslant f_*(s).$$

Consequently, the dominated convergence theorem implies

$$|f_n - f|_{p,q} \xrightarrow[n\to+\infty]{} 0\ si\ q < +\infty.$$

Exercise 11.1.9 Let $f(x) = \sum_{j=0}^{n} c_j \chi_{E_j}(x)$ with $E_i \cap E_j = \emptyset$, $i \neq j$, $c_0 > c_1 > \cdots > c_n$. If $F_j = \bigcup_{k=0}^{j} E_k$, $\alpha_j = c_j - c_{j+1}$, then $f(x) = \sum_{j=0}^{n} f_j(x)$, $f_j = \alpha_j \chi_{F_j}$ and $f_*(t) = \sum_{j=0}^{n} f_{j*}(t)$.

Solution 11.1.9 Compare the two expressions pointwise and use equimeasurability.

Exercise 11.1.10

1. Compare $L^{p_1,q_1}(\Omega)$ and $L^{p_2,q_2}(\Omega)$
2. What space is $L^{p,p}(\Omega)$?
3. For $1 \leqslant p \leqslant +\infty$, $1 \leqslant q < +\infty$, show that $L^\infty(\Omega)$ is dense in $L^{p,q}(\Omega)$.
4. Show that if $1 \leqslant q \leqslant p < +\infty$ then $f \to |f|_{p,q}$ is a norm on $L^{p,q}(\Omega)$.

Solution 11.1.10

1. To show the inclusion $L^{p,q_1}(\Omega) \subset L^{p,q_2}(\Omega)$ if $q_1 \leqslant q_2$,we have $\forall t \in \Omega_*,\ \forall f \in L^{p,q}(\Omega),\ q < +\infty$:

   (i) $|f|_{**}(t) \leqslant c\dfrac{|f|_{(p,q)}}{t^{\frac{1}{p}}}$. Indeed, since $\sigma \to |f|_{**}(\sigma)$ is decreasing

$$|f|^q_{**}(t)t^{q/p}\left(\frac{p}{q}\right)^q = |f|_{**}(t)\int_0^t \sigma^{q/p}\frac{d\sigma}{\sigma}$$

$$\leqslant \int_0^t \sigma^{q/p}|f|_{**}(\sigma)^q\frac{d\sigma}{\sigma} \leqslant |f|^q_{(p,q)},$$

   hence the result.

   (ii) We deduce $|f|_{(p,q_2)} \leqslant c\,|f|_{(p,q_1)}$, because if $q_1 \leqslant q_2$, then

$$|f|^{q_2}_{(p,q_2)} = \int_{\Omega_*} |f|_{**}(t)^{q_1}\,|f|^{q_2-q_1}_{**}(t)\cdot t^{\frac{q_2}{p}-1}dt \leqslant c\,|f|^{q_2-q_1}_{(p,q_1)}\,|f|^{q_1}_{(p,q_2)}\,.$$

   If $q_1 = q_2$ and $p_1 < p_2$ then $\left(\dfrac{t}{|\Omega|}\right)^{\frac{1}{p_2}} \geqslant \left(\dfrac{t}{|\Omega|}\right)^{\frac{1}{p_1}}$, $0<t\leqslant|\Omega|$ from which we have

$$L^{p_2,q_2}(\Omega) \hookrightarrow L^{p,1q_1}(\Omega)$$

   (in a way, this decreases with respect to the first exponent).

2. (See Bennett–Sharpley [21], theorem 4.3). The only difficulty is to show the triangle inequality. As in the proof of Lemma 4.2.2, use Proposition 1.5.2

Exercise 11.1.11 In dimension 3, calculate σ_3 (measure of the unit sector in $\mathbb{R}^3$).

Solution 11.1.11
Answer: $\sigma_3 = \dfrac{4\pi}{3}\sin^2\left(\dfrac{\alpha}{2}\right)$.

Exercise 11.1.12 Let $f,\ f \in L^p(\Omega),\ g \in L^{p'}(\Omega),\ \dfrac{1}{p}+\dfrac{1}{p'} = 1$. Show that

$$\int_{\Omega_*} f_*g_*dt = \sup\left\{\int_\Omega f\overline{g}dx,\ \overline{g}_* = g_*\right\} = L.$$

Start with the case where f does not have a plateau, and then use Exercise 11.1.3.

Solution 11.1.12 By Hardy–Littlewood's inequality, we always have

$$\int_{\Omega} f\overline{g}dx \leqslant \int_{\Omega_*} f_*\overline{g}_* : L \leqslant \int_{\Omega_*} f_*g_*.$$

Conversely, let us first consider the case where f does not have a plateau , then

$$\overline{g}(x) = g_*(|f > f(x)|) \text{ satisfies } \overline{g}_* = g_*.$$

By equimeasurability, we then have

$$\int_{\Omega_*} f_*(s)g_*(s)ds = \int_{\Omega} f(x)g_*\big(|f > f(x)|\big)dx \leqslant L$$

hence the equality. We conclude by using the density given in Exercise 11.1.3 since by hypothesis $f_*g_* \in L^1(\Omega_*)$.

Exercise 11.1.13

1. Show that if $p > N$ then $W^{1,p}(\Omega) \subset W^1L^{N,1}(\Omega)$ is a continuous inclusion.
2. If $u \in W^{1,p}(\Omega),\ x \in \Omega,\ r > 0$ s.t. $B(x,r) \subset \Omega$ then

$$\underset{B(x,r)}{\operatorname{osc}}\ u \leqslant c\,|\nabla u|_{L^p(\Omega)}\, r^{1-\frac{N}{p}} \text{ where} c = c(N,p) \text{ is to be specified.}$$

Solution 11.1.13
Hint First show that:

$$L^p(\Omega) \subset L^{N,1}(\Omega) \subset L^N(\Omega) \qquad \forall p > N.$$

Thus:

$$W^{1,p}(\Omega) \subsetneqq W^1L^{N,1}(\Omega) \subsetneqq W^{1,N}(\Omega)\,.$$

Exercise 11.1.14

- Think about the case where Ω is a bounded open Lipschitzian set s.t. $\Gamma_0 \subset \partial\Omega$ and $\Gamma_0 \cap \partial\Omega_i$ is of measure $H_{N-1} > 0$, where $\partial\Omega_i$ is the boundary of a connected component Ω_i of Ω.
- Show that $|\nabla u|_p$ is a norm equivalent to the norm of $W^{1,p}_{\Gamma_0}(\Omega)$.

Solution 11.1.14 This is a consequence of the inclusions of Theorem 4.4.8

Exercise 11.1.15 Let $(a_1, \ldots, a_n)$, $(b_1, \ldots, b_n)$ be real numbers such that

$$a_n \leqslant a_{n-1} \leqslant \cdots \leqslant a_1 \text{ and } b_n \leqslant b_{n-1} \leqslant \cdots \leqslant b_1.$$

1. Show that

$$0 \leqslant \frac{1}{2} \sum_{k=1}^{n} \sum_{j=1}^{n} (a_k - a_j)(b_k - b_j) = n \sum_{k=1}^{n} a_k b_k - \left(\sum_{k=1}^{n} a_k \right) \left(\sum_{j=1}^{n} b_j \right).$$

2. Let Ω be a measurable set of finite measure, $u : \Omega \to \mathbb{R}$ and $v : \Omega \to \mathbb{R}$ be two measurable functions and let $(s_1, \cdots, s_n)$ such that $0 < s_1 \leqslant s_2 \leqslant \cdots \leqslant s_n < |\Omega|$.
   Show that

$$\left(\frac{1}{n} \sum_{j=1}^{n} u_*(s_j) \right) \left(\frac{1}{n} \sum_{j=1}^{n} v_*(s_j) \right) \leqslant \frac{1}{n} \sum_{j=1}^{n} u_*(s_j) v_*(s_j).$$

3. We assume that Ω is a connected open set. Deduce from (2) that if $u \in L^p(\Omega)$, $v \in L^{p'}(\Omega)$ then

$$\left(\frac{1}{|\Omega|} \int_\Omega u(x)dx \right) \left(\frac{1}{|\Omega|} \int_\Omega v(x)dx \right) \leqslant \frac{1}{|\Omega|} \int_{\Omega_*} u_*(s) v_*(s) ds.$$

4. Show that if $u \in L^1(\Omega)$ then $\forall s \in [0, |\Omega|]$

$$\frac{s}{|\Omega|} \int_\Omega u(x)dx \leqslant \int_0^s u_*(\sigma) d\sigma.$$

Solution 11.1.15

1. Since the maps $j \to a_j$, $j \to b_j$ are decreasing, we deduce that for $k \neq j$

$$\frac{a_k - a_j}{k - j} \leqslant 0 \text{ and } \frac{b_k - b_j}{k - j} \leqslant 0$$

   from which we have $(a_k - a_j)(b_k - b_j) \geqslant 0$ and consequently

$$\frac{1}{2} \sum_{k=1}^{n} \sum_{j=1}^{n} (a_k - a_j)(b_k - b_j) \geqslant 0.$$

We expand this double sum to obtain the equality.

2. We apply question 1. with $a_j = u_*(s_j)$ and $b_j = v_*(s_j)$.
3. We use the fact that it is a Riemann sum and then the equimeasurability.
4. We apply 3. by choosing $v(x) = \chi_E(x)$ with $|E| = s$, or we use the decreasing of $s \to \frac{1}{s}\int_0^s u_*(\sigma)d\sigma)$

Exercise 11.1.16

1. Prove Proposition 9.4.2 (by direct calculation).
2. Show that the function k given in Theorem 9.4.6 satisfies $k \leqslant K$.

   Deduce that $\int_\Omega \Phi\big(u(t,x)\big)dx \leqslant \int_{\tilde{\Omega}} \Phi\big(U(t,x)\big)dx$, for any convex lipschitzian function $\Phi : \mathbb{R} \to \mathbb{R}_+$ with $\Phi(0) = 0$.

Solution 11.1.16 Use the Comparison Theorem 9.5.3.

Exercise 11.1.17 Let $\eta > 0$, $0 < \varepsilon < 1$, Ω be an open bounded set of measure and $f \in L^2(\Omega)$. For $v \in H_0^1(\Omega)$, we define

$$J(v) = \frac{1}{2}\int_\Omega |\nabla v|^2\,dx + \frac{1}{2\eta}\int_\varepsilon^1 [(v_*)_+]^2\,(s)\,ds - \int_\Omega f \cdot v\,dx.$$

1. Show that there exists a function $u_\eta \in H_0^1(\Omega)$ satisfying

$$J(u_\eta) = \inf\big\{J(v),\ v \in H_0^1(\Omega)\big\}.$$

2. Give the variational inequality satisfied by u_η and show that u_η remains within a bounded set of $H_0^1(\Omega)$ when η tends to 0.
3. Assume that u_η converges weakly to u in $H_0^1(\Omega)$. Show that u satisfies

$$\begin{cases} |u > 0| \leqslant \varepsilon, & (i) \\ u \in H_0^1(\Omega), & (ii) \\ -\Delta u \leqslant f \text{ in } \mathcal{D}'(\Omega). & (iii) \end{cases}$$

4. Give functions f for which we do not have the equality in (iii) and prove that if we have equality in (iii) then u is unique.

Solution 11.1.17

1. Note that $J(v) \geqslant J_0(v) = \frac{1}{2}\int_\Omega |\nabla v|^2 - \int_\Omega f\,v$, thus we have the coercivity of J as the map $v \in H_0^1(\Omega) \to \frac{1}{2\eta}\int_\varepsilon^1 (v_{*+})^2\,(\sigma)\,d\sigma$ is continuous on $H_0^1(\Omega)$-weak, and we deduce the result by following the usual arguments. Note that

$J_0(u_\eta) \leqslant J(0)$ so u_η remains bounded in $H_0^1(\Omega)$ when η varies. We will assume $u_\eta \rightharpoonup u$ in $H_0^1(\Omega)$-weak.

2. We calculate $J'(u_\eta)v = \lim\limits_{t\to 0,\, t>0} \dfrac{J(u_\eta + tv) - J(u_\eta)}{t} \geqslant 0$. Hence

$$\int_\Omega \nabla u_\eta \cdot \nabla v\, dx + \frac{1}{\eta}\int_\varepsilon^1 u_{\eta*+}(\sigma)v_{*u}(\sigma)d\sigma - \int_\Omega f\, v\, dx \geqslant 0$$

for all $v \in H_0^1(\Omega)$.

3. We interpret this inequality, knowing that if $v \in \mathcal{D}(\Omega),\ v \geqslant 0, then\ v_{*u_\eta} \geqslant 0$, thus:

$$\int_\Omega \nabla u_\eta \cdot \nabla v\, dx - \int_\Omega f\, v\, dx \geqslant 0 : -\Delta u_\eta - f \leqslant 0 : -\Delta u \leqslant f.$$

As $J(u_\eta) \leqslant J(0) \Longrightarrow \int_\varepsilon^1 \left(u_{\eta*+}\right)^2 \leqslant (constant)\eta$, we have

$$\int_\varepsilon^1 (u_{*+})^2\,(\sigma)d\sigma = 0 : u_{*+}(\sigma) = 0,\ \forall \sigma \in [\varepsilon, 1] \Longrightarrow |u > 0| \leqslant \varepsilon.$$

4. example $\begin{cases} f \geqslant 0 \\ f \not\equiv 0 \end{cases}$ if we have equality then by the maximum principle $u > 0$: $|u > 0| > \varepsilon$

Exercise 11.1.18 Let f and g be two integrable functions on a bounded interval (a, b). Assume that f is decreasing and that $0 \leqslant g \leqslant 1$ on (a, b).

1. Show that

$$\int\limits_{b-\lambda}^{b} f(t)dt \leqslant \int_a^b f(t)g(t)dt \leqslant \int\limits_{a}^{a+\lambda} f(t)dt \quad \text{where } \lambda = \int_a^b g(t)dt.$$

2. Prove from (1) the corollary of Hardy-Littlewood theorem which stipulates

$$\int_0^s u_*(\sigma)d\sigma = \mathrm{Max}\left\{ \int_\Omega u(x)z(x)dx,\ 0 \leqslant z \leqslant 1\ \int_\Omega z\, dx = s \right\}.$$

Solution 11.1.18

(1) We will only show the second inequality (which leads to 2.) the first is done in an identical way.

Let us set for $x \in (a, b)$:

$$H(x) = \int_a^{a+\int_a^x g(t)dt} f(t)dt - \int_a^x f(t)g(t)dt.$$

This is a function that vanishes at $x = a$ and its derivative is positive because

$$H'(x) = g(x)f\left(a + \int_a^x g(t)dt\right) - f(x)g(x) \geqslant 0.$$

As $0 \leqslant g \leqslant 1$ implies $\int_a^x g(t)dt \leqslant x - a$ and since f is decreasing, we have

$$f\left(a + \int_a^x g(t)dt\right) \geqslant f(x).$$

Consequently $H(x) \geqslant H(a) = 0$.

(2) We use Hardy–Littlewood inequalities and (1)

$$\int_0^s u_*(\sigma) \leqslant \text{Max}\left\{\int_\Omega u\chi_E dx : |E| = s\right\}$$

$$\leqslant \text{Max}\left\{\int_\Omega uzdx : 0 \leqslant z \leqslant 1, \int_\Omega zdx = s\right\}$$

$$\leqslant \text{Max}\left\{\int_0^{|\Omega|} u_* z_1 ds : 0 \leqslant z_1 \leqslant 1 \int_0^{|\Omega|} z_1 d\sigma = s\right\} \leqslant \int_0^s u_*(t)dt.$$

Exercise 11.1.19 Let $1 \leqslant q \leqslant p \leqslant +\infty$ and $r > N$. Find $0 < a < 1$ and c such that

$$|u|_p \leqslant c\,|\nabla u|_r^a\,|u|_q^{1-a}, \ \forall u \in W_0^{1,r}(\Omega).$$

Here Ω is a bounded open set of $\mathbb{R}^N$.

Solution 11.1.19 See the interpolation section of the course.

Exercise 11.1.20 Let $f,\ g,\ h$ be three integrable functions on a measurable domain Ω. We assume that h is bounded and that

(i) $$\int_{\{h\geqslant t\}} f(x)dx \geqslant \int_{\{h\geqslant t\}} g(x)dx\,,\ \forall t \geqslant 0,$$

(ii) $$\int_{\{h<t\}} f(x)dx \leqslant \int_{\{h<t\}} g(x)dx\,,\ \forall t < 0.$$

Show that $\displaystyle\int_\Omega gh\,dx \leqslant \int_\Omega fh\,dx.$

Solution 11.1.20 Use Fubini's theorem as in Chap. 1, Lemma 1.2.3.

Exercise 11.1.21 Let Ω be a bounded open set of $\mathbb{R}^N$. We denote by $d(x,\Gamma) = v(x)$ the distance from a point $x\in\Omega$ to the boundary $\Gamma = \partial\Omega$.

1. Show that

$$\forall s\in\Omega_*,\qquad d(\cdot,\Gamma)_*(s) \leqslant \frac{1}{\alpha_N^{\frac{1}{N}}}\Big(|\Omega|^{\frac{1}{N}} - s^{\frac{1}{N}}\Big).$$

2. Deduce that: if $\alpha\geqslant 1$ then $\displaystyle\int_\Omega d(x,\Gamma)^{-\alpha}dx = +\infty,$

   if Ω is the unit ball of $\mathbb{R}^N$, then $d(\cdot,\Gamma)_*(s) = 1-\alpha_N^{-\frac{1}{N}}s^{\frac{1}{N}}$.
3. Show that if $g\in W_0^{1,p}(0,1),\ 1<p<+\infty$ then

$$\left|\frac{g}{x(1-x)}\right|_p \leqslant \frac{2p}{p-1}|g'|_p$$

   (use a Hardy inequality).
4. Determine a constant $c>0$ such that $\forall u\in W_0^{1,p}(\Omega),\ u\geqslant 0$

$$\int_{\underset{\sim}{\Omega}} \underset{\sim}{u}(x)^p d(x,\underset{\sim}{\Gamma})^{-p}dx \leqslant c\int_{\underset{\sim}{\Omega}} \left|\nabla\underset{\sim}{u}(x)\right|^p dx.$$

   Here $1<p<+\infty,\ \underset{\sim}{\Gamma} = \partial\underset{\sim}{\Omega}$.

Solution 11.1.21

1. The function $v(x) = d(x,\Gamma)$ is in $C_+^{0,1}(\Omega)\cap H_0^1(\Omega)$ and $|\nabla v(x)|\leqslant 1$ a.e., by the (PSR) inequality, we deduce:

$$-v_*'(s) \leqslant \frac{s^{\frac{1}{N}-1}}{N\alpha_N^{\frac{1}{N}}}$$

whence, for all $s \in \overline{\Omega}_*$, we have:

$$v_*(0) - \frac{s^{\frac{1}{N}}}{\alpha_N^{\frac{1}{N}}} \leqslant v_*(s) \leqslant \frac{1}{\alpha_N^{\frac{1}{N}}}\left(|\Omega|^{\frac{1}{N}} - s^{\frac{1}{N}}\right).$$

2. By equimeasurability and the above estimate, by making a change of variables, we deduce :

$$\int_\Omega d(x,\Gamma)^{-\alpha} = \int_0^{|\Omega|} v_*(s)^{-\alpha} ds \geqslant (constant)\int_{\frac{1}{2}}^1 (1-t)^{-\alpha} dt = +\infty$$

if $\alpha \geqslant 1$.

If Ω is the unit ball of $\mathbb{R}^N$, we have $|\Omega| = \alpha_N$ and $v_*(0) = 1$ hence

$$v_*(s) = 1 - \frac{1}{\alpha_N^{\frac{1}{N}}} s^{\frac{1}{N}}.$$

3. This is Hardy's inequality, see the book J.E. Rakotoson and J.M. Rakotoson [136].
4. To simplify, we can assume that $\underset{\sim}{\Omega}$ is the unit ball, then $d(x, \underset{\sim}{\Gamma}) = 1 - |x|$. For $u \in C_c^\infty(\Omega)$, $u \geqslant 0$

$$\int_{\underset{\sim}{\Omega}} \frac{\underset{\sim}{u}(x)^p}{(1-|x|)^p} dx = N\int_0^1 \frac{u_*(\alpha_N r^N)^p r^{N-1}}{(1-r)^p} dr = N\int_0^1 \left|\frac{g(r)}{r(1-r)}\right|^p dr,$$

with $g(r) = u_*(\alpha_N r^N) r^{\frac{N-1}{p}+1}$ then $g \in W_0^{1,p}(0,1)$ (see regularity of u_*).

Conclude by Hardy's inequality from question 3 and the classic Sobolev inequality (see Chap. 4), i.e. $|\underset{\sim}{u}|_p \leqslant c_p(\Omega)|\nabla u|_p$.

Exercise 11.1.22 Let Ω be an open bounded subset of $\mathbb{R}^N$, $1 \leqslant r \leqslant p - N'$, $p < +\infty$. Determine a constant $c > 0$ such that:

$$|u|_p \leqslant c\,|\nabla u|_N^{1-\frac{r}{p}}\,|u|_r^{\frac{r}{p}}, \qquad \forall u \in W_0^{1,N}(\Omega).$$

Solution 11.1.22 See Chap. 4, Corollary 4.5.2.

Exercise 11.1.23 Let E be a measurable set in $\mathbb{R}^N$, of finite measure. We denote by χ_E the characteristic function of E. Let $(\rho_n)_{n\geqslant 0}$ be a regularizing sequence in $C_c^\infty(\mathbb{R}^N)$.

1. Show that $u_n = \chi_E * \rho_n$ satisfies

$$|Du_n|_1 \leqslant |D\chi_E|_1 \doteq P_{\mathbb{R}^N}(E).$$

2. Given that $u_n \in C_c^\infty(\mathbb{R}^N)$, show that

$$\forall x = (x_1, \cdots, x_N) \qquad |u_n(x)|^N \leqslant \prod_{j=1}^{N} \int_{-\infty}^{+\infty} \left|\partial_j u_n\right| dx_j$$

(we will write that $u_n(x) = \displaystyle\int_{-\infty}^{x_j} \partial_j u_n \, dt$).

3. Deduce for $N = 2$ that $|u_n|_2 \leqslant \prod_{j=1}^{j=2} |\partial_j u_n|_1^{\frac{1}{2}}$
4. Recall the Gagliardo–Nirenberg inequality given in the first chapter and show that

$$|u_n|_{N'} \leqslant \frac{1}{\sqrt{N}} |Du_n|_1 .$$

5. Deduce $|E|^{1-\frac{1}{N}} \leqslant \dfrac{1}{\sqrt{N}} P_{\mathbb{R}^N}(E)$.

Solution 11.1.23

1. Let $\rho_n \in C_c^\infty(\mathbb{R}^N)$ s.t. $\rho_n(-x) = \rho(x)$, $\displaystyle\int_{\mathbb{R}^N} \rho_n(x)dx = 1$, $\rho_n \geqslant 0$.

   Then, for all $\psi = (\psi_1, \cdots, \psi_N)$, with $\psi_j \in C_c^\infty(\mathbb{R}^N)$, $j = 1, \cdots, N$, we have:

$$\int_{\mathbb{R}^N} \partial_j u_n \psi_j dx = -\int_{\mathbb{R}^N} \left(\int_E \rho_n(x-y)dy \right) \partial_j \psi_j(x)dx$$

$$= \int_{\mathbb{R}^N} \chi_E(y) \left[\int_{\mathbb{R}^N} \partial_j \rho_n(x-y) \psi_j(x)dx \right] dy.$$

   Setting $\Phi_{jn}(y) = (\rho_n \star \psi_j)(y)$, we then have $||\Phi_{jn}||_\infty \leqslant 1$.

$$\int_{\mathbb{R}^N} \partial_j u_n \psi_j \, dx = \int_{\mathbb{R}^N} \chi_E(y) \cdot \partial_j \Phi_{jn}(y)dy.$$

   Since:

$$|\nabla u_n|_1 = \sup \left\{ \int_{\mathbb{R}^N} \nabla u_n \cdot \psi \, dx, \; ||\psi||_\infty \leqslant 1, \; \psi \in C_c^\infty(\mathbb{R}^N)^N \right\}$$

$$= \sup \left\{ \int_E \operatorname{div}(\Phi_{jn}), \; ||\Phi_{jn}||_\infty \leqslant 1, \; \Phi_{jn} \in C_c^\infty(\mathbb{R}^N) \right\}$$

$$\leqslant \sup \left\{ \int_E \operatorname{div}(\widetilde{\psi}) : ||\widetilde{\psi}||_\infty \leqslant 1, \; \widetilde{\psi} \in C_c^\infty(\mathbb{R}^N)^N \right\},$$

we have

$$|\nabla_{u_n}|_1 \leqslant P_{\mathbb{R}^N}(E).$$

2.

$$u_n(x_1; \ldots; x_j; \cdots; x_N) = \int_{-\infty}^{x_j} \partial_j u_n(x_1; \cdots; t; x_{j+1}; \cdots; x_N)dt$$

(Note that $|u_n|_1 \leqslant |E| < +\infty$, thus $u_n \in W^{1,1}(\mathbb{R}^N) \cap C^\infty(\mathbb{R}^N)$). Therefore,

$$|u_n(x)|^N \leqslant \prod_{j=1}^{N} \int_{-\infty}^{+\infty} |\partial_j u_n|(x_1; \cdots; t; x_{j+1}; \cdots, x_n)dt.$$

3. For $N = 2$, we then have

$$|u_n(x_1; x_2)|^2 \leqslant \int_{-\infty}^{+\infty} |\partial_1 u_n(t; x_2)|dt \cdot \int_{-\infty}^{+\infty} |\partial_2 u_n|(x_1; t)\, dt.$$

We integrate this relation on $\mathbb{R}^2$ using Fubini's theorem

$$\int_{\mathbb{R}^2} |u_n(x_1; x_2)|^2\, dx_1\, dx_2$$

$$\leqslant \left(\int_{\mathbb{R}^2} |\partial_1 u_n|(t; x_2)\, dt\, dx_2\right) \left(\int_{\mathbb{R}^2} |\partial_2 u_n|(x_1; t)\, dt\, dx_1\right)$$

which gives the result.

This is the Gagliardo–Nirenberg inequality for $N = 2$.

The general case can be done by induction.

4. From the Gagliardo–Nirenberg inequality, we have

$$|u_n|_{N'} \leqslant \prod_{j=1}^{N} |\partial_j u_n|_1^{\frac{1}{N}} \leqslant \frac{1}{N} \sum_{j=1}^{N} |\partial_j u_n]_1.$$

Since for $a_i \geqslant 0,\ i = 1, \cdots, N,\ \ (a_1 \cdots a_N)^{\frac{1}{N}} \leqslant \dfrac{1}{N}(a_1 + \cdots + a_N)$ and $a_1 + \cdots + a_N \leqslant \sqrt{N}(a_1^2 + \ldots + a_N^2)^{\frac{1}{2}}$ we get

$$|u_n|_{N'} \leqslant \frac{1}{\sqrt{N}} |\nabla u_n|_1.$$

5. We deduce from the first question

$$|u_n|_{N'} \leqslant |\frac{1}{\sqrt{N}} P_{R^N}(E).$$

Since $\chi_E \in L^{N'}(\mathbb{R}^N)$, we have

$$\lim_{n\to+\infty} |u_n - \chi_E|_{N'} = 0$$

from which we have

$$|E|^{1-\frac{1}{N}} \leqslant \frac{1}{\sqrt{N}} P_{R^N}(E).$$

This shows that

$$c_N = \sup_{|E|<+\infty} \frac{|E|^{1-\frac{1}{N}}}{P_{R^N}(E)} < +\infty.$$

This constant is called the isoperimetric constant,

$$c_N = \frac{1}{N\alpha_N^{\frac{1}{N}}}.$$

Exercise 11.1.24 Let Ω be an open bounded subset of $\mathbb{R}^N$, $a : \Omega \to]0, +\infty[$ a weight function. Define

$$W^1(\Omega, a) = \Big\{v \in L^1_{loc}(\Omega) \cap L^1(\Omega, a),\ |\nabla v| \in L^1(\Omega, a)\Big\},$$

$$L^1(\Omega, a) = \Big\{v \text{ measurable } : \int_\Omega |v(x)|\, a(x)dx < +\infty\Big\}.$$

We denote by u_*^a the rearrangement of a measurable function u with respect to the measure $d\mu = a\,dx$, and by $v_{*u,a}$ the relative rearrangement of a function v with respect to u, i-e

$$\lim_{\lambda\to 0} \frac{(u+\lambda v)_*^a - u_*^a}{\lambda} = v_{*u,a}$$

(same meaning as if $a = 1$).

We assume that $W^1(\Omega, a)$ satisfies the following property (PSR):

$$-(u_*^a)'(s) \leqslant K_a(s)\,|\nabla u|_{*u,a}(s) \qquad a.e.\ s \in \left[0,\ |\Omega|_a = \int_\Omega a(x)dx\right].$$

1. Rewrite the set of inequalities from Chap. 4 by replacing u_* with u_*^a.
2. Give interpolation inequalities relative to these weighted spaces.

Exercise 11.1.25 Let Ω be a connected open subset of $\mathbb{R}^N$. Given that $\Omega = \bigcup_{j\in\mathbb{N}} Q_j$, Q_j are cubes with pairwise disjoint interiors, i.e. $\overset{\circ}{Q}_j \cap \overset{\circ}{Q}_k = \emptyset, \quad j \neq k$, give a proof of the theorem that if $u \in W^{1,1}_{loc}(\Omega)$ then $u_* \in W^{1,1}_{loc}(\Omega_*)$.

Exercise 11.1.26 Let Ω be a subset of $\mathbb{R}^N$ of finite measure, $u \in L^1(\Omega)$, $g \in L^{p,q}(\Omega_*)$, $1 < p < +\infty$, $1 < q < +\infty$.

1. Show that $M_u(g) \in L^{p,q}(\Omega)$ and give an estimate of its norm (specifying the norms used).
2. Show that if u is plateau-free (without a flat zone) then $M_u : \left(L^{p,q}(\Omega_*), |\cdot|_{p,q}\right) \to \left(L^{p,q}(\Omega), |\cdot|_{p,q}\right)$ is an isometry.

Solution 11.1.26

1. Since the associated space of $L^{p,q}$ is $L^{p',q'}$ with $\dfrac{1}{p} + \dfrac{1}{p'} = 1 = \dfrac{1}{q} + \dfrac{1}{q'}$, (see Lemma 4.4.7), it suffices to estimate for $g \in L^{p,q}(\Omega_*)$, $h \in L^{p',q'}(\Omega)$ the integral $\int_\Omega M_u h\,dx$, for $h \geqslant 0$, $g \geqslant 0$ (because $|M_u(g)| \leqslant M_u(|g|)$). We have $\int_\Omega M_u(g)h\,dx =$

$$\int_{\Omega\backslash P(u)} g(\beta(u))h\,dx + \sum_{i\in D} |P_i| \left(\fint_{P_i^*} g\,d\sigma\right)\left(\fint_{P_i} h(x)dx\right) \tag{11.1}$$

($\fint_A$ denoting the average over A).
So by Hardy–Littlewood inequality:

$$\int_{\Omega\backslash P(u)} g(\beta(u))h\,dx \leqslant \int_{\Omega_*} g_* h_* d\sigma \leqslant |g|_{(p,q)} \cdot |h|_{p',q'}. \tag{11.2}$$

By the Hardy–Littlewood inequality, we deduce:

$$\fint_{P_i^*} g\,d\sigma \leqslant \frac{1}{t}\int_0^t g_*(\sigma)d\sigma \ si\ t \leqslant |P_i|.$$

Hence

$$\int_{P_i^*} t^{\frac{q}{p}-1} \left(\fint_{P_i^*} g \, d\sigma \right)^q \leqslant \int_{P_i^*} t^{\frac{q}{p}-1} g_{**}(t)^q \, dt.$$

Therefore

$$\fint_{P_i^*} g \, d\sigma \leqslant \left(\frac{q}{p}\right)^{\frac{1}{q}} |P_i|^{-\frac{1}{p}} \left(\int_{P_i^*} \left[t^{\frac{1}{p}} g_{**}(t) \right]^q \frac{dt}{t} \right)^{\frac{1}{q}}.$$

Likewise

$$\fint_{P_i} h(x) dx \leqslant \left(\frac{q'}{p'}\right)^{\frac{1}{q'}} |P_i|^{\frac{1}{p'}} \left(\int_{P_i^*} \left[t^{\frac{1}{p}} h_{**}(t) \right]^{q'} \frac{dt}{t} \right)^{\frac{1}{q'}}.$$

Let

$$a_i = \int_{P_i^*} \left[t^{\frac{1}{p}} g_{**}(t) \right]^q \frac{dt}{t}, \qquad b_i = \int_{P_i^*} \left[t^{\frac{1}{p}} h_{**}(t) \right]^{q'} \frac{dt}{t}. \tag{11.3}$$

Then, we deduce from relations (11.1) to (11.3)

$$\int_\Omega M_u(g) h dx \leqslant |g|_{(p,q)} \, |h|_{(p',q')} + \sum_{i \in D} a_i^{\frac{1}{q}} b_i^{\frac{1}{q'}}. \tag{11.4}$$

By the discrete Hölder inequality (D is at most countable)

$$\sum_{i \in D} a_i^{\frac{1}{q}} b_i^{\frac{1}{q'}} \leqslant \left(\sum_{i \in D} a_i \right)^{\frac{1}{q}} \left(\sum_{i \in D} b_i \right)^{\frac{1}{q'}}$$

$$= \left(\int_{\cup P_i^*} \left[t^{\frac{1}{p}} g_{**}(t) \right]^q \frac{dt}{t} \right)^{\frac{1}{q}} \cdot \left(\int_{\cup P_i^*} \left[t^{\frac{1}{p}} h_{**}(t) \right]^{q'} \frac{dt}{t} \right)^{\frac{1}{q'}} \leqslant |g|_{(p,q)} \, |h|_{(p',q')} . \tag{11.5}$$

By combining the relations (11.4) and (11.5), we deduce:

$$\int_\Omega M_u(g) h \, dx \leqslant 2 \, |g|_{(p,q)} \, |h|_{(p',q')} . \tag{11.6}$$

Hence $M_u(g) \in L^{p,q}(\Omega)$ and

$$\sup\left\{\int_\Omega (M_u(g)h)\,dx,\ |h|_{(p',q')}=1\right\} \leqslant 2\,|g|_{(p,q)}\,.$$

2. By equimeasurability, we deduce $g\big(\beta(u)\big)_* = g_*$, hence the result.

Exercise 11.1.27 Let $f\ :\ [0,+\infty[\to\ [0,+\infty[$ be measurable, $r,\ q$ two real numbers such that $r<0,\ q\geqslant 1$. Show that there exists a constant $c_1 = c_1(q,r) > 0$ such that

$$\int_0^{+\infty}\left(\int_x^{+\infty} f(t)dt\right)^q \frac{dx}{x^{r+1}} \leqslant c_1\int_0^{+\infty} f(t)^q t^q\frac{dt}{t^{r+1}}.$$

Let $W_0^1(\Omega,|\cdot|_{p,q}) = \Big\{v\in L^1(\Omega) : |\nabla v|\in L^{p,q}(\Omega)\Big\}\cap W_0^{1,1}(\Omega)$.

Show that if $1\leqslant p<N,\ 1\leqslant q\leqslant p,\ p^*=\dfrac{Np}{N-p}$ then,

$$\forall v\in W_0^1(\Omega,|\cdot|_{p,q}),\ v\geqslant 0,\ |v|_{p^*,q}\leqslant c_2\Big||\nabla v|_{*v}\Big|_{p,q}\leqslant c_2|\nabla v|_{p,q}$$

where c_2 is a constant to be specified.

Solution 11.1.27 This is a Hardy inequality, Lemma 1.4.2. We treat it like that of the first version (Chap. 1). For $f\geqslant 0$ measurable, we write for $x>0$

$$\int_x^{+\infty} f(t)dt = \int_x^{+\infty} f(t)t^{\frac{r}{q}-1}t^{1-\frac{r}{q}}dt\,.$$

By applying to this inequality the Hölder inequality relative to the measure $d\mu(t) = t^{\frac{r}{q}-1}dt$, we deduce:

$$\left(\int_x^{+\infty} f(t)dt\right)^q \leqslant \left(\int_x^{+\infty} f(t)^q t^{q-r} t^{\frac{r}{q}-1}dt\right)\left(\int_x^{+\infty} t^{\frac{r}{q}-1}dt\right)^{\frac{q'}{q}}$$

$$\left(\frac{1}{q}+\frac{1}{q'}=1\right) \qquad \leqslant \left|\frac{q}{r}\right|^{q-1} x^{\frac{r}{q'}}\int_x^{+\infty} f(t)^q t^{q-r}t^{\frac{r}{q}-1}dt\,.$$

Hence the following inequality: $\displaystyle\int_0^{+\infty}\left(\int_x^{+\infty} f(t)dt\right)^q \frac{dx}{x^{r+1}}$

$$\leqslant \left|\frac{r}{q}\right|^{q-1}\int_0^{\infty} x^{-\frac{r}{q}-1}dx \int_0^{+\infty} \chi_{[x,+\infty[}(t)h(t)dt$$

where $h(t) = f(t)^q t^{q-r} t^{\frac{r}{q}-1}$. We apply Fubini's theorem to the latter double integral

$$\int_0^{+\infty} x^{-\frac{r}{q}-1}dx \int_0^{+\infty} \chi_{[x,+\infty[}(t)h(t)dt = \int_0^{+\infty} h(t)dt \int_0^t x^{-\frac{r}{q}-1}dx$$

$$= \frac{q}{|r|}\int_0^{+\infty} h(t)t^{-\frac{r}{q}}dt$$

$$= \frac{q}{|r|}\int_0^{+\infty} f(t)^q t^{q-r-1}dt.$$

By combining the last two relations, we have:

$$\int_0^{\infty}\left(\int_x^{\infty} f(t)dt\right)^q \frac{dx}{x^{r+1}} \leqslant \left|\frac{q}{r}\right|^q \int_x^{\infty} f(t)^q t^q \frac{dt}{t^{r+1}}.$$

We apply this inequality and for this, if $v \in W_0^1(\Omega, |\cdot|_{p,q})$, $v \geqslant 0$, we know that $v_* \in W^{1,1}(s, |\Omega|)$, $s > 0$ and $v_*(s) = -\displaystyle\int_s^{|\Omega|} v_*'(t)dt$.

Let us set $f(t) = \begin{cases} -v_*'(t) & if\ t \in \overline{\Omega}_* \\ 0 & if\ t > |\Omega| \end{cases}$. We then have:

$$s^{\frac{q}{p^*}} v_*(s)^q \leqslant s^{\frac{s}{p^*}}\left(\int_s^{+\infty} f(t)dt\right)^q$$

and setting $r = -\frac{q}{p^*}$ this last inequality leads to:

$$\int_{\Omega_*} s^{\frac{q}{p^*}} v_*(s)^q \frac{ds}{s} \leqslant \int_0^{+\infty} \left(\int_s^{+\infty} f(t)dt \right)^q \frac{ds}{s^{r+1}}.$$

By application of the previous Hardy inequality, we have:

$$\int_{\Omega_*} s^{\frac{q}{p^*}} v_*(s)^q \frac{ds}{s} \leqslant \left|\frac{q}{r}\right|^q \int_x^{\infty} f(t)^q t^q \frac{dt}{t^{r+1}} = \left|\frac{q}{r}\right|^q \int_{0_*} \left|v_*'(t)\right|^q t^q \frac{dt}{t^{r+1}}.$$

By the PSR inequality $\left|v_*'(t)\right| \leqslant \frac{t^{\frac{1}{N}-1}}{N\alpha_N^{\frac{1}{N}}} |\nabla v|_{*v}(t)$ and therefore,

$$\int_{\Omega_*} s^{\frac{q}{p^*}} v_*(s)^q \frac{ds}{s} \leqslant \left|\frac{q}{r}\right|^q \int_{\Omega_*} \frac{t^{(\frac{1}{N}-1)q+q-r-1}}{\left(N\alpha_N^{\frac{1}{N}}\right)^q} \left[|\nabla v|_{*v}\right]^q(t)dt.$$

Since $r = -\frac{q}{p^*}$, by simplifying the exponent, we deduce:

$$\int_{\Omega_*} s^{\frac{q}{p^*}} v_*(s)^q \frac{ds}{s} \leqslant \left|\frac{q}{r}\right|^q \frac{1}{\left(N\alpha_N^{\frac{1}{N}}\right)^q} \int_{\Omega_*} t^{\frac{q}{p}} \left[|\nabla v|_{*v}\right]^q(t)dt.$$

Since $q \leqslant p$ the map $v \to |v|_{p,q}$ is a Fatou norm invariant under rearrangement and therefore

$$\left||\nabla v|_{*v}\right|_{p,q} \leqslant |\nabla v|_{p,q}.$$

So

$$|v|_{p^*,q} \leqslant \frac{p^*}{N\alpha_N^{\frac{1}{N}}} \left||\nabla v|_{*v}\right|_{p,q} \leqslant \frac{p^*}{N\alpha_N^{\frac{1}{N}}} |\nabla v|_{p,q}.$$

11.2 Problems

Exercise 11.2.1 Let Ω be a connected bounded open subset of $\mathbb{R}^N$ with Lipschitz boundary and let $f \in L^2(\Omega)$ be of zero mean.

1. Show that there exists a unique function $u \in H^1(\Omega)$ such that

   (a) $\displaystyle\int_\Omega u(x)\,dx = 0,$

   (b) $\displaystyle\int_\Omega \nabla u(x) \cdot \nabla v(x)\,dx = \int_\Omega f(x)v(x)\,dx, \quad \forall v \in H^1(\Omega).$

2. We now assume that Ω is a rectangle of width b and of length a. We set

$$F_u(t) = \min\left(\int_0^t |f_{*u}|_*(\sigma)d\sigma; \quad \int_0^{ab-t} |f_{*u}|_*(\sigma)d\sigma\right)$$

   $k(t) = \min\left(\sqrt{t}, \sqrt{ab-t}\right), \quad t \in [0, ab].$

   Show that $|\nabla u|_{*u}(s) \leqslant \left(\frac{a}{2b}\right)^{\frac{1}{2}} \frac{F_u(s)}{k(s)}$, for almost all $s \in [0, ab]$.

3. Show that $|\nabla u|_{*u} \in L^\infty(\Omega_*)$ and $\left||\nabla u|_{*u}\right|_\infty \leqslant \left(\frac{a}{2b}\right)^{\frac{1}{2}} |f_{*u}|_{L^2(\Omega_*)}$.
4. Derive that $u \in L^\infty(\Omega)$ and give an explicit estimate of $|u|_\infty$.

Solution 11.2.1

1. We apply the Lax-Milgram theorem (see [165]) with the subspace of $H^1(\Omega)$:

$$V = \left\{v \in H^1(\Omega), \int_\Omega v(x)dx = 0\right\}.$$

   We know that $\forall v \in V,\ |v|_{H^1(\Omega)} \leqslant c\,|\nabla v|_{L^2(\Omega)}$ where $c > 0$, and consequently, the bilinear form given by $a(v, w) = \displaystyle\int_\Omega \nabla v \cdot \nabla w\,dx$ is continuous coercive on $V \times V$, and the map $v \to \displaystyle\int_\Omega fv$ is linear continuous on V. There then exists a unique u in V such that

$$a(u, v) = \int_\Omega fv\,dx \quad \forall v \in V,$$

   hence the answer.

2. The function $v = \left(u - u_*(s)\right)_+$ is in $H^1(\Omega)$. Using the techniques of Chap. 5, we deduce that for almost every $s \in \Omega_*$

$$\left(|\nabla u|^2\right)_{*u}(s) = \frac{d}{ds}\int_\Omega f\left(u - u_*(s)\right)_+ dx = \left(\int_0^s f_{*u}(\sigma)d\sigma\right)\left(-\frac{du_*}{ds}(s)\right).$$

By applying the pointwise Poincaré–Sobolev inequality (relative to the rectangle), we know

$$-\frac{du_*}{ds}(s) \leqslant \sqrt{\frac{a}{2b}} \cdot \frac{1}{k(s)} \left|\nabla u\right|_{*u}(s)$$

with $k(s) = \min(s, |\Omega| - s)^{\frac{1}{2}}$, $|\Omega| = ab$.

Since $(|\nabla u|_{*u})^2 \leqslant \left[\,|\nabla u|^2\,\right]_{*u}$, we then have (after simplification)

$$\left|\nabla u\right|_{*u}(s) \leqslant \sqrt{\frac{a}{2b}} \cdot \frac{1}{k(s)} \left|\int_0^s f_{*u}(\sigma)d\sigma\right| \tag{11.7}$$

but as,

$$0 = \int_\Omega f dx = \int_{\Omega_*} f_{*u}(\sigma)d\sigma = \int_0^s f_{*u}(\sigma)d\sigma + \int_s^{|\Omega|} f_{*u}(\sigma)d\sigma$$

we then have according to Hardy–Littlewood:

$$\begin{cases} \left|\displaystyle\int_0^s f_{*u}(\sigma)d\sigma\right| = \left|\displaystyle\int_s^{|\Omega|} f_{*u}(\sigma)d\sigma\right| \leqslant \displaystyle\int_0^{ab-s} |f_{*u}|_*(\sigma)d\sigma \\ \left|\displaystyle\int_0^s f_{*u}(\sigma)d\sigma\right| \leqslant \displaystyle\int_0^s |f_{*u}|(\sigma)d\sigma \end{cases}. \tag{11.8}$$

By introducing

$$F_u(t) = \min\left(\int_0^t |f_{*u}|(\sigma)d\sigma;\ \int_0^{ab-t} |f_{*u}|_*(\sigma)d\sigma\right)$$

the relations (11.7) and (11.8) lead to

$$\left|\nabla u\right|_{*u}(s) \leqslant \sqrt{\frac{a}{2b}} \cdot \frac{F_u(s)}{k(s)} \quad a.e.\ s. \tag{11.9}$$

3. By the Cauchy–Schwartz inequality, $F_u(s) \leqslant k(s)\,|f_{*u}|_{L^2(\Omega_*)}$ which gives by the relation (11.9):

$$\left|\,|\nabla u|_{*u}\right|_\infty \leqslant \sqrt{\frac{a}{2b}}\,|f_{*u}|_{L^2(\Omega_*)}\,. \tag{11.10}$$

4. By taking the PSR inequality, and the inequality (11.10), we deduce:

$$-\,u_*'(s) \leqslant \frac{a}{2b}\,|f_{*u}|_{L^2(\Omega_*)}\cdot\frac{1}{k(s)}\ a.e.\ s.$$

Since $\int_\Omega u\,dx = 0$, there exists $s_0 \in \overline{\Omega}_*$ s.t. $v_*(s_0) = 0$. So,

$$u_*(0) = |u_*(0)| = -\int_0^{s_0} u_*'(t)dt \leqslant \frac{a}{2b}\,|f_{*u}|_{L^2}\cdot\int_{\Omega_*}\frac{dt}{k(t)}$$

and

$$\left|u_*(\,|\Omega|\,)\right| = -\int_{s_0}^{|\Omega|} u_*'(t)dt \leqslant \frac{a}{2b}\,|f_{*u}|_{L^2}\cdot\int_{\Omega_*}\frac{dt}{k(t)}.$$

As,

$$|u|_\infty = \mathrm{Max}\,\left(|u_*(0)|\,,\,\left|u_*(\,|\Omega|\,)\right|\right) \leqslant \frac{a}{2b}\,|f_{*u}|_{L^2}\cdot\int_{\Omega_*}\frac{dt}{k(t)}$$

and

$$\int_{\Omega_*}\frac{dt}{k(t)} = 2\int_0^{\frac{ab}{2}}\frac{dt}{\sqrt{t}} = 4\left(\frac{ab}{2}\right)^{\frac{1}{2}},$$

we deduce

$$|u|_\infty \leqslant a\left(\frac{2a}{b}\right)^{\frac{1}{2}}|f_{*u}|_{L^2} \leqslant a\left(\frac{2a}{b}\right)^{\frac{1}{2}}|f|_{L^2}\,.$$

Exercise 11.2.2 Let $W^1L^{N,1}(\Omega) = \Big\{v \in L^1(\Omega), |\nabla v| \in L^{N,1}(\Omega)\Big\}$ where Ω is an open connected bounded Lipschitzian boundary.

Q.0 Show that $u_* \in W^{1,1}(\Omega_*)$ if $u \in W^1L^{N,1}(\Omega)$ by giving an estimate. We propose to show that if $u_n \to u$ in $W^1L^{N,1}(\Omega)$ and if measure$\Big\{x : \nabla u(x){=}0\Big\}$=measure$\Big\{x : \nabla u_n(x){=}0\Big\}$=0

then $u'_{n_*} \to u'_*$ in $L^1(\Omega_*)$. We will need some of the following preliminary results:

Q.1 Show that if $u_n \to u$ in $W^1L^{N,1}(\Omega)$ then $u_n \to u$ in $W^{1,1}(\Omega)$.

Q.2 Show that $u'_{n_*} \rightharpoonup u'_*$ in $L^1(\Omega_*)$-weak.

Q.3 Show that for all $a \geqslant 0,\ b \geqslant 0$, we have:

$$\sqrt{1+b^2} \geqslant \sqrt{1+a^2} + \frac{a(b-a)}{\sqrt{1+a^2}} + \frac{1}{2}\frac{(b-a)^2}{\big[1+\max(a,b)^2\big]^{\frac{3}{2}}}\,.$$

Q.4 Show that if $\lim\limits_n \int_{\Omega_*} \sqrt{1+u'^2_{n_*}(s)}\,ds = \int_{\Omega_*} \sqrt{1+u'^2_*(s)}\,ds$ then $u'_{n_*} \to u'_*$ in $L^1(\Omega_*)$-strong.

Q.5 We denote by m (resp. m_n) the distribution function of u (resp. of u_n) and we now assume that measure$\Big\{x : \nabla u(x) = 0\Big\}$ = measure$\Big\{x : \nabla u_n(x) = 0\Big\}$ = 0.

Show that m and m_n are absolutely continuous functions from $\mathbb{R}$ to Ω_* and show that if $R(u) = [Inf_\Omega u,\ Sup_\Omega u]$ then

$$\int_{R(u)} \sqrt{1+m'^2(t)}\,dt = \int_{\Omega_*} \sqrt{1+u'^2_*}\,ds$$

(same for m_n and u'_{n_*}.)

We admit that for almost all t in $\mathbb{R}$, $\big|m'(t)\big| \leqslant \operatorname{liminf}\limits_n \big|m'_n(t)\big|$.

Q.6 Show $\lim\limits_n \int_t^{+\infty} m'_n(\sigma)d\sigma = \int_t^{+\infty} m'(\sigma)d\sigma$ for all $t \in [-\infty, +\infty]$ and $\lim\limits_n \int_{\mathbb{R}} |m'_n(t) - m'(t)|\,dt = 0$.

Q.7 Deduce from the previous questions that

$$\lim_n \int_{\Omega_*} |u'_{n_*} - u'_*|\,d\sigma = 0\,.$$

The answers to the questions in this exercise are mostly given in the course. They are briefly described. We will denote by c different constants depending only on Ω and N and on u.

Q0: According to the PSR inequality, we have for

$$u \in W^1L^{N,1}(\Omega),\ \left|u'_*(s)\right| \leqslant Q \max(s, |\Omega| - s)^{\frac{1}{N}-1} |\nabla u|_{*u}(s)$$

whence by the Hardy–Littlewood inequality

$$\int_{\Omega_*} \left|u'_*(s)\right| ds \leqslant Q2^{1-\frac{1}{N}} \int_{\Omega_*} [s^{\frac{1}{N}} (|\nabla u|_{*u})_* (s)] \frac{ds}{s}$$

$$\leqslant Q2^{1-\frac{1}{N}} |\nabla u|_{(N,1)} \leqslant \frac{NQ2^{1-\frac{1}{N}}}{N-1} |\nabla u|_{N,1}.$$

Q1: $0 < s < |\Omega|,\ \displaystyle⨍_{\Omega_*} |\nabla(u - u_n)|_* (\sigma) d\sigma \leqslant \frac{1}{s} \int_0^s |\nabla(u - u_n)|_* (\sigma) d\sigma.$

Hence

$$\int_\Omega |\nabla(u - u_n)|\, dx \leqslant \frac{|\Omega|^{1-\frac{1}{N}}}{N} |\nabla(u - u_n)|_{(N,1)}$$

$$\leqslant \frac{|\Omega|^{1-\frac{1}{N}}}{N-1} |\nabla(u - u_n)|_{N,1}.$$

Q2: Let $u_n \to u$ in $W^1L^{N,1}(\Omega)$. Then u'_{n*} remains in a bound subset of $L^1(\Omega_*)$ as $u_{n*} \to u_*$ in $L^1(\Omega_*)$ so $u'_{n*} \rightharpoonup u'_*$ in $\mathcal{D}(\Omega_*)$. Now let $E \subset \Omega_*$. By the pointwise PSR inequality and the Hardy–Littlewood's inequality, we deduce

$$\int_E \left|u'_{n*}\right| d\sigma \leqslant c \int_0^{|E|} s^{\frac{1}{N}-1} \left(|\nabla u_n|_{*u_n}\right)_* (s) ds$$

$$\leqslant c \int_0^{|E|} s^{\frac{1}{N}-1} |\nabla u_n|_{**} (s) ds$$

$$\int_E \left|u'_{n*}\right| d\sigma \leqslant c\, |\nabla(u - u_n)|_{N,1} + c \int_0^{|E|} s^{\frac{1}{N}-1} |\nabla u|_{**} (s) ds.$$

Since $\displaystyle\int_0^{|E|} s^{\frac{1}{N}-1} |\nabla u|_{**} (s) ds \xrightarrow[|E| \to 0]{} 0$, we deduce from this last inequality that (u'_{n*}) satisfies the Dunford–Pettis conditions .

Consequently $u'_{n*} \rightharpoonup u'_*$ in $L^1(\Omega_*)$-weak.

Q3: We apply the Taylor–Lagrange formula to the point a with $\Phi(t) = \sqrt{1+t^2}$, then

$$\Phi(b) = \Phi(a) + (b-a)\Phi'(a) + \frac{(b-a)^2}{2}\Phi''(c) \quad c \in]a, b[$$

$$\sqrt{1+b^2} = \sqrt{1+a^2} + \frac{a(b-a)}{\sqrt{1+a^2}} + \frac{(b-a)^2}{2(1+c^2)^{\frac{3}{2}}},\ 0 \leqslant c \leqslant \max(a,b)$$

hence

$$\sqrt{1+b^2} \geqslant \sqrt{1+a^2} + \frac{a(b-a)}{\sqrt{1+a^2}} + \frac{(b-a)^2}{2\big(1+\max(a,b)^2\big)^{\frac{3}{2}}}.$$

Q4: We apply the previous inequality with $a = |u'_*(s)| = -u'_*(s)$, $b = -u'_{n*}(s)$. So

$$\int_{\Omega_*} \left(\sqrt{1+u'_{n*}(s)^2} - \sqrt{1+u'_*(s)^2}\right) ds$$

$$\geqslant \int_{\Omega_*} \frac{u'_*}{\sqrt{1+u'^2_*}}(u'_{n*} - u'_*)ds + \int_{\Omega_*} (u'_{n*} - u'_*)^2 K_n(s)ds.$$

By weak convergence $\lim_n \int_{\Omega_*} \frac{u'_*}{\sqrt{1+u'^2_*}}(u'_{n*} - u'_*)ds = 0$ with the hypothesis of the question, we deduce:

$$\lim_n \int_{\Omega_*} (u'_{n*} - u'_*)^2 K_n(s)ds = 0$$

$$\textit{where } K_n(s) = \frac{1}{2} \cdot \frac{1}{\left[1 + \max\left(|u'_{n*}(s)|, |u'_*(s)|\right)^2\right]^{\frac{3}{2}}}.$$

Let $\theta > 0$, $b_n(s) = \max\left(|u'_{n*}(s)|, |u'_*(s)|\right)$, $s \in \Omega_*$. So,

$$|u'_{n*} - u'_*|_1 \leqslant \int_{b_n > \theta} |u'_{n*} - u'_*|\,dt + \int_{b_n \leqslant \theta} |u'_{n*} - u'_*|\,dt.$$

We then have

$$\int\limits_{b_n \leqslant \theta} \left| u'_{n*} - u'_*\right| dt \;\leqslant c\left(1+\theta^2\right)^{\frac{3}{4}} \left(\int_{\Omega_*} (u'_{n*} - u'_*)^2 K_n(s) ds\right)^{\frac{1}{2}}.$$

As

$$|b_n > \theta| \leqslant \frac{1}{\theta}\int_{\Omega_*} b_n(t)dt \leqslant \frac{1}{\theta}\left(\int_{\Omega_*} |u'_{n*}| + \int_{\Omega_*} |u'_*|\right) \leqslant \frac{c}{\theta},$$

using question Q2, we have:

$$\int\limits_{b_n > 0} \left| u'_{n*} - u'_*\right| dt \leqslant c \int\limits_0^{|b_n > \theta|} s^{\frac{1}{N}-1} \left|\nabla u\right|_{**} ds \;+ c \int\limits_0^{|b_n > \theta|} s^{\frac{1}{N}-1} \left|\nabla u_n\right|_{**} ds$$

$$\leqslant c\left|\nabla(u_n - u)\right|_{N,1} + c\int\limits_0^{\frac{c}{\theta}} s^{\frac{1}{N}-1} \left|\nabla u\right|_{**} ds\,.$$

So

$$\left|u'_{n*} - u'_*\right| \leqslant c(1+\theta^2)^{\frac{3}{4}} \left(\int_{\Omega_*} (u'_{n*} - u'_*)^2 K_n(s) ds\right)^{\frac{1}{2}}$$

$$+ c\int\limits_0^{\frac{c}{\theta}} s^{\frac{1}{N}-1} \left|\nabla u\right|_{**} ds \;+ c\left|\nabla(u_n - u)\right|_{N,1}.$$

Hence

$$\limsup_n \left|u'_{n*} - u'_*\right|_1 \leqslant c\int\limits_0^{\frac{c}{\theta}} s^{\frac{1}{N}-1} \left|\nabla u\right|_{**} ds \xrightarrow[\theta\to+\infty]{} 0.$$

According to the co-area formula, m and m_n are absolutely continuous since we have

$$\left(m(a) - m(b) = \int_a^b dt \int\limits_{u=t} \frac{dH_{N-1}(x)}{|\nabla u|\,(x)} \; if \; N \geqslant 2 \right)$$

$$\int_{-\infty}^{+\infty} dt \int\limits_{u=t} \frac{dH_{N-1}(x)}{|\nabla u|\,(x)} < +\infty, \text{ likewise for } N = 1.$$

We change the variables, $t = u_*(\sigma)$

$$\int\limits_{R(u)} \sqrt{1 + m'^2(t)dt} = \int_{\Omega_*} \sqrt{1 + u_*'^2(\sigma)d\sigma}.$$

For question Q6 see Chap. 6. Question Q7 follows from Q6, Q5 and Q4.

Exercise 11.2.3 (Plasma Problem) Let $I > 0$, $\lambda > 0$ be two given real numbers, Ω be a bounded open connected subset of $\mathbb{R}^2$ of class C^∞.

1. Show that for all $v \in L^2(\Omega)$, there exists a unique constant $C(v)$ such that

$$\int_\Omega \big(v(x) + C(v)\big)_- dx = \frac{I}{\lambda}$$

   (t_- denotes the negative part of t). Henceforth $C : L^2(\Omega) \to \mathbb{R}$ we denote the function.

   Now for $x \in \Omega$, let $A(x)$ be a matrix with coefficients in $L^\infty(\Omega)$ and coercive in the sense that:

   $\forall \xi \in \mathbb{R}^2 \; A(x)\xi \cdot \xi \geqslant |\xi|^2$, and let

$$\begin{array}{c} S : L^4(\Omega) \to L^4(\Omega) \\ \varphi \mapsto \psi \end{array} \iff \begin{cases} Av = -\text{div}\big(A(x)\nabla v\big) = -\lambda\varphi \\ v \in H_0^1(\Omega) \\ \psi = \big(v + C(v)\big)_- \end{cases} .$$

2. Show that S is well defined.
3. Let v be the solution of $Av = -\lambda\varphi$, $v \in H_0^1(\Omega)$. Show that there exists a constant $a > 0$ (depending only on Ω) to be specified, such that $|v|_\infty \leqslant \lambda a |\varphi|_{L^4(\Omega)}$.

4. Deduce that for $\psi = \big(v + C(v)\big)_-$, we have

$$|\psi|_\infty \leqslant 2\lambda a|\varphi|_{L^4(\Omega)} + \frac{I}{\lambda|\Omega|}\,.$$

What is the value of $|\psi|_{L^1(\Omega)}$?

5. Show that $|\psi|^4_{L^4} \leqslant \frac{I}{\lambda}|\psi|^3_\infty$.
6. Derive from the previous questions that for all $\varphi \in L^4(\Omega)$, we have:

$$|S\varphi|^4_{L^4} \leqslant \big(C_1|\varphi|_{L^4} + C_2\big)^3\,,$$

where C_1, C_2 are constants to be specified depending on $a, \lambda, I, |\Omega|$.

7. Let R be the largest positive root of

$$X^4 - \big(C_1 X + C_2\big)^3 = 0$$

(check that it exists).

Denote by $B_4(R)$ the ball $\Big\{v \in L^4(\Omega) : |v|_{L^4} \leqslant R\Big\}$ of $L^4(\Omega)$. Show that if $|\varphi|_{L^4} \leqslant R$ then $|S\varphi|_{L^4} \leqslant R$ and deduce that iS maps, $B_4(R)$ to $B_4(R)$.

8. By decomposing S, show that S is compact from $L^4(\Omega)$ to $L^4(\Omega)$.
9. Deduce that S has a fixed point $u \in H^1_0(\Omega) \cap L^\infty(\Omega)$.
10. Deduce that there exists $w \in H^1_0(\Omega) \oplus \mathbb{R}$ satisfying:

$$\begin{cases} Aw + \lambda w_- = 0 \\ \displaystyle\int_\Omega w_- dx = \frac{I}{\lambda}. \end{cases}$$

Solution 11.2.3

1. $t \to \displaystyle\int_\Omega (v(x) + t)_-\,dx = G_v(t)$ is strictly decreasing, moreover

$$G_v(t) \xrightarrow[t\to+\infty]{} 0, \qquad G_v(t) \xrightarrow[t\to-\infty]{} +\infty.$$

Indeed, by Fatou's theorem, we have:

$$\underline{\lim}_{t\to-\infty} G_v(t) \geqslant \int_\Omega \lim_{t\to-\infty} \big(v(x) + t\big)_- dx = +\infty$$

and

$$G_v(t) \leqslant \int_\Omega (t - v_-(x))_- dx, \qquad \lim_{t \to +\infty} (t - v_-(x))_- = 0$$

$$(t - v_-(x))_- \leqslant v_-(x),$$

hence the theorem of dominated convergence implies that

$$\lim_{t \to +\infty} \int_\Omega (t - v_-(x))_- \, dx = 0.$$

Therefore there exists a unique $C(v)$ s.t. $G_v\big(C(v)\big) = \frac{I}{\lambda}$.

2. For $\varphi \in L^4(\Omega) \subset L^2(\Omega)$, by the Lax–Milgram theorem, there exists a unique $v \in H_0^1(\Omega)$ s.t. $Av = -\lambda\varphi$. Since $C(v)$ is unique $\psi = \big(v + C(v)\big)_- = S\varphi$ is also unique.
3. Let $w = |v|$ and $\big(w - w_*(s)\big)_+ \operatorname{sign}(v) \in H_0^1(\Omega)$, then

$$\frac{d}{ds} \int_{w > w_*(s)} |\nabla w|^2 \, dx = -\frac{d}{ds} \int_{\Omega_*} \lambda\varphi\big(w - w_*(s)\big)_+ \operatorname{sign}(v),$$

$$\Big[|\nabla w|^2\Big]_{*w}(s) = +\left(\int_0^s \big(\lambda\varphi \operatorname{sign}(v)\big)_{*w}(t)dt\right)\big(w_*'(s)\big)$$

(see Chap. 5). Hence, the PSR inequality leads to

$$\left(\Big[|\nabla w|^2\Big]_{*w}\right)^{\frac{1}{2}} \leqslant \lambda Q k^{-1}(s) \left(\int_0^s |\varphi|_*(t)dt\right)$$

where $k(s) = \sqrt{s}$, $Q = \dfrac{1}{2\sqrt{\pi}}$. From this we obtain

$$-w_*'(s) \leqslant \lambda \frac{1}{4\pi} s^{-1} s^{\frac{3}{4}} |\varphi|_{L^4},$$

$$|v|_\infty = -\int_0^{|\Omega|} w_*'(s)ds \leqslant \lambda \frac{|\Omega|^{\frac{3}{4}}}{3\pi} |\varphi|_{L^4}$$

and we let

$$a = \lambda \frac{|\Omega|^{\frac{3}{4}}}{3\pi}.$$

4. Since the map $t \to G_v(t)$ is decreasing and $G_v(|v|) = 0$ we have $|v|_\infty \geqslant C(v)$.

   - If $C(v) \geqslant 0$ then $|\psi|_\infty \leqslant |v|_\infty + C(v) \leqslant 2|v|_\infty$.
   - If $C(v) < 0$ then let's distinguish two cases:
     if $|v|_\infty + C(v) \geqslant 0$ then $|C(v)| \leqslant |v|_\infty$
     and if $|v|_\infty + C(v) \leqslant 0$ then

$$\frac{I}{\lambda} = \int_\Omega (v + C(v))_- \geqslant |\Omega| \left(|v|_\infty + C(v)\right)_-$$

$$= -|\Omega| \left(|v|_\infty + C(v)\right)$$

$\dfrac{I}{\lambda|\Omega|} + |v|_\infty \geqslant |C(v)|$ then

$$|\psi|_\infty \leqslant 2|v|_\infty + \frac{I}{\lambda|\Omega|} \; and \; |\psi|_{L^1} = \frac{I}{\lambda}.$$

5.

$$|\psi|_{L^4}^4 \leqslant \int_\Omega |\psi|^4 \leqslant |\psi|_\infty^3 \int_\Omega \psi = \frac{I}{\lambda} |\psi|_\infty^3$$

$$|S\varphi|_{L^4}^4 = |\psi|_{L^4}^4 \leqslant \frac{I}{\lambda}\left(2\lambda a |\varphi|_{L^4} + \frac{I}{\lambda|\Omega|}\right)^3$$

$$|S\varphi|_{L^4}^4 \leqslant \left(2\lambda a \left(\frac{I}{\lambda}\right)^{\frac{1}{3}} |\varphi|_{L^4} + \left(\frac{I}{\lambda}\right)^{\frac{4}{3}} \frac{1}{|\Omega|}\right)^3.$$

Hence,

$$C_1 = 2\lambda a \left(\frac{I}{\lambda}\right)^{\frac{1}{3}}, \qquad C_2 = \frac{1}{|\Omega|}\left(\frac{I}{\lambda}\right)^{\frac{4}{3}}$$

$$|S\varphi|_{L^4}^4 \leqslant \left(C_1 |\varphi|_{L^4} + C_2\right)^3.$$

Let the polynomial equation $X^4 - (C_1 X + C_2)^3 = 0$ has a root $R > 0$ because

$$F(X) = X^4 - (C_1 X + C_2)^3 \xrightarrow[X \to +\infty]{} +\infty, \qquad F(0) = -C_2^3 < 0.$$

If $|\varphi|_{L^4} \leqslant R$ then $|S\varphi|_{L^4}^4 \leqslant (C_1 R + C_2)^3 = R^4$.

Let $|S\varphi|_{L^4} \leqslant R : S : B_4(R) \to B_4(R)$

$$S : L^4(\Omega) \xrightarrow{L} H_0^1(\Omega) \subset L^\infty(\Omega) \xrightarrow{i} L^4 \xrightarrow{C+I} L^4 \xrightarrow{P} L^4$$

$$\varphi \quad \longmapsto \quad v \quad \longrightarrow v \longrightarrow v + Cv \longrightarrow \big(v + C(v)\big)_-$$

then L continuous, i =compact, $C + I$ continuous, P continuous $\implies S$ is compact.

There exists $u \in H_0^1(\Omega) \subset L^\infty(\Omega)$ s.t

$$Su = u$$

(Leray–Schauder theorem).

We have:

$$Su = u \iff \begin{cases} u = \big(v + C(v)\big)_- \\ -div\big(A(x)\nabla v\big) = -\lambda u \end{cases}.$$

If we set $w = v + C(v)$ then

$$\begin{cases} Au = Aw = -\lambda w_- \\ w \in H_0^1(\Omega) \oplus \mathbb{R} \\ \displaystyle\int_\Omega w_- dx = \frac{I}{\lambda}. \end{cases}$$

Exercise 11.2.4 We recall that if $v \in W^{1,p}(0, 1)$, $1 \leqslant p \leqslant +\infty$, then

(i) $v_* \in W^{1,p}(0, 1)$, $\quad |v'_*|_{L^p(0,1)} \leqslant |v'|_{L^p(0,1)}$.
(ii) If $g \geqslant 0$ is measurable then

$$\int_0^1 |v'(x)|\, g(x)dx = \int_{-\infty}^{+\infty} \left(\sum_{x \in v^{-1}(t)} g(x) \right) dt.$$

We denote m_v as the distribution function of v or more simply m if there is no confusion.

Part I Let $u \in W^{1,p}(0, 1)$, $1 \leqslant p < +\infty$.

1. Show that for almost all $\theta \in \left(\inf_{[0,1]} u; \sup_{[0,1]} u \right)$ we have:

$$-\frac{d}{d\theta} \int_{u_* > \theta} |u'_*(\sigma) d\sigma| = 1.$$

2. Show that if $\text{measure}\Big\{x \in (0,1),\ u'(x) = 0\Big\} = 0$ then for almost all $\theta(\inf u, \sup u)$, $m'(\theta) \neq 0$ and $1 = \big(-m'(\theta)\big(u'_*(m(\theta))\big)$ where $m(\theta) = |\{u > \theta\}|$.
3. We consider the following function: for $\theta > 0$ and $h > 0$, $t \in \mathbb{R}$

$$S_{\theta,h}(t) = \begin{cases} 0 & if\ |t| \leqslant 0 \\ \text{sign}\,(t) & |t| \geqslant \theta + h \\ \dfrac{1}{h}(t-\theta) & \theta \leqslant t \leqslant \theta + h \\ \\ \dfrac{1}{h}(t+\theta) & -\theta - h \leqslant t \leqslant -\theta. \end{cases}$$

We set $v = |u|$, then show that for almost all $\theta > 0$

$$-\frac{d}{d\theta}\int_{\{v_*>\theta\}} |v'_*(\sigma)|^p\, d\sigma \leqslant -\frac{d}{d\theta}\int_{\{v>\theta\}} |v'(x)|^p\, dx\,.$$

4. Show that for almost all $\theta \in (\inf_{[0,1]} v,\ \sup_{[0,1]} v)$

$$1 \leqslant \left(-\frac{d}{d\theta}\int_{u>\theta} |v'(x)|^p\, dx\right)\big(-m'_v(\theta)\big)^{p-1}.$$

Part II We now propose to use this inequality in the evolution equations. We consider for $j \geqslant 1$, $\varphi_j(x) = \sqrt{2}\sin(j\pi x)$, $x \in (0,1)$. Let $V_m = vect\{\varphi_1, \cdots, \varphi_m\}$ (subspace generated by $\varphi_1, \cdots, \varphi_m$), P_m the orthogonal projection of $L^2(0,1)$ on V_m and $T > 0$.

1. Let $f \in L^2(]0,T[\times]0,1[)$, $u_0 \in L^2(0,1)$. Show that there exists a unique function $u_m \in H^1(0,T;V_m)$, $\frac{d}{dt}\int_0^1 u_m(t,x)\psi(x)dx + \int_0^1 \frac{\partial u_m}{\partial x}(t,x)\psi'(x)dx = \int_0^1 f(t,x)\psi(x)dx\ \forall\psi \in V_m\ and\ u_m(0) = P_m u_0$.
2. Show that u_m remains within a bound of $H^1(]0,T[\times]0,1[)$ when m varies. Show that if $u_m(t)$ is not identical zero (for fixed t). Then $\forall j \in \mathbb{N}$, $\forall\theta \in \mathbb{R}$,

$$\text{measure}\Big\{x \in (0,1),\ \frac{\partial^j}{\partial x^j}u_m(t,x) = \theta\Big\} = 0.$$

3. We set $v_m = |u_m|$. Show that for almost all $t \in (0, T)$ and for almost all $\theta > 0$

$$\int_{v_m(t)>\theta} \frac{\partial u_m}{\partial t}(t,x)dx - \frac{d}{d\theta}\int_{v_m(t)>\theta}\left(\frac{\partial u_m}{\partial x}\right)^2(t,x)dx$$

$$= \int_{v_m(t)>\theta} \big(P_m f(t)\big)(x)\,\text{sign}\,u_m dx.$$

4. For $t \in (0, T)$, $s \in (0, 1)$, $\theta > 0$, we define the functions $k_m(t,s) = \int_0^s v_{m*}(t,\sigma)d\sigma$, $g_m = |P_m f(t)|$ and $M(t,\theta) = \text{measure}\{x \in (0,1),\ v_m(t,x) > \theta\}$.
Show that $\dfrac{\partial k_m}{\partial t}(t,s) = \displaystyle\int_0^s \frac{\partial u_{m*}}{\partial t}(t,\sigma)d\sigma$ in $\mathcal{D}'(]0,T[\times]0,1[)$ and that for almost every $\theta \in \big(\inf_{[0,1]} v_m(t), \sup_{[0,1]} v_m(t)\big)$ (t being fixed),

$$1 \leqslant \left[\int_0^{M(t,\theta} g_{m*}(t)(\sigma)d\sigma - \int_0^{M(t,\theta)} \frac{\partial v_{m*}}{\partial t}(t,\sigma)d\sigma\right]\Big(-\frac{\partial M}{\partial \theta}(t,\theta)\Big).$$

5. Deduce that for fixed t, we have for almost every $s \in [0, 1]$,

$$-\frac{\partial u_{m*}}{\partial s}(t,s) \leqslant \int_0^s g_{m*}(t,s)d\sigma - \int_0^s \frac{\partial u_{m*}}{\partial t}(t,\sigma)d\sigma$$

and that in the sense of the distributions in $\mathcal{D}'(]0,T[\times]0,1[)$

$$\frac{\partial k_m}{\partial t}(t,s) - \frac{\partial^2 k_m}{\partial s^2}(t,s) \leqslant \int_0^s g_{m*}(t,\sigma)d\sigma.$$

6. Show that there is a unique function $u \in H^1(]0,T[\times]0,1[)$ s.t.

$$\frac{\partial u}{\partial t} - \frac{\partial^2 u}{\partial x^2} = f$$

$$u(0) = u_0,\ u(t,0) = u(t,1) = 0, \quad \forall t \in [0,1]$$

and if we put $k(t,s) = \int_0^1 u_*(t,\sigma)d\sigma$ then

$$\frac{\partial k}{\partial t}(t,s) - \frac{\partial^2 k}{\partial s^2}(t,s) \leqslant \int_0^s f_*(t,\sigma)d\sigma \text{ in } \mathcal{D}'(]0,T[\times]0,1[).$$

Solution 11.2.4 The solution is left to the reader.

Exercise 11.2.5 Let Ω be an open bounded subset of $\mathbb{R}^N$, $f \in L^{p'}(\Omega)$, $1 < p' < +\infty$, $f \geqslant 0$.

1. Explicitly determine the solution

$$U \in W_0^{1,p}(\underset{\sim}{\Omega})\ of\ -\Delta_p U = -\text{div}\left(|\nabla U|^{p-2}\nabla U\right) = \underset{\sim}{f}$$

   in $\underset{\sim}{\Omega}$. Here $\frac{1}{p} + \frac{1}{p'} = 1$.
   Give U'_*.
2. Let $\widehat{a} : \Omega \times \mathbb{R} \times \mathbb{R}^N \to \mathbb{R}^N$ satisfying a.e. $x \in \Omega$, $\forall(\sigma,\xi) \in \mathbb{R} \times \mathbb{R}^N$, $\widehat{a}(x,\sigma,\xi)\cdot\xi \geqslant |\xi|^p$. Show that all solutions

$$u \in W_0^{1,p}(\Omega) \text{ of } -\text{div}\big(\widehat{a}(x,u,\nabla u)\big) = f\ in\ \mathcal{D}'(\Omega)$$

   such that $\widehat{a}(x,u,\nabla u) \in L^{p'}(\Omega)^N$ satisfy
   (a) $u \geqslant 0$
   (b) $\left|u'_*(s)\right| \leqslant \left|U'_*(s)\right|$ $a.e.\ s \in \Omega_*$.
3. Show that $\underset{\sim}{u}(x) \leqslant U(x) \quad \forall x \in \underset{\sim}{\Omega}$.

Solution 11.2.5

1. If $f \in L^{p'}(\Omega)$ then $\underset{\sim}{f} \in L^{p'}(\underset{\sim}{\Omega})$. By minimizing the functional $J(v) = \frac{1}{p}\int_{\underset{\sim}{\Omega}} |\nabla v|^p\,dx - \int_{\underset{\sim}{\Omega}} \underset{\sim}{f} v$ on $W_0^{1,p}(\underset{\sim}{\Omega})$, we deduce that there exists at least one solution $U \in W_0^{1,p}(\underset{\sim}{\Omega})$ of $-\Delta_p U = \underset{\sim}{f}$. Due to the monotonicity of the p-Laplacian (which is strict) we deduce that this solution is unique. Moreover $f \geqslant 0$ implies that $\underset{\sim}{f} \geqslant 0$ and by the maximum principle we deduce that $U \geqslant 0$. Under these conditions, $J(\underset{\sim}{U}) \leqslant J(U)$ and by strict convexity of J

(or uniqueness of solution) we have: $\underset{\sim}{U} = U$. So to explicitly calculate U, it is enough to calculate U_*.
<u>Search for U</u> Let $s \in \Omega_*$ and choose $\big(U - U_*(s)\big)_+$ as a test function. We deduce

$$\int_{U>U_*(s)} |\nabla U|^p \, dx = \int_{\underset{\sim}{\Omega}} \underset{\sim}{f}\big(U - U_*(s)\big)_+ dx. \tag{11.11}$$

Since $U(x) = U_*(\alpha_N |x|^N)$ we have

$$|\nabla U(x)| = N\alpha_N \left|U_*'\right| \left(\alpha_N |x|^N\right) |x|^{N-1}.$$

Since U is radial, we have:

$$\int_{U>U_*(s)} |\nabla U|^p \, dx = (N\alpha_N)^p \int_{\alpha_N |x|^N < s} \left|U_*'\right|^p (\alpha_N |x|^N) \, |x|^{(N-1)p} \, dx$$

(by changing variables)

$$= (N\alpha_N^{\frac{1}{N}})^p \int_0^s \left|U_*'\right|^p (\sigma) \sigma^{\frac{p}{N'}} d\sigma. \tag{11.12}$$

By combining the relations (11.11) and (11.12) and by differentiating with respect to s, we deduce:

$$(N\, a_N^{\frac{1}{N}})^p \left|U_*'\right|^p (s) s^{\frac{p}{N'}} = -U_*'(s) \int_{U>U_*(s)} \underset{\sim}{f} \, dx = -U_*'(s) \int_0^s f_*(\sigma) d\sigma.$$

Since the solution is unique, we will look for U and show a posteriori that the set $\left\{s : U_*'(s) = 0\right\}$ has zero measure. We have:

$$\left|U_*'\right|^{p-1} (s) = \frac{s^{-\frac{p}{N'}}}{(N\alpha_N^{\frac{1}{N}})^p} \left(\int_0^s f_*(\sigma) d\sigma\right)$$

which gives

$$U_*'(s) = -\frac{s^{-\frac{p'}{N'}}}{(N\alpha_N^{\frac{1}{N}})^{p'}}\left(\int_0^s f_*(\sigma)d\sigma\right)^{\frac{1}{p-1}} \tag{11.13}$$

hence

$$U(x) = U_*(\alpha_N |x|^N) = -\frac{1}{(N\alpha_N^{\frac{1}{N}})^{p'}}\int_{\alpha_N|x|^N}^{|\Omega|} \sigma^{-\frac{p'}{N'}}\left[\int_0^\sigma f_*(t)dt\right]^{\frac{1}{p-1}} d\sigma\,. \tag{11.14}$$

By a direct calculation, we prove that $-\Delta_p U = \underset{\sim}{f}$, $U(x) = 0$, $x \in \partial\Omega$.

By uniqueness of solution, we conclude that U_*' is given by (11.13) and that $\left\{s : U_*'(s) = 0\right\}$ is of zero measure.

2. Let us show that $\left|u_*'(s)\right| \leqslant \left|U_*'(s)\right|$. By applying the techniques of Chap. 5, we have:

$$\left(|\nabla u|^p\right)_{*u}(s) \leqslant \left(-u_*'(s)\right)\int_0^s f_*(t)dt. \tag{11.15}$$

Note that by the maximum principle, a solution u of

$$-div\big(\widehat{a}(x,u,\nabla u)\big) = f \geqslant 0$$

is positive because

$$0 \leqslant \int_\Omega |\nabla u_-|^p\, dx \leqslant -\int_\Omega \widehat{a}(x,u,\nabla u)\cdot\nabla u_- dx = -\int_\Omega f\cdot u_- \leqslant 0$$

(here $u_-(x) = -\min\big(u(x),0\big)$).

Applying the pointwise PSR inequality we have:

$$-u_*'(s) \leqslant \frac{s^{\frac{1}{N}-1}}{N\alpha_N^{\frac{1}{N}}}\left(\left[|\nabla u|^p\right]_{*u}(s)\right)^{\frac{1}{p}}, \; s \in \Omega_*,$$

and the relation (11.15) leads to:

$$\left(\left[|\nabla u|^p\right]_{*u}(s)\right)^{\frac{1}{p}} \leqslant \frac{s^{-\frac{p'}{N'}\frac{1}{p}}}{(N\alpha_N^{\frac{1}{N}})^{\frac{1}{p-1}}}\left(\int_0^s f_*(t)dt\right)^{\frac{1}{p-1}}. \tag{11.16}$$

Since $|\nabla u|_{*u}(s) \leqslant \left(\left(|\nabla u|^p\right)_*(s)\right)^{\frac{1}{p}}$, and the expression of U'_* is given by (11.13), we deduce that:

$$|\nabla u|_{*u}(s) \leqslant \frac{s^{-\frac{p'}{N'}\frac{1}{p}}}{(N\alpha_N^{\frac{1}{N}})^{\frac{1}{p-1}}}\left(\int_0^s f_*(t)dt\right)^{\frac{1}{p-1}} = N\alpha_N^{\frac{1}{N}} s^{\frac{1}{N'}} \left|U'_*(s)\right|.$$

Hence

$$\left|u'_*(s)\right| \leqslant \frac{s^{-\frac{1}{N'}}}{N\alpha_N^{\frac{1}{N}}} |\nabla u|_{*u}(s) \leqslant \left|U'_*(s)\right|.$$

3. Since $u_*(s) = \displaystyle\int_s^{|\Omega|} \left|u'_*(t)\right| dt$ we have $u_*(s) \leqslant U_*(s)$, $\forall s \in \overline{\Omega}_*$ and hence $\underset{\sim}{u}(x) = u_*(\alpha_N |x|^N) \leqslant U_*(\alpha_N |x|^N) = U(x)$.

Exercise 11.2.6 Let Ω be an open set of measure 1, $1 < p < +\infty$, $p' = \dfrac{p}{p-1}$. For $g \geqslant 0$ measurable, we associate

$$|g|_{(p'} = \operatorname{Inf}_{g=\sum_{k=1}^{+\infty} g_k\ g_k \geqslant 0} \left\{ \sum_{k=1}^{+\infty} \inf_{0<\varepsilon<p-1} \varepsilon^{-\frac{1}{p-\varepsilon}} \left(\int_\Omega g_k^{(p-\varepsilon)'} dx\right)^{\frac{1}{(p-\varepsilon)'}} \right\}$$

where $(p-\varepsilon)'$ is the conjugate of $p-\varepsilon$.

1. Show that $\rho(g) = |g|_{(p'}$ is a Fatou norm invariant under rearrangement.
2. Let $L^{(p'}(\Omega) = \left\{g \text{ measurable } : |g|_{(p'} < +\infty\right\}$. Show that $L^{p'+\varepsilon}(\Omega) \subset L^{(p'}(\Omega) \subset L^{p'}(\Omega)\ \forall \varepsilon > 0$.

Exercise 11.2.7 Construct a map $\rho : L^0(\Omega_*) \to [0, +\infty]$ that is not a norm but that is monotonically homogeneous.

Then choose a map ρ s.t.

$$\rho\left(\left(\operatorname{Log}\frac{|\Omega|}{s}\right)^{\frac{1}{N'}}\right) < +\infty,\ N' = \frac{N}{N-1}.$$

Exercise 11.2.8 Show that the functions of $W^{1,1}_{loc}(\mathbb{R})$ are co-area regular.

Exercise 11.2.9 Let $\Omega =]0, 1[\times]0, 1[$, $u(x, y) = y - x$ and $v \in L^1(\Omega)$.
Calculate $v_{*u}(s)$, $s \in [0, 1]$ as a function of v and $u_*(s)$.
Deduce that if $v \in C(\overline{\Omega})$ then $v_{*u} \in C(\overline{\Omega}_*)$.
Show that if $v \in C^1(\overline{\Omega})$, then $v_{*u} \in W^{1,+\infty}_{loc}(\Omega_*)$.

Exercise 11.2.10 Let u, v_1, v_2 be three functions in $L^1(\Omega)$.
Show that $\forall s \in \Omega_*$

$$\int_0^s |v_{1*u} - v_{2*u}|_*(s)ds \leqslant \int_0^s |v_1 - v_2|_*(\sigma)d\sigma.$$

Solution 11.2.10 We know that for all convex and Lipschitzian $\forall \psi : \mathbb{R}_+ \to \mathbb{R}_+$, we have:

$$\int_{\Omega_*} \psi(|v_{1*u} - v_{2*u}|)d\sigma \leqslant \int_{\Omega_*} \psi(|v_1 - v_2|_*)d\sigma.$$

By taking $\psi(\sigma) = (\sigma - t)_+, \theta \in \mathbb{R}$ we deduce:

$$\int_\Omega (|v_{1*u} - v_{2*u}|(\sigma) - t) + d\sigma \leqslant \int_\Omega (|v_1 - v_2|_* - t)_+ d\sigma.$$

Proposition 1.2.1 gives the result.

11.3 New Exercises and Solutions

Exercise 11.3.1 Show that a Banach function space $L(\Omega, \rho)$ is necessarily complete (this is Proposition 1.4.5)

Solution 11.3.1 Consider a Cauchy sequence $(f_n)_n$ of $L(\Omega, \rho)$. Since there exists $c(\Omega) > 0$ such that

$$\int_\Omega |f_n - f_m| \leqslant c(\Omega)\rho(f_n - f_m), \quad \forall n,\ m$$

the sequence is then Cauchy in the Banach space $(L^1(\Omega), |\cdot|_1)$. Therefore there exists $f \in L^1(\Omega) : \lim_n \int_\Omega |f_n - f|dx = 0.$

Let us first show that $f \in L(\Omega, \rho)$. Indeed, we can extract a subsequence $(f_{\sigma(n)})_n$ that converges pointwise on Ω

$$\lim_n |f_{\sigma(n)}(x) - f(x)| = 0 \; a.e. \; in \; \Omega.$$

Consider $g_n = \inf_{k \geqslant n} |f_{\sigma(k)}|$. Then $\lim_n g_n(x) = |f(x)|$ and $g_n \leqslant g_{n+1}$. Thus, since ρ is a Fatou norm,

$$\lim_n \rho(g_n) = \rho(f).$$

Since $\rho(g_n) \leqslant \sup_{j \in \mathbb{N}} \rho(f_j) < +\infty$ we have $\rho(f) < +\infty$.

Let us show $\rho(f - f_m) \xrightarrow[n \to +\infty]{} 0$. Let $n,\ m$ be finite. Consider the function

$$\delta_{nm} = \inf_{k \geqslant n} |f_{\sigma(k)} - f_m|.$$

Then

$$\delta_{nm} \leqslant \delta_{(n+1)m}, \qquad \lim_n \delta_{n,m}(x) = |f - f_m|(x).$$

As $\delta_{nm} \leqslant |f_{\sigma(n)} - f_m|$ we deduce

$$\rho(f - f_m) = \lim_{n \to +\infty} \rho(\delta_{nm}) \leqslant \lim_n \inf \rho(f_{\sigma(n)} - f_m).$$

Let $\varepsilon > 0$, $then\ \exists m_\varepsilon \in \mathbb{N} such that\ \forall n \geqslant m_\varepsilon,\ \forall m \geqslant m_\varepsilon$, we have

$$\rho(f_{\sigma(n)} - f_m) \leqslant \varepsilon.$$

By letting $n \to +\infty$ we deduce

$$\rho(f - f_m) \leqslant \varepsilon.$$

◇

Exercise 11.3.2 Let $L(\Omega, \rho)$ be a Banach function space, invariant under rearrangement. Let u and v in $L(\Omega, \rho)$ s.t. $\rho(v_{*u}) = \rho(v)$ and $v \geqslant 0$.

1. Show that

$$\lim_{\lambda \to 0,\, \lambda > 0} \rho\left(\frac{(u + \lambda v)_* - u_*}{\lambda}\right) = \rho(v_{*u}).$$

2. We assume that $L(\Omega, \rho) = L^p(\Omega),\ 1 < p < +\infty$. Show that in this case

$$\lim_{\lambda \to 0} \rho\left(\frac{(u + \lambda v)_* - u_*}{\lambda} - v_{*u}\right) = 0.$$

3. Then prove the result of Theorem 7.3.4, ii).

Solution 11.3.2

1. Let $\dfrac{dw_\lambda}{ds} = \dfrac{(u + \lambda v)_* - u_*}{\lambda}, \quad \lambda > 0.$

We have from Corollary 1.3.3, Chap. 1:

$$\int_0^s |(u+\lambda v)_* - u_*|_* dt \leqslant \lambda \int_0^s v_*(t)dt \ \ \forall s \in \Omega_*.$$

We deduce, since $\frac{dw_\lambda}{ds} \geqslant 0$,

$$\int_0^s \left(\frac{dw_\lambda}{ds}\right)_*(t)dt \leqslant \int_0^s v_*(t)dt, \ \forall s.$$

By applying Theorem 2.2.1 of Lorentz–Luxemburg,

$$\rho\left(\frac{dw_\lambda}{ds}\right) = \rho\left(\left(\frac{dw_\lambda}{ds}\right)_*\right) \leqslant \rho(v) = \rho(v_{*u}).$$

As $\frac{dw_\lambda}{ds} \underset{\lambda\to 0}{\rightharpoonup} v_{*u}$ in $L^1(\Omega_*)$-weak, for $g \in L(\Omega_*, \rho')$ s.t.
$\rho'(g) \leqslant 1$, we then have

$$0 \leqslant \int_{\Omega_*} \frac{dw_\lambda}{ds} \cdot T_k(|g|)ds \leqslant \rho\left(\frac{dw_\lambda}{ds}\right) \leqslant \rho(v_{*u}).$$

When $\lambda \to 0$, $0 \leqslant \int_{\Omega_*} v_{*u} T_k(|g|)ds \leqslant \underline{\lim}_{\lambda\to 0} \rho\left(\frac{dw_\lambda}{ds}\right)$. When $k \to +\infty$, we deduce

$$0 \leqslant \int_{\Omega_*} v_{*u}|g|ds \leqslant \liminf_{\lambda\to 0} \rho\left(\frac{dw_\lambda}{ds}\right).$$

Hence

$$\rho(v_{*u}) \leqslant \liminf_{\lambda\to 0} \rho\left(\frac{dw_\lambda}{ds}\right) \leqslant \limsup_{\lambda\to 0} \rho\left(\frac{dw_\lambda}{ds}\right) \leqslant \rho(v_{*u}),$$

which proves

$$\lim_{\lambda\to 0} \rho\left(\frac{dw_\lambda}{ds}\right) = \rho(v_{*u}).$$

2. If $L(\Omega, \rho) = L^p(\Omega)$ then $\frac{dw_\lambda}{ds}$ converges to v_{*u} in $L^p(\Omega_*)$-weak when $\lambda \to 0$ as

$$\lim_{\lambda\to 0} \left|\frac{dw_\lambda}{ds}\right|_p = |v_{*u}|_p.$$

The uniform convexity of $L^p(\Omega_*)$ then implies that

$$\lim_{\lambda \to 0} \left| \frac{dw_\lambda}{ds} - v_{*u} \right|_p = 0.$$

3. According to Lemma 7.2.1, if $v \in L^p(\Omega)$ and u is a simple function then $(v_{*u})_* = v_*$. In particular,

$$|v_{*u}|_{L^p(\Omega_*} = |v|_{L^p(\Omega)}.$$

The result then follows from the second question. ◇

Here are two useful exercises for computing K-functionals in space interpolation problems.

The first is Hunt's lemma

Exercise 11.3.3 Let v_1, v_2 be two measurable functions on Ω (always of finite measure).

Then $\forall \sigma_1 \in [0, |\Omega|]$, $\forall \sigma_2 \in [0, |\Omega|]$

$$(v_1 + v_2)_*(\sigma_1 + \sigma_2) \leqslant v_{1*}(\sigma_1) + v_{2*}(\sigma_2).$$

Solution 11.3.3 We put $\lambda = v_{1*}(\sigma_1) + v_{2*}(\sigma_2)$, $t = |v_1 + v_2 > \lambda|$. Then we have the following inclusion

$$\left\{x \in \Omega : (v_1 + v_2)(x) > \lambda\right\} \subset \bigcup_{i=1}^{2} \left\{x : v_i(x) > v_{i*}(\sigma_i)\right\}$$

hence $t = v_1 + v_2 > \lambda|$

$$\leqslant \text{measure}\left\{v_1 > v_{1*}(\sigma_1)\right\} + \text{measure}\left\{v_2 > v_{2*}(\sigma_1)\right\} \leqslant \sigma_1 - \sigma_2.$$

Then $t \leqslant \sigma_1 + \sigma_2$ implies that

$$(v_1 + v_2)_*(\sigma_1 + \sigma_2) \leqslant (v_1 + v_2)_*(|v_1 + v_2 > \lambda|) \leqslant \lambda.$$

◇

Exercise 11.3.4 Let f be a positive measurable function on Ω and λ a real number. We define $g = (f - \lambda)\chi_{\{f>\lambda\}}$.

1. Calculate g_*.
2. We choose $\lambda = f_*(a)$, $a \in]0, |\Omega|[$, and we let $h = f - g$

   (a) Calculate $f_*(s)$, $s \in]0, |\Omega|[$
   (b) What is the relation between $g_* + h_*$ and f_*?

Solution 11.3.4

1. Since $g = (f - \lambda)_+$ and the map $\sigma \xrightarrow{\Phi} (\sigma - \lambda)_+$ is increasing, we have

$$g_* = \Phi(f)_* = \Phi(f_*),$$

see Lemma 1.1.1, hence

$$g_* = (f_* - \lambda)_+ \,.$$

2. (a)

$$\chi_{\{f > f_*(a)*\}}(s) = \chi_{(0,a)}(s) \;\; a.e.$$

and

$$h_*(s) = \begin{cases} f_*(a) = \lambda & if \;\, s \in (0, a) \\ f_*(s) & if \;\, s \in (a, |\Omega|) \end{cases}$$

i.e.

$$h_*(s) = f_*(a)\chi_{(0,a)}(s) + f_*(s)\chi_{(a,|\Omega|)}(s).$$

2. (b) We see that

$$f_* = h_* + g_*.$$

Exercise 11.3.5 Let $f \geqslant 0, \;\; f \in L^1(\Omega) + L^\infty(\Omega)$ and $t \in [0, |\Omega|]$.

1. Recall the proof of

$$\text{Min}_{\;\lambda \geqslant 0} \left\{ t\lambda + \int_\Omega (f(x) - \lambda) dx \right\} = \int_0^t f_*(s) ds$$

(use Proposition 1.2.1, given that

$$t f_*(t) + \int_\Omega (f(x) - f_*(t))_+ dx = \int_0^t f_*(s) ds \geqslant 0)$$

2. Let $K(t, f) = \inf\limits_{f=g+h} \{|g|_1 + t|h|_\infty\}\}$.

   (a) Show that

$$\inf_{\lambda \geqslant 0} \left\{ t\lambda + \int_\Omega (f(x) - \lambda)_+ dx \right\} \leqslant K(t, f).$$

(b) Show that $K(t, f) = \int_0^t f_*(s)ds.$

Solution 11.3.5

1. Repeat the proof of Proposition 1.2.1.
2. (a) Let $\varepsilon > 0$. Then there exists $g \in L^1(\Omega),\ h \in L^\infty(\Omega)$ s.t. $f = g + h$

$$|g|_1 + t|h|_\infty \leqslant K(t, f) + \varepsilon.$$

As $g = f - h \geqslant f - |h|_\infty, \qquad \lambda_\varepsilon = |h|_\infty.$
So

$$(f - \lambda_\varepsilon)_+ \leqslant g_+ : \int_\Omega (f(x) - \lambda_\varepsilon)_+ dx \leqslant |g|_1.$$

Then

$$t\lambda_\varepsilon + \int_\Omega (f(x) - \lambda_\varepsilon)_+ dx \leqslant K(t, f) + \varepsilon.$$

Therefore Min $_{\lambda>0}\left\{t\lambda + \int_\Omega (f(x) - \lambda)_+ dx\right\} \leqslant K(t, f) + \varepsilon, \forall \varepsilon > 0$, hence the result.

(b) Let $\lambda \geqslant 0,\ g = (f - \lambda)_+ \in L^1(\Omega)$. Then $h = f - g \in L^\infty(\Omega)$ and $|h|_\infty = \lambda$ because $h(x) = \begin{cases} \lambda & if\ f(x) > \lambda, \\ varphi(x) & if\ f(x) \leqslant \lambda. \end{cases}$. So

$$K(t, f) \leqslant |g|_1 + t|h|_\infty = t\lambda + \int_\Omega (f(x) - \lambda)_+ dx,$$

and we have

$$K(t, f) = \text{Min}_{\lambda\geqslant 0}\left\{t\lambda + \int_\Omega f(x) - \lambda)_+ dx\right\} = \int_0^t f_*(s)ds.$$

◇

For simplicity, we adopt the following notations for two quantities A_1 and A_2. We will write $A_1 \lesssim A_2$ if there exists a (universal) constant $c > 0$ such that $A_1 \leqslant cA_2$.

We will write $A_1 \simeq A_2$ (A_1 is equivalent to A_2) if $A_1 \lesssim A_2,\ A_2 \lesssim A_1$.

(universal in general means that it does not depend on variables but only on data like the domain Ω and the dimension N of the space.

Exercise 11.3.6 We recall the definition of the large Lebesgue space $L^{p)}(\Omega)$ as follows: for $1 < p < +\infty$, $|\Omega| = 1$, $L^{p)}(\Omega)$

$$= \left\{ f \in L^0(\Omega) : |f|_{p),1} = \sup_{0<t<1} (-\operatorname{Log} t)^{-\frac{1}{p}} \left(\int_t^1 |f|_*(s) ds \right)^{\frac{1}{p}} < \infty \right\}.$$

We then define the following K-functional for $0 < t < 1$,
$f \in L^{p)}(\Omega)$, $f \geqslant 0$,

$$K(f, t; L^{p)}, L^p) = \inf_{f=g+h} \left\{ |g|_{p),1} + t|h|_p \right\}.$$

$$Let \varphi(t) = e^{1-\frac{1}{t^p}} and\ K_1(t) = \sup_{0<s<\varphi(t)} (1 - \operatorname{Log} s)^{-\frac{1}{p}} \left(\int_s^1 |f|_*^p(x) dx \right)^{\frac{1}{p}}.$$

1. Show using Hunt's lemma from Exercise 11.3.3 that

$$K_1(t) \lesssim K(f, t; L^{p)}, L^p)$$

(as an exercise we will estimate the universal constant $c > 0$).
2. We consider the decomposition of f given in Exercise 11.3.4 with $\lambda = f_*\big(\varphi(t)\big)$, i.e we consider $g = (f - \lambda)\chi_{\{f>\lambda\}}$, $h = f - g$.
   (a) Show that $|g|_{p),1} \leqslant K_1(t)$.
   (b) Show that $t|h|_p \lesssim K_1(t)$.
   (c) Derive that $K\Big(f, t; L^{p)}, L^p\Big) \approx K_1$.

Solution 11.3.6

1. Let $f = g + h$ be an arbitrary decomposition of f with $g \in L^{p)}$, $h \in L^p$. According to Exercise 11.3.3 we then have

$$f_*(x) \leqslant g_*\left(\frac{x}{2}\right) + h_*\left(\frac{x}{2}\right), \quad \forall x \in [0, 1].$$

Replacing this in the expression of $K_1(t)$, since

$$\left(\int_s^1 f_*^p(x) dx \right)^{\frac{1}{p}} \leqslant \left(\int_s^1 g_*^p \left(\frac{x}{2}\right) dx \right)^{\frac{1}{p}} + \left(\int_s^1 h_*^p \left(\frac{x}{2}\right) dx \right)^{\frac{1}{p}}$$

we deduce the existence of a constant $c > 0$ independent of $t,\ g,\ h$, such that

$$K_1(t) \leqslant c\Big[\,|g|_{p),1} + |h|_p \sup_{0<s<\varphi(t)} (1-\text{Log } s)^{-\frac{1}{p}}\Big].$$

As the function $s \to (1-\text{Log } s)^{-\frac{1}{p}}$ is increasing

$$K_1(t) \leqslant c\Big[\,|g|_{p),1} + t|h|_p\Big]$$

hence

$$K_1(t) \leqslant c \inf_{f=g+h} \Big\{|g|_{p),1} + t|h|_p\Big\} = c\,K(f, t; L^{p)}, L^p).$$

2. (a) We have by definition

$$|g|_{p),1} = \sup_{0<s<1} \Big(1-\text{Log } s\Big)^{-\frac{1}{p}} \left(\int_s^1 (f_* - \lambda)_+^p(x)dx\right)^{\frac{1}{p}}$$
$$\leqslant \sup_{0<s<\varphi(t)} (1-\text{Log}\,(s))^{-\frac{1}{p}} \left(\int_s^1 f_*^p(x)dx\right)^{\frac{1}{p}} = K_1(t)$$

(because if $\varphi(t) < s$ then
$\int_s^1 (f_* - \lambda)_+^p(x)dx = 0, \lambda = f_*(\varphi(t))$).

(b) Since, as previously noted that $t = (1-\text{Log } \varphi(t))^{-\frac{1}{p}}$, $t \in [0, 1]$ and

$$h_*(s) = \lambda\chi_{(0,\varphi(t))}(s) + f_*(s)\chi_{(\varphi(t),1)}(s)$$

we deduce

$$t|h|_p \leqslant t\lambda\varphi(t)^{\frac{1}{p}} + t\left(\int_{\varphi(t)}^1 f_*^p(x)dx\right)^{\frac{1}{p}}$$
$$\lesssim \left(1-\text{Log } \frac{\varphi(t)}{2}\right)^{-\frac{1}{p}} \lambda\varphi(t)^{\frac{1}{p}} +$$
$$\sup_{0<s<\varphi(t)} (1-\text{Log } s)^{-\frac{1}{p}} \left(\int_{\varphi(t)}^1 f_*^p(x)ds\right)^{\frac{1}{p}}$$
$$\lesssim \left(1-\text{Log } \frac{\varphi(t)}{2}\right)^{-\frac{1}{p}} \left(\int_{\frac{\varphi(t)}{2}}^{\varphi(t)} f_*^p(x)dx\right)^{\frac{1}{p}} + K_1(t)$$
$$\lesssim K_1(t).$$

(c) We deduce that

$$K\Big(t, f; L^{p)}, L^p\Big) \leqslant |g|_{p),1} + t|h|_p \lesssim K_1(t).$$

Note that for f of any sign,

$$K\Big(t, |f|; L^{p)}, L^p\Big) = K(t, f; L^{p)}, L^p\Big).$$

we can use this result in the previous exercise.

Exercise 11.3.7 Let $(X_i, ||\cdot||_i)$, $i = 0, 1$ be two normalized subspaces of $L^0(\Omega)$ s.t.

$$\forall f \in X_i,\ \ |f| \in X_i \Longleftrightarrow f \in X_i, \quad i = 0, 1$$

Let $t \in [0, |\Omega|]$, $f \in X_0 + X_1 \subset L^0(\Omega)$, and

$$K(f, t; X_0, X_1) = \inf_{f=h+g} \Big\{||h||_{X_0} + t||g||_{X_1}\Big\}.$$

Show that $K(f, t; X_0, X_1) = K(|f|, t; X_0, X_1)$.

Solution 11.3.7 Note that $f = \text{sign}\,(f)|f|$ and $|f| = \text{sign}\,(f)f$, $\forall f \in L^0(\Omega)$.

So if $f = h + g$ then $|f| = h\,\text{sign}\,(f) + g\,\text{sign}\,(f)$ and we have that $\widetilde{h} = h\,\text{sign}\,(f)$ satisfies $|\widetilde{h}| = |h|$ and $\widetilde{g} = g\,\text{sign}\,(f)$, $satisfies|\widetilde{g}| = |g|$.

Likewise for $|f|$ if $|f| = h + g$ we have $f = h\,\text{sign}\,|f| + g\,\text{sign}\,(f)$, hence the result. ◇

Exercise 11.3.8

1. Let u and v be in $L^2(\Omega) and\ \Phi : \mathbb{R} \to \mathbb{R}$ be strictly increasing such that $\Phi(u) \in L^1(\Omega)$. Show that

$$\int_\Omega uv dx = \int_{\Omega_*} u_* v_{*u} d\sigma \ and\ v_{*\Phi(u)} = v_{*u}.$$

2. Let $L(\Omega, \rho)$ a rearrangement-invariant Banach function space. Show that if $v \in L(\Omega, \rho)$, $u \in L(\Omega, \rho')$ (associated space) then

   (a) $\left|\int_\Omega uv dx\right| \leqslant \rho'(u)\rho(v_{*|u|})$

   (b) $\rho(v) = \sup_{u \in L(\Omega, \rho'),\ u \neq 0} \rho(|v|_{*|u|})$

Solution 11.3.8

1. By equimeasurability we have $\int_\Omega u^2 = \int_{\Omega_*} u_*^2$ and

$$\int_\Omega (u+\lambda v)^2 = \int_\Omega (u+\lambda v)_*^2, \quad \forall\,\lambda > 0.$$

Thus,

$$\int_\Omega \frac{(u+\lambda v)^2 - u^2}{\lambda} = \int_{\Omega_*} \frac{(u+\lambda v)_*^2 - u_*^2}{\lambda}$$

and letting $\lambda \to 0$ we find the equality We can also use the mean value formula. The equality $v_{*\Phi(u)} = v_{*u}$ follows from the fact that

$$\Phi(u)_* = \Phi(u_*), \quad w(s) = F(s,u;v) = F(s,\Phi(u);v) \text{ and } v_{*u} = \frac{dw}{ds}.$$

(a) Since $\left|\int uvdx\right| \leqslant \int_\Omega |u|\,|v|dx$ we can assume that $u \geqslant 0$ and $v \geqslant 0$. For $k \in \mathbb{N}$, consider the strictly increasing and bounded function $\Phi_k : \mathbb{R}_+ \to \mathbb{R}_+$ defined as follows:

$$\Phi_k(\sigma) = \begin{cases} \sigma & if\ 0 \leqslant \sigma \leqslant k \\ k + \frac{1}{2} - e^{-\frac{\mathrm{Log}\,2}{k}\sigma} & otherwise. \end{cases}$$

Then $\forall \sigma \in [0,+\infty[$, $0 \leqslant \Phi_k(\sigma) \leqslant \Phi_{k+1}(\sigma)$, $\Phi_k(\sigma)$ tends to σ as $k \to +\infty$. Reasoning as above, we have

$$\int_\Omega \Phi_k(u)v = \int_{\Omega_*} \Phi_k(u_*)v_{*\Phi_k(u)}, \quad \Phi_k(u_*) = \Phi_k(u)_*.$$

As $v_{*\Phi_k(u)} = v_{*u}$ we deduce that

$$0 \leqslant \int_\Omega \Phi_k(u)v \leqslant \rho'\big(\Phi_k(u)_*\big)\cdot\rho(v_{*u}) = \rho'\big(\Phi_k(u)\big)\rho(v_{*u}).$$

When $k \to +\infty$ the Beppo Levi theorem as well as the fact that ρ' is a Fatou norm implies

$$0 \leqslant \int_0 uvdx \leqslant \rho'(u)\rho(v_{*u}).$$

As

$$\rho(v) = \sup\left\{\int_\Omega uvdx,\ \ \rho'(u) \leqslant 1\right\} \leqslant \sup_u \rho(v_{*u}) \leqslant \rho(v).$$

◇

Exercise 11.3.9 (Dini Continuity for $W^1L^\alpha_{exp}(\Omega)$) Let Ω be a bounded open subset of measure 1 of $\mathbb{R}^N$, $N \geqslant 2$, and for $\alpha > 0$

$$W^1L^\alpha_{exp}(\Omega) = \left\{v \in W^{1,1}(\Omega) : \sup_{0<t<1} \frac{|\nabla v|_{**}(t)}{(1 - \text{Log }t)^\alpha} < +\infty\right\}.$$

Let $C_D(\Omega) = \Big\{v \in C(\Omega)\ s.t.$ there exists a function $\omega : \mathbb{R}_+ \to \mathbb{R}_+$ continuous at $0, \omega(0) = 0, \displaystyle\int_0^1 \frac{\omega(t)}{t}dt < +\infty,\ \text{osc }_{Br} v \leqslant \omega(r),\ B(r)$ contained in $\Omega\Big\}$.

1. Show that $W^1L^\alpha_{exp}(\Omega) \subset C_D(\Omega)$
2. Determine a function $\omega_\alpha : \mathbb{R}_+ \to \mathbb{R}_+$ s.t.

$$\text{osc }_{B(r)} v \leqslant \omega_\alpha(r),\ \forall\, B(r) \text{ ball of radius} r \text{contained in} \Omega.$$

3. Check that $\displaystyle\int_0^\varepsilon \frac{\omega_\alpha(r)}{r}dr < +\infty$, for $\varepsilon > 0$ small.

Solution 11.3.9

1. Let $||\nabla v||_{,\alpha} = \sup_{0<t<1} \dfrac{|\nabla v|_{**}(t)}{(1 - \text{Log }t)^\alpha}$ $for\ v \in W^1L^\alpha L_{exp}(\Omega)$. Since $\displaystyle\int_0^1 t^{\frac{1}{N}}(1 - \text{Log }t)^\alpha \frac{dt}{t} < +\infty$ we have

   $W^1L^\alpha_{exp}(\Omega) \subset W^1L^{N,1}(\Omega)$ and we saw in Chap. 4 that for $B(r) \subset \Omega$ ($B(r)$ ball)

$$\text{osc }_{B(r)} v \leqslant \frac{\alpha_N^{1-\frac{1}{N}}}{\alpha_{N-1}} \int_0^{\alpha_N r^N} t^{\frac{1}{N}} |\nabla v|_*(t)\frac{dt}{t}$$

   and $|\nabla v|_*(t) \leqslant |\nabla v|_{**}(t) \leqslant ||\nabla v||_{,\alpha}(1 - \text{Log }t)^\alpha$. Hence the estimation by setting $c_N = \dfrac{\alpha_N^{1-\frac{1}{N}}}{\alpha_{N-1}}$

$$\text{osc }_{B(r)} v \leqslant \left[c_N \int_0^{\alpha_N r^N} t^{\frac{1}{N}}(1 - \text{Log }t)^\alpha \frac{dt}{t}\right] ||\nabla v||_{,\alpha} = \omega_\alpha(r).$$

2. By change of variable, we therefore have the following equivalent of $\omega_\alpha(r)$

$$\omega_\alpha(r) \simeq e^{\frac{1}{N}} \int_{1-\mathrm{Log}\,\alpha_N - N\,\mathrm{Log}\,r}^{+\infty} e^{-\frac{\sigma}{N}} \sigma^\alpha d\sigma \doteq \overline{\omega}_\alpha(r).$$

Case Where $\alpha = 1$ Integrating by parts, we obtain

$$\overline{w}_1(r) \leqslant N\alpha_N^{\frac{1}{N}} \mathrm{Max}\,(N;\, 1 + N - \mathrm{Log}\,\alpha_N) r(1 - \mathrm{Log}\,r)$$

and since $\displaystyle\int_0 \frac{\overline{\omega}_1(r)}{r} < +\infty$ we have

$$\int_0 \frac{\overline{\omega}_\alpha(r)}{r} dr < +\infty.$$

General Case We use gamma functions because

$$\overline{\omega}_\alpha(r) = e^{\frac{1}{N}} N^{\alpha+1} \int_{\gamma_N - \mathrm{Log}\,r}^{+\infty} e^{-t} t^\alpha dt \doteq \Gamma(\alpha+1;\, \gamma_N - \mathrm{Log}\,r)$$

where $\gamma_N = \dfrac{1 - \mathrm{Log}\,(\alpha_N)}{N}$.

The asymptotic behavior of $\Gamma(a : x)$ is given in [135]

$$\Gamma(a; x) \underset{x \to +\infty}{\sim} e^{-x} x^{a-1}$$

From this, we obtain

$$\underset{B(r)}{\mathrm{osc}}\, v \leqslant c_N\, r(1 + \log r|)^\alpha ||\nabla v||_{,\alpha} \quad and \quad \int_0 \frac{\overline{\omega}_\alpha(t)}{t} dt < +\infty.$$

◇

Exercise 11.3.10 Deduce from the previous exercise that there exists a constant K_0 (depending only on N and Ω) (we will assume $|\Omega| = 1$ for simplicity) s.t. $\forall\, v \in W^{2,N}(\Omega)$ (Sobolev space $W^2 L^N(\Omega)$)

$$\mathrm{osc}_{\,B(r)}(v) \leqslant K_N ||v||_{W^{2,N}} r(1 + |\,\mathrm{Log}\,r|)^{\frac{N-1}{N}}$$

Solution 11.3.10 We use the injection of $W^1 L^N$ into L^α_{exp} of Proposition 4.4.1, to deduce that $W^2 L^N \hookrightarrow W^1 L^{1/N'}_{exp}(\Omega)$; then we apply the previous result with $\alpha = \dfrac{1}{N'}$.

Exercise 11.3.11 We assume that Ω is a connected Lipschitzian bounded open set. Show, using the methods developed in Chap. 4, that

$$W^1L^{N',\infty}(\Omega) \hookrightarrow L^{\frac{N}{N-2},\infty}(\Omega)\ if\ N \geqslant 3$$

and

$$W^1L^{2,\infty}(\Omega \hookrightarrow L^{\frac{1}{2}}_{exp}(\Omega)\ if\ N = 2$$

by giving an estimate of the inclusion constants.

Solution 11.3.11 The proof is identical to that of Proposition 4.4.1.

Exercise 11.3.12 Show that $W^1L^{p,1}(\Omega)$ in $L^{p*}(\Omega)$ is continuous is not compact, $1 < p < N$.

Hint Consider $u \in C_c^\infty(B(1))$, $B(1)$ unit ball s.t. $u = 1$ on $B\left(\frac{1}{2}\right)$ (ball of radius $\frac{1}{2}$).

Show that the sequence $u_m(x) = 2^{m(\frac{N}{p}-1)}u(2^m x)$ is not Cauchy in $L^{p*}(B(1))$ and show $(u_m)_m$ remains bounded in $W^1L^{p,1}(\Omega)$.

Solution 11.3.12 We prove that for m and n in $\mathbb{N}$

$$|u_m - u_n|_{p*} = |u_{|m-n|} - u|_{p*} \geqslant 2^{1-\frac{N}{p}}(1 - 2^{1-\frac{N}{p}})|B|^{\frac{1}{p*}}$$

(this inequality is apparently due to Sobolev). We calculate for $D_j = \dfrac{\partial}{\partial x_j}$ $|D_j u_m|_{L^{p,1}(B)} = |D_j u|_{L^{p,1}(B)}$ and $|u_m|_{p*} = |u|_{p*}$. Hence the result. ◇

Note that

$$\textit{the injection}\ W^1L^{(p}(\Omega) \hookrightarrow L^{p*}(\Omega),\ p^* = \frac{Np}{N-p} 1 < p < N \textit{is compact.}$$

Exercise 11.3.13 Let Ω be a set of finite measure, $f : \Omega \longrightarrow [0, +\infty[$ measurable. Show that $\forall t \in [0, |\Omega|]$

$$f_*(t) = \sup_{|E|=t}\left(\operatorname*{ess\,inf}_{x\in E} f(x)\right) \doteq F(t).$$

Solution 11.3.13 Let $t \in [0, |\Omega|]$. By Lyapunov's theorem there exists $E_t \subset \Omega$ s.t.

$$|E_t| = t \quad \text{and} \quad \left\{f > f_*(t)\right\} \subset E_t \subset \left\{f \geqslant f_*(t)\right\}$$

$$F(t) \geqslant \operatorname*{ess\,inf}_{E_t} f(x) \geqslant f_*(t).$$

Conversely, if $\varepsilon > 0$, there exists $E_\varepsilon \subset \Omega$ s.t.

$$|E_\varepsilon| = t \quad \text{and} \quad F(t) \leqslant \operatorname*{ess\,inf}_{x \in E_\varepsilon} f(x) + \varepsilon.$$

As

$$\left(f\Big|_{E_\varepsilon}\right)_*\left(|E_\varepsilon|\right) = \operatorname*{ess\,inf}_{x \in E_\varepsilon} f(x) \text{ and} \left(f\Big|_{E_\varepsilon}\right)_*(t) \leqslant f_*(t),$$

we deduce the result

$$F(t) \leqslant f_*(t) + \varepsilon \quad \forall \varepsilon. \qquad \diamond$$

Exercise 11.3.14 (Laplacian Eigenfunctions and Critical Points) Let Ω be a bounded open Lipschitzian set and $(\varphi_j)_{j \geqslant 1}$ be the sequence of eigenfunctions s.t.

$$-\Delta \varphi_j = \lambda_j \varphi_j, \quad \varphi_j \in C_0^1(\overline{\Omega}) \cap H^k(\Omega).$$

Show $\forall u \in Vect\left\{\varphi_1, \cdots, \varphi_m\right\} = V_m, \ u \not\equiv 0$

$$\text{the set} C_u = \left\{x \in \Omega : \nabla u(x) = 0\right\} \text{ is of zero measure.}$$

Solution 11.3.14 Let $u \in V_m \backslash \{0\}$. Suppose measure$(C_u) > 0$. According to a property of G. Stampacchia (see H. Brezis [28, 29]) we then have for almost all $x \in C_u$

$$\Delta u(x) = 0 : \text{measure}\left\{x \in C_u : \Delta u(x) = 0\right\} > 0.$$

Since $u \in V_m$ there exists $(\alpha_1, \cdots, , a_m \neq (0, \cdots 0)\, s.t. \sum_{j=0}^{m} \alpha_j \varphi_j = u.$

Then

$$\sum_{j=0}^{m} \lambda_j \alpha_j \varphi_j(x) = \Delta u(x) = 0 \quad \forall x \in C_u \backslash N \quad (mes(N) = 0).$$

But $\left\{\varphi_j, j = 0, \cdots, m\right\}$ is free and $\lambda_j \geqslant \lambda_1 > 0$so that $\alpha_j = 0 \,\forall j$, a contradiction.

Exercise 11.3.15 Let u be a C^1-function on $\mathbb{R}$.

Assume that $D_u = \left\{t \in \mathbb{R} : |u = t| > 0\right\}$ is finite and the points $x \in \mathbb{R} \backslash I\!P(u)$ such that$u'(x) = 0$ are isolated.

Show that $dm_{ou}(t) = -\sum_{t \in Du} \beta_t \delta_t$ where δ_t denotes the Dirac measure at t and $\beta_t > 0$ is to be specified. Here as before for $t \in \mathbb{R}$,

$$m_{ou}(t) = meas\Big\{x \in \mathbb{R} : u(x) > t,\ u'(x) = 0\Big\}.$$

Show that u is a co-area regular function.

Solution 11.3.15 Since D_u is finite, we may find $\text{ess inf}\, u \leqslant t_1 < t_2 < \cdots < t_k \leqslant \text{ess sup}\, u$ real numbers such that $D_u = \{t_1; \ldots; t_k\}$.

Let $I\!P(u) = \bigcup_{u=1}^{k} \{u = t_i\}$, $\alpha_k = meas(I\!P(u))$.

More generally $\alpha_i = meas\Big(\bigcup_{j=1+1}^{k} \{u = t_j\}\Big)$ $1 \leqslant i \leqslant k$. Then we have:

$$m_{ou}(t) = \begin{cases} \alpha_k & if\ t \leqslant t_1 \\ \alpha_{k-1} & if\ t_1 < t \leqslant t_2 \\ \vdots & \\ \alpha_{k-i} & if\ t_i < t \leqslant t_{i+1} \\ \vdots & \\ \alpha_1 & if\ t_{k-1} < t \leqslant t_k \\ \alpha_0 = 0 & if\ t > t_k. \end{cases}$$

Therefore the derivative of m_{ou} in the sense of distribution is then:

$$dm_{ou} = -\sum_{i=0}^{k-1} (\alpha_{k-i} - \alpha_{k-i-1})\delta_{t_{i+1}}.$$

The regular part is then zero so u is a co-area-regular function.

Remark Let $B = \Big\{x \in \mathbb{R} : u'(x) = 0\Big\}$. Under the hypothesis that the points of $A = B - I\!P(u)$ are isolated, A is at most countable according the Cantor–Berdixon's theorem. Therefore it is of measure zero and the expression of $m_{ou}(t)$ is reduced to

$$m_{ou}(t) = meas\Big\{x \in I\!P(u),\ u(x) > t : u'(x) = 0\Big\} = meas \bigcup_{\{i : t_i > t\}} \{u = t_i\}$$

where $D_u = \Big\{t_i : i \in \{1, \cdots, k\}\Big\}$.

Bibliography

1. Adams, R.: Sobolev Spaces. Academic Press, Cambridge (1975)
2. Ahmed, I., Fiorenza, A., Formica, M.R., Gogastishvili, A., Rakotoson, J.M.: Some new results related to Lorentz G-$Gamma$-spaces and interpolation. J. Math. Anal. Al. **483**(2), 123623 (2020)
3. Ahmed, I., Fiorenza, A., Formica, M.R., Gogastishvilli, A., El Hamidi, A., Rakotoson, J.M.: Quasilinear PDEs, interpolations spaces and Hölderian mappings. Anal. Math. **49**(4), 895–950 (2023). https://doi.org/10.1007/s10470-023-0245-z
4. Almgren, F., Lieb, E.: Symmetric rearrangement is sometimes continuous. J. Am. Math. Soc. **2**, 683–773 (1989)
5. Alvino, A.: Sulla disegualianza di Sobolev in spazi di Lorentz. Boll. Un. Mat. Ital. **14**, 148–156 (1977)
6. Alvino, A.: Un caso limite della diseguafliana di Sobolev in spazi di Lorentz. Rend. Acad. Sci. Napoli **XLIV**, 105–112 (1978)
7. Alvino, A., Trombetti, G.: Sulle migliori costanti di maggiorazione per una classe di equationi ellittiche degeneri. Ricerche Mat. **27**, 413–428 (1978)
8. Alvino, A., Trombetti, G.: Sulle migliori costanti di maggiorazione per una classe di equationi ellittiche degeneri e non. Ricerche Mat. **30**, 15–33 (1981)
9. Alvino, A., Trombetti, G.: A lower bound for the first eigenvalue of an elliptic operator. J. Math Anal. Appl. **94**, 328–337 (1983)
10. Alvino, A., Lions, P.L., Trombetti, G.: On optimization problems with prescribed rearrangements. Nonlinear Anal. Theory Methods Appl. **13**(2), 185–220 (1989)
11. Alvino, A., Lions, P.L., Trombetti, G.: Comparison results for elliptic and parabolic equation via Schwarz symmetrization. Ann. Inst. Henri Poincaré. **7**, 37–65 (1990)
12. Alvino, A., Lions, P.L., Trombetti, G.: Comparison results for elliptic and parabolic equation via symmetrization: a new approach. Differ. Int. Eq. **4**, 25–50 (1991)
13. Alvino, A., Matarasso, S., Trombetti, G.: Variational inequalities and rearrangements. Rend. Mat. Acc. Lincei. **9**, 271–285 (1992)
14. Alvino, A., Díaz J.I., Lions, P.L., Trombetti, G.: Elliptic equations and steiner symmetrization. Commun. Pure Appl. Math. **XLIX**, 217–236 (1996)
15. Alvino, A., Ferone, V., Trombetti, G.: Estimates for the gradient of solutions of nonlinear elliptic equations with L^1. Ann. Mat. Pura Appl. **178**, 129–142 (2000)
16. Baernstein, A.: A unified approach to symmetrization. Partial Differ. Equ. Elliptic Type **XXXV**, 47–91 (1995). A. Alvino et al, Symposia matematica
17. Baernstein, A., II, Drasin, D., Laugesen, R.: Symmetrization in Analysis, vol. 36. Cambridge University Press, Cambridge (2019)

J. M. Rakotoson, *Relative Rearrangement*, Lecture Notes in Mathematics 2376,
https://doi.org/10.1007/978-3-032-02228-8

18. Bandle, C.: Isoperimetric Inequalities and Applications. Pitman, London (1980)
19. Bandle, J., Mossino, J.: Rearrangements in variational inequalities. Ann. Mat. Pura Appl. **138**, 1–14 (1984)
20. Bénilan, P., Crandall, M.L.: Semigroup theory and evolution equations. Clements, P., et al. (eds.), pp. 41–76. Marcel Decker, New York (1991)
21. Bennett, C., Sharpley, R.: Interpolation of Operators. Academic Press, Cambridge (1983)
22. Berestycki, H., Brézis, H.: On a free boundary problem arising in plasma physics. Nonlinear Anal. **4**, 415–436 (1980)
23. Bermúdes, A., Seoane, M.L.: Numerical solution of a nonlocal problem arising in plasma physics. Math. Comput. Model. **27**, 45–59 (1998)
24. Betta, F., Mercaldo, A.: Existence and regularity results for a nonlinear elliptic equation. Rend. Mat. **VII**(11), 737–759 (1991)
25. Betta, F., Mercaldo, A.: Comparison and regularity results for a nonlinear elliptic equation. Nonlinear Anal. Theory Methods Appl. **20**(1), 63–77 (1993)
26. Betta, F., Mercaldo, A.: Geometric inequalities related to Steiner symmetrization. Differ. Integral Equ. **10**(3), 473–486 (1997)
27. Blum, J., Gallouet, T., Simon, J.: Equilibrium of a Plasma in a Tokamak. Pitman, London (1985)
28. Brézis, H.: Functional Analysis, Sobolev Spaces and Partial Differentiel Equations. Masson, Issy-les-Moulineaux (1983)
29. Brézis, H.: Analyse Fonctionnelle Théorie et Applications. Springer, Berlin (2010)
30. Brézis , H., Wainger, S.: A note on limiting cases of Sobolev embedding and convolution inequalities. Commun. Partial Differ. Equ. **5**, 773–789 (1980)
31. Brock, F.: Continuous Steiner-symmetrization. Math. Nachrichten **172**, 25–48 (1995)
32. Brothers, J., Ziemer, W.P.: Minimal rearrangements of Sobolev functions. J. Reine Angew. Math. **384**: 153–179 (1988)
33. Burchard, A.: Steiner symmetrization is continuous in $W^{1,p}$. Geom. Funct. Anal. **7**, 823–860 (1997)
34. Burton, G.R.: Rearrangement of functions, maximization of convex functionals, and vortex rings. Math. Anal. **276**(2), 225–253 (1987)
35. Chabi, A., Haraux, A.: Un théorème de valeurs intermédiaires dans les espaces de Sobolev et applications. Ann. Faculty Sci. Toulouse **VII**, 87–100 (1985)
36. Chavel, I.: Isoperimetric Inequalities. Cambridge Tracts in Mathematics, vol. 145. Cambridge University Press, Cambridge (2001)
37. Chiacchio, F., Blasio, Gdi.: Isoperimetric inequalities for the first Newmann eigenvalue in Gauss space. Ann. Inst. Poincaré **29**, 199–216 (2012)
38. Chiti, G.: Rearrangement of functions and convergence on Orlicz spaces. Appl. Anal. **9**(1), 23–27 (1979)
39. Chong, K.M., Rice, N.M.: Equimeasurable Rearrangements of Functions. Queen's University, Kingston (1971)
40. Choquet, G.: Theory of capacities. Ann. Inst. Fourier **5**, 131–395 (1955)
41. Chrayteh, H., Rakotoson, J.M.: Eigenvalue problems with fully discontinuous operators and critical exponents. Nonlinear Anal. **73**, 2036–2055 (2010)
42. Cianchi, A.: On relative isoperimetric inequalities in the plane. Bolletino della Unione Math. Ital. **7**(3-B), 289–325 (1989)
43. Cianchi, A.: Optimal gradient bounds and heat equation. Differ. Int Equ. **6**(5), 1079–1088 (1993)
44. Cianchi, A., Maz'ja, V.: Gradient regularity via rearrangement for p-Laplacian type elliptic boundary value problems. J. Eur. Math. Soc. **16**, 571–595 (2014)
45. Cianchi, A., Pick, L.: Sobolev embeddings into BMO, VMO, and L^{∞}. Arch. fur Math. **36**(2), 317–340 (1998)
46. Cianchi, A., Edmunds, D.E., Gurka, P.: On weighted Poincaré inequalities. Math. Nachr. **180**, 15–41 (1996)

47. Corona, J.M.: The continuity of the rearrangement in $W^{1,p}(\mathbb{R})$. Ann. della Scuola Normale Superiore di Pisa. Série IV **11**(1), 57–85 (1984)
48. Crandall, M.G., Tartar, L.: Some relations between nonexpansive and ordre preserving maps. Proc. Am. Math. Soc. **78**(3), 385–390 (1980)
49. Cruz-Uribe, D.: The Hardy-Littlewood maximal operator on variable-$L^{p(\cdot)}$ spaces. In: Seminar of Mathematical Analysis (Malaga/Seville, 2002/2003) 147–156, Colecc. Abierta, 64, Univ. Sevilla Secr. Publ., Seville, (2003)
50. Cruz-Uribe, D., Fiorenza, A.: Variable Lebesgue spaces Foundation and Harmonic Analysis Applied and Numerical Harmonic Analysis. Springer, Berlin (2013)
51. De Giorgi, E.: Su una teoria generale della misura $(r-1)$-dimensionale in uno spazio ad r-dimension. Ann. Mat. Pura Appl. **36**, 191–213 (1952)
52. Díaz, J.I.: Applications of Symmetric Rearrangement to Certain Nonlinear Elliptic Equations with a Free Boundary. Pitman, London (1985)
53. Díaz, J.I.: Nonlinear Partial Differential Equations and Free Boundaries. Elliptic Equations, vol. 1. Pitman, London (1985)
54. Díaz, J.I.: Desigualdades de tipo isoperimétrico para problemas de Plateau y capilaridad. Rev. Acad. Canaria Ciencas **III**(1), 127–166 (1991)
55. Díaz, J.I.: Symmetrization of nonlinear elliptic and parabolic problems and applications: a particular overview. In: Progress in Partial Differential Equations Elliptic and Parabolic Problems. Pitman Research Notes in Mathematics, vol. 266, pp. 1–16. Longman Harlow, London (1992)
56. Díaz, J.I.: Qualitative study of nonlinear parabolic equations: an introduction. Extract. Math. **16**(2), 303–341 (2001)
57. Díaz, J.I.: Modelos bidimensionales de equilibrio magnetohidrodinámico para Stellarators, Informe 3 Formulaci. Euroatom CIEMAT Associaton Reports, Madrid (1991)
58. Díaz, J.I., Mossino, J.: Isoperimetric inequalities in the parabolic obstacle problems. J. Math. Pures Math. **71**(9), 233–266 (1992)
59. Díaz, J.I., Nagai, T.: Symmetrization in a parabolic-elliptic system related to chemotaxis. Adv. Math. Sc. Appl. **5**, 659–680 (1995)
60. Díaz, J.I., Rakotoson, J.M.: On a two–dimensional stationary free boundary problem arising in the confinement of a plasma in a Stellarator. C.R. Acad. Sci. Paris I **317**, 353–358 (1993)
61. Díaz, J.I., Rakotoson, J.M.: On a nonlocal stationary free boundary problem arising in the confinement of a plasma in a Stellarator geometry. Arch. Rat. Mech. Anal. **134**(1), 53–95 (1996)
62. Díaz, J.I., Nagai, T., Rakotoson, J.M.: Symmetrization techniques on unbounded domains: application to a chemotaxis system on $I\!R^N$. J. Differ. Equ. **145**(1), 156–183 (1998)
63. Díaz, J.I., Padial, J.F., Rakotoson, J.M.: Mathematical treatement of the magnetic confinement in a current carrying Stellerator. Nonlinear Anal. TMA **34**, 857–887 (1998)
64. Díaz, J.I., Lerena, M., Padial, J.F., Rakotoson, J.M.: Nonlocal elliptic-parabolic equation arising in the stability of a confined plasma in a Stellerator. C.R. Acad. Sci. Paris **329**, 773–777 (1999)
65. Díaz, J.I., Gómez-Castro, D., Rakotoson, J.M., Temam, R.: Linear diffusion with singular absorption potential and/or unbounded convective flow: the weighted space approach. Discrete Continuous Dynam. Syst. **28**(2), 509–546 (2018)
66. Diening, L., Harjulehto, P., Hasto, P., Ruzicka, M.: Lebesgue and Sobolev spaces with variables exponents. Lecture Notes in Mathematics. Springer, Berlin (2011)
67. Di Blasia, G., Volzone, B.: Comparison and regularity results for the fractional Laplacian via symmetrization methods. J. Differ. Equ. **253**, 2593–2615 (2012)
68. Di Fratta, G., Fiorenza, A.: A direct approach to the duality of grand and small Lebesgue spaces. Nonlinear Anal. **70**, 2582–2592 (2009).
69. Douglas, R.J.: Rearrangement of functions on unbounded domains. Proc. R. Soc. Edinburgh **124**(A), 623–644 (1994)
70. Ekeland, I., Temam, R.: Convex Analysis and Variational Problems. North-Holland, Amsterdam (1976)

71. Federer, H.: Geometric Measure. Springer, Berlin (1989)
72. Feo, F., Martin, J., Posteraro, R.: Sobolev anisotropic inequalities with monomial weights (2022). ArXiv:1910.08996v1 [Math.FA], vol. 505
73. Ferone, A., Volpicelli, R.: Some relations between pseudo-rearrangement and relative rearrangement. Nonlinear Anal. **41**(7–8), 855–869 (2000)
74. Ferone, V., Posteraro, M.R.: A remark on a comparison theorem. Commun. Partial Differ. Equ. **16**, 1255–1262 (1991)
75. Ferone, A., Volpicelli, R.: Polar factorization and pseudo-rearrangement : applications to Polyà-Szegö type inequalities. Nonlinear Anal. **53**, 929–949 (2003)
76. Ferone, V., Posteraro, M.R., Volpicelli, R.: An inequality concerning rearrangements of functions and Hamilton-Jacobi equations. Arch. Rational Mech. Anal. **125**(3), 257–269 (1993)
77. Ferone, A., Jalal, M., Rakotoson, J.M., Volpicelli, R.: Nonlocal generalized models for a confined plasmas in a Tokamak. Appl. Math. Lett. **12**(1), 43–46 (1999)
78. Ferone, V., Posteraro, M.R., Rakotoson, J.M.: L^{∞}-estimates for nonlinear elliptic problems with p-growth in the gradient*. J. Inequal. Appl. **3**, 109–125 (1999)
79. Ferone, A., Jalal, M., Rakotoson, J.M., Volpicelli, R.: A topological approach for generalized nonlocal models for a confined plasmas in a Tokamak. Commun. Appl. Anal. **5**(2), 159–182 (2001)
80. Fiorenza, A.: Duality and reflexivity in Grand Lebesgue spaces. Collect. Math. **51**(2), 131–148 (2000)
81. Fiorenza, A., Karadzhov,G.E.: Grand and small Lebesgue spaces and their analogs. Z. Anal. Anwendungen **23**, 657–681 (2004)
82. Fiorenza, A., Rakotoson, J.M.: Petits espaces de Lebesgue et quelques applications. Comptes Rendus Math. **334**, 23–26 (2002)
83. Fiorenza, A., Rakotoson, J.M.: New proprieties of small Lebesgue spaces and their applications. Math. Ann. **326**, 543–561 (2003)
84. Fiorenza, A., Rakotoson, J.M.: Compactness, interpolation inequalities for small Lebesgue-Sobolev spaces and applications. Calculus Var. PDE **25**(2), 187–203 (2005)
85. Fiorenza, A., Rakotoson, J.M.: Relative rearrangement and Lebesgue spaces $L^{p(\cdot)}$ with variable exponents. J. Math. Pures Appl. **88**, 506–521 (2007)
86. Fiorenza, A., Rakotoson, J.M., Sbordone, C.: Variable exponents and grand Lebesgue spaces: some optimal results. Commun. Contemp. Math. **17**(6), 1550023 (2015)
87. Fiorenza, A., Formica, M.R., Gogatishvili, A., Kopaliani, T., Rakotoson, J.M.: Characterization of interpolation between Grand, small or classical Lebesgue spaces. Nonlinear Anal. **177**, 422–453 (2018)
88. Fiorenza, A., Gogatishvilli, A., Nekvinda, A., Rakotoson, J.M.: Remarks on compactness results for variable exponent spaces $L^{p(\cdot)}$. J. Math. Pures Appl. **1**(57), 136–144 (2022)
89. Fiorenza, A., Rakotoson, J.M., Zitouni, L.: Relative rearrangement method for estimating dual norms. Ind. Univ. Math. J. **58**(3), 1127–1149 (2009)
90. Fleming, R., Rishel, F.W.: An integral formula for total gradient variation. Arch. Math. **11**, 218–222 (1960)
91. Freidberg, J.P.: Ideal Magnetohydrodynamics. Plenum Press, New York (1987)
92. Giarusso, E., Nunziante, D.: Comparison theorems for a class of first-order Hamilton-Jacobi equations. Ann. Toulouse **7**, 57–75 (1985)
93. Giarusso, E., Nunziante, D.: Symmetrization in a class of first-order Hamilton-Jacobi equations. Nonlinear Anal. **8**(4), 289–299 (1984)
94. Gilbarg, D., Trudinger, N.S.: Elliptic Partial Differential Equations of Second Order. Springer, Berlin (1983)
95. Gogatishviki, A., Pick, L., Soudsky, A.: Characterization of associate spaces of weighted Lorentz spaces with applications. Stud. Math. **224**(1), 1–23 (2014)
96. Gustafsson, B., Mossino, J.: Isoperimetric inequalities for Stefan problem. SIAM J. Math. Anal. **20**(5), 1095–1108 (1989)

97. Hardy, G.H., Littlewood, J.E., Polyà, G.: Inequalities. Cambridge University Press, Cambridge (1964)
98. Hewitt, E., Stromberg, K.: Real and Abstract Analysis. Springer, New York (1965)
99. Hunt, R.: On $L(p, q)$ spaces. L'enseignement Math. **12**, 249–276 (1966)
100. Ioku, N.: Sharp Sobolev inequalities in Lorentz spaces for a mean oscillation. J. Funct. Anal. **266** 2944–2958 (2014)
101. Iwaniec, T., Sbordone, C.: On the integrability of the Jacobian under minimal hypotheses. Arch. Rat. Mech. Anal. **119**, 129–143 (1992)
102. Kawohl, B.: Rearrangements and Convexity of Level Sets in PDE. Springer Lecture Notes, vol. 1150. Springer, Berlin (1985)
103. Kesavan, S.: Symmetrization & Applications. Series in Analysis, vol. 3. World Scientific, Hackensack (2006)
104. Kokilashvili, V., Samko, S.: Singular integrals and potential in some banach function spaces with variable exponent. J. Funct. Spaces Appl. **1**, 45–59 (2003)
105. Lieb, E.: Sharp constants in the Hardy-Littlewood-Sobolev and related inequalities. Ann. Math. **118**(2), 349–374 (1983)
106. Lions, P.L., Pacella, F.: Isoperimetric inequalities for convex cones. Proc. Am. Math. Soc. **109**(2), 477–485 (1990)
107. Lions, P.L., Pacella, F., Tricarico, M.: Best constants on Sobolev inequalities for functions vanishing on some part of the boundary and related question. Ind. Univ. Math. J. **37**, 301–324 (1988)
108. Liu, W. Barrett, J.W.: Finite element approximation of some degenerate monotone quasilinear elliptic systems. SIAM J. Num. Anal. **33**, 88–106 (1996)
109. Ljusternik, L.A.: Brun-Minkowski inequality for arbitrary sets. Dokl. Akad. Nausk SSSR **3**, 55–58 (1935)
110. Maderna, C., Salsa, S.: Some special properties of solutions to obstacle problems. Rend. Sem. Mat. Un. Padova **71**, 121–129 (1984)
111. Martin, J., Milman, M.: Pointwise symmetrization inequalities for Sobolev functions and applications. Adv. Math. **225**, 121–199 (2010)
112. Maz'ja, V.: Sobolev Spaces. Springer, Berlin (1985)
113. Mercaldo, A.: Boundedness of minimizers of degerate functionals. Differ. Int. Equ. **9**(3), 541–546 (1996)
114. Moser, J.: A sharp form of an inequality by N. Trudinger. Ind. Univ. Math. J. **20**, 1077–1092 (1971)
115. Mossino, J.: Inégalités Isopérmétriques et Applications en Physique. Hermann, Paris (1984)
116. Mossino, J., Rakotoson, J.M.: Isoperimetric inequalities in parabolic equations. Ann. della Scuola Normale Superiore di Pisa IV **13**(1), 51–73 (1986).
117. Mossino, J., Temam, R.: Directional derivative of the increasing rearrangement mapping and application to a queer differential equation in plasma physics. Duke Math. J. **48**(3), 475–495 (1981)
118. O'Neil, R.: Convolution operators and $L(p, q)$ spaces. Duke. Math. J. **80**, 129–142 (1986)
119. Polyà, G., Szegö, W.N.: Isoperimetric Inequalities in Mathematical Physics. Annals of Mathematics Studies, vol. 27. Princeton University Press, Princeton (1951)
120. Puel, J.P.: A non linear eigenvalue problem with free boundary. Comptes Rendus de l' Académie des Sciences (CRAS) **284**(A), 861–863 (1977)
121. Rakotoson, J.M.: Résultat de régularité et d'existence pour certaines équations elliptiques quasilinéaires. Comptes Rendus de l' Académie des Sciences (CRAS) **302**(1), 567–570 (1986)
122. Rakotoson, J.M.: Réarrangement relatif dans des équations quasilinéaires avec un second membre distribution. Application à un théorème d'existence et de régularité. J. Differ. Equ. **66**(3), 391–419 (1987)
123. Rakotoson, J.M.: Some properties of the relative rearrangement. J. Math. Anal. Appl. **135**(2), 488–500 (1988)

124. Rakotoson, J.M.: A differentiability result for the relative rearrangement. Differ. Integral Equ. **2**, 363–377 (1989)
125. Rakotoson, J.M.: Relative rearrangement for highly nonlinear equations. Nonlinear Anal. Theory Methods Appl. **24**(4), 493–507 (1995)
126. Rakotoson, J.M.: Strong continuity of the relative rearrangement maps and application to a Galerkin approach for nonlocal problems. Appl. Math Lett. **8**(6), 61–63 (1995)
127. Rakotoson, J.M.: Galerkin approximations, strong continuity of the relative rearrangement map and application to plasma physics equations. Differ. Integral Equ. **12**(1), 67–81 (1999)
128. Rakotoson, J.M.: General pointwise relations for the relative rearrangement and applications. Appl. Anal. **80**(1–2), 201–232 (2001).
129. Rakotoson, J.M.: Some new applications of the pointwise relations for the relative rearrangement. Adv. Differ. Equ. **7**(5), 617–640 (2002)
130. Rakotoson, J.M.: Relative rearrangement and interpolations. Rev. R. Acad. Ciencas Mad. **97**(1), 133–145 (2003)
131. Rakotoson, J.M.: Multivalued fixed point index and non local problems involving relative rearrangement. Nonlinear Anal. **66**, 2470–2499 (2007). https://doi.org/10.1016/j.na(2006.03.32)
132. Rakotoson, J.M.: Réarrangement Relatif : Un Instrument d'estimation dans les Problèmes aux Limites. Springer, Berlin (2008)
133. Rakotoson, J.M.: Lipschitz properties in variable exponent problem via relative rearrangement. Chin. Ann. Math. B **6**, 991–1006 (2010)
134. Rakotoson, J.M.: Regularity of a very weak solution for parabolic equation and applications. Adv Differ. Equ. **16**(9–10), 867–894 (2011)
135. Rakotoson, J.M.: New Hardy inequalities and behaviour of linear elliptic equations. J. Funct. Anal **263**, 2893–2920 (2012)
136. Rakotoson J.E., Rakotoson, J.M.: Analyse fonctionnelle appliquée aux équations dérivées partielles. Presses Universitaires de France PUF, Paris (1999). ISBN 2130498388
137. Rakotoson, J.E., Rakotoson, J.M.: Local regularity of the monotone rearrangement. Appl. Math. Lett. **17**(3), 353–355 (2004)
138. Rakotoson, J.M., Seoane, M.L.: Numerical approximations of the Relative rearrangement. The piecewise linear case: application to some nonlocal problems. Math. Model. Numer. Anal. **34**(2), 477–499 (2000)
139. Rakotoson, J.M., Seoane, M.L.: Strong convergence of the directional derivative of the decreasing rearrangement mapping and related questions. Differ. Integral Equ. **17**(11–12), 1347–1358 (2004)
140. Rakotoson, J.M., Serre, D.: Un problème d'optimisation lié aux équations de Navier-Stokes. Ann. Della Scuola Norm. di Pisa,**XX**, 633–649 (1993)
141. Rakotoson, J.M., Simon, B.: Relative rearrangement on a measure space. Application to the regularity of weighted monotone rearrangement. Part I–II. Appl. Math. Lett. **6**(1), 75–78, 79–92 (1993)
142. Rakotoson, J.M., Simon, B.: Relative rearrangement on a finite measure space. Application to the regularity of weighted monotone rearrangement. Rev. R. Acad. Cienc. Exactas Fís. Nat. (Esp.) **91**(1), 17–31 (1997)
143. Rakotoson, J.M., Simon, B.: Relative rearrangement on a finite measure space. Application to weighted spaces and to P.D.E. Rev. R. Acad. Cienc. Exactas Fís. Nat. (Esp.) **91**(1), 33–45 (1997)
144. Rakotoson, J.M., Temam, R.: Une formule intégrale du type Federer et applications. Comptes Rendus Acad. Sci. **304**, 443–446 (1987)
145. Rakotoson, J.M., Temam, R.: Une nouvelle méthode d'estimation L^{∞}. Application aux inéquations variationnelles. Compt. Rendus Acad. Sci. **304**(17), 527–530 (1987)
146. Rakotoson, J.M., Temam, R.: A co–area formula with applications to monotone rearrangement and to regularity. Arch. Rational Mech. Anal. **109**, 213–238 (1990)
147. Rockafellar, R.T.: Convex Analysis. Princeton University Press, Princeton (1970)

148. Rodemich, E.: The Sobolev inequality with best possible constants. Analysis Seminar at California Institute of Technology (1966)
149. Rodrigues, J.F.: Strong Solutions for Quasi-Linear Elliptic-Parabolic Problems with Time-Dependent Obstaclesm, p. 266. Pitman, London (1987)
150. Rudin W.: Analyse Réelle et Complexe: Cours et Exercices. Dunod, Paris (1998). ISBN:978-2-10-004004-9
151. Saramito, B.: Stabilité d'un Plasma: Modélisation Mathématique et Simulation Numérique, volume RMA, 34. Masson, Paris (1994)
152. Senba, T., Suzuki,T.: Weak solutions to a parabolic-elliptic systems of chemotaxis. J. Funct. Anal. 191, 17–51 (2002)
153. Serre, D.: Sur le principe variationnel des équations de la mécanique des fluides parfaits. Mod. Math. Anal. Num. **27**(6), 739–758 (1993)
154. Shafranov, V.D.: On magneto-hydrodynamical equilibrium configurations. Sov. Phys. JETP **6**(33), 545–554 (1958)
155. Simon, J.: Asymptotic behavior of a plasma induced by an electric current. Nonlinear Anal. T.M.A. **9**(2), 149–169 (1985)
156. Simon, B.: Réarrangement relatif sur un espace mesuré. Thèse Université de Poitiers (1994)
157. Stein, E.M.: The differentiability of functions in $\mathbb{R}^n$. Ann. Math. **113**, 383–385 (1981)
158. Strichartz, R.S.: A note on Trudinger's extension of Sobolev inequalities. Ind. Univ. Math. J. **21**, 841–841 (1972)
159. Takasi, S., Takashi, S.: Weak solutions to Parabolic-Elliptic system of chemotaxis. J. Funct. Anal. **191**, 17–51 (2002)
160. Talenti, G.: Elliptic equations and rearrangements. Ann. Scuola Norm. Sup. Pisa Cl. Sci. **3**, 697–718 (1976)
161. Talenti, G.: Best constant in Sobolev inequality. Ann. Mat. Pura Appl. **110**, 353–372 (1976)
162. Talenti, G.: Rearrangements of functions and partial differential equations. In: Fasano, A., Primicerio, M. (eds.) Nonlinear Diffusion Problems, pp. 153–178. Springer, Berlin (1986)
163. Talenti, G.: Rearrangements and PDE. In: Everitt, W.N. (ed.) Inequalities, Fifty Years on from Hardy, Littlewood and Pòlya, pp. 211–230. Marcel Dekker, New York (1991)
164. Talenti, G.: The art of rearranging. Milan J. Math. **84**, 105–157 (2016)
165. Temam, R.: Analyse Numérique. Presses Universitaires de France, Paris (1970)
166. Temam, R.: A non-linear eigenvalue problem: the shape equilibrium of a confined plasma. Arch. Rat. Mech. Anal. **60**, 51–73 (1976)
167. Temam, R.: Remarks on a free boundary problem arising in plasma physics. Commun. Partial Differ. Equ. **2**(6), 563–585 (1977)
168. Temam, R.: Monotone rearrangement of functions and the Grad-Mercier equation of plasma physics. In: de Giogi, E., Mosco U. (eds.) Proceedings of the International Meeting on Recent Methods in Nonlinear Analysis (Rome, 1978), pp. 83–98. Pitagora, Bologna (1979)
169. Temam, R.: Infinite-Dimensional Dynamical Systems in Mechanics and Physics, vol. 68. Springer, Berlin (1988)
170. Tian, Y., Li, F.: Comparison results for nonlinear degenerate Dirichlet and Neumann problems with general growth in the gradient. J. Math. Anal. Appl. **378**, 749–763 (2011)
171. Ziemer, W.: Weakly Differentiable Functions. Springer, Berlin (1989)

Index

J. M. Rakotoson, *Relative Rearrangement*, Lecture Notes in Mathematics 2376,
https://doi.org/10.1007/978-3-032-02228-8

LECTURE NOTES IN MATHEMATICS Springer

Editors in Chief: J.-M. Morel, B. Teissier;

Editorial Policy

1. Lecture Notes aim to report new developments in all areas of mathematics and their applications – quickly, informally and at a high level. Mathematical texts analysing new developments in modelling and numerical simulation are welcome.

   Manuscripts should be reasonably self-contained and rounded off. Thus they may, and often will, present not only results of the author but also related work by other people. They may be based on specialised lecture courses. Furthermore, the manuscripts should provide sufficient motivation, examples and applications. This clearly distinguishes Lecture Notes from journal articles or technical reports which normally are very concise. Articles intended for a journal but too long to be accepted by most journals, usually do not have this "lecture notes" character. For similar reasons it is unusual for doctoral theses to be accepted for the Lecture Notes series, though habilitation theses may be appropriate.

2. Besides monographs, multi-author manuscripts resulting from SUMMER SCHOOLS or similar INTENSIVE COURSES are welcome, provided their objective was held to present an active mathematical topic to an audience at the beginning or intermediate graduate level (a list of participants should be provided).

   The resulting manuscript should not be just a collection of course notes, but should require advance planning and coordination among the main lecturers. The subject matter should dictate the structure of the book. This structure should be motivated and explained in a scientific introduction, and the notation, references, index and formulation of results should be, if possible, unified by the editors. Each contribution should have an abstract and an introduction referring to the other contributions. In other words, more preparatory work must go into a multi-authored volume than simply assembling a disparate collection of papers, communicated at the event.

3. Manuscripts should be submitted either online at www.editorialmanager.com/lnm to Springer's mathematics editorial in Heidelberg, or electronically to one of the series editors. Authors should be aware that incomplete or insufficiently close-to-final manuscripts almost always result in longer refereeing times and nevertheless unclear referees' recommendations, making further refereeing of a final draft necessary. The strict minimum amount of material that will be considered should include a detailed outline describing the planned contents of each chapter, a bibliography and several sample chapters. Parallel submission of a manuscript to another publisher while under consideration for LNM is not acceptable and can lead to rejection.

4. In general, **monographs** will be sent out to at least 2 external referees for evaluation.

   A final decision to publish can be made only on the basis of the complete manuscript, however a refereeing process leading to a preliminary decision can be based on a pre-final or incomplete manuscript.

   Volume Editors of **multi-author works** are expected to arrange for the refereeing, to the usual scientific standards, of the individual contributions. If the resulting reports can be forwarded to the LNM Editorial Board, this is very helpful. If no reports are forwarded or if other questions remain unclear in respect of homogeneity etc, the series editors may wish to consult external referees for an overall evaluation of the volume.

5. Manuscripts should in general be submitted in English. Final manuscripts should contain at least 100 pages of mathematical text and should always include
   - a table of contents;
   - an informative introduction, with adequate motivation and perhaps some historical remarks: it should be accessible to a reader not intimately familiar with the topic treated;
   - a subject index: as a rule this is genuinely helpful for the reader.
   - For evaluation purposes, manuscripts should be submitted as pdf files.

6. Careful preparation of the manuscripts will help keep production time short besides ensuring satisfactory appearance of the finished book in print and online. After acceptance of the manuscript authors will be asked to prepare the final LaTeX source files (see LaTeX templates online: https://www.springer.com/gb/authors-editors/book-authors-editors/manuscriptpreparation/5636) plus the corresponding pdf- or zipped ps-file. The LaTeX source files are essential for producing the full-text online version of the book, see http://link.springer.com/bookseries/304 for the existing online volumes of LNM). The technical production of a Lecture Notes volume takes approximately 12 weeks. Additional instructions, if necessary, are available on request from lnm@springer.com.

7. Authors receive a total of 30 free copies of their volume and free access to their book on SpringerLink, but no royalties. They are entitled to a discount of 33.3 % on the price of Springer books purchased for their personal use, if ordering directly from Springer.

8. Commitment to publish is made by a *Publishing Agreement*; contributing authors of multiauthor books are requested to sign a *Consent to Publish form*. Springer-Verlag registers the copyright for each volume. Authors are free to reuse material contained in their LNM volumes in later publications: a brief written (or e-mail) request for formal permission is sufficient.

Addresses:
Professor Jean-Michel Morel, CMLA, École Normale Supérieure de Cachan, France
E-mail: moreljeanmichel@gmail.com

Professor Bernard Teissier, Equipe Géométrie et Dynamique,
Institut de Mathématiques de Jussieu – Paris Rive Gauche, Paris, France
E-mail: bernard.teissier@imj-prg.fr

Springer: Ute McCrory, Mathematics, Heidelberg, Germany,
E-mail: lnm@springer.com